Biological and Chemical Oceanography

Biological and Chemical Oceanography

Editor: Austin Brennan

R CALLISTO REFERENCE

www.callistoreference.com

Callisto Reference,
118-35 Queens Blvd., Suite 400,
Forest Hills, NY 11375, USA

Visit us on the World Wide Web at:
www.callistoreference.com

ISBN: 978-1-64116-582-2 (Hardback)

Cataloging-in-Publication Data

Biological and chemical oceanography / edited by Austin Brennan.
    p. cm.
Includes bibliographical references and index.
ISBN 978-1-64116-582-2
1. Oceanography. 2. Marine biology. 3. Chemical oceanography. 4. Marine ecology. I. Brennan, Austin.
GC11.2 .B56 2022
551.46--dc23

Table of Contents

Preface

The world is advancing at a fast pace like never before. Therefore, the need is to keep up with the latest developments. This book was an idea that came to fruition when the specialists in the area realized the need to coordinate together and document essential themes in the subject. That's when I was requested to be the editor. Editing this book has been an honour as it brings together diverse authors researching on different streams of the field. The book collates essential materials contributed by veterans in the area which can be utilized by students and researchers alike.

The study of the ocean and its biological and physical aspects is known as oceanography. It is an earth science that includes a wide range of topics such as ocean current, ecosystem, and geophysical fluid dynamics. It also encompasses the study of plate tectonics as well as the geology of the sea floor. It examines different physical properties and chemical substances found in the ocean and across its boundaries. It blends the understanding of the processes within a number of disciplines like biology, chemistry, climatology, geology, geography, hydrology, physics and astronomy in order to acquire an in-depth knowledge of the oceans. Biological oceanography and chemical oceanography are two primary branches of oceanography. Biological oceanography includes the ecology of marine organisms. The study is done on the basis of the ecological characteristics of an individual organism and the physical, chemical and geological aspects of its ocean environment. The chemistry of the ocean is studied under chemical oceanography. It is concerned with the understanding of seawater properties. This book covers in detail some existent theories and innovative concepts revolving around biological and chemical oceanography. It includes contributions made by international experts. It is meant for students who are looking for an elaborate reference text on these disciplines.

Each chapter is a sole-standing publication that reflects each author's interpretation. Thus, the book displays a multi-facetted picture of our current understanding of application, resources and aspects of the field. I would like to thank the contributors of this book and my family for their endless support.

Editor

Picophytoplankton biomass distribution in the global ocean

E. T. Buitenhuis[1], W. K. W. Li[2], D. Vaulot[3], M. W. Lomas[4], M. R. Landry[5], F. Partensky[3], D. M. Karl[6], O. Ulloa[7], L. Campbell[8], S. Jacquet[9], F. Lantoine[10], F. Chavez[11], D. Macias[12], M. Gosselin[13], and G. B. McManus[14]

[1]Tyndall Centre for Climate Change Research and School of Environmental Sciences, University of East Anglia, Norwich NR4 7TJ, UK

[2]Fisheries and Oceans Canada, Bedford Institute of Oceanography, Dartmouth, Nova Scotia, Canada

[3]CNRS and UPMC, Paris 06, UMR7144, Station Biologique, 29680 Roscoff, France

[4]Bermuda Institute of Ocean Sciences, St. George's GE01, Bermuda, USA

[5]Scripps Institution of Oceanography, University of California San Diego, La Jolla, California, USA

[6]Department of Oceanography, University of Hawaii, Honolulu, HI 96822, USA

[7]Department of Oceanography, University of Concepción, Casilla 160-C, Concepción, Chile

[8]Department of Oceanography, Texas A&M University, College Station, TX 77843, USA

[9]INRA, UMR CARRTEL, 75 Avenue de Corzent, 74200 Thonon-les-Bains, France

[10]UPMC Univ Paris 06, CNRS, LECOB, Observatoire Océanologique, 66650, Banyuls/Mer, France

[11]MBARI, 7700 Sandholdt Rd, Moss Landing, CA 95039, USA

[12]Department of Coastal Ecology and Management, Instituto de Ciencias Marinas de Andalucía (ICMAN-CSIC), Avd. Republica Saharaui s/n, CP11510, Puerto Real, Cádiz, Spain

[13]Institut des sciences de la mer de Rimouski, Université du Québec à Rimouski, 310 Allée des Ursulines, Rimouski, Québec G5L 3A1, Canada

[14]Department of Marine Sciences, University of Connecticut, Groton, CT 06340, USA

Correspondence to: E. T. Buitenhuis

Abstract. The smallest marine phytoplankton, collectively termed picophytoplankton, have been routinely enumerated by flow cytometry since the late 1980s during cruises throughout most of the world ocean. We compiled a database of 40 946 data points, with separate abundance entries for *Prochlorococcus*, *Synechococ-* and picoeukaryotes. We use average conversion factors for each of the three groups to convert the abundance data to carbon biomass. After gridding with 1° spacing, the database covers 2.4 % of the ocean surface area, with the best data coverage in the North Atlantic, the South Pacific and North Indian basins, and at least some data in all other basins. The average picophytoplankton biomass is $12 \pm 22\,\mu g\,C\,l^{-1}$ or $1.9\,g\,C\,m^{-2}$. We estimate a total global picophytoplankton biomass of 0.53–$1.32\,Pg\,C$ (17–39 % *Prochlorococcus*, 12–15 % *Synechococcus* and 49–69 % picoeukaryotes), with an intermediate/best estimate of $0.74\,Pg\,C$. Future efforts in this area of research should focus on reporting calibrated cell size and collecting data in undersampled regions.

1 Introduction

Picophytoplankton are usually defined as phytoplankton less than 2 or 3 µm diameter (e.g. Sieburth et al., 1978; Takahashi and Hori, 1984; Vaulot et al., 2008). They are the smallest class of phytoplankton and are composed of both prokaryotes and eukaryotes. The eukaryotes (0.8–3 µm) are a taxonomically diverse group that includes representatives from four algal phyla: the Chlorophyta, Haptophyta, Cryptophyta and Heterokontophyta (Vaulot et al., 2008). The prokaryotes belong to the phylum cyanobacteria and are subdivided into the genera *Prochlorococcus* (~ 0.6 µm) and *Synechococcus* (~ 1 µm), with each group having many ecotypes that dominate in different ocean regions (Johnson et al., 2006).

Picophytoplankton tend to dominate the phytoplankton biomass under oligotrophic conditions such as in the subtropical gyres (Alvain et al., 2005), where their high surface-to-volume ratio makes them the best competitors for low nutrient concentrations (Raven, 1998). The abundance of the prokaryotes is often inversely related with the eukaryotes, which are favoured by more physically active mixed layers (e.g. Boumann et al., 2011). Furthermore, with warming of the temperate to subpolar North Atlantic and the Canadian high Arctic, picophytoplankton (specifically picoeukaryotes) have been found to become an increasingly large fraction of the total chlorophyll (Li et al., 2009; Moran et al., 2010).

As part of the marine ecology data synthesis effort (MAREDAT, this special issue), we compiled a database on picophytoplankton in the global ocean. MAREDAT is a community effort to synthesise abundance and carbon biomass data for the major lower trophic level taxonomic groups in the marine ecosystem. It addresses both autotrophs and heterotrophs and covers the size range from bacteria to macrozooplankton.

2 Data

We compiled data for picophytoplankton abundance in three taxonomic groups: *Prochlorococcus*, *Synechococcus*, and picoeukaryotes (Table 1). We used the size range of picoeukaryotes as defined by the contributing researchers. The size range has a large impact on the resulting biomass (see Discussion). All of the data were obtained by flow cytometry. Both the raw data and the gridded data are available from PANGAEA (http://doi.pangaea.de/10.1594/PANGAEA.777385) and the MAREDAT webpage (http://maremip.uea.ac.uk/.maredat.html).

2.1 Conversion factors

Conversion factors from cell abundance to carbon biomass for the three picophytoplankton groups were compiled from the literature (Table 2). Conversion factors were either measured directly on unialgal cultures in the laboratory or derived from indirect methods on in situ samples. Most of the

Figure 1. Horizontal distribution of the number of observations. Data points have been enlarged to $5° \times 5°$.

indirect measures were derived from cell sizes that were estimated from average forward-angle light scatter (FALS) multiplied by a carbon content per biovolume. The conversion factors of Veldhuis et al. (1997) were based on nitrate uptake in incubated in situ samples and assuming a C : N ratio of 6. Since the biggest source of variability in the other indirect measures is the carbon content per biovolume, which was measured on laboratory cultures, the advantage of using in situ biovolume to determine conversion factors does not seem to improve the local applicability of these data, and we therefore used the directly measured conversion factors as the standard.

2.2 Quality control

Contributed data were assumed to have undergone the contributing researchers' own internal quality control procedures. As a statistical filter for outliers, we applied the Chauvenet criterion (Buitenhuis et al., 2012b) to the total carbon data. The data were not normally distributed, so we log-transformed them, excluding zero values. No high outliers were found by this criterion. The highest picophytoplankton biomass in the database is 575 µg C l^{-1}, measured near the coast of Oman (Indian Ocean).

3 Results

The database contains 40 946 data points (Fig. 1). Data are included from a number of stations that have been sampled repeatedly over many years or programs where measurements have been made on a fine-resolution grid. Therefore, after gridding, we obtained 10 747 data points on the World Ocean Atlas grid ($1° \times 1° \times 33$ vertical layers $\times 12$ months), representing a coverage of vertically integrated and annually averaged biomass for 2.4 % of the ocean surface. For further details on the gridding, see Buitenhuis et al. (2012b). To limit

Table 1. Data sources.

Cruise	Date	Area	Reference/Investigator
Li87022	Jun 1987	North Atlantic	Li and Wood (1988); Li et al. (1992)
CHLOMAX	Sep–Oct 1987	Sargasso Sea	Neveux et al. (1989)
Endeavour177	May–Jun 1988	Sargasso Sea	Olson et al. (1990)
Li88026	Sep 1988	North Atlantic	Li et al. (1992)
Bermuda	1988–1989	Sargasso Sea	Olson et al. (1990)
EROSDISCO89	Jan 1989	Mediterranean Sea	Vaulot et al. (1990)
Li89003	Apr 1989	North Atlantic	Li et al. (1992)
Oceanus206	May 1989	Sargasso Sea	Olson et al. (1990)
EROSBAN	Jul 1989	Mediterranean Sea	Partensky (unpublished data)
NIOZNatl89	Aug–Sep 1989	North Atlantic	Veldhuis and Kraay (1990); Veldhuis et al. (1993)
Palau	Aug–Sep 1990	Tropical Pacific West	Shimada et al. (1993)
NOPACCS	Aug–Oct 1990	Pacific Ocean	Ishizaka (unpublished data)
Australia	Nov–Dec 1990	Tropical Pacific West	Shimada et al. (1993)
HOT	1990–2008	Tropical Pacific	Campbell et al. (1997); Karl (unpublished data)
BATS	1990–2010	North Atlantic	DuRand et al. (2001); Lomas et al. (2010)
Iselin 9102	Feb 1991	Carribean Sea	McManus and Dawson (1994)
Li91001	Apr 1991	North Atlantic	Li (unpublished data)
BOFS	Jul 1991	North Atlantic	BODC (British Oceanographic Data Centre)
POEM91	Oct 1991	Mediterranean Sea	Li et al. (1993)
EUMELI3	Oct 1991	Tropical Atlantic	Partensky et al. (1996)
EQPACTT007	Feb–Mar 1992	Equatorial Pacific	Landry et al. (1996)
Eddy92	Mar 1992	Mediterranean Sea	Yacobi et al. (1995)
EROSVALD	Mar 1992	Mediterranean Sea	Vaulot, Marie (unpublished data)
EQPACTT008	Mar–Apr 1992	Equatorial Pacific	Binder et al. (1996)
EQPACTT008D	Mar–Apr 1992	Equatorial Pacific	DuRand and Olson (1996)
NIOZIndian	May 1992–Feb 1993	Indian Ocean/Red Sea	Veldhuis and Kraay (1993)
SurugaBay	May 1992–Oct 1993	Japan	Shimada et al. (1995)
EUMELI4	Jun 1992	Tropical Atlantic	Partensky et al. (1996)
Surtropac17	Aug 1992	Equatorial Pacific	Blanchot and Rodier (1996)
EQPACTT011	Aug–Sep 1992	Equatorial Pacific	Landry et al. (1996)
Li92037	Sep 1992	North Atlantic	Li (1994, 1995)
EQPACTT012	Sep–Oct 1992	Equatorial Pacific	DuRand and Olson (1996)
EUMELI5	Dec 1992	Tropical Atlantic	Partensky et al. (1996)
Aquaba	1992–1993	Red Sea	Lindell and Post (1995)
Malaga93	Jan 1993	Mediterranean Sea	Garcia et al. (1994)
Li93002	May 1993	North Atlantic	Li (1994, 1995)
EROSDISCO93	Jul 1993	Mediterranean Sea	Simon, Barlow, Marie (unpublished data)
NOAA93	Jul–Aug 1993	North Atlantic	Buck et al. (1996)
Flupac	Sep–Oct 1994	Equatorial Pacific	Blanchot et al. (2001)
OLIPAC	Nov 1994	Equatorial Pacific	Neveux et al. (1999)
ArabianTTN043	Jan 1995	Arabian Sea	Campbell et al. (1998)
ArabianTTN045	Mar–Apr 1995	Arabian Sea	Campbell et al. (1998)
Delaware95	Apr 1995	North Atlantic	Li (1997)
MINOS	Jun 1995	Mediterranean Sea	Vaulot, Marie, Partensky (unpublished data)
Chile95	Jun 1995	South Pacific	Li (unpublished data)
Lopez96	Jun 1995	Sargasso Sea	Li (unpublished data)
Li95016	Jul 1995	North Atlantic	Li and Harrison (2001)
Ictio-Alborán Cadiz 95	Jul 1995	North Atlantic	Echevarría et al. (2009)
ArabianTTN049	Jul–Aug 1995	Arabian Sea	Olson (unpublished data)
ArabianTTN050	Aug–Sep 1995	Arabian Sea	Campbell et al. (1998)
NOAA95	Sep–Oct 1995	Indian Ocean	Buck (unpublished data)
ArabianTTN053	Nov 1995	Arabian Sea	Olson (unpublished data)
ArabianTTN054	Dec 1995	Arabian Sea	Campbell et al. (1998)
AZOMP	1995–2009	North Atlantic	Li (2002, 2009); Li et al. (2009)

Table 1. Continued.

Cruise	Date	Area	Reference/Investigator
OMEX/D1221	Jun 1996	North Atlantic	BODC
AZMP	1997–2009	North Atlantic	Li (2002, 2009); Li et al. (2009)
Kiwi6	Oct–Nov 1997	Antarctica	Landry (unpublished data)
Kiwi7	Dec 1997	Antarctica	Landry (unpublished data)
Almo-1	Dec 1997	Mediterranean Sea	Jacquet, Marie (unpublished data)
AESOPS/NBP97-1	1997	Ross Sea	Olson, Sosik (unpublished data)
Almo-2	Jan 1998	Mediterranean Sea	Jacquet et al. (2010)
Kiwi8	Jan–Feb 1998	Antarctica	Landry (unpublished data)
Kiwi9	Feb–Mar 1998	Antarctica	Landry (unpublished data)
Southwest Pacific	Mar–Apr 1998	South Pacific	Campbell et al. (2005)
PROSOPE99	Sept 1999	Mediterranean Sea	Marie et al. (2006)
GLOBEC LTOP	Mar 2001–Sep 2003	North Pacific	Sherr et al. (2005)
JOIS	2002–2009	North Atlantic, Arctic	Li (2002, 2009); Li et al. (2009)
NP	Feb 2004–Mar 2005	North Atlantic	Lomas et al. (2009)
BIOSOPE	Oct–Dec 2004	South East Pacific	Grob et al. (2007)
ArcticNet2005	Aug–Sep 2005	Arctic, North Atlantic	Tremblay et al. (2009)
DOP	May 2006–May 2008	North Atlantic	Lomas (unpublished data)
C3O	2007–2008	North Atlantic, Arctic	Li (2002, 2009); Li et al. (2009)
Bering Sea	Mar 2008–May 2010	North Pacific	Moran et al. (2012)
Line P	Aug 2010–Jun 2011	North Pacific	Lomas (unpublished data)
FOODWEB	Feb–Aug 2011	North Atlantic	Lomas (unpublished data)

Table 2. Cell abundance to carbon biomass conversion factors [fg C cell^{-1}].

	Prochlorococcus	*Synechococcus*	picoeukaryotes	reference
Direct, from cultures		250		Kana and Glibert (1987)
		600	3800 ± 100	Verity et al. (1992)
			800, 1360	Montagnes et al. (1994)
	49 ± 9			Cailliau et al. (1996)
		350 (200–500)		Liu et al. (1999)
			4400	Llewellyn and Gibb (2000)
	27 ± 6			Claustre et al. (2002)
	53 ± 9	170 ± 65		Bertilsson et al. (2003)
	16 ± 1	249 ± 21		Fu et al. (2007)
average	36	255*	2590	
Indirect, mostly from culture C per volume × in situ volume	92	175		Veldhuis et al. (1997)
	53	246	2108	Campbell et al. (1994)
	56	112		DuRand et al. (2001)
	39 ± 1	82 ± 8	530 ± 185	Worden et al. (2004)
average	60	154	1319	

* Excluding Verity et al. (1992), 324 fg C cell^{-1} including Verity et al. (1992).

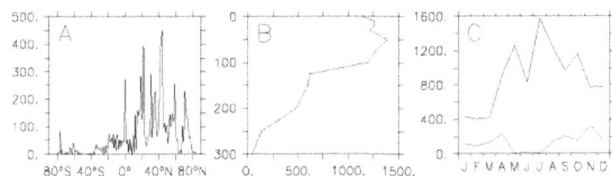

Figure 2. Number of grid points with data, as a function of (**A**) latitude. (**B**) Depth. Observations below 300 m are not shown (1.4 % of the data). The deepest observation is at 3000 m and the deepest non-zero observation at 1100 m. (**C**) Time. Red: Southern Hemisphere, black: total.

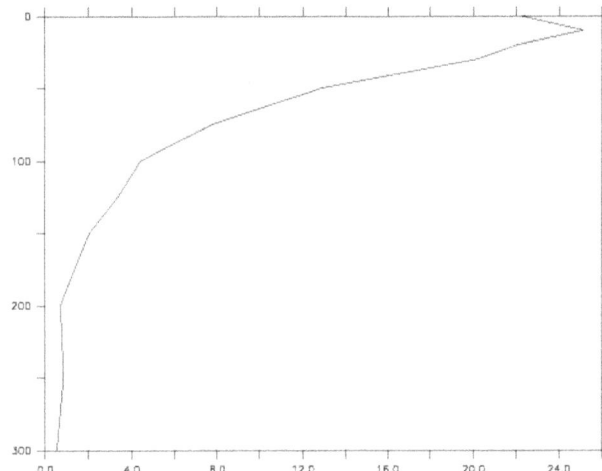

Figure 3. Average picophytoplankton biomass [μg C l^{-1}] as a function of depth [m].

the overrepresentation of well sampled locations, we present results of the gridded data. Only 15 % of the data are from the Southern Hemisphere (Fig. 2a), 33 % are from the tropics (43 % of the ocean surface), while 13 % are from the polar oceans (5 % of the ocean surface). Observations in the upper 112.5 m make up 81 % of the data (Fig. 2b), but the number of observations decreases more slowly than biomass (Fig. 3), and there are still 480 observations at 200 m depth (Fig. 2b), thus defining the vertical biomass profile fairly well. Zero values make up 1.6 % of the data, and 95 % of those are from below 62.5 m depth. There is some sampling bias towards the growing season, with 67 % of the data sampled in the spring and summer months (Fig. 2c).

The average picophytoplankton biomass is $12 \pm 22\,\mu$g C l^{-1} (Fig. 4) or $1.9\,$g C m^{-2}. Of the vertically integrated biomass, 54 % occurs in the upper 40 m and 93 % in the upper 112.5 m (Fig. 2). *Synechococcus* is found at the most shallow depths (97 % above 112.5 m, Fig. 5), followed by picoeukaryotes (92 % above 112.5 m), while *Prochlorococcus* biomass decreases more slowly with depth (87 % above 112.5 m).

The average biomass is slightly higher in the tropics and considerably lower in the Arctic (Figs. 4, 6), but the standard deviation within latitudinal bands is high, so that none of the differences are significant. Antarctica: $11 \pm 8\,\mu$g C l^{-1} or $1.2\,$g C m^{-2}, south temperate zone (67–23° S): $13 \pm 23\,\mu$g C l^{-1} or $2.2\,$g C m^{-2}, tropics: $15 \pm 24\,\mu$g C l^{-1} or $2.2\,$g C m^{-2}, north temperate zone: $12 \pm 22\,\mu$g C l^{-1} or $1.9\,$g C m^{-2}, and Arctic: $6 \pm 8\,\mu$g C l^{-1} or $0.6\,$g C m^{-2}. We calculate the global picophytoplankton biomass from the zonal and time-averaged concentration filled by interpolation across up to 22° latitude (Fig. 6) multiplied by the volume at each latitude and depth, integrating to the bottom and counting missing values as 0. We thus estimate a total global picophytoplankton biomass of 0.74 Pg C (17 % *Prochlorococcus*, 15 % *Synechococcus* and 69 % picoeukaryotes). Interpolation across up to 10° latitude only leaves a few missing values and estimates 0.73 Pg C. If we use the indirect in situ conversion factors for each of the three groups (Table 2), the total biomass (with up to 22° interpolation) is 0.53 Pg C (39 % *Prochlorococcus*, 12 % *Synechococcus*, 49 % picoeukaryotes).

Picoeukaryotes tend to dominate by > 75 % poleward of 40°, and dominate below 62.5 m depth in the tropics and below 225 m everywhere (Fig. 7). *Prochlorococcus* tends to dominate above 225 m between 20–40° N and shares dominance with picoeukaryotes between 10–30° S and at the surface in the tropics. *Synechococcus* only dominates around 50° S and is relatively abundant above 62.5 m between 10–40° N. This is consistent with the community structure of picophytoplankton that has been analysed by Bouman et al. (2011).

4 Discussion

Although data coverage, at 2.4 % of the ocean surface, is by no means complete, if we randomly select half of the depth profiles in 10 random samples, the average integrated biomass varies between 96 and 104 % of the value for the whole dataset, while the averages from the Southern and Northern Hemispheres are 119 % and 96 %, respectively. On the other hand, the average using the indirect in situ conversion factors is 72 % of the value estimated using the direct conversion factors. Thus, the main uncertainty in determining the global picophytoplankton biomass in this analysis is the conversion from cell abundance to carbon biomass. There is a fairly tight relationship between forward-angle light scatter (FALS; Cavender-Bares et al., 2001; DuRand et al., 2002) or right-angle light scatter (RALS; Simon et al., 1994; Worden et al., 2004), as measured by flow cytometry, and cell size, which is probably the main source of uncertainty in the conversion factor. Only about a third of our data came with FALS or RALS data, and even in those cases these were in arbitrary units. We recommend the routine measurement of calibrated size as the additional measurement that would do most to improve our knowledge of global picophytoplankton biomass distribution.

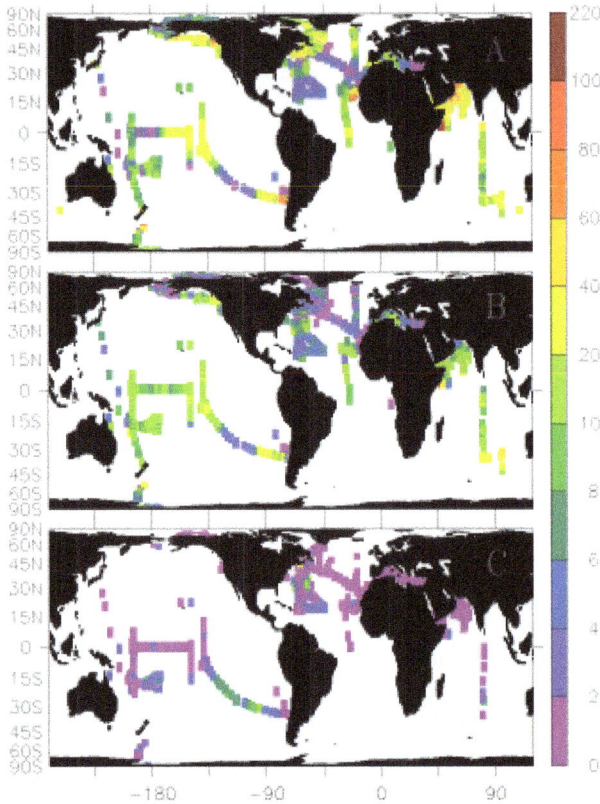

Figure 4. Picophytoplankton biomass [μg C l⁻¹]. **(A)** 0–40 m, **(B)** 40–112.5 m, **(C)** 112.5–225 m.

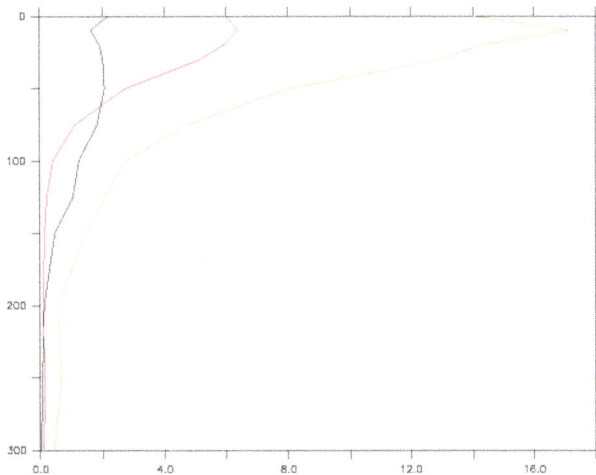

Figure 6. Zonal and time-averaged biomass [μg C l⁻¹] of **(A)** *Prochlorococcus*, **(B)** *Synechococcus*, **(C)** picoeukaryotes. Data have been filled by latitudinal interpolation of up to 22°.

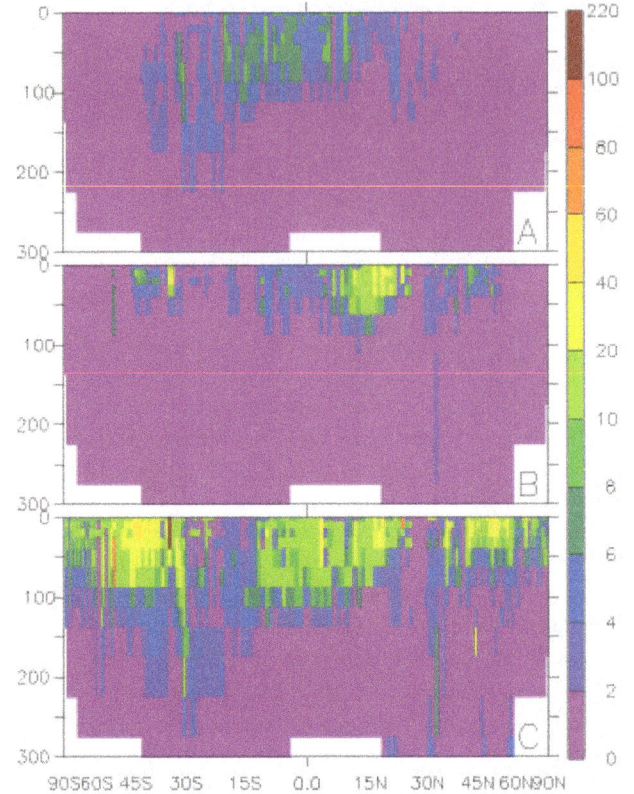

Figure 5. Average depth profiles of *Prochlorococcus* (black), *Synechococcus* (red) and picoeukaryotes (green) biomass [μg C l⁻¹].

In addition to the uncertainty in the carbon conversion factor, there is uncertainty about the abundance of *Prochlorococcus* in near-surface oligotrophic waters, where the cellular chlorophyll content, and thus the ability to detect them as algae from their red fluorescence, is at its minimum and near

the detection limit of standard flow cytometers (Dusenberry and Frankel, 1994).

It has been repeatedly shown that *Prochlorococcus* and *Synechococcus* increase in cell size with depth up to ∼ 150 m. In contrast, previously published results for picoeukaryotes showed little variation in size as a function of depth (Li et al., 1993; DuRand et al., 2001; Grob et al., 2007). We compared the increase in size for the three groups at two locations. At BATS (Bermuda Atlantic Timeseries Station; which includes the data of DuRand et al., 2001), we also find an increase in cell size of *Prochlorococcus* and *Synechococcus* but not picoeukaryotes (Fig. 8a). However, in the Western Mediterranean (Almo-1 and -2, Jacquet et al., 2010), we find a similar increase in cell size of *Prochlorococcus* and *Synechococcus*, but a much larger increase of picoeukaryotes (Fig. 8b). The difference this could make to the global picophytoplankton biomass is large. If we use the standard conversion factors in the surface and increase these linearly up to a factor 3 below 150 m depth (blue lines in Fig. 8), then the global biomass becomes 1.32 Pg C (+78 %), or if we only apply this increasing conversion factor to *Prochlorococcus* and *Synechococcus*, we estimate a global biomass of 0.93 Pg C (+25 %). Our standard conversion factors are taken from laboratory studies. Conversion factors for heterotrophic bacteria

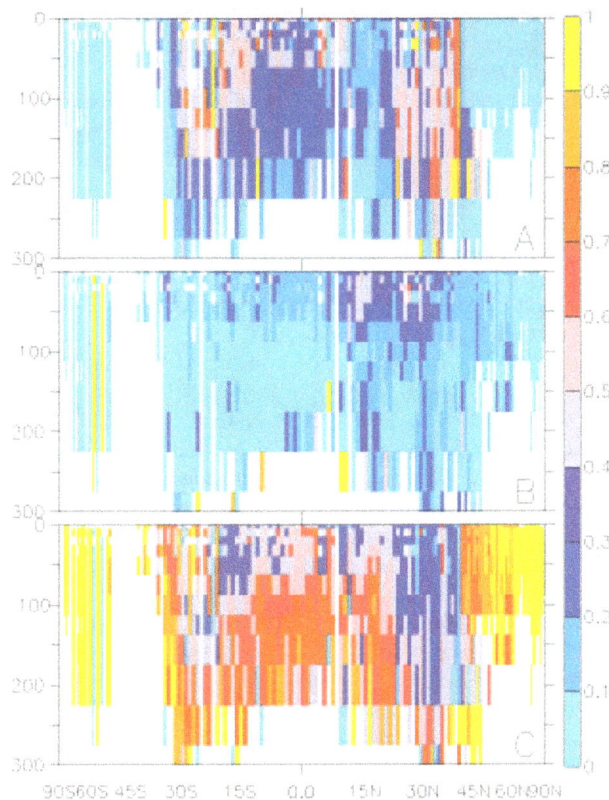

Figure 7. Zonal and time-averaged fraction of total picophyto-plankton (**A**) *Prochlorococcus*, (**B**) *Synechococcus*, (**C**) picoeukary-otes.

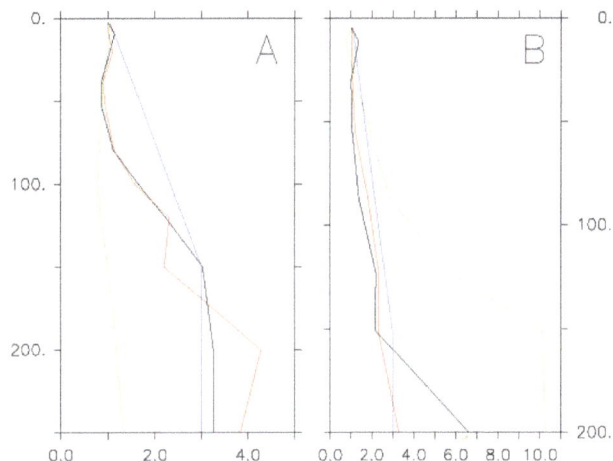

Figure 8. Cell size as a function of depth, normalised to cell size at the surface, (black) *Prochlorococcus*, (red) *Synechococcus*, (green) picoeukaryotes, (blue) exploratory conversion factor that increases up to a factor 3 below 150 m depth. (**A**) At BATS, (**B**) in the Western Mediterranean (Almo-1 and -2).

from laboratory studies tend to be higher than from in situ measurements (Buitenhuis et al., 2012a). Indeed, even if we do not account for an increase of cell size with depth, the laboratory conversion factors lead to a higher biomass estimate than the indirect conversion factors. Other sources of variability are seasonal variations of cell size (DuRand et al., 2001) of all picophytoplankton and increasing cell size of *Prochlorococcus* with latitude towards the equator (Viviani et al., 2011). Thus, it is clear that there is considerable uncertainty in the conversion factors, but in the absence of general trends for the cell size variability of each group under all conditions, our estimate of 0.74 Pg C represents our best estimate of the global picophytoplankton biomass.

Le Quéré et al. (2005) estimated that the global picophytoplankton biomass, including nitrogen fixers, is 0.28 Pg C. Our estimate, excluding nitrogen fixers, is considerably higher at 0.74 Pg C, and even our estimate using the indirect conversion factors is still almost double at 0.53 Pg C. Le Quéré et al. (2005) suggested that a third of global phytoplankton biomass is in the pico size class. Therefore, a 2–3-fold difference in the estimated picophytoplankton biomass

would not only be important for calculating the relative contribution that picophytoplankton make to the phytoplankton but also for calculating the total biomass of phytoplankton as the base of the ocean ecosystem.

For picoeukaryotes, the definition of the size range to be included is a major source of ambiguity. Whether phytoplankton between 2 and 3 μm diameter are included as picophytoplankton not only affects the abundance of the picoeukaryotes, but also which conversion factor is applicable. Here, we have included measurements of cells up to 3 μm diameter in the carbon conversion factor (Table 2). As a consequence, our conclusion that picoeukaryotes constitute 69 % of global picophytoplankton biomass critically depends on the definition of the size cut-off.

In summary, thanks to the routine use of flow cytometry for measurement of picophytoplankton abundance, we obtained a global dataset with reasonable coverage. The two main issues that deserve future attention are better resolution of cell sizes and better sampling coverage in the Southern Hemisphere.

Acknowledgements. We thank Claude Belzile, Jacques Neveux and Geneviève Tremblay for their comments on a draft manuscript, the EU (CarboChange, contract 264879) for financial support to ETB, and the Networks of Centres of Excellence of Canada-ArcticNet for financial support to MG. We thank the reviewers for their helpful comments.

Edited by: S. Pesant

References

Alvain, S., Moulin, C., Dandonneau, Y., and Breon, F. M.: Remote sensing of phytoplankton groups in case 1 waters from global SeaWiFS imagery, Deep-Sea Res. Pt. I, 52, 1989–2004, 2005.

Bertilsson, S., Berglund, O., Karl, D. M., and Chisholm, S. W.: Elemental composition of marine Prochlorococcus and Synechococcus: Implications for the ecological stoichiometry of the sea, Limnol. Oceanogr., 48, 1721–1731, 2003.

Binder, B. J., Chisholm, S. W., Olson, R. J., Frankel, S. L., and Worden, A. Z.: Dynamics of picophytoplankton, ultraphytoplankton and bacteria in the central equatorial Pacific, Deep-Sea Res. Pt. II, 43, 907–931, 1996.

Blanchot, J. and Rodier, M.: Picophytoplankton abundance and biomass in the western tropical Pacific Ocean during the 1992 El Nino year: Results from flow cytometry, Deep-Sea Res. Pt. I, 43, 877–895, 1996.

Blanchot, J., Andre, J. M., Navarette, C., Neveux, J., and Radenac, M. H.: Picophytoplankton in the equatorial Pacific: vertical distributions in the warm pool and in the high nutrient low chlorophyll conditions, Deep-Sea Res. Pt. I, 48, 297–314, 2001.

Bouman, H. A., Ulloa, O., Barlow, R., Li, W. K. W., Platt, T., Zwirglmaier, K., Scanlan, D. J., and Sathyendranath, S.: Water-column stratification governs the community structure of subtropical marine picophytoplankton, Environmental Microbiology Reports, 3, 473–482, 2011.

Buck, K. R., Chavez, F. P., and Campbell, L.: Basin-wide distributions of living carbon components and the inverted trophic pyramid of the central gyre of the North Atlantic Ocean, summer 1993, Aquat. Microb. Ecol., 10, 283–298, 1996.

Buitenhuis, E. T., Li, W. K. W., Lomas, M. W., Karl, D. M., Landry, M. R., and Jacquet, S.: Bacterial biomass distribution in the global ocean, Earth Syst. Sci. Data Discuss., 5, 301–315, doi:10.5194/essdd-5-301-2012, 2012a.

Buitenhuis, E. T., Vogt, M., Bednarsek, N., Doney, S. C., Leblanc, K., Le Quéré, C., Luo, Y.-W., Moriarty, R., O'Brien, C., O'Brien, T., Peloquin, J., and Schiebel, R.: MAREDAT: Towards a World Ocean Atlas of MARine Ecosystem DATa, Earth Syst. Sci. Data Discuss., in preparation, 2012b.

Cailliau, C., Claustre, H., Vidussi, F., Marie, D., and Vaulot, D.: Carbon biomass, and gross growth rates as estimated from C-14 pigment labelling, during photoacclimation in Prochlorococcus CCMP 1378, Mar. Ecol.-Prog. Ser., 145, 209–221, 1996.

Campbell, L., Nolla, H. A., and Vaulot, D.: The Importance of Prochlorococcus to Community Structure in the Central North Pacific-Ocean, Limnol. Oceanogr., 39, 954–961, 1994.

Campbell, L., Liu, H. B., Nolla, H. A., and Vaulot, D.: Annual variability of phytoplankton and bacteria in the subtropical North Pacific Ocean at Station ALOHA during the 1991–1994 ENSO event, Deep-Sea Res. Pt. I, 44, 167–192, 1997.

Campbell, L., Landry, M. R., Constantinou, J., Nolla, H. A., Brown, S. L., Liu, H., and Caron, D. A.: Response of microbial community structure to environmental forcing in the Arabian Sea, Deep-Sea Res. Pt. II, 45, 2301–2325, 1998.

Campbell, L., Carpenter, E. J., Montoya, J. P., Kustka, A. B., and Capone, D. G.: Picoplankton community structure within and outside a Trichodesmium bloom in the southwestern Pacific Ocean, Vie et Milieu, 55, 185–195, 2005.

Cavender-Bares, K. K., Rinaldo, A., and Chisholm, S. W.: Microbial size spectra from natural and nutrient enriched ecosystems, Limnol. Oceanogr., 46, 778–789, 2001.

Claustre, H., Bricaud, A., Babin, M., Bruyant, F., Guillou, L., Le Gall, F., Marie, D., and Partensky, F.: Diel variations in Prochlorococcus optical properties, Limnol. Oceanogr., 47, 1637–1647, 2002.

DuRand, M. D. and Olson, R. J.: Contributions of phytoplankton light scattering and cell concentration changes to diel variations in beam attenuation in the equatorial Pacific from flow cytometric measurements of pico-, ultra- and nanoplankton, Deep-Sea Res. Pt. II, 43, 891–906, 1996.

DuRand, M. D., Olson, R. J., and Chisholm, S. W.: Phytoplankton population dynamics at the Bermuda Atlantic Time-series station in the Sargasso Sea, Deep-Sea Res. Pt. II, 48, 1983–2003, 2001.

DuRand, M. D., Green, R. E., Sosik, H. M., and Olson, R. J.: Diel variations in optical properties of Micromonas pusilla (Prasinophyceae), J. Phycol., 38, 1132–1142, 2002.

Dusenberry, J. A. and Frankel, S. L.: Increasing the Sensitivity of a Facscan Flow Cytometer to Study Oceanic Picoplankton, Limnol. Oceanogr., 39, 206–209, 1994.

Echevarría, F., Zabala, L., Corzo, A., Navarro, G., Prieto, L., and Macias, D.: Spatial distribution of autotrophic picoplankton in relation to physical forcings: the Gulf of Cadiz, Strait of Gibraltar and Alboran Sea case study, J. Plankton Res., 31, 1339–1351, 2009.

Fu, F.-X., Warner, M. E., Zhang, Y., Feng, Y., and Hutchins, D. A.: Effects of increased temperature and CO_2 on photosynthesis, growth, and elemental ratios in marine Synechococcus and Prochlorococcus (Cyanobacteria), J. Phycol., 43, 485–496, 2007.

Garcia, C. M., Jimenez-Gomez, F., Rodriguez, J., Bautista, B., Estrada, M., Garcia Braun, J., Gasol, J. M., Figueiras, F. G., and Guerrero, F.: The size structure and functional composition of ultraplankton and nanoplankton at a frontal station in the Alboran Sea. Working groups 2 and 3 report, Sci. Mar., 58, 43–52, 1994.

Grob, C., Ulloa, O., Claustre, H., Huot, Y., Alarcón, G., and Marie, D.: Contribution of picoplankton to the total particulate organic carbon concentration in the eastern South Pacific, Biogeosciences, 4, 837–852, doi:10.5194/bg-4-837-2007, 2007.

Jacquet, S., Prieur, L., Nival, P., and Vaulot, D.: Structure and variability of the microbial community associated to the Alboran Sea frontal system (Western Mediterranean) in winter, J. Oceanogr., Research and Data, 3, 47–75, 2010.

Johnson, Z. I., Zinser, E. R., Coe, A., McNulty, N. P., Woodward, E. M. S., and Chisholm, S. W.: Niche partitioning among Prochlorococcus ecotypes along ocean-scale environmental gradients, Science, 311, 1737–1740, 2006.

Kana, T. M. and Glibert, P. M.: Effect of Irradiances up to 2000 $\mu E\,m^{-2}\,s^{-1}$ on Marine Synechococcus WH7803. 1. Growth, Pigmentation, and Cell Composition, Deep-Sea Res., 34, 479–495, 1987.

Landry, M. R., Kirshtein, J., and Constantinou, J.: Abundances and distributions of picoplankton populations in the central equatorial Pacific from 12° N to 12° S, 140° W, Deep-Sea Res. Pt. II, 43, 871–890, 1996.

Le Quéré, C., Harrison, S. P., Prentice, I. C., Buitenhuis, E. T., Aumont, O., Bopp, L., Claustre, H., Da Cunha, L. C., Geider, R., Giraud, X., Klaas, C., Kohfeld, K. E., Legendre, L., Manizza, M., Platt, T., Rivkin, R. B., Sathyendranath, S., Uitz, J., Watson, A.

J., and Wolf-Gladrow, D.: Ecosystem dynamics based on plankton functional types for global ocean biogeochemistry models, Glob. Change Biol., 11, 2016–2040, 2005.

Li, W. K. W.: Phytoplankton biomass and chlorophyll concentration across the North Atlantic, Sci. Mar., 58, 67–79, 1994.

Li, W. K. W.: Composition of Ultraphytoplankton in the Central North-Atlantic, Mar. Ecol.-Prog. Ser., 122, 1–8, 1995.

Li, W. K. W.: Cytometric diversity in marine ultraphytoplankton, Limnol. Oceanogr., 42, 874–880, 1997.

Li, W. K. W.: Macroecological patterns of phytoplankton in the northwestern North Atlantic Ocean, Nature, 419, 154–157, 2002.

Li, W. K. W.: From cytometry to macroecology: a quarter century quest in microbial oceanography, Aquat. Microb. Ecol., 57, 239–251, 2009.

Li, W. K. W. and Harrison, W. G.: Chlorophyll, bacteria and picophytoplankton in ecological provinces of the North Atlantic, Deep-Sea Res. Pt. II, 48, 2271–2293, 2001.

Li, W. K. W. and Wood, A. M.: Vertical Distribution of North-Atlantic Ultraphytoplankton – Analysis by Flow-Cytometry and Epifluorescence Microscopy, Deep-Sea Res., 35, 1615–1638, 1988.

Li, W. K. W., Dickie, P. M., Irwin, B. D., and Wood, A. M.: Biomass of Bacteria, Cyanobacteria, Prochlorophytes and Photosynthetic Eukaryotes in the Sargasso Sea, Deep-Sea Res., 39, 501–519, 1992.

Li, W. K. W., Zohary, T., Yacobi, Y. Z., and Wood, A. M.: Ultraphytoplankton in the Eastern Mediterranean-Sea – Towards Deriving Phytoplankton Biomass from Flow Cytometric Measurements of Abundance, Fluorescence and Light Scatter, Mar. Ecol.-Prog. Ser., 102, 79–87, 1993.

Li, W. K. W., McLaughlin, F. A., Lovejoy, C., and Carmack, E. C.: Smallest Algae Thrive As the Arctic Ocean Freshens, Science, 326, 539–539, 2009.

Lindell, D. and Post, A. F.: Ultraphytoplankton Succession Is Triggered by Deep Winter Mixing in the Gulf-of-Aqaba (Eilat), Red-Sea, Limnol. Oceanogr., 40, 1130–1141, 1995.

Liu, H. B., Bidigare, R. R., Laws, E., Landry, M. R., and Campbell, L.: Cell cycle and physiological characteristics of Synechococcus (WH7803) in chemostat culture, Mar. Ecol.-Prog. Ser., 189, 17–25, 1999.

Llewellyn, C. A. and Gibb, S. W.: Intra-class variability in the carbon, pigment and biomineral content of prymnesiophytes and diatoms, Mar. Ecol.-Prog. Ser., 193, 33–44, 2000.

Lomas, M. W., Roberts, N., Lipschultz, F., Krause, J. W., Nelson, D. M., and Bates, N. R.: Biogeochemical responses to late-winter storms in the Sargasso Sea. IV. Rapid succession of major phytoplankton groups, Deep-Sea Res. Pt. I, 56, 892–908, 2009.

Lomas, M. W., Steinberg, D. K., Dickey, T., Carlson, C. A., Nelson, N. B., Condon, R. H., and Bates, N. R.: Increased ocean carbon export in the Sargasso Sea linked to climate variability is countered by its enhanced mesopelagic attenuation, Biogeosciences, 7, 57–70, doi:10.5194/bg-7-57-2010, 2010.

Marie, D., Zhu, F., Balague, V., Ras, J., and Vaulot, D.: Eukaryotic picoplankton communities of the Mediterranean Sea in summer assessed by molecular approaches (DGGE, TTGE, QPCR), FEMS Microb. Ecol., 55, 403–415, 2006.

McManus, G. B. and Dawson, R.: Phytoplankton Pigments in the Deep Chlorophyll Maximum of the Caribbean Sea and the Western Tropical Atlantic-Ocean, Mar. Ecol.-Prog. Ser., 113, 199–206, 1994.

Montagnes, D. J. S., Berges, J. A., Harrison, P. J., and Taylor, F. J. R.: Estimating Carbon, Nitrogen, Protein, and Chlorophyll-a from Volume in Marine-Phytoplankton, Limnol. Oceanogr., 39, 1044–1060, 1994.

Moran, S. B., Lomas, M. W., Kelly, R. P., Gradinger, R., Iken, K., and Mathis, J. T.: Seasonal succession of net primary productivity, particulate organic carbon export, and autotrophic community composition in the eastern Bering Sea, Deep-Sea Res. Pt. II, 65–70, 84–97, 2012.

Moran, X. A. G., Lopez-Urrutia, A., Calvo-Diaz, A., and Li, W. K. W.: Increasing importance of small phytoplankton in a warmer ocean, Glob. Change Biol., 16, 1137–1144, 2010.

Neveux, J., Vaulot, D., Courties, C., and Fukai, E.: Green Photosynthetic Bacteria Associated with the Deep Chlorophyll Maximum of the Sargasso Sea, CR. Acad. Sci. III-Vie., 308, 9–14, 1989.

Neveux, J., Lantoine, F., Vaulot, D., Marie, D., and Blanchot, J.: Phycoerythrins in the southern tropical and equatorial Pacific Ocean: Evidence for new cyanobacterial types, J. Geophys. Res.-Oceans, 104, 3311–3321, 1999.

Olson, R. J., Chisholm, S. W., Zettler, E. R., Altabet, M. A., and Dusenberry, J. A.: Spatial and Temporal Distributions of Prochlorophyte Picoplankton in the North-Atlantic Ocean, Deep-Sea Res., 37, 1033–1051, 1990.

Partensky, F., Blanchot, J., Lantoine, F., Neveux, J., and Marie, D.: Vertical structure of picophytoplankton at different trophic sites of the tropical northeastern Atlantic Ocean, Deep-Sea Res. Pt. I, 43, 1191–1213, 1996.

Raven, J. A.: The twelfth Tansley Lecture. Small is beautiful: the picophytoplankton, Funct. Ecol., 12, 503–513, 1998.

Sherr, E. B., Sherr, B. F., and Wheeler, P. A.: Distribution of coccoid cyanobacteria and small eukaryotic phytoplankton in the upwelling ecosystem off the Oregon coast during 2001 and 2002, Deep-Sea Res. Pt. II, 52, 317–330, 2005.

Shimada, A., Hasegawa, T., Umeda, I., Kadoya, N., and Maruyama, T.: Spatial Mesoscale Patterns of West Pacific Picophytoplankton as Analyzed by Flow-Cytometry – Their Contribution to Subsurface Chlorophyll Maxima, Mar. Biol., 115, 209–215, 1993.

Shimada, A., Nishijima, M., and Maruyama, T.: Seasonal appearance of Prochlorococcus in Suruga Bay, Japan in 1992–1993, J. Oceanogr., 51, 289–300, 1995.

Sieburth, J. M., Smetacek, V., and Lenz, J.: Pelagic Ecosystem Structure – Heterotrophic Compartments of Plankton and Their Relationship to Plankton Size Fractions – Comment, Limnol. Oceanogr., 23, 1256–1263, 1978.

Simon, N., Barlow, R. G., Marie, D., Partensky, F., and Vaulot, D.: Characterization of Oceanic Photosynthetic Picoeukaryotes by Flow-Cytometry, J. Phycol., 30, 922–935, 1994.

Takahashi, M. and Hori, T.: Abundance of Picophytoplankton in the Subsurface Chlorophyll Maximum Layer in Sub-Tropical and Tropical Waters, Mar. Biol., 79, 177–186, 1984.

Tremblay, G., Belzile, C., Gosselin, M., Poulin, M., Roy, S., and Tremblay, J.-E.: Late summer phytoplankton distribution along a 3500 km transect in Canadian Arctic waters: strong numerical dominance by picoeukaryotes, Aquat. Microb. Ecol., 54, 55–70, 2009.

Vaulot, D., Partensky, F., Neveux, J., Mantoura, R. F. C., and Llewellyn, C. A.: Winter Presence of Prochlorophytes in Surface Waters of the Northwestern Mediterranean-Sea, Limnol.

Oceanogr., 35, 1156–1164, 1990.

Vaulot, D., Eikrem, W., Viprey, M., and Moreau, H.: The diversity of small eukaryotic phytoplankton (<=3 μm) in marine ecosystems, FEMS Microb. Rev., 32, 795–820, 2008.

Veldhuis, M. J. W. and Kraay, G. W.: Vertical-Distribution and Pigment Composition of a Picoplanktonic Prochlorophyte in the Subtropical North-Atlantic – a Combined Study of Hplc-Analysis of Pigments and Flow-Cytometry, Mar. Ecol.-Prog. Ser., 68, 121–127, 1990.

Veldhuis, M. J. W. and Kraay, G. W.: Cell Abundance and Fluorescence of Picoplankton in Relation to Growth Irradiance and Nitrogen Availability in the Red-Sea, Neth. J. Sea Res., 31, 135–145, 1993.

Veldhuis, M. J. W., Kraay, G. W., and Gieskes, W. W. C.: Growth and Fluorescence Characteristics of Ultraplankton on a North South Transect in the Eastern North-Atlantic, Deep-Sea Res. Pt. II, 40, 609–626, 1993.

Veldhuis, M. J. W., Kraay, G. W., VanBleijswijk, J. D. L., and Baars, M. A.: Seasonal and spatial variability in phytoplankton biomass,

productivity and growth in the northwestern Indian Ocean: The southwest and northeast monsoon, 1992–1993, Deep-Sea Res. Pt. I, 44, 425–449, 1997.

Verity, P. G., Robertson, C. Y., Tronzo, C. R., Andrews, M. G., Nelson, J. R., and Sieracki, M. E.: Relationships between Cell-Volume and the Carbon and Nitrogen-Content of Marine Photosynthetic Nanoplankton, Limnol. Oceanogr., 37, 1434–1446, 1992.

Viviani, D. A., Bjoerkman, K. M., Karl, D. M., and Church, M. J.: Plankton metabolism in surface waters of the tropical and subtropical Pacific Ocean, Aquat. Microb. Ecol., 62, 1–12, 2011.

Worden, A. Z., Nolan, J. K., and Palenik, B.: Assessing the dynamics and ecology of marine picophytoplankton: The importance of the eukaryotic component, Limnol. Oceanogr., 49, 168–179, 2004.

Yacobi, Y. Z., Zohary, T., Kress, N., Hecht, A., Robarts, R. D., Waiser, M., and Wood, A. M.: Chlorophyll Distribution Throughout the Southeastern Mediterranean in Relation to the Physical Structure of the Water Mass, J. Marine Syst., 6, 179–190, 1995.

A high-frequency atmospheric and seawater $p\mathrm{CO}_2$ data set from 14 open-ocean sites using a moored autonomous system

A. J. Sutton[1,2], C. L. Sabine[2], S. Maenner-Jones[2], N. Lawrence-Slavas[2], C. Meinig[2], R. A. Feely[2], J. T. Mathis[2], S. Musielewicz[1,2], R. Bott[2], P. D. McLain[2], H. J. Fought[3], and A. Kozyr[4]

[1]Joint Institute for the Study of the Atmosphere and Ocean, University of Washington, Seattle, Washington, USA
[2]Pacific Marine Environmental Laboratory, National Oceanic and Atmospheric Administration, Seattle, Washington, USA
[3]Battelle Memorial Institute, Columbus, Ohio, USA
[4]Carbon Dioxide Information Analysis Center, Oak Ridge National Laboratory, Department of Energy, Oak Ridge, Tennessee, USA

Correspondence to: A. J. Sutton (adrienne.sutton@noaa.gov)

Abstract. In an intensifying effort to track ocean change and distinguish between natural and anthropogenic drivers, sustained ocean time series measurements are becoming increasingly important. Advancements in the ocean carbon observation network over the last decade, such as the development and deployment of Moored Autonomous $p\mathrm{CO}_2$ (MAPCO2) systems, have dramatically improved our ability to characterize ocean climate, sea–air gas exchange, and biogeochemical processes. The MAPCO2 system provides high-resolution data that can measure interannual, seasonal, and sub-seasonal dynamics and constrain the impact of short-term biogeochemical variability on carbon dioxide (CO_2) flux. Overall uncertainty of the MAPCO2 using in situ calibrations with certified gas standards and post-deployment standard operating procedures is $< 2\,\mu$atm for seawater partial pressure of CO_2 ($p\mathrm{CO}_2$) and $< 1\,\mu$atm for air $p\mathrm{CO}_2$. The MAPCO2 maintains this level of uncertainty for over 400 days of autonomous operation. MAPCO2 measurements are consistent with shipboard seawater $p\mathrm{CO}_2$ measurements and GLOBALVIEW-CO2 boundary layer atmospheric values. Here we provide an open-ocean MAPCO2 data set including over 100 000 individual atmospheric and seawater $p\mathrm{CO}_2$ measurements on 14 surface buoys from 2004 through 2011 and a description of the methods and data quality control involved. The climate-quality data provided by the MAPCO2 have allowed for the establishment of open-ocean observatories to track surface ocean $p\mathrm{CO}_2$ changes around the globe.

1 Introduction

The global ocean as well as its interactions with the atmosphere, climate, and marine ecosystem is undergoing a rapid and dramatic transition as it responds to multiple drivers on timescales from days to decades. Sustained observations guide our understanding of this ever-evolving earth system, which, in turn, informs the development of solutions for human societies to cope with global change. The iconic Mauna Loa atmospheric carbon dioxide (CO_2) time series, or "Keeling curve", is an example of how observations gain importance with time, as they provide the basis for understanding future changes to the earth system in the context of current and historical observations (Keeling et al., 1976; Thoning et al., 1989; Hofmann et al., 2009). Similar "ocean observatories" must be sustained in order to track ocean carbon uptake and ocean acidification in the midst of the large natural temporal and spatial variability in the marine environment. These observations will provide a record of past and current behavior of the ocean carbon system and are central to predicting its future.

While high-quality ocean carbon measurements collected on global hydrographic surveys have been carried out approximately once a decade since the 1980s, the scientific community identified that constraining ocean biogeochemical models would require much greater temporal and spatial resolution of field data (Sabine et al., 2010). Autonomous technology to measure surface ocean carbon was developed to address this need and has undergone rapid advancement in the last three decades (Takahashi, 1961; Weiss et al., 1982; Wanninkhof and Thoning, 1993; Feely et al., 1998; Pierrot et al., 2009). Autonomous underway systems that can measure the partial pressure of CO_2 (pCO_2) on ships were the first major breakthrough in our ability to collect high-frequency observations in the global ocean. These systems are designed to produce climate-quality data sets with measurements accurate to within $1\,\mu$atm for atmospheric CO_2 and $2\,\mu$atm for surface seawater pCO_2. This level of accuracy has allowed the scientific community to constrain regional sea–air CO_2 fluxes to $0.2\,\mathrm{Pg\,C\,yr^{-1}}$, a level of resolution necessary to test process-based models and predict the future behavior of the carbon cycle (Bender et al., 2002; Pierrot et al., 2009).

While underway pCO_2 observations have greatly enhanced our understanding of the spatial variability in sea–air CO_2 fluxes (Takahashi et al., 2009; Wanninkhof et al., 2013), they have not solved the problem of quantifying temporal variability at a given point in space. In highly variable regions such as the equatorial Pacific and coastal systems, fixed, high-frequency observations can improve our understanding of how short-term variability impacts CO_2 flux. Episodic phenomena are important drivers of biogeochemical variability, and mooring time series of pCO_2 and related properties provide the ability to assess the controls and impacts at these short timescales. Seawater pCO_2 observations that fully capture diurnal variations at a fixed site can also be used to test parameterizations of carbon cycle processes used in ocean biogeochemical models. The Moored Autonomous pCO_2 (MAPCO2) system was developed to address this need by autonomously measuring surface ocean pCO_2 and marine boundary layer (MBL) atmospheric CO_2 every 3 h on surface buoys at approximately the same level of accuracy as underway pCO_2 systems. With this recent development of mooring autonomous pCO_2 technology, the combination of all three monitoring approaches (i.e., hydrographic surveys, underway, and buoy measurements) has improved our understanding of the spatial and temporal variability of ocean carbon at the sea surface. For the first time, ocean pCO_2 observations from multiple platforms have been incorporated into the most recent update (v2.0) of the Surface Ocean CO_2 Atlas (SOCAT), a data synthesis effort aimed at bringing together all available CO_2 data in the surface ocean in a common

format (Bakker et al., 2014). The data presented here are identical to those in SOCATv2.0.

Here we describe the methods, data quality control (QC), and data access for an open-ocean MAPCO2 data set collected on 14 surface buoys from 2004 through 2011. These surface ocean pCO_2 observatories are critical for characterizing the natural variability of the ocean carbon cycle, contributing to our understanding of secular trends in ocean chemistry, validating and interpreting modeling results, and developing more sophisticated global carbon models.

2   Methods and data quality control

In 2004, the National Oceanographic and Atmospheric Administration's (NOAA) Pacific Marine Environmental Laboratory (PMEL) began to work with the Monterey Bay Aquarium Research Institute to improve the accuracy, reliability, and ease of use of an early moored pCO_2 system developed for buoys in the equatorial Pacific. Like the well-established underway pCO_2 method (Wanninkhof and Thoning, 1993; Feely et al., 1998; Pierrot et al., 2009), this early moored system described by Friederich et al. (1995) and the MAPCO2 system described in Sect. 2.1 combine air–water equilibrators with an infrared (IR) analyzer for CO_2 gas detection. In 2009, the MAPCO2 technology was transferred to Battelle Memorial Institute and is commercially available as the Sealogy® pCO_2 monitoring system. This system is now accessible to the larger scientific community and deployed at over 50 locations in open-ocean, coastal, and coral reef environments, including on NOAA's global moored CO_2 network (www.pmel.noaa.gov/co2/story/Buoys+and+ Autonomous+Systems) and Australia's Integrated Marine Observing System (http://imos.org.au).

2.1   Description of MAPCO2 system

The MAPCO2 system includes four separate watertight cases that house the electronics, battery, transmitter, and a reference gas cylinder. The reference gases used on all the PMEL systems are traceable to World Meteorological Organization (WMO) standards and are provided by NOAA's Earth System Research Laboratory (ESRL). In the electronics case are the controls for the system, a memory flash card for data storage, a LI-COR LI-820 CO_2 gas analyzer, and a Sensirion SHT71 relative humidity and temperature sensor. The MAPCO2 also includes an oxygen sensor for internal diagnostic purposes. The LI-820 determines the CO_2 gas concentration by measuring the absorption of IR energy as a sample gas flows through an optical path. The CO_2 concentration is based on the difference ratio in the IR absorption between a reference and a sample optical path. The MAPCO2 uses temperature and relative humidity (RH) to calculate the mole fraction of CO_2 (xCO_2) in air in equilibrium with surface seawater. The LI-820 is calibrated before every measurement using a zero-CO_2 reference and an ESRL standard gas that

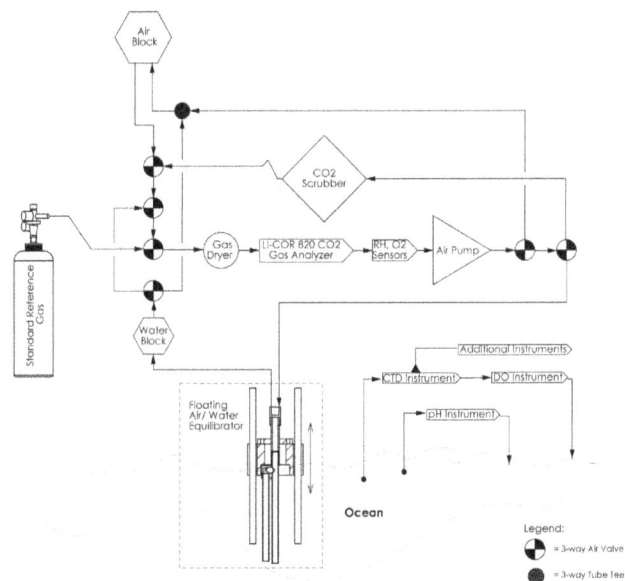

Figure 1. Schematic diagram of main components and sampling paths within the MAPCO$_2$ system. The floating air–water equilibrator is shown in more detail in Fig. 2.

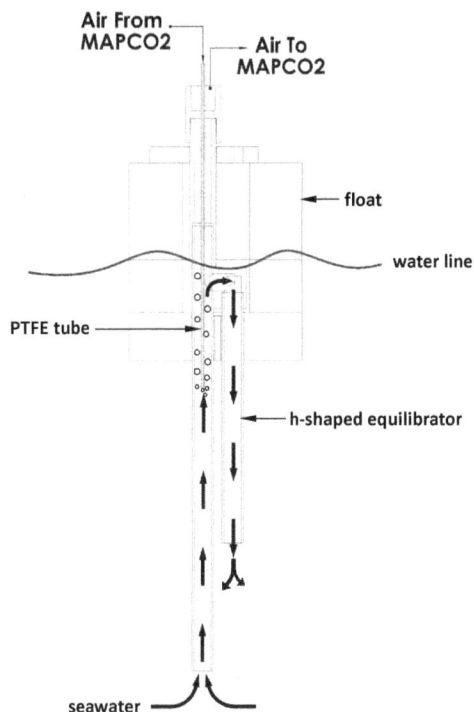

Figure 2. Schematic diagram of the floating air–water equilibrator assembly in the MAPCO$_2$ system during the seawater equilibration cycle. Air is pumped from the MAPCO$_2$ through a PTFE tube and bubbled into the equilibrator. As the bubbles rise through the water, the air comes into equilibrium with the dissolved gases in the surface seawater. The rising air bubbles in the equilibrator also create circulation by pushing water up and over the horizontal leg of the h-shaped equilibrator and out the short leg of the equilibrator. Image is not to scale.

spans the ocean pCO$_2$ values where the system is deployed. The system also includes a GPS for accurate position and time, an iridium satellite communication link, an airblock deployed approximately 1 m above the ocean surface for atmospheric sampling, and an "h"-shaped bubble equilibrator assembly described by Friederich et al. (1995) (Figs. 1, 2). The equilibrator is the only part of the system in seawater and is made of copper–nickel alloy to prevent bio-fouling.

A schematic diagram of the main components and sampling paths in the MAPCO$_2$ system is shown in Fig. 1. A typical measurement cycle, including in situ calibration and the atmospheric and seawater measurements, takes approximately 20 min. At the beginning of each cycle, the system generates a zero standard by cycling a closed loop of air through a soda lime tube to remove all of the CO$_2$. This scrubbed air establishes the zero calibration. Next, the system is calibrated with a high standard reference gas, or "span" gas. The value of this gas is set in the MAPCO$_2$ system before deployment (typically $\sim 500\,\mu$mol mol^{-1}). The gas flows through the detector for CO$_2$ analysis and is vented to the atmosphere through the airblock. Once the detector is fully flushed, the flow is stopped and the system returns to atmospheric pressure. Using a two-point calibration from the zero and span values, the LI-820 is optimized for making surface ocean CO$_2$ measurements.

To make the seawater xCO$_2$ measurement, the MAPCO$_2$ system equilibrates a closed loop of air with surface seawater in the h-shaped equilibrator, which is mounted in a float designed by PMEL to ensure the optimum depth for equilibration (Fig. 2). The air cycles through the system by pumping air out of flexible polytetrafluoroethylene (PTFE) tubing

to approximately 14 cm beneath the surface of the seawater. While the air bubbles through the column of water, the air comes into equilibrium with the dissolved gases in the surface seawater. This air then returns to the system, passing through a silica gel drying agent and the relative humidity sensor. The drying agent is used to prevent condensation in the LI-820 detector and is replaced after each deployment. The air then circulates through the equilibrator again. The closed loop of air repeats this cycle for 10 min. The rising air bubbles in the equilibrator create seawater circulation in the equilibrator by pushing the water up and over the horizontal leg of the equilibrator and out the short leg of the equilibrator (Fig. 2). This draws new water into the long leg of the equilibrator, ensuring that the recirculated air is always in contact with new seawater. After 10 min of equilibration, the pump is stopped and the LI-820 values are read on the air sample at 2 Hz for 30 s and averaged to give the seawater xCO$_2$ measurement. This is a measurement of integrated seawater CO$_2$ levels during the 10 min equilibration time.

After the equilibrator reading, a MBL air reading is made by drawing air in through the airblock, partially drying it, and

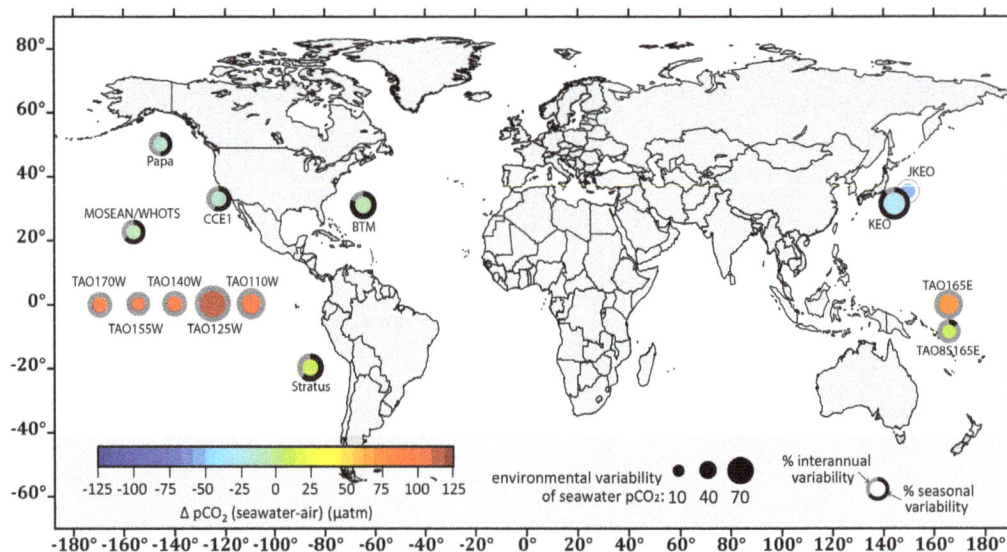

Figure 3. Location of open-ocean moorings in the $MAPCO_2$ data set. Inner circle color illustrates the mean ΔpCO_2 of the finalized data at that location. Inner circle size is relative to the environmental variability in the time series defined here as the standard deviation of seawater pCO_2 values. The outer ring shows the proportion of environmental variability in seawater pCO_2 due to the seasonal cycle (black) and interannual variability (gray). Seasonal variability is defined as the mean seasonal peak amplitude, and interannual variability is the mean Δ of annual mean values. Seasonal and interannual variability cannot be quantified at JKEO with a time series of < 1 year and is represented here by an outer ring with no color.

passing it through the LI-820. Once the LI-820 path has been flushed, the flow is stopped and a 30 s average reading is collected. All measurements and calibrations are made at atmospheric pressure. The seawater CO_2 measurement occurs approximately 17 min after the start of the measurement cycle followed by the air CO_2 measurement 2 min later. Response time of the $MAPCO_2$ is dictated by the length of the full 20 min measurement cycle, and in fast mode the $MAPCO_2$ system can measure atmospheric and seawater CO_2 once every 30 min.

Different types of sensors are used throughout the system for analytical, troubleshooting, and data quality control purposes. Additional parameters measured in each cycle (i.e., zero, span, equilibrator, and air) include temperature, pressure, relative humidity, and oxygen. Other sensors can also be integrated into the $MAPCO_2$, including CTD (conductivity, temperature, and depth) instruments with auxiliary sensors attached (e.g., dissolved oxygen, fluorescence, turbidity) and pH sensors. The raw data collected by the $MAPCO_2$ and integrated sensors are stored on a memory flash card, and averaged data from each 3-hourly cycle are telemetered from the buoy via the iridium satellite communications system. This communications system also enables the user to control the $MAPCO_2$ remotely. The user can determine the sampling frequency and other variables, but the $MAPCO_2$ is nominally designed to make CO_2 measurements every 3 h with daily data transmissions for at least 400 days.

PMEL's $MAPCO_2$ systems have been deployed on open-ocean buoys starting in 2004 with the establishment of

NOAA's global moored CO_2 network and the efforts of numerous partners (see Acknowledgements). Table 1 lists the mooring coordinates and dates of CO_2 time series operation; Fig. 3 illustrates the locations, number of measurements, and average ΔpCO_2 (sea–air) from the 14 surface CO_2 buoys included in this data set. These mooring ΔpCO_2 observations are consistent with results of a synthesis of underway observations reported by Takahashi et al. (2009). Other than the Bermuda Testbed Mooring (BTM) and Japanese Kuroshio Extension Observatory (JKEO) time series, which have been discontinued, and the Multi-disciplinary Ocean Sensors for Environmental Analyses and Networks (MOSEAN) buoy, which was moved approximately 20 km to the new Woods Hole Oceanographic Institution (WHOI) Hawaii Ocean Time-Series Station (WHOTS) location, the $MAPCO_2$ time series shown in Fig. 3 and Table 1 continue to be maintained. Seven of the 14 CO_2 buoys are located in the equatorial Pacific on the Tropical Atmosphere Ocean (TAO) array. Additional open-ocean $MAPCO_2$ sites maintained by PMEL now exist in the North Atlantic, northern Indian, and Southern oceans (see http://www.pmel.noaa.gov/co2/story/Buoys+and+Autonomous+Systems); however, they have been deployed since 2011 and are not included in the finalized data set presented here.

2.2 Data reduction and processing

The IR analyzer has a nonlinear response to CO_2, but that response is very well characterized by the manufacturer.

Table 1. Details of each CO_2 mooring time series including name, coordinates (decimal degrees), and dates of operational CO_2 measurements.

Abbreviation	Full Name	Latitude	Longitude	Year established	Current Status
MOSEAN	Multi-disciplinary Ocean Sensors for Environmental Analyses and Networks	22.8	−158.1	2004	moved to WHOTS in 2007
WHOTS	WHOI Hawaii Ocean Time-Series Station	22.7	−158.0	2007	ongoing
BTM	Bermuda Testbed Mooring	31.5	−64.0	2005	discontinued in 2007
Papa	Papa	50.1	−144.8	2007	ongoing
KEO	Kuroshio Extension Observatory	32.3	144.6	2007	ongoing
JKEO	Japanese Kuroshio Extension Observatory	37.9	146.6	2007	discontinued in 2007
CCE1	California Current Ecosystem 1	33.5	−122.5	2008	ongoing
Stratus	Stratus	−19.7	−85.6	2006	ongoing
TAO110W	Tropical Atmosphere Ocean 0°, 110° W	0.0	−110.0	2009	ongoing
TAO125W	Tropical Atmosphere Ocean 0°, 125° W	0.0	−125.0	2004	ongoing
TAO140W	Tropical Atmosphere Ocean 0°, 140° W	0.0	−140.0	2004	ongoing
TAO155W	Tropical Atmosphere Ocean 0°, 155° W	0.0	−155.0	2010	ongoing
TAO170W	Tropical Atmosphere Ocean 0°, 170° W	0.0	−170.0	2005	ongoing
TAO165E	Tropical Atmosphere Ocean 0°, 165° E	0.0	165.0	2010	ongoing
TAO8S165E	Tropical Atmosphere Ocean 8° S, 165° E	−8.0	165.0	2009	ongoing

LI-COR has a function built into their firmware that accounts for the nonlinear response and linearizes the output data. The linear function is calibrated prior to each atmospheric and seawater measurement with the zero- (intercept) and high-CO_2 standard reference gas (slope). The accuracy of the linearized, calibrated output is confirmed prior to deployment by analyzing a range of intermediate-CO_2 standards in our laboratory.

The primary check of accuracy before and after deployment is a comparison to ESRL CO_2 standards traceable to WMO standards, typically six standards that range from 0 to $< 800\,\mu\text{mol mol}^{-1}$. Systems are not certified for deployment until values are within the expected range of the standards that span the typical seawater CO_2 values at the mooring location (typically within $2\,\mu\text{mol mol}^{-1}$). A comparison to the underway pCO_2 system in the lab is then done to assess stability of the measurements over at least 1 week. During this test, each MAPCO$_2$ is tested in a seawater tank in the lab against another MAPCO$_2$ system and a General Oceanics 8050 underway pCO_2 system that are permanently mounted for continuous sampling in the seawater tank. The standard MAPCO$_2$ is regularly compared to the underway system, which is calibrated every 8 h using four standard reference gases from approximately 0 to $1000\,\mu\text{mol mol}^{-1}$. Laboratory testing of the MAPCO$_2$ systems suggests instrument precision is $< 0.6\,\mu\text{mol mol}^{-1}$ for xCO_2 values between 100 and $600\,\mu\text{mol mol}^{-1}$.

When the MAPCO$_2$ is recovered from the field, the system is compared against six gas standards to verify accuracy, and the high-frequency raw data stored on the internal memory flash card are downloaded to a local database. The high-frequency raw data from each 3-hourly cycle are then used for final processing of each data set. Averaged xCO_2 (wet) seawater and atmospheric measurements (defined in Table 2)

from each cycle are calculated starting with the raw detector counts using the published LI-COR function. The span gas coefficients used in the function during post-processing are derived from the linear regression between the calibration coefficients and the corresponding LI-820 temperature measurements acquired during the span cycle over the course of the deployment. This post-deployment reprocessing facilitates the accurate calculation of xCO_2 (wet) values from the raw detector counts when rare miscalibrations occur, resulting in erroneous coefficients during the deployment. Since the LI-820 is calibrated prior to each cycle of xCO_2 (wet) measurements using the zero- and high-CO_2 standard reference gas, detector drift is negligible. This is confirmed by a mean difference between corrected and original raw data of $-0.02\,\mu\text{mol mol}^{-1}$.

Data are quality-controlled and flagged according to the SOCAT guidelines (Pfeil et al., 2013). For pCO_2 mooring purposes, we use three quality flags (QFs): a flag value of 2 represents an acceptable measurement, 3 is a questionable measurement, and 4 is a bad measurement. A measurement can be questionable for a variety of reasons often revealed by MAPCO$_2$ system diagnostic information (e.g., low equilibrator pressure causing incomplete seawater equilibration), and the reasoning for each flag is included in the metadata QC log so the end user can decide whether or not to use questionable data. Prior to a data QC software update in June 2013, xCO_2 values flagged as bad (QF$=4$) were still included in the published data sets, but after the software update bad values are replaced with -999. Other parameters published in the data sets that do not have an associated flag, such as sea surface temperature (SST) and sea surface salinity (SSS), are given a value of -999 or -9.999 when the measurement is missing or bad.

Table 2. Final data variable names and descriptions.

Variable name	Description	Units	Equation (if applicable)
Mooring	mooring name as shown in Fig. 2 and Table 1	character string	
Latitude	average latitude during deployment	decimal degrees	
Longitude	average longitude during deployment	decimal degrees	
Date	date of measurement in UTC	MM/DD/YYYY	
Time	time of measurement in UTC	HH:MM	
xCO2_SW_wet	mole fraction of carbon dioxide in air in equilibrium with surface seawater at SST and humidity	μmol mol^{-1}	
xCO2_SW_QF	primary flag associated with seawater xCO$_2$ measurement	WOCE standards[a]	
H2O_SW	mole fraction of water in gas from equilibrator	μmol mol^{-1}	
xCO2_Air_wet	mole fraction of carbon dioxide in air at ~ 1.5 m above the sea surface at sample humidity	μmol mol^{-1}	
xCO2_SW_QF	primary flag associated with air xCO$_2$ measurement	WOCE standards[a]	
H2O_Air	mole fraction of water in air	μmol mol^{-1}	
Licor_Atm_Pressure	atmospheric pressure at ~ 1.5 m above the sea surface	hPa	
Licor_Temp	licor temperature	°C	
Percent_O2[b]	% oxygen in surface seawater divided by % oxygen in air at ~ 1.5 m above the sea surface	%	
SST[c]	sea surface temperature	°C	
SSS[c]	sea surface salinity		
xCO2_SW_dry	mole fraction of carbon dioxide in dry air in equilibrium with surface seawater	μmol mol^{-1}	1
xCO2_Air_dry	mole fraction of carbon dioxide in dry air at ~ 1.5 m above the sea surface	μmol mol^{-1}	1
fCO2_SW_sat	fugacity of carbon dioxide in wet air (100 % humidity) in equilibrium with surface seawater	μatm	4
fCO2_Air_sat	fugacity of carbon dioxide in wet air (100 % humidity) at ~ 1.5 m above the sea surface	μatm	4
dfCO2	fCO2_SW_sat – fCO2_Air_sat	μatm	
pCO2_SW_sat[d]	partial pressure of carbon dioxide in wet air (100 % humidity) in equilibrium with surface seawater	μatm	5
pCO2_Air_sat[d]	partial pressure of carbon dioxide in wet air (100 % humidity) at ~ 1.5 m above the sea surface	μatm	5
dpCO2[d]	pCO2_SW_sat – pCO2_Air_sat	μatm	

Notes: [a] SOCAT flags used in this data set: 2 = acceptable measurement; 3 = questionable measurement; 4 = bad measurement (note: bad data values are reported in the final data file submitted to CDIAC prior to QC software upgrade in June 2013 but reported as −999 in files submitted after the upgrade). [b] Oxygen measured in the MAPCO$_2$ system is exposed to air and likely modified within the system prior to measurement. Rapid changes in oxygen are not properly captured using this method. This data should not be used as a quantitative measure of oxygen. [c] Usually measured by other academic partners at each site. See metadata for each deployment for details on SST and SSS measurements. [d] pCO2 only presented in data sets submitted to CDIAC after June 2013 when QC software was upgraded to include this calculation. Data users of earlier data sets can calculate pCO2 as defined in Eq. (4).

As a final check of the data QC process, atmospheric xCO$_2$ (dry) data are compared to MBL data from the GLOBALVIEW-CO2 product and the MAPCO$_2$ systems deployed before and after the deployment of interest (GLOBALVIEW-CO2, 2013). When a MAPCO$_2$ system is recovered and a new system deployed, there is typically some overlap in measurements at each location. In cases when there is an offset in air xCO$_2$ values between systems at the same location, which is often corroborated by an offset from the GLOBALVIEW-CO2 MBL time series as well, a correction (typically $\leq 3\,\mu$mol mol^{-1}) is applied to the atmospheric and seawater xCO$_2$ (wet) values. This correction is noted in the metadata and can be removed by the data user if desired. The GLOBALVIEW-CO2 MBL data set serves as a useful and unifying comparison data set, especially since other in situ comparison data are often lacking. As we build MAPCO$_2$ time series at each of these locations, we start to build an understanding of how the MAPCO$_2$ observations typically compare to the MBL data set. For example, winter atmospheric xCO$_2$ values measured by our MAPCO$_2$ systems at Papa are consistently lower than MBL values (Fig. 4a).

Post-QC calculation of pCO$_2$ and fCO$_2$ (fugacity of CO$_2$) are made according to recommendations of the underway pCO$_2$ community (Pierrot et al., 2009). However, MAPCO$_2$ measurements of xCO$_2$ vary from the underway pCO$_2$ method. The MAPCO$_2$ system uses the LI-820 and

the RH to report the mole fraction of CO$_2$ in air in equilibrium with surface seawater, called xCO$_2$ (wet). This "partially wet" measurement typically has a RH of ~ 75 % (seawater and atmospheric samples), which is not completely dried as in the underway pCO$_2$ method, due to lack of drying methods available for extended autonomous operation. However, since we measure RH and temperature of the sample air stream exiting the LI-820, we can calculate xCO$_2$ (dry) using Eqs. (1)–(3). First, xCO$_2$ in dry air is calculated by

$$x\text{CO}_2\,(\text{dry}) = x\text{CO}_2\,(\text{wet}) \times \frac{P_{\text{Licor}}}{P_{\text{Licor}} - VP_{\text{Licor}}}, \quad (1)$$

where xCO$_2$ (wet) is the LI-820 measured concentration (μmol mol^{-1}), P_{Licor} is the pressure of the atmospheric and seawater samples measured in the LI-820 (kPa) and considered atmospheric pressure, and VP_{Licor} is the vapor pressure in the LI-820 (kPa). RH measurements of the air samples exiting the LI-820 are used to calculate VP_{Licor} in Eq. (1) using the following as defined by Buck (1981) and LI-COR for the IR analyzers:

$$VP_{\text{sat}} = (0.61121)(1.004)e^{\left(\frac{17.502 \times T_{\text{RH}}}{240.97 + T_{\text{RH}}}\right)} \quad (2)$$

$$VP_{\text{Licor}} = (\text{RH}_{\text{sample}} - \text{RH}_{\text{span}}) \times \frac{VP_{\text{sat}}}{100}, \quad (3)$$

where VP_{sat} is the saturation vapor pressure of the RH sensor cell (kPa); T_{RH} is the temperature of the RH sensor (°C);

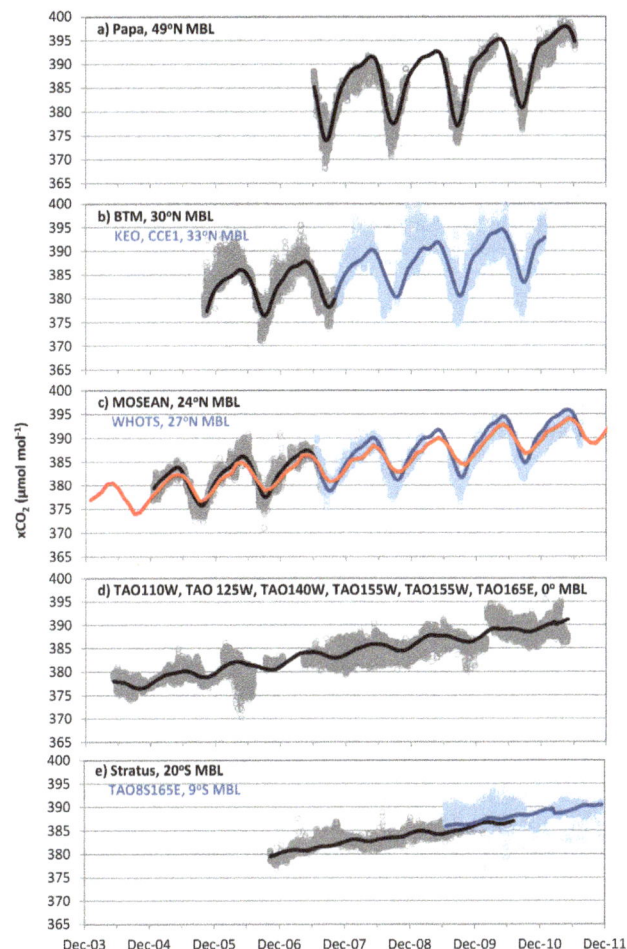

Figure 4. MAPCO$_2$ and GLOBALVIEW-CO2 MBL atmospheric xCO$_2$ (μmol mol^{-1}) presented by latitude: (**a**) Papa MAPCO$_2$ (gray points) and MBL at 49° N (black line); (**b**) BTM MAPCO$_2$ (gray points) and MBL at 30° N (black line), KEO and CCE1 MAPCO$_2$ (blue points) and MBL at 33° N (blue line); (**c**) MOSEAN MAPCO$_2$ (gray points) and MBL at 24° N (black line), WHOTS MAPCO$_2$ (blue points) and MBL at 27° N (blue line), and Mauna Loa Observatory atmospheric xCO$_2$ (red line); (**d**) six equatorial MAPCO$_2$ buoys (gray points) and MBL at 0° (black line); and (**e**) Stratus MAPCO$_2$ (gray points) and MBL at 20° S (black line), TAO8S165E MAPCO$_2$ (blue points) and MBL at 9° S (blue line). MBL data from GLOBALVIEW-CO2 (2013). Mauna Loa Observatory monthly mean data from Pieter Tans, NOAA/ESRL (http://www.esrl.noaa.gov/gmd/ccgg/trends/), and Ralph Keeling, Scripps Institution of Oceanography (http://scrippsco2.ucsd.edu/).

RH$_{sample}$ is the RH of the air sample (%); and RH$_{span}$ is the RH of the span (%), i.e., the background RH level for the system. Equation (2) is a calculation of vapor pressure optimized for the temperature interval of -20 to $50\,^{\circ}$C as defined by Buck (1981). This equation includes coefficients for calculating VP_{sat} with an enhancement factor (a correction for dealing with moist air as a function of temperature

and pressure) of 1.004 for 20 °C and 1000 mb (Buck, 1981). VP_{sat} and RH of the air sample are then used to calculate VP_{Licor}. Once the VP_{Licor} is known, the dilution effect can then be removed from the partially wet xCO$_2$ measurement using Eq. (1) to calculate xCO$_2$ (dry).

Since the MAPCO$_2$ equilibration occurs directly in the ocean, it does not require the warming correction necessary for underway pCO$_2$ systems. Therefore, pCO$_2$ in wet air (100 % saturation) in equilibrium with the surface seawater is calculated by

$$p\text{CO}_2(\text{sat}) = x\text{CO}_2(\text{dry}) \times (P_{\text{Licor}} - p\text{H}_2\text{O}), \qquad (4)$$

where P_{Licor} is atmospheric pressure for the atmospheric and surface seawater samples (atm) and pH$_2$O is the water vapor pressure (atm) at equilibrator temperature as defined by Weiss and Price (1980). fCO$_2$ in wet air (100 % saturation) in equilibrium with the surface seawater is calculated by

$$f\text{CO}_2(\text{sat}) = p\text{CO}_2(\text{sat}) \times e^{\left[\frac{P_{\text{Licor}} \times (B_{11} + 2\delta_{12})}{R \times T}\right]}, \qquad (5)$$

where the ideal gas constant $R = 82.0578\,\text{cm}^3\,\text{atm}\,\text{mol}^{-1}\,\text{K}^{-1}$, T is SST (K) from the CTD, and the B_{11} virial coefficient and δ_{12} cross-virial coefficient for CO$_2$ are as defined by Weiss (1974). The raw CO$_2$ data, temperature, salinity, and pressures are included in all published MAPCO$_2$ data sets so other data users can recalculate xCO$_2$, fCO$_2$, and pCO$_2$. Additional parameters included with the pCO$_2$ mooring data set are listed and described in Table 2.

2.3 Uncertainty of pCO$_2$ measurements

Precision and accuracy of the MAPCO$_2$ measurements have been assessed in both laboratory and field settings. As stated in Sect. 2.2, the precision of the MAPCO$_2$ system in a laboratory setting is $0.6\,\mu$mol mol^{-1}. Standard deviation of the high-frequency raw data ($\sim$ 58 repeated measurements over 30 s) in the field is a good assessment of the in situ precision of the MAPCO$_2$ system. Mean standard deviation of the raw data from the 14 buoy time series presented here is $0.7\,\mu$mol mol^{-1} for seawater xCO$_2$ and $0.6\,\mu$mol mol^{-1} for air xCO$_2$, which is similar to precision measured in the laboratory. While estimating accuracy in a laboratory setting is feasible, the more-desired estimate of in situ accuracy is difficult to obtain due to the limited availability of validation samples for comparison and the mismatch in space and time of these validation samples compared to the MAPCO$_2$ measurements. These issues related to accuracy will be discussed in more detail below. In this section, we present MAPCO$_2$-estimated in situ precision, accuracy, and uncertainty, which we define as the overall error of the measurement encompassing instrument precision and accuracy as well as propagation of error.

Propagation of error must be considered when calculations are based on variables with individual uncertainties. These

Table 3. Sources of error for the calculation of $x\mathrm{CO_2}$ (dry) at atmospheric pressure $= 101\,\mathrm{kPa}$, $\mathrm{RH_{sample}} = 75\,\%$, $\mathrm{RH_{span}} = 30\,\%$, $\mathrm{SST} = 25\,^{\circ}\mathrm{C}$, $\mathrm{SSS} = 35$, and $x\mathrm{CO_2}$ (wet) $= 375\,\mu\mathrm{mol\,mol^{-1}}$. Total estimated precision and accuracy are calculated using the root-sum-of-squares method (RSS): $\mathrm{RSS} = \left(\sum a^2\right)^{1/2}$.

Sources of error	Variable precision ($\pm$)	Effect on precision of final calculation (a)	Variable accuracy ($\pm$)	Effect on accuracy of final calculation (a)
VP_{Licor} calculation				
VP_{sat} (kPa)	0.019[a]	0.009	0.057[a]	0.026
$\mathrm{RH_{sample}}$	0.1[b]	negligible	3.0[b]	0.1
$\mathrm{RH_{span}}$	0.1[b]	negligible	3.0[b]	0.1
Assumption that $VP_{\mathrm{RH}} = VP_{\mathrm{Licor}}$				0.052
Total estimated error: VP_{Licor}		0.009		0.153
$x\mathrm{CO_2}$ (dry) calculation				
$x\mathrm{CO_2}$ (wet) ($\mu\mathrm{mol\,mol^{-1}}$)	0.7[c]	0.7	1.5[c]	1.5
P_{Licor} (kPa)	0.001[b]	negligible	0.010[b]	negligible
VP_{Licor} (kPa) (calculated above)	0.009	0.034	0.153	0.585
Total estimated error: $x\mathrm{CO_2}$ (dry)		0.7		1.6

Notes: [a] Error calculated using manufacturer-estimated error for T_{RH} of $\pm 0.1\,^{\circ}\mathrm{C}$ precision and $\pm 0.3\,^{\circ}\mathrm{C}$ accuracy (see Eq. 2). [b] Error reported by manufacturer. [c] Precision estimate based on standard deviation of the high-frequency raw data (~ 58 repeated measurements over 30 s) in the field; accuracy estimate based on pre-deployment testing in the laboratory. Negligible indicates value $<$ significant digits of variable.

types of errors that impact the calculated $p\mathrm{CO_2}$ and $f\mathrm{CO_2}$ values have been assessed for underway $p\mathrm{CO_2}$ systems and are typically small ($< 0.1\,\mu\mathrm{atm}$) with minimal impact to the overall uncertainty when combined with the larger uncertainty ($< 2\,\mu\mathrm{atm}$) in the actual $x\mathrm{CO_2}$ measurement (Feely et al., 1998; Wanninkhof and Thoning, 1993; Pierrot et al., 2009). However, we utilize a different method to calculate $x\mathrm{CO_2}$ (dry) for the MAPCO$_2$ system, as discussed in Sect. 2.1, so it is important to address the potential error in this new method. The RH measurements used to calculate $x\mathrm{CO_2}$ (dry) have separate precisions and accuracies that can propagate through Eqs. (1)–(3) (Table 3). The total estimated precision and accuracy of $x\mathrm{CO_2}$ (dry) are calculated by summing each variable's precision and accuracy using the root-sum-of-squares method. As presented in Table 3, propagation of all the errors from the separate variables does not cause the precision of calculated $x\mathrm{CO_2}$ (dry) to differ from measured $x\mathrm{CO_2}$ (wet) and results in a small impact to the accuracy ($0.1\,\mu\mathrm{mol\,mol^{-1}}$).

In addition to the propagation of error, an estimate of in situ accuracy is key to determining the overall uncertainty of the MAPCO$_2$ system. The GLOBALVIEW-CO2 data product maintained by NOAA ESRL can be used as one data set for comparison to the MAPCO$_2$ air $x\mathrm{CO_2}$ (dry) measurements (GLOBALVIEW-CO2, 2013). Figure 4 shows 3-hourly atmospheric MAPCO$_2$ measurements and biweekly atmospheric CO$_2$ values from the MBL layer of GLOBALVIEW-CO2 at the latitude closest to each MAPCO$_2$ location. Atmospheric MAPCO$_2$ data presented here are in the finalized, processed form as described in

Sect. 2.2. Both MBL and MAPCO$_2$ data capture seasonal variability and long-term trends, but, as expected, high-frequency MAPCO$_2$ measurements show short-term variability typically deviating from the smoothed MBL data product by $< 5\,\mu\mathrm{mol\,mol^{-1}}$ (Fig. 4). The Mauna Loa atmospheric CO$_2$ record is also shown in Fig. 4c and provides a reference for illustrating the larger seasonal variability in the lower atmosphere directly influenced by the presence of the ocean's surface. For the time series longer than 2 years, growth rates of the 3-hourly MAPCO$_2$ and biweekly MBL atmospheric CO$_2$ are presented in Table 4. Atmospheric CO$_2$ growth rates observed by five of the seven mooring time series differ from the MBL data by $\leq 0.1\,\mu\mathrm{mol\,mol^{-1}\,yr^{-1}}$, suggesting that the finalized MAPCO$_2$ observations are consistent with other atmospheric data products generated using different methods.

MAPCO$_2$ and MBL data are compared in more detail in Table 5. This includes descriptive statistics of the finalized, processed atmospheric data in addition to pre-finalized data prior to any adjustments or offsets. The 3-hourly MAPCO$_2$ measurement that is closest in time to the biweekly MBL estimate is used to calculate the Δ (MAPCO$_2$–MBL). Pre-QC MAPCO$_2$ data show a slight negative bias ($-1.5 \pm 2.4\,\mu\mathrm{mol\,mol^{-1}}$) to MBL values (Table 5). The mean difference between finalized MAPCO$_2$ data and MBL values is smaller ($-0.3 \pm 1.7\,\mu\mathrm{mol\,mol^{-1}}$) due to the application of occasional offsets during data QC described in Sect. 2.2. Standard deviations likely reflect the natural variability in atmospheric CO$_2$ at the sea surface illustrated in Fig. 4. Low standard error of the mean and low confidence

Table 4. Growth rate of GLOBALVIEW-CO2 MBL and MAPCO$_2$ atmospheric xCO$_2$ time series over the time period of the data sets listed in Table 5 (GLOBALVIEW-CO2, 2013). For mooring time series locations see Fig. 3.

Time series > 2 years	Growth rate (μmol mol^{-1} yr^{-1})	
	MBL	MAPCO$_2$
Papa	3.1	3.1
KEO	1.7	1.1
MOSEAN/WHOTS	2.0	1.6
TAO125W	1.9	2.0
TAO140W	1.9	1.9
TAO170W	1.9	1.9
Stratus	1.8	1.9

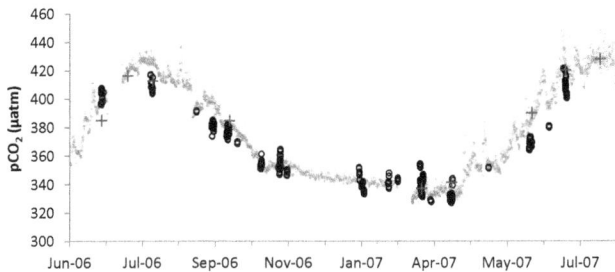

Figure 6. **(a)** TAO125W surface seawater MAPCO$_2$ observations (gray points) for the entire time series at this location with average R/V *Ka'imimoana* underway pCO$_2$ data within 10 km and 10 min of the MAPCO$_2$ measurements (black open circles). Two examples of comparison data over 1-week time series are shown in panels **(b)** and **(c)**, with MAPCO$_2$ measurements corresponding to the average underway observations illustrated in gray open circles. Selection boxes in **(a)** are not to scale of actual axes in **(b)** and **(c)** panels. *Ka'imimoana* data from NOAA PMEL, http://cdiac.ornl.gov/oceans/VOS_Program/kaimimoana.html.

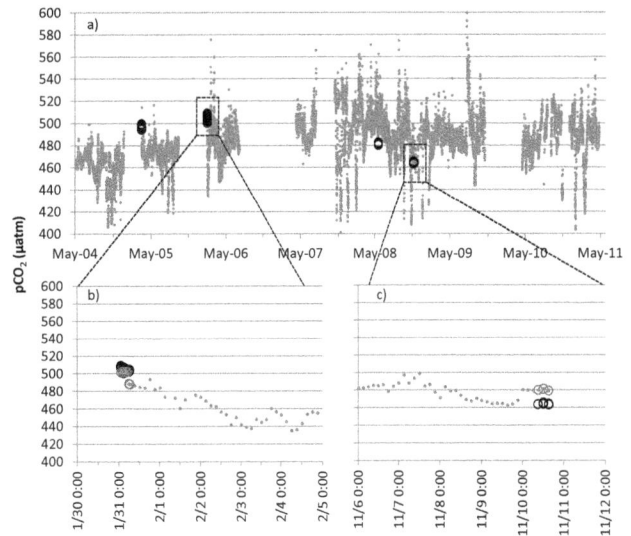

Figure 5. Seawater pCO$_2$ values from BTM MAPCO$_2$ (gray points), Bermuda Atlantic Time-series Study (BATS) discrete (plus signs), and R/V *Atlantic Explorer* underway (open circles) used in the Table 5 statistics. BATS data from Bermuda Institute of Ocean Sciences, bats.bios.edu. *Atlantic Explorer* data from Bermuda Institute of Ocean Sciences, http://cdiac.ornl.gov/oceans/CARINA/.

level values reported for the atmospheric comparison in Table 5 suggest strong statistical significance in the mean MAPCO$_2$–MBL values.

While environmental variability may introduce some error to the MAPCO$_2$ and MBL air comparison, the resulting mean differences in the atmospheric data are likely due primarily to uncertainty in the measurements, which in this case we associate with the MAPCO$_2$ system. However, surface ocean pCO$_2$ exhibits large temporal and spatial variability. For example, it is common to observe variability in underway pCO$_2$ measurements from the R/V *Atlantic Explorer* of approximately 10 μatm within 10 km of BTM over a period of 3 h (Fig. 5). We observe even larger variability in the eastern equatorial Pacific, with changes up to 50 μatm over a period of 3 h and > 100 μatm over the course of a day (Fig. 6a). This patchiness can create errors in comparing MAPCO$_2$ measurements to ship-based measurements made at safe distance from the surface buoy. In Fig. 6b, for example, the difference between the TAO125W MAPCO$_2$ and underway measurements from the R/V *Ka'imimoana* (made within 10 km and 10 min of the MAPCO$_2$ measurement) start

at ± 2 μatm on 31 January 2006 at 14:00:00, but as the ship begins to leave the surface buoy 6 h later the measurements diverge as the MAPCO$_2$ starts to detect a decreasing trend in surface seawater pCO$_2$ values at the buoy location that persists for the next 8 days. In another example shown in Fig. 6c, the 15 μatm difference between the MAPCO$_2$ and underway system observed on 10 November 2008 is similar to the daily variability observed at the buoy in the 4 days prior to arrival of the *Ka'imimoana* and could reflect true differences observed by the underway and MAPCO$_2$ systems located 1–7 km apart. These examples highlight the difficulty of separating environmental variability and instrument uncertainty in these types of comparison exercises.

In order to minimize environmental variability while maximizing sample size for descriptive statistics, we use discrete measurements made within 10 km and 1.5 h and averaged underway pCO$_2$ measurements made within 10 km and 10 min of the MAPCO$_2$ system measurements for the seawater pCO$_2$ comparison analysis. While underway and MAPCO$_2$ systems utilize similar methodology, discrete pCO$_2$ presented in Table 5 is calculated from measurements of dissolved inorganic carbon (DIC) and total alkalinity (TA) using the program CO2SYS developed by Lewis and Wallace (1998) with the constants of Lueker et al. (2000). Typical error in calculated pCO$_2$ using this method is < 5 %. Only finalized seawater MAPCO$_2$ data are used for the descriptive

Table 5. Descriptive statistics of Δ (MAPCO$_2$ measurement – comparison measurement). The MAPCO$_2$ measurements (both pre- and post-offset if applied during data QC) are compared to biweekly GLOBALVIEW-CO2 MBL values from the latitude nearest to average buoy location, single discrete measurements made within 10 km and 1.5 h, and averaged underway pCO$_2$ measurements made within 10 km and 10 min of the MAPCO$_2$ system measurement. Standard error is the standard error of the mean, and confidence intervals illustrate that with a 95 % probability the actual population mean = sample mean $\pm$ confidence interval.

	n	Mean	Standard error	Standard deviation	Confidence interval (95 %)
MAPCO$_2$ air xCO$_2$ (dry) comparison to MBL[a] air (μmol mol^{-1})					
Data prior to QC (estimate of MAPCO$_2$ system in situ accuracy)	1823	-1.5	0.1	2.4	0.1
Finalized data (estimate of finalized MAPCO$_2$ data accuracy)	1823	-0.3	< 0.1	1.7	0.1
MAPCO$_2$ seawater pCO$_2$ comparison to calculated pCO$_2$ (μatm) from discrete DIC, TA					
WHOTS vs. HOTS[b]	7	0.1	1.4	3.7	3.4
BTM vs. BATS[c]	9	1.3	1.9	5.6	4.3
Papa vs. Station P[d]	10	-0.4	2.0	6.2	4.5
MAPCO$_2$ seawater pCO$_2$ comparison to underway pCO$_2$ (μatm)					
BTM vs. *Atlantic Explorer*[e]	76	1.8	0.5	4.8	1.1
TAO125W vs. *Ka'imimoana*[f]	16	-3.3	3.8	15.2	8.1
TAO140W vs. *Ka'imimoana*[f]	13	2.1	2.3	8.3	5.0

Notes on data sources and archives: [a] GLOBALVIEW-CO2 marine boundary layer (MBL) data source: NOAA Earth System Research Laboratory, http://www.esrl.noaa.gov/gmd/ccgg/globalview/co2/co2_intro.html (GLOBALVIEW-CO2, 2013). [b] Hawaii Ocean Time-Series (HOTS) data source: University of Hawaii, hahana.soest.hawaii.edu/hot. [c] Bermuda Atlantic Time-series Study (BATS) data source: Bermuda Institute of Ocean Sciences, http://bats.bios.edu. [d] Station P data source: University of Washington and NOAA PMEL. [e] *Atlantic Explorer* data source: Bermuda Institute of Ocean Sciences, http://cdiac.ornl.gov/oceans/CARINA/. [f] *Ka'imimoana* data source: NOAA PMEL, http://cdiac.ornl.gov/oceans/VOS_Program/kaimimoana.html.

statistics presented in Table 5. Unlike the descriptive statistics for the MAPCO$_2$ air comparisons, the statistics that result from using MAPCO$_2$ seawater measurements pre-MBL offset are not statistically different than the finalized, post-MBL offset statistics presented in Table 5. This could be due to the large natural variability in seawater pCO$_2$ compared to atmospheric CO$_2$.

Agreement between discrete and mooring surface ocean pCO$_2$ measurements is within 1.3 μatm (mean Δ in Table 5; BTM example in Fig. 5). Although more discrete measurements have been made at these and other mooring locations, this comparison is based on discrete samples restricted to within 10 km and 1.5 h of the MAPCO$_2$ system measurements with $n > 5$. Even with these restrictions, it is likely that environmental variability is not completely removed and is reflected in the mean Δ standard deviations of 3.7–6.2 μatm (Table 5). The small sample sizes (≤ 10 at each site) also resulting from these restrictions create large uncertainty in mean Δ values, with standard error and confidence levels exceeding mean Δ values. This analysis shows promising results with a close agreement between discrete and MAPCO$_2$ measurements; however, more discrete samples will need to be collected within 10 km and 1.5 h of MAPCO$_2$ system measurements in order to improve the statistical significance of the seawater pCO$_2$ comparison.

Sample sizes are larger ($13 \leq n \leq 76$) for the comparison between underway and MAPCO$_2$ measurements at the BTM, TAO125W, and TAO140W locations. While underway measurements exist at other equatorial Pacific mooring locations, comparisons within 10 km and 10 min are restricted to TAO125W and TAO140W due to the large gaps in pCO$_2$ mooring data, the infrequent mooring-servicing ship visits to each site ($\sim$ once every 1–1.5 years), and the necessity for the mooring-servicing ship to leave for the next station before the MAPCO$_2$ system has gone through a few cycles and measurements have stabilized. Even with these challenges, there are 76 comparison samples at BTM during the two buoy deployments in 2006–2007 (Fig. 5). These measurements show a mean Δ of 1.8 ± 4.8 μatm with a low confidence interval of 1.1, indicating strong statistical significance ($p < 0.05$) that the actual mean Δ is between 0.7 and 2.9 μatm (Table 5). Standard deviations of the difference between the BTM versus discrete (5.6) and underway (4.8) measurements are similar, which may be reflective of the environmental variability in this region of the surface ocean. Mean Δ in the equatorial Pacific is higher (-3.3 ± 15.2 μatm at TAO125W and 2.1 ± 8.3 μatm at TAO140W), but statistical significance of these values is low due to the lower sample sizes and higher environmental variability in this region (Fig. 3). The largest standard deviation in mean Δ of 15.2 is at TAO125W, which is the site that exhibits the largest natural variability (i.e., total range of $\sim 200\,\mu$atm, Fig. 6a) in surface seawater pCO_2 of the open-ocean mooring data sets compared in Table 5.

The MAPCO$_2$ system has also been involved in two independent ocean pCO_2 instrument intercomparisons. During an Alliance for Coastal Technologies demonstration project, the difference between the MAPCO$_2$ system and an underway pCO_2 system was $-9\pm 8\,\mu$atm in coastal Washington, USA waters and $-3\pm 9\,\mu$atm in coral reef waters of Kaneohe Bay, Hawaii, USA (Schar et al., 2010). Separating environmental variability from instrument uncertainty in this case is challenging. Small-scale environmental variability (i.e., meters) due to natural spatial patchiness of pCO_2 was determined to be 10–15 μatm at the coastal site and $< 2\,\mu$atm at the coral site and may account for much of the difference observed between the MAPCO$_2$ and reference measurements. An intercomparison between buoy and underway pCO_2 systems held at the National Research Institute of Fishery Engineering in Hasaki, Kamisu city, Ibaraki, Japan, was done in the more controlled environment of an indoor seawater pool (UNESCO, 2010). In this intercomparison, the MAPCO$_2$ was within 1 μatm compared to the underway pCO_2 reference system in conditions within the calibration gas range.

In summary, the MAPCO$_2$ system performs very well in laboratory and field settings in comparison to a variety of other methods. Considering the precision estimate of the MAPCO$_2$ measurements in the field ($< \pm 0.7\,\mu$mol mol^{-1}), the statistically strong ($p < 0.05$) mean differences in MAPCO$_2$ versus comparison measurements in Table 5 ($< \pm 1.8\,\mu$atm), and the small propagation of error resulting from the xCO$_2$ (dry) calculation ($< \pm 0.1\,\mu$mol mol^{-1}), we estimate in situ MAPCO$_2$ precision at $< \pm 0.7\,\mu$mol mol^{-1} and accuracy at $< \pm 2.0\,\mu$mol mol^{-1} for xCO$_2$ (dry) measurements. Overall uncertainty of pCO_2 and fCO_2 observations from the MAPCO$_2$ system is estimated to be $< 2.0\,\mu$atm for values between 100 and 600 μatm for over 400 days of autonomous operation. However, the uncertainty of finalized, quality-controlled data is likely better for atmospheric pCO_2 and fCO_2 observations at $< 1.0\,\mu$atm when following the post-deployment standard operating procedures described in Sect. 2.2.

3 Data description and access

Finalized MAPCO$_2$ data are reported to the Carbon Dioxide Information Analysis Center (CDIAC; http://cdiac.ornl.gov/oceans/Moorings) and archived at additional data centers such as the National Oceanographic Data Center (http://www.nodc.noaa.gov). The archived data are organized by site and deployment date. The numeric data package (NDP) associated with this publication includes the 56 deployments listed in Table 6 and is available at doi:10.3334/CDIAC/OTG.TSM_NDP092 or http://cdiac.ornl.gov/oceans/Moorings/ndp092. The methods described here are associated with the mooring pCO_2 data included in

this NDP. These data are made freely available to the public and the scientific community in the belief that their wide dissemination will lead to greater understanding and new scientific insights. Users of the data are requested to cite this publication when using the entire open-ocean mooring data set or cite according to the CDIAC data archive when using individual mooring data sets. When preparing manuscripts using these data, users are asked to invite lead pCO_2 mooring investigators to coauthor or to send draft manuscripts using these data to the lead investigators to ensure that the quality and limitations of the data are accurately represented.

The mooring data set includes 3-hourly seawater and atmospheric CO$_2$ observations from 14 moorings since 2004, encompassing over 100 000 individual measurements. As presented in Fig. 3, climatological means of surface ocean pCO_2 measured on moorings are consistent with observations from other platforms (Bakker et al., 2014; Takahashi et al., 2009); however, much of the value in high-frequency mooring observations is demonstrated at shorter timescales. Figure 3 shows that short-term (≤ 2 years) variability at the subtropical sites tends to be dominated by the seasonal cycle, and tropical sites tend to be dominated by interannual variability. At the subtropical sites, seawater CO$_2$ is typically highest in the summer and lowest in the winter. The Papa site is the highest-latitude mooring in this data set and exhibits approximately equal short-term variation driven by the seasonal cycle and interannual variability caused by strong weather events in this region of the North Pacific. The highest interannual variability is observed in the equatorial Pacific driven by El Niño and La Niña events (Fig. 3) and dominates any small seasonal signal that may exist in this region (Sutton et al., 2014). In the most extreme conditions, seawater pCO_2 values can vary over 100 μatm within 24 h at 0°, 125° W (Fig. 6a). Variability of 100–150 μatm is also common in the equatorial Pacific during the extension of the warm water pool during El Niño events on timescales of months and the passing of tropical instability waves on timescales of weeks (e.g., Fig. 4 in Sutton et al., 2014). Sustained, long-term mooring time series also provide the opportunity to identify and remove the short-term variability from the time series and investigate long-term trends. For example, in a synthesis of equatorial Pacific mooring data, Sutton et al. (2014) found that the uptake of anthropogenic CO$_2$ and an acceleration in equatorial upwelling since the shift in the Pacific Decadal Oscillation in 1998 has led to high rates of pCO_2 change of $+2.3$ to $+3.3\,\mu$atm yr^{-1} in this region. This decadal shift in CO$_2$ outgassing is consistent with underway pCO_2 observations made in this region since 1982 (Feely et al., 2014).

Mooring data from most of the deployments through 2010 listed in Table 6 are also included in the most recent version of SOCAT (Bakker et al., 2014). This SOCATv2.0 synthesis involves a standardized, second-level quality control of 10.1 million surface seawater fCO_2 measurements from many different sources, including underway and mooring

Table 6. List of open-ocean mooring deployments in the open-ocean MAPCO$_2$ data set. n is the total number of measurements collected at each mooring location during these deployments.

Mooring	Start date MM/DD/YYYY	End date MM/DD/YYYY	Mooring	Start date MM/DD/YYYY	End date MM/DD/YYYY
MOSEAN/WHOTS	12/19/2004	05/23/2005	TAO110W	09/19/2009	11/03/2009
	05/29/2005	01/20/2006		03/15/2010	07/14/2010
	06/18/2006	12/21/2006		07/22/2010	10/28/2010
	01/28/2007	07/30/2007	n	2148	
	06/26/2007	06/05/2008			
	06/05/2008	02/12/2009	TAO125W	05/08/2004	12/20/2004
	07/11/2009	08/01/2010		03/16/2005	09/15/2005
	08/01/2010	07/13/2011		01/31/2006	07/08/2006
n	17 645			04/13/2007	07/17/2007
				10/16/2007	11/10/2008
BTM	10/02/2005	07/03/2006		11/13/2008	10/21/2009
	07/14/2006	03/02/2007		04/22/2010	11/06/2010
	03/13/2007	10/01/2007	n	13 609	
n	5354				
			TAO140W	05/23/2004	09/12/2004
Papa	06/08/2007	06/10/2008		09/13/2004	03/01/2005
	06/11/2008	11/11/2008		03/02/2005	09/22/2005
	06/13/2009	03/27/2010		01/17/2006	05/14/2006
	06/16/2010	06/13/2011		09/14/2006	12/18/2006
n	9235			05/31/2007	11/20/2007
				05/10/2008	09/03/2009
KEO	09/28/2007	08/08/2008		09/04/2009	01/30/2010
	09/13/2008	09/04/2009		11/26/2010	03/23/2011
	09/05/2009	09/24/2010	n	14 276	
	09/30/2010	12/24/2010			
n	9182		TAO155W	01/13/2010	08/25/2010
			n	1791	
JKEO	02/18/2007	10/03/2007			
n	1837		TAO170W	07/04/2005	06/23/2006
				07/31/2007	08/13/2008
CCE1	11/11/2008	02/06/2009		08/26/2008	06/01/2009
	05/19/2009	12/14/2009		06/02/2009	12/12/2009
	12/15/2009	09/01/2010		02/03/2010	02/04/2011
	09/02/2010	10/11/2010	n	12 528	
n	4775				
			TAO165E	02/23/2010	02/27/2011
Stratus	10/16/2006	10/29/2007	n	2955	
	10/27/2007	10/27/2008			
	10/26/2008	01/18/2010	TAO8S165E	06/22/2009	09/19/2010
	01/19/2010	07/07/2010		10/18/2010	11/15/2011
n	10 889		n	6720	

systems. SOCAT also produces a gridded surface ocean fCO$_2$ data product in a uniform format available at http://www.socat.info. Rödenbeck et al. (2013) compared the previous version of SOCAT (v1.5), which did not include mooring data, to some of the open-ocean MAPCO$_2$ time series in Table 6. In a comparison between seawater pCO$_2$ data from the TAO170W MAPCO$_2$ and data-driven model estimates based on SOCATv1.5, Rödenbeck et al. (2013) find that seawater pCO$_2$ estimates in the tropics are unrelated, or even opposite, to the mooring observations. This discrepancy arises because that particular location is not well constrained by the SOCATv1.5 data set. We expect the recent mooring additions to SOCATv2.0 and the open-ocean MAPCO$_2$ data set presented here to make a large impact on our efforts to model and understand the global carbon cycle in the coming years.

4 Conclusion

Mooring observations can play a critical role in improving our ability to model, understand, and describe the ocean carbon cycle on all timescales. In particular, time series from remote, data-sparse areas of the ocean collected on moorings fulfill a unique niche by providing the high-resolution data necessary to explore questions about short-term variability at fixed locations. Here we provide a data set of 3-hourly surface seawater and marine boundary layer atmospheric pCO_2 observations on 14 open-ocean moorings in the Pacific and Atlantic from 2004 to 2011. When using the in situ and post-calibration methods described here, overall uncertainty for the MAPCO$_2$ data is $< 2\,\mu$atm for seawater pCO_2 and $< 1\,\mu$atm for air pCO_2, making the MAPCO$_2$ system a climate-quality method for tracking surface ocean pCO_2. These types of sustained, temporally resolved observations allow us to improve our understanding of the role of shorter-term variability and key biogeochemical processes on the global carbon system. Potential uses of these data to inform our understanding of a changing ocean include investigating high-frequency variability in surface ocean biogeochemistry, developing seasonal CO_2 flux maps for the global oceans (e.g., Takahashi climatology and SOCAT), studying ocean acidification, and evaluating regional and global carbon models.

Acknowledgements. This work was funded by the Climate Observation Division within NOAA's Climate Program Office. The authors thank Gernot Friederich, Peter Brewer, and Francisco Chavez at the Monterey Bay Aquarium Research Institute for their efforts in developing the early mooring pCO_2 system, and the Battelle Memorial Institute for their later investments in this technology that have made the MAPCO$_2$ systems available to the broader research and ocean-observing community. The NOAA network of pCO_2 moorings would not be possible without the industrious efforts of PMEL technical and engineering staff as well as our partners and their funders who support the maintenance of the open-ocean buoys: Nick Bates (BTM), Meghan Cronin (Papa and KEO), Michael McPhaden (TAO array), Tommy Dickey (MOSEAN), Al Plueddemann and Robert Weller (WHOTS and Stratus), and Uwe Send (CCE1). PMEL contribution 4061 and JISAO contribution 2254.

Edited by: D. Carlson

References

Bakker, D. C. E., Pfeil, B., Smith, K., Hankin, S., Olsen, A., Alin, S. R., Cosca, C., Harasawa, S., Kozyr, A., Nojiri, Y., O'Brien, K. M., Schuster, U., Telszewski, M., Tilbrook, B., Wada, C., Akl, J., Barbero, L., Bates, N. R., Boutin, J., Bozec, Y., Cai, W.-J., Castle, R. D., Chavez, F. P., Chen, L., Chierici, M., Currie, K., de Baar, H. J. W., Evans, W., Feely, R. A., Fransson, A., Gao, Z., Hales, B., Hardman-Mountford, N. J., Hoppema, M., Huang, W.-J., Hunt, C. W., Huss, B., Ichikawa, T., Johannessen, T., Jones, E. M., Jones, S. D., Jutterström, S., Kitidis, V., Körtzinger, A., Landschützer, P., Lauvset, S. K., Lefèvre, N., Manke, A. B., Mathis, J. T., Merlivat, L., Metzl, N., Murata, A., Newberger, T., Omar, A. M., Ono, T., Park, G.-H., Paterson, K., Pierrot, D., Ríos, A. F., Sabine, C. L., Saito, S., Salisbury, J., Sarma, V. V. S. S., Schlitzer, R., Sieger, R., Skjelvan, I., Steinhoff, T., Sullivan, K. F., Sun, H., Sutton, A. J., Suzuki, T., Sweeney, C., Takahashi, T., Tjiputra, J., Tsurushima, N., van Heuven, S. M. A. C., Vandemark, D., Vlahos, P., Wallace, D. W. R., Wanninkhof, R., and Watson, A. J.: An update to the Surface Ocean CO_2 Atlas (SOCAT version 2), Earth Syst. Sci. Data, 6, 69–90, doi:10.5194/essd-6-69-2014, 2014.

Bender, M., Doney, S., Feely, R. A., Fung, I. Y., Gruber, N., Harrison, D. E., Keeling, R., Moore, J., Sarmiento, J., Sarachik, E., Stephens, B., Takahashi, T., Tans, P. P., and Wanninkhof, R.: A Large Scale Carbon Observing Plan: In Situ Oceans and Atmosphere (LSCOP), Nat. Tech. Info. Service, Springfield, 201 pp., 2002.

Buck, A. L.: New equations for computing vapor pressure and enhancement factor, J. Appl. Meteorol., 20, 1527–1532, 1981.

Feely, R. A., Wanninkhof, R., Milburn, H. B., Cosca, C. E., Stapp, M., and Murphy, P.: A new automated underway system for making high precision pCO_2 measurements onboard research ships, Anal. Chim. Acta, 377, 185–191, doi:10.1016/S0003-2670(98)00388-2, 1998.

Feely, R. A., Cosca, C. E., Sutton, A. J., Sabine, C. L., Wanninkhof, R., and Mathis, J. T.: Decadal changes of air-sea CO_2 fluxes in the Equatorial Pacific Ocean, Geophys. Res. Lett., in preparation, 2014.

Friederich, G. E., Brewer, P. G., Herlien, R., and Chavez, F. P.: Measurement of sea surface partial pressure of CO_2 from a moored buoy, Deep-Sea Res. Pt. I, 42, 1175–1186, doi:10.1016/0967-0637(95)00044-7, 1995.

GLOBALVIEW-CO2: Cooperative Global Atmospheric Data Integration Project, 2013, updated annually, Multi-laboratory compilation of synchronized and gap-filled atmospheric carbon dioxide records for the period 1979–2012 (obspack_co2_1_GLOBALVIEW-CO2_2013_v1.0.4_2013-12-23), compiled by NOAA Global Monitoring Division: Boulder, Colorado, USA Data product accessed at: doi:10.3334/OBSPACK/1002, 2013.

Hofmann, D. J., Butler, J. H., and Tans, P. P.: A new look at atmospheric carbon dioxide, Atmos. Environ., 43, 2084–2086, doi:10.1016/j.atmosenv.2008.12.028, 2009.

Keeling, C. D., Bacastow, R. B., Bainbridge, A. E., Ekdahl, C. A., Guenther, P. R., Waterman, L. S., and Chin, J. F. S.: Atmospheric carbon dioxide variations at Mauna Loa Observatory, Hawaii, Tellus, 28, 538–551, doi:10.1111/j.2153-3490.1976.tb00701.x, 1976.

Lewis, E. and Wallace, D. W. R.: Program Developed for CO2 System Calculations, Carbon Dioxide Information Analysis Center, Oak Ridge National Laboratory, U.S. Department of Energy, Oak Ridge, Tennessee, 1998.

Lueker, T. J., Dickson, A. G., and Keeling, C. D.: Ocean pCO_2 calculated from dissolved inorganic carbon, alkalinity, and equations for K1 and K2: validation based on laboratory measurements of CO_2 in gas and seawater at equilibrium, Mar. Chem., 70, 105–119, doi:10.1016/S0304-4203(00)00022-0, 2000.

Pfeil, B., Olsen, A., Bakker, D. C. E., Hankin, S., Koyuk, H., Kozyr, A., Malczyk, J., Manke, A., Metzl, N., Sabine, C. L., Akl, J.,

Alin, S. R., Bates, N., Bellerby, R. G. J., Borges, A., Boutin, J., Brown, P. J., Cai, W.-J., Chavez, F. P., Chen, A., Cosca, C., Fassbender, A. J., Feely, R. A., González-Dávila, M., Goyet, C., Hales, B., Hardman-Mountford, N., Heinze, C., Hood, M., Hoppema, M., Hunt, C. W., Hydes, D., Ishii, M., Johannessen, T., Jones, S. D., Key, R. M., Körtzinger, A., Landschützer, P., Lauvset, S. K., Lefèvre, N., Lenton, A., Lourantou, A., Merlivat, L., Midorikawa, T., Mintrop, L., Miyazaki, C., Murata, A., Nakadate, A., Nakano, Y., Nakaoka, S., Nojiri, Y., Omar, A. M., Padin, X. A., Park, G.-H., Paterson, K., Perez, F. F., Pierrot, D., Poisson, A., Ríos, A. F., Santana-Casiano, J. M., Salisbury, J., Sarma, V. V. S. S., Schlitzer, R., Schneider, B., Schuster, U., Sieger, R., Skjelvan, I., Steinhoff, T., Suzuki, T., Takahashi, T., Tedesco, K., Telszewski, M., Thomas, H., Tilbrook, B., Tjiputra, J., Vandemark, D., Veness, T., Wanninkhof, R., Watson, A. J., Weiss, R., Wong, C. S., and Yoshikawa-Inoue, H.: A uniform, quality controlled Surface Ocean CO_2 Atlas (SOCAT), Earth Syst. Sci. Data, 5, 125–143, doi:10.5194/essd-5-125-2013, 2013.

Pierrot, D., Neill, C., Sullivan, K., Castle, R., Wanninkhof, R., Lüger, H., Johannessen, T., Olsen, A., Feely, R. A., and Cosca, C. E.: Recommendations for autonomous underway pCO_2 measuring systems and data-reduction routines, Deep-Sea Res. Pt. II, 56, 512–522, doi:10.1016/j.dsr2.2008.12.005, 2009.

Rödenbeck, C., Keeling, R. F., Bakker, D. C. E., Metzl, N., Olsen, A., Sabine, C., and Heimann, M.: Global surface-ocean pCO_2 and sea–air CO_2 flux variability from an observation-driven ocean mixed-layer scheme, Ocean Sci., 9, 193-216, doi:10.5194/os-9-193-2013, 2013.

Sabine, C., Ducklow, H., and Hood, M.: International carbon co-ordination: Roger Revelle's legacy in the Intergovernmental Oceanographic Commission, Oceanography, 23, 48–61, 2010.

Schar, D., Atkinson, M., Johengen, T., Pinchuk, A., Purcell, H., Robertson, C., Smith, G. J., and Tamburri, M.: Performance demonstration statement PMEL MAPCO2/Battelle Seaology pCO_2 Monitoring System, UMCES Technical Report Series: Ref. No. [UMCES]CBL 10-092, 2010.

Sutton, A. J., Feely, R. A., Sabine, C. L., McPhaden, M. J., Takahashi, T., Chavez, F. P., Friederich, G. E., and Mathis, J. T.: Natural variability and anthropogenic change in equatorial Pacific surface ocean pCO_2 and pH, Global Biogeochem. Cy., 2013,

GB004679, doi:10.1002/2013gb004679, 2014.

Takahashi, T.: Carbon dioxide in the atmosphere and in Atlantic Ocean water, J. Geophys. Res., 66, 477–494, 1961.

Takahashi, T., Sutherland, S. C., Wanninkhof, R., Sweeney, C., Feely, R. A., Chipman, D. W., Hales, B., Friederich, G., Chavez, F., Sabine, C., Watson, A., Bakker, D. C. E., Schuster, U., Metzl, N., Yoshikawa-Inoue, H., Ishii, M., Midorikawa, T., Nojiri, Y., Körtzinger, A., Steinhoff, T., Hoppema, M., Olafsson, J., Arnarson, T. S., Tilbrook, B., Johannessen, T., Olsen, A., Bellerby, R., Wong, C. S., Delille, B., Bates, N. R., and de Baar, H. J. W.: Climatological mean and decadal change in surface ocean pCO_2, and net sea-air CO_2 flux over the global oceans, Deep-Sea Res. Pt. II, 56, 554–577, doi:10.1016/j.dsr2.2008.12.009, 2009.

Thoning, K. W., Tans, P. P., and Komhyr, W. D.: Atmospheric carbon dioxide at Mauna Loa Observatory: 2. Analysis of the NOAA GMCC data, 1974–1985, J. Geophys. Res.-Atmos., 94, 8549–8565, doi:10.1029/JD094iD06p08549, 1989.

UNESCO: SOCAT Equatorial Pacific, North Pacific, and Indian Ocean Regional Workshop, Tokyo, Japan, 8–11 February 2010, IOC Workshop Report No. 229, available at: http://www.ioccp.org/images/D3meetingReports/WR229_eo.pdf (last access: 29 October 2014), 2010.

Wanninkhof, R. and Thoning, K.: Measurement of fugacity of carbon dioxide in surface water and air using continuous sampling methods, Mar. Chem., 44, 189–205, 1993.

Wanninkhof, R., Park, G.-H., Takahashi, T., Sweeney, C., Feely, R., Nojiri, Y., Gruber, N., Doney, S. C., McKinley, G. A., Lenton, A., Le Quéré, C., Heinze, C., Schwinger, J., Graven, H., and Khatiwala, S.: Global ocean carbon uptake: magnitude, variability and trends, Biogeosciences, 10, 1983–2000, doi:10.5194/bg-10-1983-2013, 2013.

Weiss, R. F.: Carbon dioxide in water and seawater: the solubility of a non-ideal gas, Mar. Chem., 2, 203–215, doi:10.1016/0304-4203(74)90015-2, 1974.

Weiss, R. F. and Price, B. A.: Nitrous oxide solubility in water and seawater, Mar. Chem., 8, 347–359, doi:10.1016/0304-4203(80)90024-9, 1980.

Weiss, R. F., Jahnke, R. A., and Keeling, C. D.: Seasonal effects of temperature and salinity on the partial pressure of CO_2 in seawater, Nature, 300, 511–513, 1982.

Distribution of mesozooplankton biomass in the global ocean

R. Moriarty[1] and T. D. O'Brien[2]

[1]School of Earth, Atmospheric and Environmental Sciences, University of Manchester, Williamson Building, Oxford Road, Manchester M13 9PL, UK

[2]National Marine Fisheries Service, 1315 East-West Highway, Silver Spring, Maryland, USA

Correspondence to: T. D. O'Brien (todd.obrien@noaa.gov)

Abstract. Mesozooplankton are cosmopolitan within the sunlit layers of the global ocean. They are important in the pelagic food web, having a significant feedback to primary production through their consumption of phytoplankton and microzooplankton. In many regions of the global ocean, they are also the primary contributors to vertical particle flux in the oceans. Through both they affect the biogeochemical cycling of carbon and other nutrients in the oceans. Little, however, is known about their global distribution and biomass. While global maps of mesozooplankton biomass do exist in the literature, they are usually in the form of hand-drawn maps for which the original data associated with these maps are not readily available. The dataset presented in this synthesis has been in development since the late 1990s, is an integral part of the Coastal and Oceanic Plankton Ecology, Production, and Observation Database (COPEPOD), and is now also part of a wider community effort to provide a global picture of carbon biomass data for key plankton functional types, in particular to support the development of marine ecosystem models. A total of 153 163 biomass values were collected, from a variety of sources, for mesozooplankton. Of those 2 % were originally recorded as dry mass, 26 % as wet mass, 5 % as settled volume, and 68 % as displacement volume. Using a variety of non-linear biomass conversions from the literature, the data have been converted from their original units to carbon biomass. Depth-integrated values were then used to calculate an estimate of mesozooplankton global biomass. Global epipelagic mesozooplankton biomass, to a depth of 200 m, had a mean of $5.9 \, \mu g \, C \, L^{-1}$, median of $2.7 \, \mu g \, C \, L^{-1}$ and a standard deviation of $10.6 \, \mu g \, C \, L^{-1}$. The global annual average estimate of mesozooplankton in the top 200 m, based on the median value, was 0.19 Pg C. Biomass was highest in the Northern Hemisphere, and there were slight decreases from polar oceans (40–90°) to more temperate regions (15–40°) in both hemispheres. Values in the tropics (15° N–15° S) were intermediate between those at the northern and southern temperate latitudes.

1 Introduction

Mesozooplankton are found throughout the world's oceans. They are defined as zooplankton ranging from 200 μm to 2 cm (Sieburth et al., 1978), consisting primarily of crustacean plankton (copepods), meroplanktonic larva and smaller individual gelatinous zooplankton. Mesozooplankton are traditionally sampled by towed nets with mesh sizes ranging from 200 to 333 μm (Harris et al., 2000). They feed directly on phytoplankton, microzooplankton, other mesozooplankton and detritus, and have a significant feedback to primary production (Buitenhuis et al., 2006). In the global ocean they are one of the primary contributors to vertical particle flux in the oceans. Thus they are important in both the pelagic food web and export production, affecting the biogeochemical cycling of carbon and other nutrients in the oceans.

While global maps of mesozooplankton biomass exist in the literature (Bogorov et al., 1968; Reid Jr., 1962), they exist only in the form of hand-drawn maps, and the original data compiled for creating these maps are not widely available, if at all. Volume 5 of the World Ocean Atlas (WOA) 2001 (O'Brien et al., 2002) was one of the first freely available, global data compilations of zooplankton biomass created. Since then, this dataset has been expanded upon in method and data content at fairly regular intervals (O'Brien, 2005, 2007, 2010). For this synthesis, data from O'Brien (2010), along with additional new data, have been processed through the new and hybrid techniques outlined in this document.

Mesozooplankton are an important group within the plankton community. While mesozooplankton and microzooplankton collection methods and biogeochemical contribution differ greatly, a distinction is not always made between the two groups in biogeochemical models that represent all zooplankton as one box, e.g., nutrient-phytoplankton-detritus-zooplankton (NPDZ) models. NPDZ models have been shown to underestimate the interannual variability of chlorophyll a, which suggests these models also underestimate decadal- and century-scale sensitivity of climate variability (Buitenhuis et al., 2006). Models that more closely represent our current understanding of the marine ecosystem are being built in an effort to address this issue. Mesozooplankton communities have shown to exhibit decadal-scale variability with climate (Beaugrand et al., 2003), and as they have an effect on both primary production and carbon export they need to be explicitly represented in biogeochemical models (Le Quéré et al., 2005). Including mesozooplankton sensitivity to climate variability on a decadal scale, in models that capture important marine ecosystem processes, should bring us closer to modeling the response and feedbacks between marine ecosystems and climate variability, which are largely unknown at present. There is a pressing need for observations that allow the development and validation of these models, and mesozooplankton constitute a group of significant importance in this regard.

The data presented in this paper are part of a wider community effort known as MARine Ecosystem DATa (MAREDAT). MAREDAT is a collection of global biomass datasets. It contains data on the global distribution of a variety of the major plankton functional types (PFTs) currently represented in marine ecosystem models. These include picophytoplankton, diazotrophs, coccolithophores, *Phaeocystis*, diatoms, picoheterotrophs, microzooplankton, mesozooplankton, pteropods and macrozooplankton. MAREDAT is part of the MARine Ecosystem Model Inter-comparison Project (MAREMIP) that led to this compilation of observation-based global biomass datasets. The biomass data that populate MAREDAT are freely available for use in model evaluation and development, and to the scientific community as a whole.

The original mesozooplankton biomass data extracted from COPEPOD were run through standard COPEPOD

translation and standardization routines (Sect. 2.1), converted to common biomass units, sampling mesh sizes, and depth intervals (Sect. 2.2), and run through standard COPEPOD quality control routines and secondary quality control measures (Sect. 2.3). The results of the quality control routines and the gridded mesozooplankton carbon biomass data are examined and discussed in Sect. 3.

2 Data and methods

2.1 Origin of data

Mesozooplankton biomass data were extracted from the Coastal and Oceanic Plankton Ecology, Production, and Observation Database (COPEPOD, http://www.st.nmfs.noaa.gov/copepod), a global plankton database project of the US National Marine Fisheries Service (NMFS). COPEPOD's data content comes from ongoing and historical NMFS ecosystem surveys and monitoring projects, from data rescued by COPEPOD's Historical Plankton Data Search and Rescue project (COPEPOD-SAR), from international institutional and project-based sampling programs, and from individual investigators (e.g., thesis data, individual cruises).

COPEPOD's data, including mesozooplankton data, come from a wide variety of sources and in a wide variety of formats. There is a two-phase process that allows data to be translated faithfully from original file format and variables to the COPEPOD variable definition set and data structure. In the first phase, there are procedures in place that allow the original methods and metadata documentation to be reviewed ensuring accurate representation during translation. Once the original values are available in standard COPEPOD electronic format, there are two issues: (1) original units are not always comparable, and (2) taxonomic resolution is not always uniform. During the second phase, common base unit values are transformed into standard units and all taxonomic data are standardized and classified into groupings. For the purposes of this synthesis all mesozooplankton biomass values have also been converted to $\mu g\,C\,L^{-1}$. For more information in relation to the treatment and standardization of data in COPEPOD, see O'Brien (2010) (http://www.st.nmfs.noaa.gov/copepod/2010).

A total of 110 datasets were used in the global mesozooplankton biomass compilation. Table 1 lists the first 30 of these datasets, ranked in order of their spatial contribution, which represent 80 % of the spatial data coverage and 80 % of the total observations. The remaining 80 datasets individually contribute less than 1 % each to the spatial coverage. The datasets in Table 1 were ranked and sorted by the number of monthly 1×1 degree grid cells (Mcells) of spatial coverage, which they contributed to the global gridded fields, as opposed to ranking by number of observations. This method gives a higher ranking to the most spatially visible members in the global grid, such as the International Indian Ocean Expedition (IIOE), which is the most visible and dominant

Table 1. Sources for COPEPOD mesozooplankton biomass and biovolume data.

Dataset Title (as used in data files)	# of Mcells	Mcell Ranking	% Contribution to Global Field	Cumulative % Contribution	# of Observations	Observation Ranking	Abbreviated Information
CalCOFI	1952	1	7.5	7.5	38 548	1	California Cooperative Oceanic Fisheries Investigations (CalCOFI)
Odate Collection	1695	2	6.5	14.1	16 395	3	Dataset of Zooplankton Biomass in the Western North Pacific Ocean (1951–1990, K. Odate Collection)
IIOE	1409	3	5.4	19.5	1826	15	International Indian Ocean Expedition (IIOE)
HUFO-DAT	1347	4	5.2	24.7	3783	7	Hokkaido University Long-term Fisheries and Oceanographic Database (HUFO-DAT)
EASTROPAC	1287	5	5.0	29.7	3497	8	Eastern Tropical Pacific (EASTROPAC: 1967–1968) project
CSK	1207	6	4.7	34.3	2462	9	Cooperative Study of the Kuroshio and adjacent regions (CSK)
NMFS Marine Mammal Surveys	871	7	3.4	37.7	977	23	NMFS Southwest Fisheries Science Center (SWFSC) Marine Mammal surveys
IBSS Biomass Collection	825	8	3.2	40.9	1324	21	Institute of Biology of the Southern Seas (IBSS)
EcoFOCI	785	9	3.0	43.9	8803	5	Ecosystems and Fisheries-Oceanography Coordinated Investigations (EcoFOCI)
INODC Zooplankton	715	10	2.8	46.6	1851	14	National Institute of Oceanography (NIO) zooplankton database.
SEAMAP	704	11	2.7	49.4	9019	4	Southeast Monitoring and Assessment Program (SEAMAP)
Vityaz Pacific Ocean and Indian Ocean Cruises	594	12	2.3	51.7	1971	13	Institute of Oceanology/USSR Academy of Sciences – Vityaz Data Archive
EcoMon-RV (continuation of *MARMAP*)	593	13	2.3	53.9	18 749	2	NMFS Northeast Fisheries Science Center (NEFSC) Ecosystem Monitoring (EcoMon) Research Vessels division (EcoMon-RV)
VITYAZ Zooplankton	580	14	2.2	56.2	3948	6	Institute of Oceanology/USSR Academy of Sciences – Vityaz Data Archive
North Pacific Survey	518	15	2.0	58.2	944	25	North Pacific Survey 1955–1958
Institute of Marine Research – JAKARTA	503	16	1.9	60.1	1327	20	Institute of Marine Research – Jakarta, National Institute of Oceanology, Indonesian Institute of Sciences
BCF – POFI	494	17	1.9	62.0	1155	22	Bureau of Commercial Fisheries (BCF) – Pacific Oceanic Fisheries Investigations
PINRO Collection	491	18	1.9	63.9	2432	10	Knipovich Polar Research Institute of Marine Fisheries and Oceanography (PINRO)
CSIRO Australia	476	19	1.8	65.8	1496	17	Commonwealth Scientific and Industrial Research Organization (CSIRO)
R/V *ELTANIN*	463	20	1.8	67.5	972	24	United States Antarctic Research Project (USAP/USARP)
JMA North Pacific Surveys	461	21	1.8	69.3	1806	16	Japan Meteorological Agency (JMA)
NORWESTLANT	444	22	1.7	71.0	835	27	International Commission for the Northwest Atlantic Fisheries (ICNAF) – Northwest Atlantic project
EQUALANT	369	23	1.4	72.5	737	28	Equatorial Atlantic Surveys (EQUALANT) I, II, III
JARE	359	24	1.4	73.8	712	29	Japanese Antarctic Research Expedition (JARE) database – National Institute of Polar Research (NIPR)

Table 1. Continued.

Dataset Title (as used in data files)	# of Mcells	Mcell Ranking	% Contribution to Global Field	Cumulative % Contribution	# of Observations	Observation Ranking	Abbreviated Information
NMFS-SWFSC Surveys	347	25	1.3	75.2	2050	11	NMFS Southwest Fisheries Science Center (SWFSC) near shore surveys
Foxton 1956	341	26	1.3	76.5	1354	19	P. Foxton Discovery Reports Volume XXVIII (1956)
IMR Norwegian Sea Survey	337	27	1.3	77.8	878	26	Institute of Marine Research (IMR)
EASTROPIC	264	28	1.0	78.8	551	30	Eastern Tropical Pacific project (EASTROPIC: 1955)
IMECOCAL	233	29	0.9	79.7	2048	12	Investigaciones Mexicanas de la Corriente de California (IMECOCAL)
R/V *Dolphin* Cruise	226	30	0.9	80.6	1463	18	R/V *Dolphin* cruises (1965–1968)
Sum of the 30 datasets listed above	20 890		80.6	79.7	133 913		* Datasets 31 through 110 individually contributed less than 1 % each to the Global Field, but when combined
80 Additional Data Sets*	5033	31–110	19.4	21.2	22 761	31–110	together contributed 19.4 % to the Global Field and 21.2 % to the total number of Observations.
Grand Total	25 923	Mcells			153 163	records	

data source in the Indian Ocean. While this dataset is spatially ranked 3rd, it would be ranked 15th based only on observations. In contrast, the long-running EcoMon/MARMAP dataset ranks 2nd in observations but ranks 13th spatially, because those 35 yr of repeat sampling in the same 1 × 1 degree grid cell actually only contribute 12 monthly means (12 Mcells) each to the global grids created by this synthesis.

2.2   Data conversion

2.2.1   Biomass conversion

There are four different types of biomass within the COPEPOD mesozooplankton dataset: wet mass, dry mass, displacement volume, and settled volume (see Fig. 1). The determination of total sample biomass or biovolume, as compared to microscope-based full sample identification and enumeration, is relatively fast and simple and is therefore the most prevalent zooplankton measurement type and method found in both historical and ongoing mesozooplankton monitoring and survey programs (O'Brien, 2010; O'Brien et al., 2011). Of the largest data contributors to the database, the ongoing NMFS survey projects EcoFOCI, CalCOFI, SEAMAP, and EcoMon/MARMAP exclusively use displacement volume; Japanese survey programs almost exclusively use wet mass, and historical sampling by Russian/former Soviet Union (FSU) surveys uses a mixture of wet mass, displacement volume, and settled volume. Dry mass data are rare, coming primarily from the most recent sampling programs (e.g., JGOFS, GLOBEC, Norwegian Sea Survey). Published equations allow these four biomass types to be converted to carbon biomass. Total carbon mass was selected as the common zooplankton biomass proxy because of its fundamental

use in food chain and energy flow applications (Harris et al., 2000; Wiebe et al., 1975) and the abundance of published conversion equations to this biomass type (e.g., Cushing et al., 1958; Balvay, 1987; Wiebe, 1988; Bode et al., 1998; Harris, 2000). The non-linear biomass conversion equations of Balvay (1987), Wiebe (1988) and Bode et al. (1998) are used (Table 2).

2.2.2   Sampling mesh sizes

The mesozooplankton size fraction was extracted from COPEPOD by selecting only data from mesh sizes 150 to 650 μm. Three general mesh groups occur centered on 200 μm, 333 μm, and 505 μm (Fig. 2a). Historically, the most common mesh size was 333 μm (Fig. 2b), used by large, and often continuous, monitoring programs carried out by the US and Japan and by historical multi-national projects such as IIOE and NORWESTLANT. Recently sampled data, as well as the historical Russian/FSU data, focus more on data in 200 μm mesh data (Fig. 2c). Finally, large areas of the eastern Pacific used 505 μm mesh nets for their ichthyoplankton-focused surveys (Fig. 2d). A mesh category, mCAT, was assigned to each of these groupings, labeled m200, m333, m505. The original values and the assigned mCAT values are both documented in the original mesozooplankton dataset.

Mesh size affects what components of the zooplankton population are actually caught (Landry et al., 2001; Hernroth, 1987; DeVries and Stein, 1991; Colton et al., 1980), with smaller mesh nets generally collecting more biomass than larger mesh nets due to their better capture of the smaller taxa species and smaller life stages. As each mesh size does not offer a complete geographic coverage (333 μm is absent in the mid-Atlantic and Southern Ocean, and 200 μm is

Table 2. Biomass conversion equations.

Original Biomass Measure	Equation	Reference
Displacement Volume (DV) to Carbon Mass (CM)	$\log CM = (\log DV + 1.434)/0.820$	Wiebe (1988)
Wet Mass (WM) to Carbon Mass	$\log CM = (\log WM + 1.537)/0.852$	Wiebe (1988)
Dry Mass (DM) to Carbon Mass	$\log CM = (\log DM - 0.499)/0.991$	Wiebe (1988)
Settled Volume (SV) to Dry Mass	$\log DM = 0.843 \cdot \log SV + 1.417$	Balvay (1987)
Dry Mass to Carbon Mass	$\log CM = (\log DM - 0.499)/0.991$	Wiebe (1988)
Ash-free Dry Mass (AFDM) to Carbon Mass	$\log CM = (\log AFDM - 0.410)/0.963$	Bode et al. (1998)

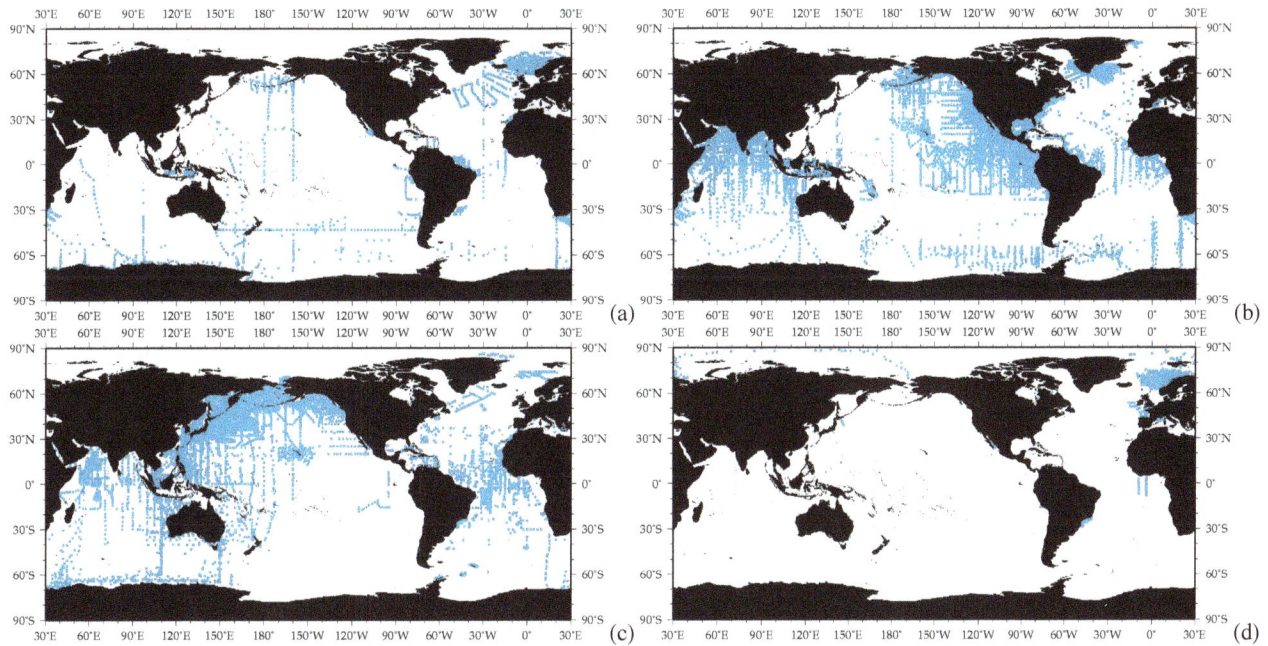

Figure 1. Distribution of the different types of biovolume and biomass samples: **(a)** settled volume, **(b)** displacement volume, **(c)** wet mass and **(d)** dry mass.

absent in the equatorial and eastern Pacific), the mesh conversion equations used in O'Brien (2005) (http://www.st.nmfs.noaa.gov/copepod/2005) were calculated using the updated mesozooplankton biomass data presented here in Table 3. As 333 μm was the most numerically abundant data type, and co-sampled 333 and 505 μm data were more prevalent than 200 and 500 μm co-sampled data, all sizes were calculated to their equivalent 333 μm values. In general, smaller mesh nets capture a larger portion of the smaller species and smaller life stages, while larger mesh nets capture less of the smaller species and life stages (Harris et al., 2000). The equations in Table 3 reduce the biomass values from 200 μm mesh nets, and increase the biomass values from 505 μm nets, to make them reasonably equivalent to data sampled with a 333 μm mesh net.

2.2.3 Depth intervals

Zooplankton and mesozooplankton alike are unevenly distributed with depth. Unlike the discrete depths of bottle-

sampled plankton (e.g., 10 m, 25 m), over 95 % of the available mesozooplankton data in COPEPOD were sampled with a single net towed over a single depth interval that generally runs from a target depth to the surface, e.g., 0–50 m, 0–100 m, with 0–150 m and 0–200 m being the most common (Fig. 3). Zooplankton data sampled from these depth intervals can be used to describe the average population throughout that interval, but they cannot be used to discuss data at an individual depth level, e.g., 20 m. A small handful of data were sampled at multiple depth intervals using a multiple net sampler; e.g., the Russian Juday multi-net frequently samples at depths 0–10 m, 10–25 m, 25–50 m, 50–100 m and 100–200 m. By adding these pieces together, it was possible to build standard depths, e.g., 0–25 m from 0–10 m and 10–25 m or 0–200 m from 0–10 m, 10–25 m, 25–50 m, 50–100 m and 100–200 m. The mesozooplankton biomass data presented here have been organized into 11 depth categories, which allow the data to be selected at a variety of different depths (see Table 4). Within the standard 33 level WOA data

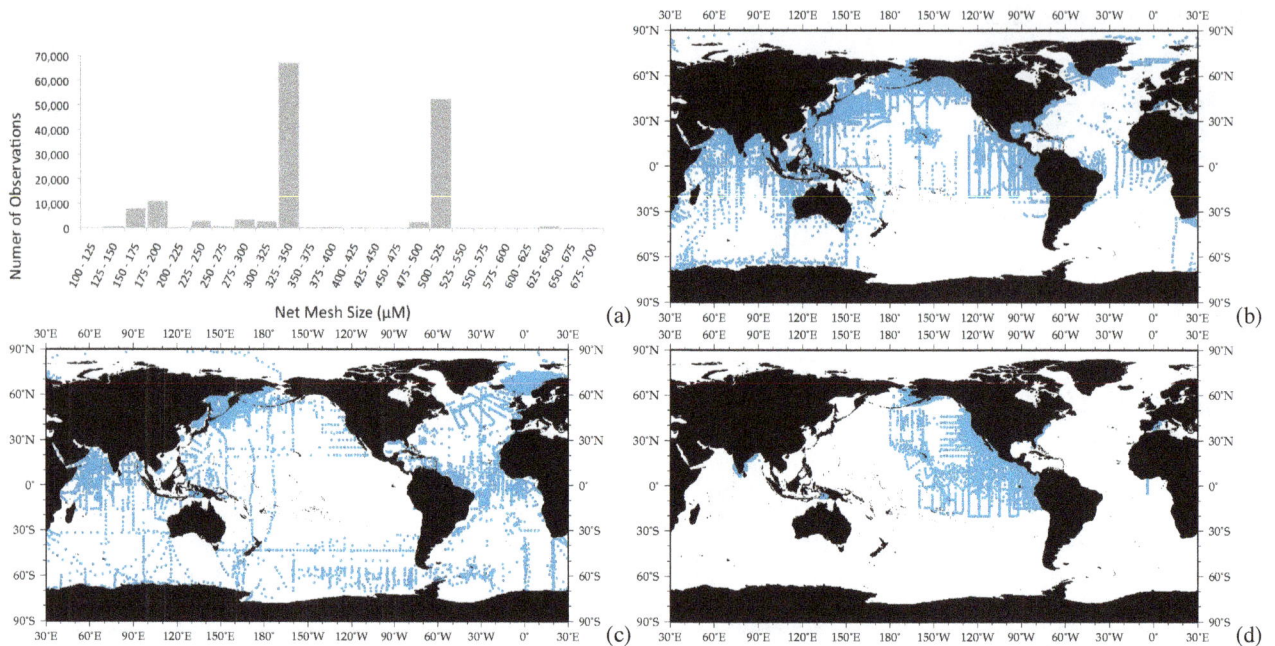

Figure 2. Biomass and biovolume sampling mesh distribution: (**a**) frequency distribution of mesh size (μm), (**b**) distribution of 333 μm, (**c**) 200 μm and (**d**) 505 μm mesh catches.

Table 3. Sampling mesh conversion equations.

Original Mesh Size	Equation	Reference
Mesozooplankton carbon mass sampled via 200 μm mesh net (CM_{M200}) to 333 μm mesh equivalent (CM_{M333})	$\text{Log } CM_{M333} = 0.6195 \cdot \log CM_{M200}$	O'Brien (2005)
Mesozooplankton carbon mass sampled via 505 μm mesh net (CM_{M505}) to 333 μm mesh equivalent (CM_{M333})	$\text{Log } CM_{M333} = 1.2107 \cdot \log CM_{M505}$	O'Brien (2005)

grid used in the MAREMIP database, the mesozooplankton were stored at the WOA depth level representing the midpoint of the tow interval. For example, the 0–40 m interval (zCat i040) was stored as 20 m (WOA level 2) while the 0–200 m interval (zCAT i200) was stored as 100 m (WOA level 7) (see Table 4).

The data were gridded using the original entries for latitude, longitude and month from all datasets. Mesozooplankton concentrations in $\mu g\,C\,L^{-1}$ were binned on the 4-dimensional WOA grid. This is a monthly grid with horizontal resolution of 1×1 degree and 33 vertical depth levels, with the first ten levels representing depths 0, 10, 20, 30, 50, 75, 100, 125, 150, and 200 m. Depth intervals were assigned to represent WOA levels, as described above. Only data that were gridded in the top 200 m of the ocean were used for calculation of global epipelagic mesozooplankton annual average biomass.

2.3 Quality control

Numerical range-based quality control of zooplankton data is complicated because of differences in sampling method, mesh size, seasonality and diurnal vertical migration (O'Brien, 2007). The mesozooplankton data acquired from COPEPOD already have quality control flags assigned to each value by COPEPOD. The COPEPOD quality control method (O'Brien, 2007, http://www.st.nmfs.noaa.gov/copepod/2007) for zooplankton biomass data divides the world into 15 major geographic basins, six mesh size categories, 12 months, four seasons and four biomass types.

The COPEPOD 2007 quality control system has three different types of outlier warning flags that are assigned based on three n-dependent ranging tiers. If a data value falls outside of 99 %, 99.9 % or 99.99 % of all other available same-category data present within the COPEPOD database, they are flagged. Using the COPEPOD quality control system, an individual mesozooplankton wet mass collected with a 333 μm mesh size in the North Pacific is compared to (1) the full numeric range of all wet mass data present within

Table 4. Description of COPEPOD depth interval criteria and World Ocean Atlas equivalents.

COPEPOD zCAT	Upper	Lower	Max upper	Min lower	Max lower	Min zdiff*	World Ocean Atlas	
	z-target (m)		z-allowed				z-ID	z-LAYER
i010	0	10	3	–	15	8	1	0
i020	0	20	5	15	30	16	2	10
i040	0	40	5	30	50	32	3	20
i060	0	60	10	50	80	48	4	30
i100	0	100	10	80	125	80	5	50
i150	0	150	10	125	175	120	6	75
i200	0	200	15	175	225	160	7	100
i250	0	250	15	225	275	200	8	125
i300	0	300	15	275	350	240	9	150
i400	0	400	20	350	450	320	10	200
i500	0	500	20	450	600	400	11	250

Notes:

* 80 % of interval

COPEPOD zCAT is the COPEPOD four-character token used to represent each depth interval, e.g., i010 = 0–10 m, i100 = 0–100 m, i200 = 0–200 m. *Upper z* and *Lower z-target* are the ideal depth intervals desired by this (zCAT) category. *Max upper z* is the maximum non-surface interval allowed by this zCAT. (This really applies more to deeper depth intervals and multi-net tows, i.e., 0–25 m, 25–50 m). *Min* and *Max lower z-allowed* are the allowed range above and below the lower z-target. (They keep the individual COPEPOD zCATs from overlapping with each other.) *Min zdiff allowed* is important if a tow is shorter than (found within) the min and max depths; this makes sure it has at least an 80 % coverage of the interval. (This is to prevent a "0–500 m" tow from being comprised of a 400–500 m-only depth fragment.)

Supplementary note: in any given tow interval within the COPEPOD dataset, a bottom depth correction flag (BDCF) will be set if the bottom depth at the sampling location is less than the lower target range for a given zCAT. This means that a 0–100 m tow in a 110 m bottom depth area would qualify as a i100, i150, i200, i250, i300, i400, and i500 value. Except for the i100, the other depths would include a "BDCF" marked in the data file. This allows a user to use all data from a single depth category, i.e., i200, or to combine multiple depth categories, i100, i150, i200 – by excluding any BDCF flags to remove duplicated data between the multiple depth files.

COPEPOD, e.g., all other wet mass data sampled in any oceanic region in any month (F1); (2) the basin-specific annual range, e.g., wet mass data sampled only in the North Pacific in any month (F2); and (3) the seasonal range, e.g., wet mass data sampled only in the North Pacific in June, July or August (F3). For the purposes of compiling the mesozooplankton biomass data, COPEPOD mesozooplankton biomass values were excluded if their flagging indicated that they fell outside of 99.9 % of same category data from any region and any month (F1), fell outside of 99 % of same category data from the same region regardless of season (F2) and during the same season (F3). The stricter criterion used for the F1 flag's range checking is intended to detect extreme outliers at the global (and any season) level without excluding reasonable differences due to geographic sub-regions and/or season (which are tested by the F2 and F3 range flagging).

The suggested minimum quality control for the MARE-DAT datasets was to apply Chauvenet's criterion for data rejection (Glover et al., 2011; Buitenhuis et al., 2012). Chauvenet's criterion was applied only to the log-transformed mesozooplankton biomass data, which are normally distributed. The mean $\bar{x}$ and the standard deviation σ of the log-transformed data were calculated and used to calculate the critical value x_c. One half of $1/(2n)$ was used as Chauvenet's criterion in a two-tailed test; however, only data on one tail, the high one, $\bar{x} + x_c$, were rejected.

Table 5. Global and latitudinal band values for the gridded mesozooplankton biomass data.

Latitude	Biomass (μg C L^{-1})					
	n	Min.	Max.	Mean	Median	±std.
Global	42 245	0.017	345.4	5.91	2.68	10.57
90–40° N	13 539	0.019	302.6	7.76	3.61	11.59
40–15° N	14 247	0.017	345.4	6.34	2.81	11.43
15° N–15° S	8825	0.057	240.5	4.63	2.67	9.33
15–40° S	2230	0.029	44.93	1.66	0.98	2.35
40–90° S	3404	0.020	177.3	2.91	1.48	5.93

3 Results and discussions

3.1 Results of quality control

The mesozooplankton data coming from COPEPOD had already undergone rigorous in-house quality control criteria (e.g., O'Brien, 2007). Out of the 156 380 originally collected mesozooplankton biomass data points, 3217 were then excluded based on COPEPOD's outlier detection flagging (quality control): 2 % of these outliers were flagged as > 99.99 % outliers; 19 % were flagged as > 99.9 % outliers; and 79 % were flagged as > 99 % outliers. Chauvenet's criterion was applied to all remaining 153 163 data points of log-transformed mesh corrected carbon biomass values. No data points from the biomass dataset were rejected as outliers

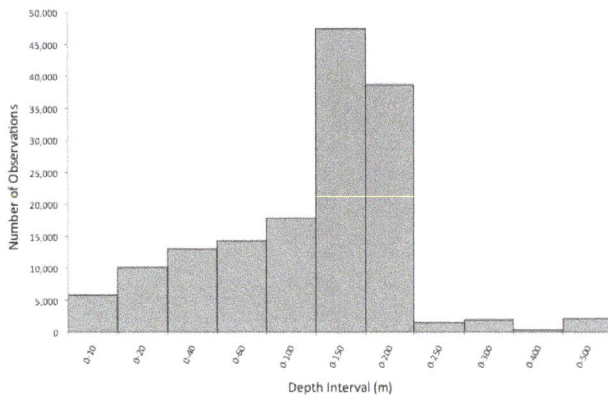

Figure 3. Distribution of original sampling depth. Depth interval 0–10 corresponds to zCAT i010; depth interval 0–20 corresponds to zCAT i020, etc. (see Table 4).

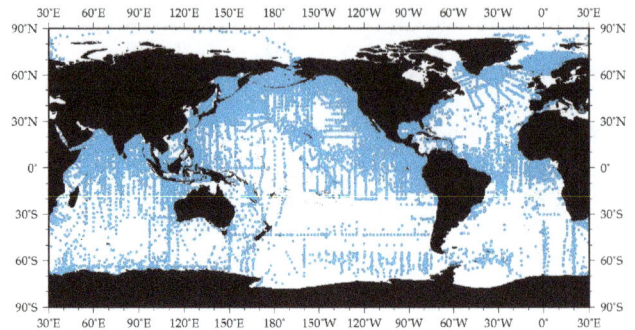

Figure 4. Global distribution of all mesozooplankton biomass data (converted to carbon and a common 333 μm equivalent mesh size). Each point represents a station where mesozooplankton were recorded.

using Chauvenet's criterion; all values being lower than the critical value of the mean $+4.6534 \times$ standard deviation.

Sampling protocols, handling, preservation and measurement techniques were not considered when removing outliers. These variables are assumed reasonably consistent within COPEPOD, but are most likely not uniform across datasets and projects. Issues related to sampling such as the inherent variability of field populations (Landry et al., 2001), mesh size, type of net, gear avoidance, seasonal/diel vertical migrations, sample handling, e.g., sample splitting, size fractionation and sample analysis, all sources of random sampling error, were considered to have a greater effect than the sampling bias issues found across projects/datasets.

3.2 Biomass description

The mesozooplankton biomass database contains 153 163 data points. Data from a number of stations that have been sampled repeatedly over many years, or programs where measurements have been made on a fine-resolution grid have been included. Therefore, after gridding, we obtained 42 245 data points on the WOA grid ($1° \times 1° \times 12$ months $\times 33$ depths), representing coverage of annually averaged biomass for 20 % of the ocean surface. To limit the overrepresentation of well-sampled locations, we present results of the gridded data.

The gridded data were split between regions as follows: 46 % of the data were found in the Pacific Ocean, 16 % in the Atlantic Ocean, 16 % in the Indian Ocean and 14 % in the polar oceans. The tropics, including the equatorial Atlantic, equatorial Pacific, Indian Ocean, which represent 43 % of the ocean surface, accounted for 39 % of the data. In contrast 14 % of the data came from the polar oceans, which represent 5 % of the ocean surface. Only 22 % of the data were found in the Southern Hemisphere (Fig. 4). There is some sampling bias towards the local summer season (Fig. 5e and f), with peak cells found in summer months in both hemispheres.

The distribution of biomass values between open water and shelf water was also examined. "Shelf water" was defined as a 1-degree grid cell in an area with a bottom depth of less than 200 m or adjacent to a grid containing land. Globally, the ratio of open vs. shelf water mesozooplankton biomass values was exactly 50 %. However, when Northern and Southern hemispheres are compared, the partitioning between open and shelf water was 47 % to 53 % in the north and 80 % to 20 % in the south; i.e., the Southern Hemisphere data were dominated by open water values. These values reflect the asymmetry in the proportion of samples collected in both hemispheres. Greater shelf water area and greater sampling effort (in terms of samples collected) in the Northern Hemisphere is important to consider when comparing these values. Although open water values seem to dominate the Southern Hemisphere data, biomass values for the region may not necessarily reflect de facto open water environment. Ice cover in the Southern Ocean means that although many samples are collected along the ice edge, the effective coastline, depending upon the season, these are labeled as "open" water by the criteria stated above.

3.3 Global estimates

Global estimates of mesozooplankton biomass were calculated from the gridded data in the top 200 m of the global ocean (see Table 5). Global mesozooplankton biomass had a mean of $5.9\,\mu g\,C\,L^{-1}$, a median of $2.7\,\mu g\,C\,L^{-1}$ and a standard deviation of $10.6\,\mu g\,C\,L^{-1}$. Biomass was highest in the Northern Hemisphere, and there were slight decreases from polar oceans (40–90°) to more temperate regions (15–40°) in both hemispheres. Values in the tropics (15° N–15° S) were intermediate between those at the northern and southern temperate latitudes. The standard deviation within the latitude bands was high so the differences in the mean were not significant. The global total of mesozooplankton carbon biomass in the top 200 m of the ocean was estimated at 0.19 PgC. This total was

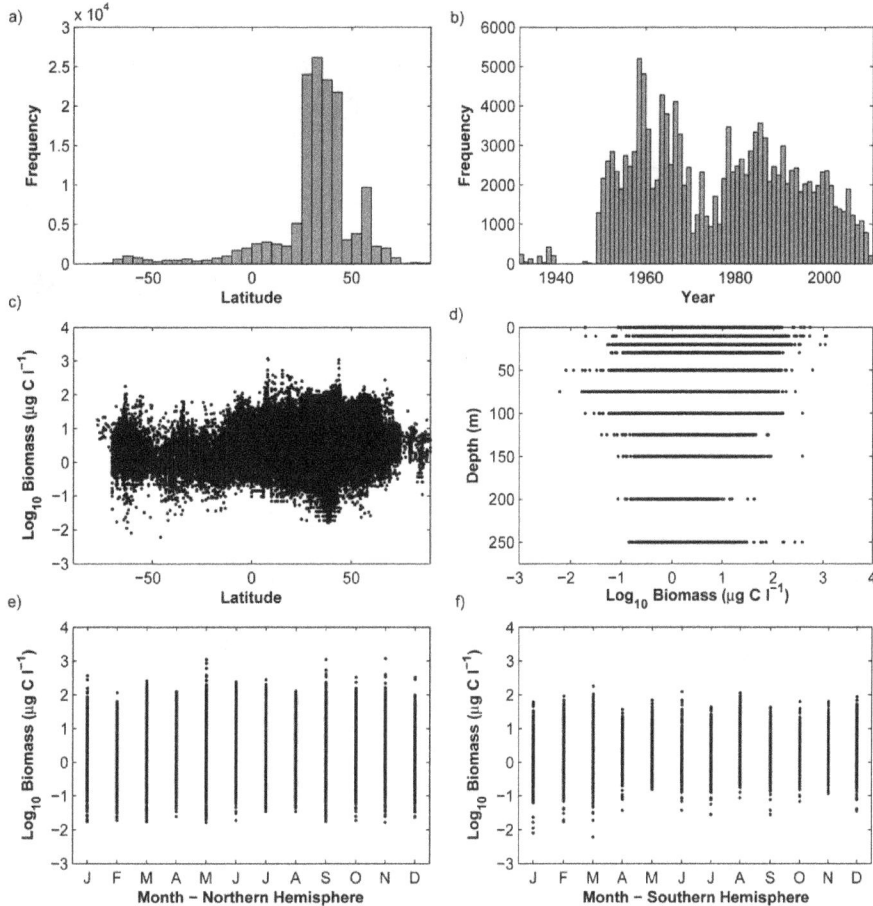

Figure 5. Description of mesozooplankton biomass observations: (**a**) latitudinal distribution, (**b**) yearly distribution, (**c**) latitudinal depth distribution, (**d**) depth distribution, (**e**) monthly distribution in the Northern and (**f**) Southern hemispheres.

calculated by multiplying the median mesozooplankton biomass value $(2.7 \, \mu g \, C \, L^{-1} = 2700 \, \mu g \, C \, m^{-3}) \times$ the area of the global ocean $(3.56 \times 10^{14} \, m^2) \times$ the upper 200 m surface layer $(200 \, m) \times 1 \times 10^{-21} \, \mu g \, Pg^{-1}$, which gave a value of 0.19 Pg C.

An overview of all 11 PFT groups currently included in the MAREDAT project is given in the Introduction to the MAREDAT ESSD Special Issue (see Buitenhuis et al., 2012). A comparison of all PFT biomasses, e.g., picophytoplankton, diazotrophs, coccolithophores, *Phaeocystis*, diatoms, picoheterotrophs, microzooplankton, mesozooplankton, pteropods and macrozooplankton, is also presented. It is important to be aware that the majority of the other plankton groups only have a small fraction of the data coverage seen in the mesozooplankton data of this paper. For these groups, the spatial and temporal coverage were limited such that only a basic comparison of "latitudinal ranges" and "annual averages" was possible.

4   Conclusions and recommendations

A coherent map of mesozooplankton global distribution and biomass is presented. Global mesozooplankton biomass was estimated from the median biomass value of $2.7 \, \mu g \, C \, L^{-1}$ ($= 0.19$ Pg C annual average mesozooplankton biomass in the top 200 m) and a standard deviation of $10.6 \, \mu g \, C \, L^{-1}$. The global, latitudinal and depth estimates of biomass concentrations will be useful for understanding ocean biogeochemistry, and for evaluating global models that include mesozooplankton. Although less developed versions of the mesozooplankton data have been published before as part of the regular COPEPOD database report series (O'Brien, 2005, 2007, 2010), this is the first time individual mesh categories (mCATs) and depth intervals (zCAT) have been distributed. This is also the first time these data have been collected together as a whole for publication in a journal together with the publication of the associated dataset. The dataset description and methods should act as a guide to those interested in using this dataset. It is important when using a dataset such as this that the associated caveats are understood and should be considered when drawing conclusions based on these data.

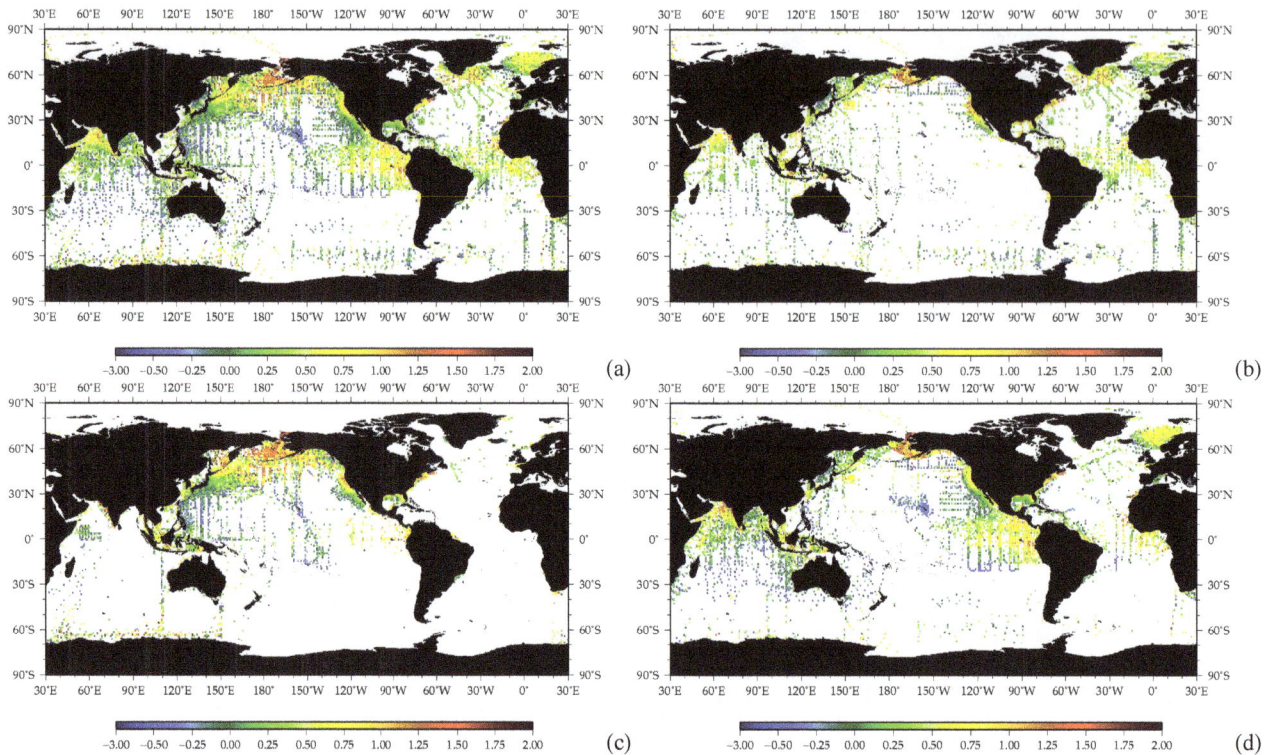

Figure 6. Annual mean mesozooplankton biomass (μg C L^{-1}): **(a)** combined 0–100 m, 0–150, and 0–200 m sampling depth intervals (zCAT i100 + i150 + i200); **(b)** 0–100 m sampling depth interval (zCAT i100) **(c)** 0–150 m sampling depth interval (zCAT i150) and **(d)** 0–200 m sampling depth interval (zCAT i200). Sampling mesh of ~ 333 μm in all.

The data compiled for this effort represent over 50 years of sampling effort made by institutions and scientists from around the world. While combined depth, mesh, and method maps such as Figs. 4 and 6a show a nearly global distribution of data, the extent of this coverage disappears quickly when one looks only at data from a specific month. Detailed maps of the monthly data distribution by depth, mesh, and original biomass type are available online at the COPEPOD project website (http://www.st.nmfs.noaa.gov/copepod). Mesozooplankton investigators and policy makers are encouraged to view these maps, as they show clearly that not all regions of the ocean are adequately sampled and the maps may provide guidance for future mesozooplankton monitoring or process studies to fill in these gaps.

Communication between biogeochemical modelers, data managers and experimentalists is at an all time high. There is an increasing interest to combine expertise from the modeling and experimental communities to produce and share the data products necessary to parameterize and validate marine ecosystem models. COPEPOD regularly interacts with scientific projects such as MAREMIP and international working groups such as the ICES Working Group on Zooplankton Ecology (WGZE). Through collaboration with the scientists and user community, COPEPOD strives to constantly improve its data content and to ensure data products, such as

the biomass fields in this paper, are available and useful to the scientific community.

Acknowledgements. A significant portion of the historical plankton data content present within COPEPOD is possible through data rescue, digitization, and funding provided by NOAA's Climate Data Modernization Program (CDMP). We thank Erik Buitenhuis, Meike Vogt and Stéphane Pesant for their support for the duration of this project.

Edited by: S. Pesant

References

Balvay, P. G.: Equivalence entre quelques parametres estimatifs de l'abondance du zooplankton total, Schweiz. Z. Hydrol., 49, 75–83, 1987.

Beaugrand, G., Brander, K. M., Lindley, J. A., Souissi, S., and Reid, P. C.: Plankton effect on cod recruitment in the North Sea, Nature, 426, 661–664, doi:10.1038/nature02164, 2003.

Bode, A., Álvarez-Ossorio, M. T., and Gonzáles, N.: Estimation of mesozooplankton biomass in a coastal upwelling area off NW Spain, J. Plankton Res., 20, 1005–1014, 1998.

Bogorov, V. G., Vinograd, M. E, Voronina, N. M., Kanaeva, I. P., and Suetova, I. A.: Distribution of zooplankton biomass within the superficial layer of the world ocean, Dokl. Akad. Nauk SSSR, 182, 1205–1207, 1968.

Buitenhuis, E., Le Quéré, C., Aumont, O., Beaugrand, G., Bunker, A., Hirst, A., Ikeda, T., O'Brien, T., Piontkovski, S., and Straile, D.: Biogeochemical fluxes through mesozooplankton, Global Biogeochem. Cy., 20, GB2003, doi:10.1029/2005gb002511, 2006.

Buitenhuis, E. T., Vogt, M., Moriarty, R., Bednaršek, N., Doney, S. C., Leblanc, K., Le Quéré, C., Luo, Y.-W., O'Brien, C., O'Brien, T., Peloquin, J., Schiebel, R., and Swan, C.: MAREDAT: towards a World Ocean Atlas of MARine Ecosystem DATa, Earth Syst. Sci. Data Discuss., 5, 1077–1106, doi:10.5194/essdd-5-1077-2012, 2012.

Colton, J. B., Green, J. R., Byron, R. R., and Frisella, J. L.: BONGO net retention rates as effected by towing speed and mesh size, Can. J. Fish. Aquat. Sci., 37, 606–623, doi:10.1139/f80-077, 1980.

Cushing, D. H., Humprey, G. H., Banse, K., and Laevastui, T.: Report of the committee on terms and equivalents. Rapp. P.-V. Reun. Cons. Int. Explor. Mer, 144, 15–16, 1958.

DeVries, D. R. and Stein, R. A.: Comparison of three zooplankton samplers: a taxon-specific assessment, J. Plankton Res., 13, 53–59, 1991.

Glover, D. M., Jenkins, W. J., and Doney, S. C.: Modeling Methods for Marine Science, Cambridge University Press, Cambridge, 588 pp., 2011.

Harris, R. P., Wiebe, P. H., Lenz, J., Skjldal, H. R., and Huntley, M.: ICES Zooplankton Methodology Manual, Academic Press, 684 pp., 2000.

Hernroth, L.: Sampling and filtration efficency of 2 commonly used plankton nets – a comparitive study of the Nansen net and the UNESCO WP-2 net, J. Plankton Res., 9, 719–728, doi:10.1093/plankt/9.4.719, 1987.

Landry, M. R., Al-Mutairi, H., Selph, K. E., Christensen, S., and Nunnery, S.: Seasonal patterns of mesozooplankton abundance and biomass at Station ALOHA, Deep-Sea Res. Pt. II, 48, 2037–2061, doi:10.1016/s0967-0645(00)00172-7, 2001.

Le Quéré, C., Harrison, S. P., Prentice, I. C., Buitenhuis, E. T., Aumont, O., Bopp, L., Claustre, H., Da Cunha, L. C., Geider, R., Giraud, X., Klaas, C., Kohfeld, K. E., Legendre, L., Manizza, M., Platt, T., Rivkin, R. B., Sathyendranath, S., Uitz, J., Watson, A. J., and Wolf-Gladrow, D.: Ecosystem dynamics based on plankton functional types for global ocean biogeochemistry models, Glob. Change Biol., 11, 2016–2040, 2005.

O'Brien, T. D.: COPEPOD: A Global Plankton Database, US Dep. Commerce, NOAA Tech. Memo, 136 pp., 2005.

O'Brien, T. D.: COPEPOD: The Global Plankton Database. A review of the 2007 database contents and new quality control methodology, US Dep. Commerce, NOAA Tech. Memo, 28 pp., 2007.

O'Brien, T. D.: COPEPOD: The Global Plankton Database. An overview of the 2010 database contents, processing methods, and access interface, US Dep. Commerce, NOAA Tech. Memo NMFS-F/ST-36, 28 pp., 2010.

O'Brien, T. D., Conkright, M. E., Boyer, T. P., Stephens, C., Antonov, J. I., Locarnini, R. A., and Garcia, H. E.: World Ocean Atlas 2001, Volume 5: Plankton, NOAA Atlas NESDIS 53, edited by: Levitus, S., US Government Printing Office, Washington DC, 89 pp., 2002.

O'Brien, T. D., Wiebe, P. H., and Hay, S.: ICES Zooplankton Status Report 2008/2009, 152 pp., 2011.

Reid Jr., J. L.: On Circulation, Phosphate-Phosphorus Content, and Zooplankton Volumes in the Upper Part of the Pacific Ocean, Limnol. Oceanogr., 7, 287–306, 1962.

Sieburth, J. M., Smetacek, V., and Lenz, J.: Pelagic Ecosystem Structure: Heterotrophic Compartments of the Plankton and Their Relationship to Plankton Size Fractions, Limnol. Oceanogr., 23, 1256–1263, 1978.

Wiebe, P. H.: Functional regression equations for zooplankton displacement volume, wet weight, dry weight, and carbon. A correction, Fish. Bull., 86, 833–835, 1988.

Wiebe, P. H., Boyd, S., and Cox, J. L.: Relationships between zooplankton displacement volume, wet weight, dry weight, and carbon, Fish. Bull., 73, 777–786, 1975.

4

Gas phase acid, ammonia and aerosol ionic and trace element concentrations at Cape Verde during the Reactive Halogens in the Marine Boundary Layer (RHaMBLe) 2007 intensive sampling period

R. Sander[1], A. A. P. Pszenny[2], W. C. Keene[3], E. Crete[2,*], B. Deegan[4,**], M. S. Long[3,***], J. R. Maben[3], and A. H. Young[2]

[1]Air Chemistry Department, Max-Planck Institute of Chemistry, P.O. Box 3060, 55020 Mainz, Germany
[2]University of New Hampshire, Durham, NH, USA
[3]Department of Environmental Sciences, University of Virginia, Charlottesville, VA 22904, USA
[4]Mount Washington Observatory, North Conway, New Hampshire, USA
*now at: The Earth Institute, Columbia University, NY, USA
**now at: 97 Raymond St., Fairhaven, MA, USA
***now at: Harvard University, Cambridge, MA, USA

Correspondence to: R. Sander (rolf.sander@mpic.de)

Abstract. We report mixing ratios of soluble reactive trace gases sampled with mist chambers and the chemical composition of bulk aerosol and volatile inorganic bromine (Br_g) sampled with filter packs during the Reactive Halogens in the Marine Boundary Layer (RHaMBLe) field campaign at the Cape Verde Atmospheric Observatory (CVAO) on São Vicente island in the tropical North Atlantic in May and June 2007. The gas-phase data include HCl, HNO_3, HONO, HCOOH, CH_3COOH, NH_3, and volatile reactive chlorine other than HCl (Cl^*). Aerosol samples were analyzed by neutron activation (Na, Al, Cl, V, Mn, and Br) and ion chromatography (SO_4^{2-}, Cl^-, Br^-, NH_4^+, Na^+, K^+, Mg^{2+}, and Ca^{2+}).

1 Introduction

Multiphase halogen chemistry impacts important, interrelated chemical processes in marine air. Bromine activation chemistry leads to catalytic ozone destruction and modification of oxidation processes including HO_x and NO_x cycling. Spatiotemporal variability in many reactants, products and reaction pathways are poorly characterized, rendering uncertain the global significance of tropospheric halogen chemistry.

2 Data set description and access

RHaMBLe was a large-scale investigation of reactive halogen cycling and associated impacts on oxidation processes in the marine boundary layer over the eastern North Atlantic Ocean (Lee et al., 2010). As part of RHaMBLe, (Reactive Halogens in the Marine Boundary Layer) an intensive 25 day process study was conducted during spring 2007 at the Cape Verde Atmospheric Observatory (Fig. 1) on the windward shore of São Vicente Island (16.8° N, 24.9° W) in the tropical, eastern North Atlantic (Fig. 2). Data from this field intensive campaign coupled with measurements within the surrounding region and at the observatory during other periods have been used to address a range of topics including

Figure 1. The sampling tower at the Cape Verde Atmospheric Observatory in 2007.

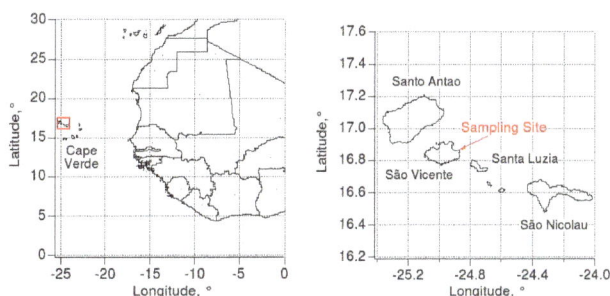

Figure 2. Location of the sampling site.

halogen-mediated destruction of ozone (Read et al., 2008), pollution-enhanced production of Cl radicals and associated influences on oxidation processes (Lawler et al., 2009), the cycling of reactive nitrogen oxides (Lee et al., 2009), and the composition and processing of aerosols (Allan et al., 2009), among others. Herein, we report a suite of soluble reactive trace gases, volatile inorganic bromine, and ionic and elemental aerosol constituents measured from the top of the observatory's 30 m tower during the RHaMBLe campaign.

The complete data set is available in NASA Ames format[1] from the Zenodo repository service under doi:10.5281/zenodo.6956. In addition, we also provide comma-separated-values (csv) files of the full data set in the Supplement.

3   Instruments and methods

All air volumes reported here are normalized to standard temperature and pressure (273 K and 1.013×10^5 Pa).

---

[1]http://badc.nerc.ac.uk/help/formats/NASA-Ames

Figure 3. Mist chambers at the Cape Verde sampling tower.

3.1   Mist chambers

Water-soluble, volatile inorganic chlorine and nitrate (dominated by and hereafter referred to as HCl and HNO_3, respectively), NH_3, HCOOH, and CH_3COOH were sampled over 2 h intervals at nominal flow rates of $20 \, L \, min^{-1}$ with a single set of tandem mist chambers (Figs. 3, 4), each of which contained 20 mL deionized water (Lawler et al., 2009). To minimize artifact phase changes caused by mixing chemically distinct aerosol size fractions on bulk prefilters, air was sampled through a size-fractionating inlet that inertially removed super-μm aerosols from the sample stream. Sub-μm aerosol was removed downstream by an in-line 47 mm Teflon filter (Zefluor 2 μm pore diameter). In-line filters were changed daily. Samples were analyzed on site by ion chromatography (IC) usually within a few hours after recovery. Data were corrected based on dynamic handling blanks that were loaded, briefly (few seconds) exposed to ambient air flow, recovered, processed, and analyzed using procedures identical to those for samples. Collection efficiencies for all species were greater than 95 % and, consequently, corrections for inefficient sampling were not necessary. Detection limits (DLs; estimated following Keene et al., 1989) for HCl, HNO_3, NH_3, HCOOH, and CH_3COOH were 26, 12, 3, 29, and $44 \, pmol \, mol^{-1}$, respectively. The corresponding precision for each analyte is approximately one half of the estimated DL.

Reactive inorganic chlorine gases (Cl^*) were sampled in parallel through an identical inlet with similar sets of tandem mist chamber samplers (Keene et al., 1993; Maben et al., 1995; Lawler et al., 2009). The upstream chamber contained acidic solution (37.5 mM H_2SO_4 and 0.042 mM $(NH_4)_2SO_4$), which removed HCl quantitatively but efficiently passed other forms of volatile Cl, and the downstream chamber contained alkaline solution (30.0 mM $NaHCO_3$ and 0.408 mM $NaHSO_3$), which sampled Cl^* (including Cl_2 HOCl, and

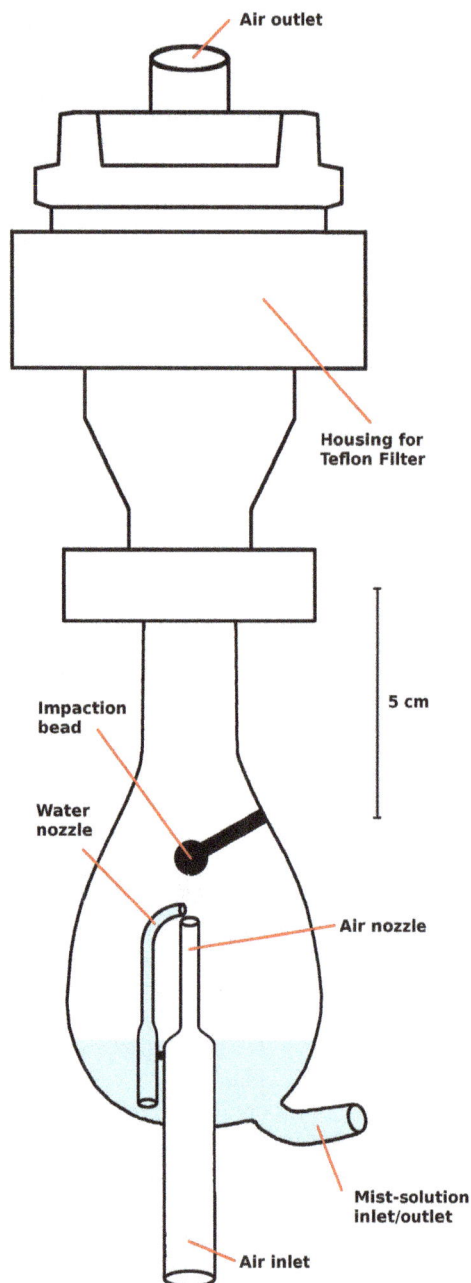

Figure 4. Schematic of a mist chamber (to scale). The teflon filter (only housing is shown) serves as a liquid barrier.

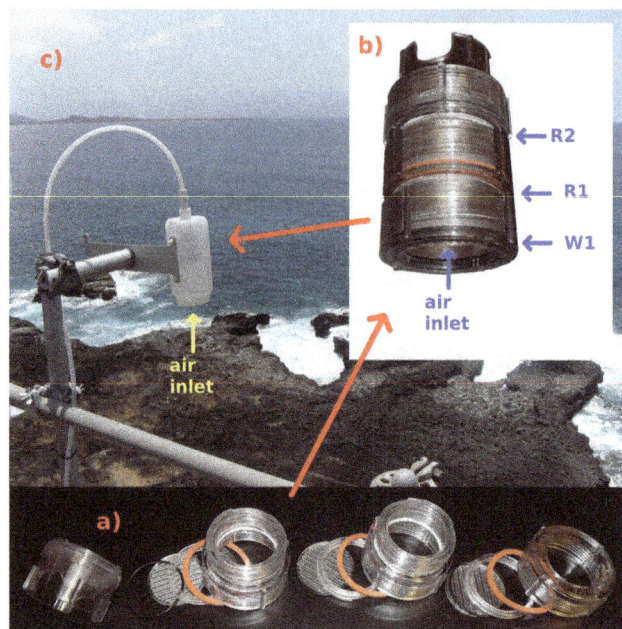

Figure 5. Filter packs: (**a**) individual pieces (without filters); (**b**) assembled filter pack with positions of the Whatman (W1) and Rayon (R1, R2) filters indicated; (**c**) deployment at the Cape Verde sampling tower.

The average precision for Cl* was approximately $\pm 15\%$ or $\pm 7\,\text{pmol}\,\text{mol}^{-1}$ Cl, whichever was the greater absolute value, and the corresponding average DL based on Keene et al. (1989) was $14\,\text{pmol}\,\text{mol}^{-1}$ Cl.

3.2 Filter-pack sampling and chemical analysis

Using a modification of the technique by Rahn et al. (1976), total aerosol and inorganic gases were sampled using filter packs with 3 filters (Fig. 5). Particles were sampled on dry 47 mm diameter Whatman 41 cellulose filters that were positioned upstream (W1 in Fig. 5). Each filter was precleaned with deionized water (DIW) and dried prior to the campaign. Alkaline-reactive trace gases were sampled on moist tandem Rayon filters (SKU 64007, Leader Evaporator), impregnated with an alkaline solution (10 g LiOH and 10 mL glycerol per 100 mL), positioned in tandem downstream (R1 and R2 in Fig. 5). Filters from the same lot were used for all samples, blanks, and analytical standards. Sample and field blank filters were mounted in 47 mm diameter Nuclepore polycarbonate cassettes. Seven samples were collected each day with sample changes keyed to sunrise and sunset and other change times adjusted such that sampling interval durations were similar (3 to 4 h). The airflow was about 80 standard liters per minute (SLPM) and monitored with a mass flow meter. Field blanks were obtained once per day at different times. Each field blank was loaded, deployed, exposed briefly (few seconds) to ambient air, recovered, processed,

probably contributions from ClNO₃, ClNO₂, BrCl, ClO, and Cl). Available evidence (Keene et al., 1993; Maben et al., 1995) indicates that this sampling technique reliably discriminates volatile inorganic Cl from Cl associated with both particles and organic gases and that it quantitatively differentiates between HCl and other forms of volatile inorganic Cl. However, the speciation of Cl* cannot be determined unequivocally. Mist solutions were analyzed on site by IC.

Table 1. Data summary. Here, DL is the detection limit. $N(\text{tot})$ and $N(<\text{DL})$ are the total number of data points and the number below the DL, respectively.

	minimum	maximum	median	DL*	$N(\text{tot})$	$N(<\text{DL})$	unit
			mist chamber				
HCl	<DL	613	206	26	212	13	pmol mol^{-1}
Cl*	<DL	222	21	14	212	79	pmol mol^{-1}
HNO_3	<DL	124	14	12	212	90	pmol mol^{-1}
NH_3	<DL	651	18	3	212	51	pmol mol^{-1}
HCOOH	<DL	796	128	29	212	68	pmol mol^{-1}
CH_3COOH	<DL	550	78	44	212	68	pmol mol^{-1}
			filter pack (NAA)				
Na	1.12	7.69	3.3	0.031	147	0	$\mu\text{g m}^{-3}$
Al	<DL	0.559	0.095	0.013	147	2	$\mu\text{g m}^{-3}$
Cl	0.39	6.84	2.6	0.079, 0.13	147	0	$\mu\text{g m}^{-3}$
EF(Cl)	0.194	0.752	0.432		147		
Cl deficit	1.21	6.96	3.17		147		$\mu\text{g m}^{-3}$
V	<DL	2.21	0.68	0.26	147	16	ng m^{-3}
Mn	<DL	4.98	1.12	0.52	147	17	ng m^{-3}
Br	<DL	26.3	8.39	4.2	147	35	ng m^{-3}
EF(Br)	0.0524	1.06	0.392		147		
Br deficit	−1.33	36.7	11.3		147		ng m^{-3}
Br_g	3.26	41.7	16.8		147		ng m^{-3}
ΔBr	−20.8	29.2	5.48		147		ng m^{-3}
			filter pack (IC)				
Cl^-	0.168	7.66	2.65	0.14, 0.2	147	1	$\mu\text{g m}^{-3}$
$EF(Cl^-)$	0.0723	0.763	0.447		147		
Cl^- deficit	1.31	6.53	2.96		147		$\mu\text{g m}^{-3}$
Br^-	<DL	16.7	4.59	1.3, 8.2	147	115	ng m^{-3}
$EF(Br^-)$	0.0605	0.963	0.227		147		
Br^- deficit	0.404	40.2	14.3		147		ng m^{-3}
SO_4^{2-}	0.866	6.26	2.36	0.09, 0.036	147	0	$\mu\text{g m}^{-3}$
Na^+	1.11	7.91	3.2	0.069, 0.14	147	0	$\mu\text{g m}^{-3}$
NH_4^+	<DL	1.11	0.292	0.49, 0.19	147	51	$\mu\text{g m}^{-3}$
K^+	<DL	0.31	0.146	0.064, 0.11	147	32	$\mu\text{g m}^{-3}$
Mg^{2+}	0.143	0.979	0.418	0.046, 0.006	147	0	$\mu\text{g m}^{-3}$
Ca^{2+}	<DL	0.632	0.19	0.35, 0.15	147	59	$\mu\text{g m}^{-3}$

* For IC and for chloride from NAA, the blanks' analyses indicated two distinct subsets with quite different concentrations of several ions. Individual detection limits were calculated for these subsets.

and analyzed using procedures identical to those for samples. Between each use filter cassettes were completely disassembled, cleaned with dilute Liquinox solution, rinsed with DIW and dried in a clean bench. Exposed filters were transferred to clean polyethylene envelopes, frozen, and transported to and stored frozen at the University of New Hampshire (UNH) prior to preparation for analysis.

Using tandem impregnated filters allowed to calculate the collection efficiency ε for volatile inorganic bromine Br_g:

$$\varepsilon = \frac{m_1}{m_{\text{tot}}} = \frac{m_2}{(m_{\text{tot}} - m_1)}. \tag{1}$$

Here, m_1 and m_2 are the masses deposited on the first and second impregnated filter, respectively, and m_{tot} is the total

mass that would be deposited at 100 % collection efficiency. Rearranging the equations and inserting the known values of m_1 and m_2 yields a collection efficiency of

$$\varepsilon = 1 - m_2/m_1 \tag{2}$$

and a total mass of

$$m_{\text{tot}} = \frac{m_1}{1 - m_2/m_1}. \tag{3}$$

All filters were analyzed by neutron activation analysis (NAA) using a procedure based on that described by Uematsu et al. (1983). Standards were prepared by spotting aliquots of a NIST-traceable mixed element standard solution (Ultra Scientific, North Kingstown, RI) on blank filter circles. Standards, samples and field blanks were each

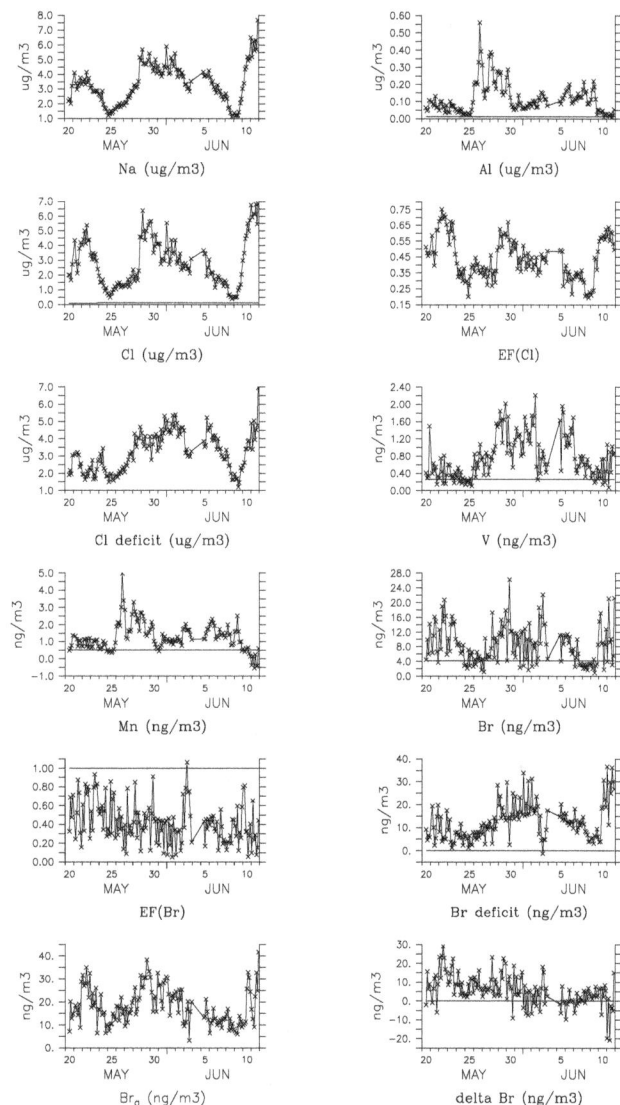

Figure 6. Time series of the filter pack data, analyzed by NAA. The red lines denote the detection limit and the blue lines denote sea water composition, i.e., EF = 1 and deficit = 0.

spiked with 20 ng of indium in dilute nitric acid (as an aliquot of a NIST-traceable standard solution; Ultra Scientific) as internal flux monitor, sealed in a clean polyethylene envelope and subsequently irradiated at the Rhode Island Nuclear Science Center (RINSC) for 300 s at a nominal flux of 4×10^{12} cm^{-2} s^{-1} thermal neutrons. Following irradiation, samples were allowed to decay for approximately 5 min during which they were transferred to non-irradiated envelopes, and counted for 900 s live time on a Ge(Li) gamma-ray spectrometer. From the particle filters, data were obtained for six elements: Na, Al, Cl, Mn, V, and Br. Using Eq. (3), gaseous Br data were obtained from the impregnated filters. All laboratory manipulations of cassettes and filters prior to irradia-

tion were carried out in class 100 clean benches by personnel wearing unpowdered plastic gloves.

About three years after the NAA analysis, the radioactivity had decreased to a safe level, and it was possible to handle the samples again. The aerosols on the particle filters were extracted under sonication in DIW. The ionic species SO_4^{2-}, Cl^-, Br^-, NH_4^+, Na^+, K^+, Mg^{2+}, and Ca^{2+} were determined by high-performance ion chromatography (IC) using procedures similar to those described by Keene et al. (2009). Data for samples were corrected based on median concentrations of analytes recovered from handling blanks ($n = 10$).

To estimate the detection limits for the filter pack aerosol samples, we used Eq. (14) of Currie (1995). An element or ion was deemed detected if its mass (NAA) or extract concentration (IC) exceeded its minimal detectable value estimated via

$$DL = 2 t_{1-\alpha,\nu} s_0, \qquad (4)$$

where DL is the detection limit (minimal detectable value), t is the value of Student's t statistic for probability level α and ν degrees of freedom (i.e., number of blanks), and s_0 is the standard deviation of mass or concentration values determined in the N blanks analyzed. The choice of $\alpha = 0.025$ corresponds to rejecting a null hypothesis of "not detected" at the 95 % confidence level. The average sampled air volume for all samples of 12.4 m^3 STP (0 °C, 1 atm) was assumed for all calculations.

For IC, the blanks' analyses indicated two distinct subsets of blanks of $N = 3$ and $N = 20$, respectively, with quite different concentrations of several ions. For NAA, chlorine appeared to have the same two subsets. For other elements determined no distinct subsets were apparent.

The phase partitioning of HCl and NH$_3$ with aerosol solutions is pH dependent, and aerosol pH typically varies as a function of size. Because the pH of aerosols sampled in bulk on a filter may diverge from the pH of the aerosol size fractions with which most Cl^- and NH_4^+ is associated in ambient air, these species are subject to artifact phase changes when sampled in bulk (e.g., Keene et al., 1990). Consequently, the absolute concentrations and associated enrichment factors and deficits relative to sea salt that are reported herein may not be representative of those for ambient aerosols during the campaign.

4   Data summary

A detailed analysis of the data is beyond the scope of a paper in this journal. Here, we only present the data and draw some basic conclusions. Table 1 lists minimum, maximum, average, and median values, and Figs. 6, 7, and 8 show time series of the measurements.

Regarding the analytical techniques applied here, we found very good correlations between Na and Cl measured by NAA vs. IC (Fig. 10) coupled with slopes and intercepts

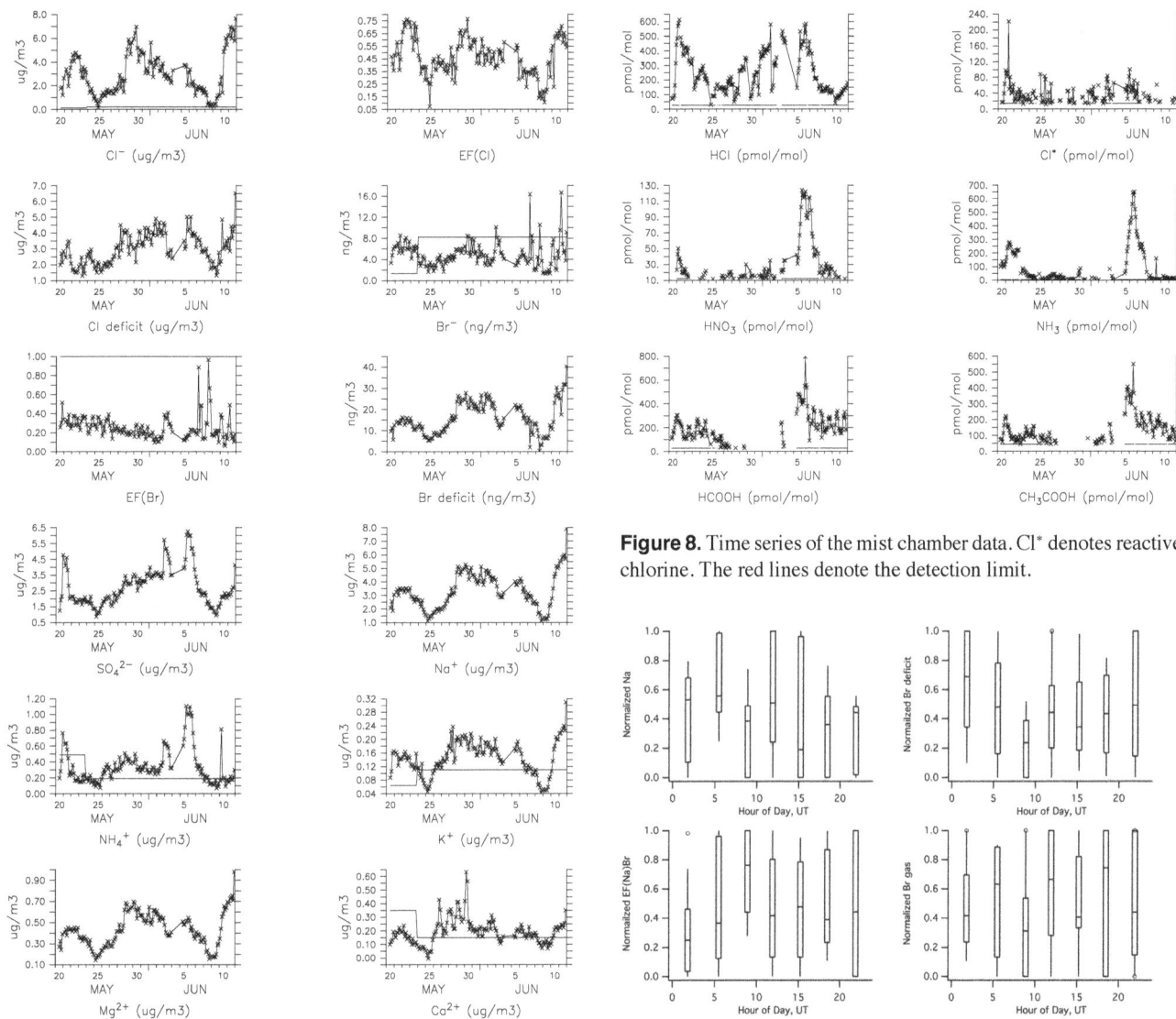

Figure 7. Time series of the filter pack data, analyzed by IC. The blue lines denote sea water composition, i.e., EF = 1 and deficit = 0. The red lines denote the detection limit. The different detection limits reflect different blank mean amounts and variances in filters used during early vs. later portions of the campaign.

Figure 8. Time series of the mist chamber data. Cl* denotes reactive chlorine. The red lines denote the detection limit.

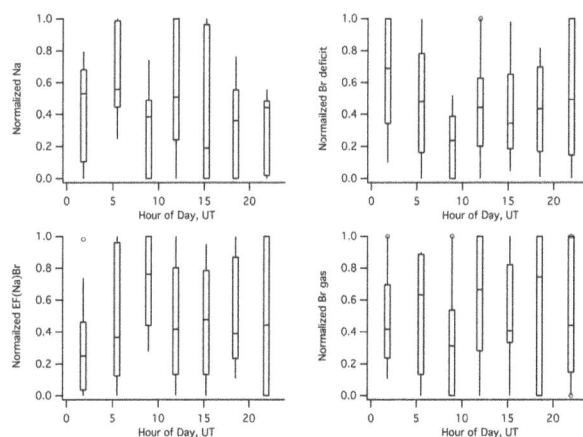

Figure 9. Box-and-whisker plots of the diel variability of normalized Na and Br data. In these plots the daily maximum concentration was set to 1 and the minimum concentration was set to 0. The solid line near the middle of each box represents the median. The bottom of the box represents the 25th percentile and the top of the box represents the 75th percentile. The whisker ends mark the 10th and 90th percentiles. Outliers are displayed for values below the 1st or above the 99th percentile.

that are statistically indistinguishable from 1.0 and 0.0, respectively. This indicates that (1) virtually all particulate Na and Cl were in the forms Na^+ and Cl^-, respectively, and (2) Na^+ and Cl^- in samples did not deteriorate while stored for 3 yr.

Correlations between selected data are shown in Fig. 10. A good correlation can indicate a common origin. The relative concentrations of Na, Mg and K are the same as for sea salt. The correlation between Al and Mn points towards crustal dust. The presence of volatile acids at detectable mixing ratios indicates that aerosols were either acidic or rapidly acidified throughout the campaign. A good correlation be-

tween the acids HCOOH and CH_3COOH (Fig. 10) suggests common sources and/or processing.

In the context of reactive halogens, the main focus of the campaign, a few definitions are needed for the evaluation of the results. Adopting the equations from Sander et al. (2003) and using the seawater mass ratios $([Cl]/[Na])_{seawater} = 1.8 \, kg \, kg^{-1}$ and $([Br]/[Na])_{seawater} = 0.0062 \, kg \, kg^{-1}$, we define the enrichment factors EF(X) for X = Cl and Br as

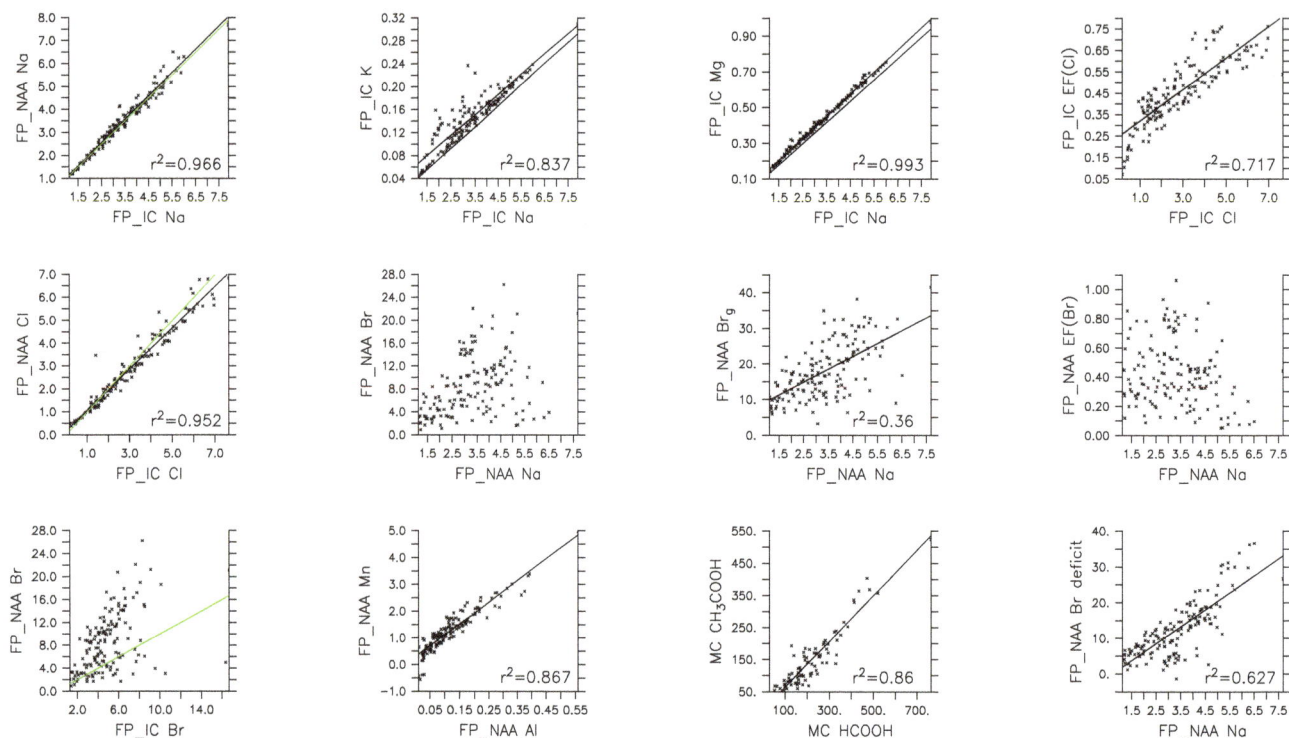

Figure 10. Scatter plots. The first column shows comparisons between filter pack samples analyzed by neutron activation (prefix FP_NAA) and ion chromatography (prefix FP_IC), respectively, with a 1 : 1 line in green. The second and third columns show correlations between selected data, where the prefix MC refers to mist chamber data, and a blue line indicates the sea water composition. Finally, the last column shows correlations to the derived quantities enrichment factor and deficit, as defined in Sect. 4. The units of all plots are the same as listed in Table 1. Regression lines in red are only shown if $r^2 > 0.3$.

$$EF(X) = \frac{[X]/[Na]}{([X]/[Na])_{seawater}}, \qquad (5)$$

where square brackets denote mass concentrations. Another useful quantity is the absolute deficit, which is defined as

$$deficit = \left[[Na] \times \left(\frac{[X]}{[Na]} \right)_{seawater} \right] - [X]. \qquad (6)$$

We also define the quantity ΔBr, which is the difference between gas-phase bromine and the bromine deficit. The following are our main conclusions regarding measured halogens.

- Values for total bromine analyzed by NAA are higher than those for bromide analyzed by IC. This indicates that some aerosol bromine is in a form other than Br^-.

- EF(Br) was not correlated with particulate Na^+. Absolute deficits of particulate Br and mixing ratios of gasphase bromine (Br_g) increased with increasing sea-salt concentrations. These relationships together with the presence of acidic gases at detectable levels indicate that sufficient acidity was available to titrate sea-salt alkalinity and sustain halogen activation over the full range of sea-salt loadings during the experiment.

- Br_g concentrations are almost always higher than absolute deficits of particulate Br (the difference is shown as "delta Br" in Fig. 6). This implies that Br_g has a longer atmospheric lifetime than the parent aerosol. This has also been observed in a previous campaign (Keene et al., 2009).

- Systematic diel variability was not evident in either the aerosol Br enrichment factor nor inorganic gaseous Br (Fig. 9). This is in contrast to previous measurements in the MBL at Hawaii (Sander et al., 2003) and over the eastern Atlantic (Keene et al., 2009).

Acknowledgements. We thank B. Faria (Instituto Nacional de Meteorologia e Geofísica, São Vicente) and technical staff at the Cape Verde Atmospheric Observatory for outstanding logistical support. K. Read, J. Plane, D. Heard, G. McFiggans, L. Carpenter, E. Saltzman and their respective students and staff collaborated in the research effort. This research was funded by the US National Science Foundation via awards AGS-0646864 to the University of New Hampshire, and AGS-0646854 and AGS-0541570 to the University of Virginia.

Edited by: V. Sinha

References

Allan, J. D., Topping, D. O., Good, N., Irwin, M., Flynn, M., Williams, P. I., Coe, H., Baker, A. R., Martino, M., Niedermeier, N., Wiedensohler, A., Lehmann, S., Müller, K., Herrmann, H., and McFiggans, G.: Composition and properties of atmospheric particles in the eastern Atlantic and impacts on gas phase uptake rates, Atmos. Chem. Phys., 9, 9299–9314, doi:10.5194/acp-9-9299-2009, 2009.

Currie, L. A.: Nomenclature in evaluation of analytical methods including detection and quantification capabilities, Pure Appl. Chem., 67, 1699–1723, 1995.

Keene, W. C., Talbot, R. W., Andreae, M. O., Beecher, K., Berresheim, H., Castro, M., Farmer, J. C., Galloway, J. N., Hoffmann, M. R., Li, S.-M., Maben, J. R., Munger, J. W., Norton, R. B., Pszenny, A. A. P., Puxbaum, H., Westberg, H., and Winiwarter, W.: An intercomparison of measurement systems for vapor and particulate phase concentrations of formic and acetic acids, J. Geophys. Res., 94D, 6457–6471, 1989.

Keene, W. C., Pszenny, A. A. P., Jacob, D. J., Duce, R. A., Galloway, J. N., Schultz-Tokos, J. J., Sievering, H., and Boatman, J. F.: The geochemical cycle of reactive chlorine through the marine troposphere, Global Biogeochem. Cycles, 4, 407–430, 1990.

Keene, W. C., Maben, J. R., Pszenny, A., and Galloway, J. N.: Measurement technique for inorganic chlorine gases in the marine boundary layer, Environ. Sci. Technol., 27, 866–874, 1993.

Keene, W. C., Long, M. S., Pszenny, A. A. P., Sander, R., Maben, J. R., Wall, A. J., O'Halloran, T. L., Kerkweg, A., Fischer, E. V., and Schrems, O.: Latitudinal variation in the multiphase chemical processing of inorganic halogens and related species over the eastern North and South Atlantic Oceans, Atmos. Chem. Phys., 9, 7361–7385, doi:10.5194/acp-9-7361-2009, 2009.

Lawler, M. J., Finley, B. D., Keene, W. C., Pszenny, A. A. P., Read, K. A., von Glasow, R., and Saltzman, E. S.: Pollution-enhanced reactive chlorine chemistry in the eastern tropical Atlantic boundary layer, Geophys. Res. Lett., 36, L08810, doi:10.1029/2008GL036666, 2009.

Lee, J. D., Moller, S. J., Read, K. A., Lewis, A. C., Mendes, L., and Carpenter, L. J.: Year-round measurements of nitrogen oxides and ozone in the tropical North Atlantic marine boundary layer, J. Geophys. Res., 114D, D21302, doi:10.1029/2009JD011878, 2009.

Lee, J. D., McFiggans, G., Allan, J. D., Baker, A. R., Ball, S. M., Benton, A. K., Carpenter, L. J., Commane, R., Finley, B. D., Evans, M., Fuentes, E., Furneaux, K., Goddard, A., Good, N., Hamilton, J. F., Heard, D. E., Herrmann, H., Hollingsworth, A., Hopkins, J. R., Ingham, T., Irwin, M., Jones, C. E., Jones, R. L., Keene, W. C., Lawler, M. J., Lehmann, S., Lewis, A. C., Long, M. S., Mahajan, A., Methven, J., Moller, S. J., Müller, K., Müller, T., Niedermeier, N., O'Doherty, S., Oetjen, H., Plane, J. M. C., Pszenny, A. A. P., Read, K. A., Saiz-Lopez, A., Saltzman, E. S., Sander, R., von Glasow, R., Whalley, L., Wiedensohler, A., and Young, D.: Reactive Halogens in the Marine Boundary Layer (RHaMBLe): the tropical North Atlantic experiments, Atmos. Chem. Phys., 10, 1031–1055, doi:10.5194/acp-10-1031-2010, 2010.

Maben, J. R., Keene, W. C., Pszenny, A. A. P., and Galloway, J. N.: Volatile inorganic Cl in surface air over eastern North America, Geophys. Res. Lett., 22, 3513–3516, 1995.

Rahn, K. A., Borys, R. D., and Duce, R. A.: Tropospheric halogen gases: Inorganic and organic components, Science, 192, 549–550, 1976.

Read, K. A., Mahajan, A. S., Carpenter, L. J., Evans, M. J., Faria, B. V. E., Heard, D. E., Hopkins, J. R., Lee, J. D., Moller, S. J., Lewis, A. C., Mendes, L., McQuaid, J. B., Oetjen, H., Saiz-Lopez, A., Pilling, M. J., and Plane, J. M. C.: Extensive halogen-mediated ozone destruction over the tropical Atlantic Ocean, Nature, 453, 1232–1235, 2008.

Sander, R., Keene, W. C., Pszenny, A. A. P., Arimoto, R., Ayers, G. P., Baboukas, E., Cainey, J. M., Crutzen, P. J., Duce, R. A., Hönninger, G., Huebert, B. J., Maenhaut, W., Mihalopoulos, N., Turekian, V. C., and Van Dingenen, R.: Inorganic bromine in the marine boundary layer: a critical review, Atmos. Chem. Phys., 3, 1301–1336, doi:10.5194/acp-3-1301-2003, 2003.

Uematsu, M., Duce, R. A., Prospero, J. M., Chen, L., Merrill, J. T., and McDonald, R. L.: Transport of mineral aerosol from Asia over the North Pacific Ocean, J. Geophys. Res., 88C, 5343–5352, 1983.

5

Measurements of total alkalinity and inorganic dissolved carbon in the Atlantic Ocean and adjacent Southern Ocean between 2008 and 2010

U. Schuster[1,*], A. J. Watson[1,*], D. C. E. Bakker[2], A. M. de Boer[3,*], E. M. Jones[4,*], G. A. Lee[2], O. Legge[2], A. Louwerse[1,*], J. Riley[5], and S. Scally[*]

[1]College of Life and Environmental Sciences, University of Exeter, Exeter, EX4 4PS, UK

[2]Centre for Ocean and Atmospheric Science, School of Environmental Sciences, University of East Anglia, Norwich Research Park, Norwich, NR4 7TJ, UK

[3]Department of Geological Sciences and Bolin Centre for Climate Research, Stockholm University, Stockholm, Sweden

[4]Alfred Wegener Institute for Polar and Marine Research, Climate Sciences, Postfach 120161, 27515 Bremerhaven, Germany

[5]International CLIVAR Project Office, National Oceanography Centre, Southampton, Waterfront Campus, European Way, Southampton, SO14 3ZH, UK

[*]formerly at: School of Environmental Sciences, University of East Anglia, Norwich Research Park, Norwich, NR4 7TJ, UK

Correspondence to: U. Schuster (u.schuster@exeter.ac.uk)

Abstract. Water column dissolved inorganic carbon and total alkalinity were measured during five hydrographic sections in the Atlantic Ocean and Drake Passage. The work was funded through the Strategic Funding Initiative of the UK's Oceans2025 programme, which ran from 2007 to 2012. The aims of this programme were to establish the regional budgets of natural and anthropogenic carbon in the North Atlantic, the South Atlantic, and the Atlantic sector of the Southern Ocean, as well as the rates of change of these budgets. This paper describes in detail the dissolved inorganic carbon and total alkalinity data collected along east–west sections at 47° N to 60° N, 24.5° N, and 24° S in the Atlantic and across two Drake Passage sections. Other hydrographic and biogeochemical parameters were measured during these sections, and relevant standard operating procedures are mentioned here.

Over 95 % of dissolved inorganic carbon and total alkalinity samples taken during the 24.5° N, 24° S, and the Drake Passage sections were analysed onboard and subjected to a first-level quality control addressing technical and analytical issues. Samples taken along 47° N to 60° N were analysed and subjected to quality control back in the laboratory. Complete post-cruise second-level quality control was performed using crossover analysis with historical data in the vicinity of measurements, and data were submitted to the CLIVAR and Carbon Hydrographic Data Office (CCHDO), the Carbon Dioxide Information Analysis Center (CDIAC) and and will be included in the Global Ocean Data Analyses Project, version 2 (GLODAP 2), the upcoming update of Key et al. (2004).

Table 1. Data coverage and parameter measured.

Geographical region, EXPO code	Repository reference	Coverage	Dates
47 to 60° N, 74DI20080820	doi:10.3334/CDIAC/OTG.CLIVAR_AR07W_74DI20080820	47.51° N to 60.6° N/55.5° W to 11.1° W	20 August to 25 September 2008
24.5° N, 74DI20100106	doi:10.3334/CDIAC/OTG.CLIVAR_A05_2010	23.3° N to 27.9° N/79.9° W to 13.4° W	5 January to 19 February 2010
24° S, 740H20090307	doi:10.3334/CDIAC/OTG.CLIVAR_A9.5_2009	37.4° S to 22.2° S/53.5° W to 13.7° E	7 March and 21 April 2009
Drake Passage, 740H20090203	doi:10.3334/CDIAC/OTG.CLIVAR_A21_JC031_2009	A21: 57.1° S to 64.1° S/68.3° W to 63.1° W, SR01b: 61.1° S to 54.7° S/54.6 to 58.0° W, with an extension to 53.14° S 58.00° W	3 February and 3 March 2009

1 Introduction

The world oceans had taken up approximately half of the anthropogenic carbon dioxide (CO_2) released by the burning of fossil fuel and producing cement between the Industrial Revolution and 1994 (Sabine et al., 2004). Without this sink of CO_2, the atmospheric content of CO_2 would be much higher than it is today, by approx. 80 ppm, depending on the variability of the source/sink balance since the industrial revolution. Moreover, the oceanic uptake of atmospheric CO_2 is not uniform in either time or space, so that the future concentration of atmospheric CO_2 is highly dependent on the behaviour and variability of this sink.

The control of the spatial and temporal variability of oceanic CO_2 uptake is not well understood. One of the major influences is the Meridional Overturning Circulation (MOC) and in particular the Atlantic MOC, which have significant impact on climate and their influence on the carbon cycle (Peréz et al., 2013; Broecker and Peng, 1992; Watson et al., 1995).

In order to predict the future behaviour of the ocean sink, it is crucial to first get a better understanding of its present behaviour. To that end, sustained observations are needed of interior ocean biogeochemical and hydrographic parameters.

Between 2007 and 2012, a strategic research programme of the UK's National Environmental Research Council (NERC) Marine Centres, Oceans2025 (http://www.oceans2025.org/) was funded. Theme 1 – Ocean circulation, sea level and climate change – aimed to establish regional budgets of heat, freshwater, and carbon via the measurements of physical, chemical and biological parameters on a set of hydrographic sections in the Atlantic and the Atlantic sector of the Southern Ocean. To ensure high-quality measurements of carbon and transient tracers on these sections, a Strategic Funding Initiative (SOFI) was funded entitled "A carbon and transient tracer measurement programme in the Atlantic and Southern Ocean under Oceans2025". In this paper we describe the measurements of dissolved inorganic carbon (DIC) and total alkalinity (TA) performed on these sections.

These measurements were a UK-funded contribution to the international repeat hydrography effort (CLIVAR/GO-SHIP programmes, http://www.go-ship.org/).

2 Common sampling and analyses

As part of SOFI, DIC and TA were analysed in seawater samples on five hydrographic sections in the Atlantic Ocean and the Atlantic sector of the Southern Ocean; Table 2 lists in north to south direction, geographical region (WOCE section), Expocode, cruise name, year, ship, and carbon PI along 47° N to 60° N in 2008, 24.5° N in 2010, 24° S in 2009, and Drake Passages in 2009. The sections and positions of stations at which DIC and TA samples were taken are indicated in Fig. 1. Hereafter, the cruises are referred to by cruise name (i.e. DI332, DI346, JC032, and JC031).

All seawater samples for DIC and TA were drawn from the Niskin bottles of the conductivity/temperature/depth (CTD) rosette into either 500 mL or 250 mL borosilicate ground glass bottles according to the Standard Operating Procedure (SOP) #1 (Dickson et al., 2007). In short, sample bottles are rinsed and then overfilled from the bottom. Subsequently during DI332, DI346, and JC032, samples were routinely poisoned by removing 1 % of sample volume and adding 0.02 % saturated mercuric chloride solution. Finally, the samples were stoppered and stored in a cool, dark place, until analysed either onboard or back in the laboratory. During JC031, samples were only poisoned upon collection if the analysis did not take place within 8 h for surface samples and within 20 h for deep samples. Approximately 30 % of JC031 samples were poisoned.

DIC was analysed by coulometry (Johnson et al., 1985, 1987, 1993) following SOP #2 (Dickson et al., 2007) and TA by open cell potentiometric titration (Mintrop et al., 2000) following SOP #3b (Dickson et al., 2007). Two types of instruments were used: (i) for DIC only, a stand-alone extraction unit (Robinson and Williams, 1991) connected to a coulometer (UIC, USA, model 5011), and (ii) for both DIC and TA, two Versatile INstrument for the Determination of Titration Alkalinity (VINDTA version 3C, Marianda,

Table 2. Geographic region (WOCE section), Expocode, cruise name, year, ship, and carbon PI for each cruise.

Geographical region (WOCE section)	Expocode	Cruise name	Year	Ship	Carbon PI
47° N to 60° N (AR7W and AR7E)	74DI20080820	DI332	2008	RRS *Discovery*	Ute Schuster
24.5 °N (A05)	74DI20100106	DI346	2010	RRS *Discovery*	Ute Schuster
24 °S (A09.5)	740H20090307	JC032	2009	RRS *James Cook*	Ute Schuster
Drake Passage (A21 and SR01b)*	740H20090203	JC031	2009	RRS *James Cook*	Dorothee Bakker

* The denotation A21/SR01 has been used both for cruises in the western Drake Passage (e.g. the western section on JC031 and 06MT11_5, Chipman et al., 1994) or in the central Drake Passage (e.g. LMG200603 and LMG200909).

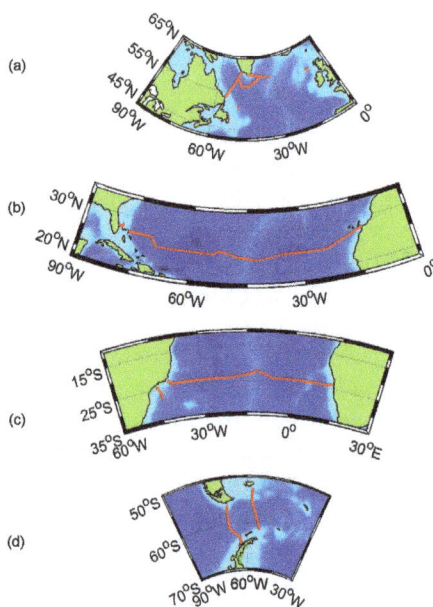

Figure 1. Positions of stations at which DIC and TA samples were taken in the Atlantic and Drake Passage during (**a**) DI332, (**b**) DI346, (**c**) JC032, and (**d**) JC031 (A21 in the west and SR1b in the east).

Germany, SN #004 and #007), connected to a coulometer (UIC, USA, model 5011) and a Titrino (Metrohm UK Ltd.).

For the analysis, seawater is allowed into the stand-alone extractor by gravity, whilst it is drawn into the VINDTAs by slow peristaltic pump. Glass pipettes, approx. 20 mL for DIC and approx. 100 mL for TA, are first rinsed with new seawater sample before being filled to overflowing. The volumes of these pipettes were accurately calibrated before and after the cruises and the laboratory analysis.

The VINDTAs' sample pipettes and alkalinity titration cell are thermostated at 25 °C, and seawater samples were brought up to 25 °C prior to analysis. Conversely, the stand-alone extractor's DIC sample pipette is insulated but not water jacketed, and seawater samples were not warmed up prior to analyses.

During DIC analysis, all inorganic dissolved carbon is converted to CO_2 by addition of excess phosphoric acid (1 M, 8.5 %) to a calibrated volume of seawater sample. Oxygen-free-nitrogen gas (OfN, BOC, UK), after passing through soda lime to remove any traces of CO_2, is used to carry the evolving CO_2 to the coulometer cell, where all CO_2 is quantitatively absorbed, forming an acid that is coulometrically titrated.

During TA analysis, aliquots of 0.1 M hydrochloric acid are added to the seawater sample, and the electromotive force is measured by a pH and reference electrode assembly. TA is calculated using a Gran plot and curve fit (Mintrop et al., 2000).

A 500 mL bottle allows both DIC and TA to be analysed twice per sample (as a successive in-bottle duplicate), thereby providing information on the precision of measurements. A 250 mL bottle only allows DIC and TA to be analysed once, but as done during DI332, duplicate samples from the same Niskin allow for same-depth duplicates.

The VINDTAs, and the stand-alone extractor during JC031 and JC032, were installed in a seagoing laboratory container of the Laboratory for Global Marine and Atmospheric Chemistry (LGMAC) of UEA, UK, on the ships' aft decks. During DI346, JC031, and JC032, DIC and TA analyses and first-level quality control were performed at sea, and data were submitted, together with other cruise parameters, at end of the cruise to the CLIVAR and Carbon Hydrographic Data Office (CCHDO). Samples were collected and stored during DI332 for VINDTA analysis back in the laboratory. Data submitted after second-level quality control are available at Carbon Dioxide Information Analysis Center

Table 3. Total number of (i) stations sampled, (ii) depths sampled, (iii) DIC and TA samples left after first- and second-level quality control for each cruise.

Cruise name	Total number of			
	stations sampled	depths sampled	depths reported to data centres	
			of which TA values	of which DIC values
DI332	74	297	239	288
DI346	135	1427	1226	1322
JC032	116	1606	1504	1475
JC031	63	1380	1044	1060

(CDIAC) and are included in the GLODAP 2 (Global Ocean Data Analyses Project, version 2) effort.

Details of the measurements of temperature, salinity, and nutrients are not included in this paper as its emphasis is on DIC and TA, but sampling, analysis, and quality control procedures have followed recommended standard operating procedures developed during the era of the World Ocean Circulation Experiment (WOCE) in the 1990s, which are now maintained by GO-SHIP (Hood et al., 2010). WOCE and GO-SHIP SOPs are found for salinity in Stalcup (1991) and Kawano (2010), for nutrients in Gordon et al. (1993) and Hydes et al. (2010), and for dissolved oxygen in Culberson (1991) and Langdon (2010). Details of the cruises covered in this paper are given in Bacon (2010) for DI332, King and Hamersley (2012) for DI346, King and Hamersley (2010) for JC032, and McDonagh and Hamersley (2009) for JC031.

3 Specific cruises' sampling and analyses

Table 3 lists the total number of stations sampled, depths sampled, and DIC and TA values reported for each cruise.

3.1 DI332, "Arctic gateway"

RRS *Discovery* cruise DI332, between 20 August and 25 September 2008 from Canada, via Greenland, to Scotland, was planned as one cruise in the vicinity of WOCE section AR7, as an occupation of both the western Labrador Sea part (AR7W) and the eastern part across the Irminger Basin, Iceland Basin, Rockall–Hatton Plateau and Rockall Trough (AR7E) (cruise track in Fig. 1a; Bacon, 2010). AR7W was completed, whilst AR7E could not completed due to mechanical failures and foul weather.

A total of 74 CTD stations were occupied with a 24×20 litre Niskin bottle rosette. West of Greenland (AR7W), Niskins of 28 stations were sampled into 250 mL bottles for DIC and TA analyses. Depths throughout the water column were sampled, with a minimum of three duplicate depths at each of the sampled stations (one set of duplicates at the

deepest depth, one at the shallowest depth, and one set at an intermediate depth), which served as same-depth duplicates.

A total of 297 depths were sampled and stored for later analysis back in the laboratory, which was carried out between 30 July and 8 October 2009 using VINDTA SN #007. Instrument calibration was done throughout the analysis using certified reference materials (CRMs, batch 90). Following first- and second-level quality control (see Sect. 4), 290 sample values of 21 stations were left, which contained 239 TA and 288 DIC values.

Final second-level quality-controlled data have been submitted to British Oceanographic Data Centre (BODC), and are included in the GLODAP 2 effort, via CDIAC.

3.2 DI346, 24.5° N in 2010

RRS *Discovery* cruise DI346 was between 5 January and 19 February 2010 from The Bahamas to Lisbon, Portugal, along nominal latitude 24.5° N in the Atlantic (cruise track in Fig. 1b; King and Hamersley, 2012). It was a repeat occupation of the section, previously occupied including carbon measurements in 2004 (e.g. Brown et al., 2010), 1998 (e.g. Macdonald et al., 2003), and 1992 (e.g. Rosón et al., 2003).

A total of 135 CTD stations were occupied with a 24×20 litre Niskin bottle rosette. Samples for DIC and TA were not taken from all depths at each station. Generally, 16 depths were sampled from each station, including the shallowest and deepest ones with the other depths selected to allow for optimum interpolation across the whole section. Initially, all samples were taken in 500 mL bottles. From Station 34 until Station 129, every third station was fully sampled in 250 mL bottles and initially stored, with four depths of each of these stations sampled in duplicate bottles. All other stations, sampled in 500 mL bottles, were analysed as a priority, and once profiles for these stations had been obtained, selected 250 mL bottles were analysed in order to strengthen areas of missing or suspect data.

Instrument calibration was done throughout the cruise using CRMs (batch 97), and first-level quality-controlled DIC and TA data were submitted, together with other cruise

Table 4. Precision of DIC and TA measurements, defined as the standard deviation of in-bottle duplicate measurements during DI346, JC032, and JC031, and of same-depth duplicate measurements during DI332.

	DIC precision [µmol kg⁻¹]			TA precision [µmol kg⁻¹]	
	VINDTA SN #004	VINDTA SN #007	Stand-alone extractor	VINDTA SN #004	VINDTA SN #007
DI332	NA	±1.4	NA	NA	±0.9
DI346	±1.4	±1.8	NA	±1.2	±1.1
JC032	NA	±1.4	±1.7	±1.7	±1.8
JC031	NA	±2.7	±1.4	±2.3	±2.7

Table 5. Accuracy of DIC and TA measurements for each instrument and section. The accuracy is defined as the standard deviation of CRM values around the mean, after first-level QC and second-level QC using the CARINA dataset. A second-level QC with new data in GLODAP 2 recommends to lower TA during JC031 by 10 µmol kg⁻¹ (R. Key, personal communication, 2014).

	DIC accuracy [µmol kg⁻¹]			TA accuracy [µmol kg⁻¹]	
	VINDTA SN #004	VINDTA SN #007	Stand-alone extractor	VINDTA SN #004	VINDTA SN #007
DI332	NA	±2.1	NA	NA	±2.1
DI346	±3.0	±3.0	NA	±2.1	±3.0
JC032	NA	±3.1	±3.2	±3.2	±3.4
JC031	NA	±3.0	±3.0	±3.0	±3.0

parameters, at end of cruise to the CLIVAR and Carbon Hydrographic Data Office (CCHDO), and final second-level quality-controlled data (1226 TA and 1322 DIC) have been resubmitted to CCHDO and will be included in the GLODAP 2 effort, via CDIAC.

3.3 JC032, 24° S in 2009

RRS *James Cook* cruise JC032 was between 7 March and 21 April 2009 from Montevideo, Uruguay, to Walvis Bay, Namibia, across the Brazil Current and along nominal latitude 24° S in the Atlantic Ocean (cruise track in Fig. 1c; King and Hamersley, 2010).

A total of 118 CTD stations were occupied with 24 rosettes of which 4 were 20 L Niskin bottles for near-surface sampling and the remaining depths sampled using 10 L Niskin bottles. For DIC and TA, 116 stations were sampled. At each station sampled, the top two and bottom two were always sampled for DIC and TA, with in-between depths sampled alternatively for optimum interpolation across the whole section.

All samples were taken in 500 mL bottles with duplicate analyses DIC and TA done of each bottle, first DIC by the stand-alone extractor and VINDTA SN #007, followed by TA by the VINDTA SN #004 and #007 (the DIC on VINDTA SN #004 was inoperable during JC032).

Instrument calibration was done throughout the cruise using CRMs (bath 90), and first-level quality-controlled DIC

and TA data were submitted, together with other cruise parameters, at end of cruise to the CLIVAR and Carbon Hydrographic Data Office (CCHDO), and final second-level quality-controlled data (1504 TA and 1475 DIC) have been re-submitted to CCHDO and will be included in the GLODAP 2 effort, via CDIAC.

3.4 JC031, Drake Passage in 2009

RRS *James Cook* cruise JC031 took place between 3 February and 3 March 2009 from Punta Arenas, Chile, to Montevideo, Uruguay, across Drake Passage (WOCE sections A21 and SR01b, cruise track in Fig. 1d; McDonagh and Hamersley, 2009). Section A21 (also known as SR01) had been occupied before in 1990 and 1999 with DIC measurements made in 1990 (Chipman et al., 1994), and Sect. SR01b, located further to the east, has been occupied every year, except two, since 1993, with the 2009 occupation being the first time DIC and TA measurements were made here.

A total of 84 CTD stations were occupied with a 24-bottle Niskin bottle rosette with 10 L and 20 L bottles. For DIC and TA, a total of 63 stations were sampled. Samples for DIC and TA were taken from depths throughout the water column at each station, with more samples taken in the top 1000 m to aid resolving vertical structure.

All samples were taken in 500 mL bottles with duplicate analyses for DIC and TA done of each bottle, first DIC by the stand-alone extractor and VINDTA SN #007, followed by TA by the VINDTA SN #004 and #007 (the DIC part of VINDTA SN #004 was inoperable during JC031).

Instrument calibration was done throughout the cruise using CRMs (batch 90 and 92); data quality assurance and initial first-level data quality control was done throughout the cruise, with full first-level and second-level quality control done post cruise. The final data (1044 TA and 1060 DIC) will be included in the GLODAP 2 effort, via CDIAC.

4 Quality control

Quality control (QC) of DIC and TA measurements was done in two distinct steps: during first-level QC, data are checked for obvious outliers, and technical or analytical problems during measurements; during second-level QC, cross-over analysis is performed with other sections and corrections identified where necessary. Essentially, first-level QC addresses precision whilst second-level QC addresses accuracy.

4.1 First-level QC

Throughout DIC and TA analyses, regular 500 mL certified reference materials (CRMs) were analysed as in-bottle duplicates. Generally during one day's analyses, one CRM was run after the coulometer cell had stabilised, one mid-cell, and one at the end. This resulted in generally three, occasionally two CRMs being run per CTD cast. CRM batches used were #90 during DI332, #97 during DI346, #90 during JC032, and #90 and #92 during JC031.

Initial DIC and TA calibrations were done onboard (DI346, JC032, and JC031) or in the laboratory (DI332) by correcting all DIC and TA values by the difference between the mean of all CRM measurements and the CRM values of the respective batches used.

WOCE quality flags (Joyce and Corry, 1994) were then assigned to each sample, initially flag 2 for all measurements. All DIC and TA were then checked for obvious outliers, identified by unusually high differences between duplicates, unusually high differences to neighbouring Niskins after optimum interpolation, unusually long TA or DIC titration times, non-smooth titration curves, unusually high residuals in calculated TA. All such outliers were then flagged as 4 when identified as a bad measurement, and flagged as 3 when uncertain. Figure 2 shows the CRM values for DIC and TA during the analyses of samples from the four cruises. Included are samples with WOCE flag 2 or 3 only, but prior to final recalibration; hence the plots indicate the variability of the instruments' response over time. Calibrations were checked and subsequently re-done for different TA acid batches, for different CRMs (e.g. JC031), and different sample pipette volumes (e.g. DI346). Finally, when duplicates' flags were 2 or 3, the mean DIC or TA of the two was reported with the highest WOCE flag of the duplicates for DI346, JC032, and JC031. The precision of DIC and TA measurements is given in Table 4, defined for DI346, JC032, and JC031 as the standard deviation of in-bottle duplicate measurements, and for DI332 as same-depth duplicate measurements (Dickson et al., 2007; SOP #23).

4.2 Second-level QC

Second-level QC was carried out using the Matlab crossover analysis toolbox (Tanhua, 2010). This method evaluates the consistency of deep-water measurements by comparing them with data from a reference data set. The reference data used were the CARINA Atlantic data set (updated August 2012) at http://cdiac.ornl.gov/ftp/oceans/2nd_QC_Tool/refdata/. Station profiles from the cruises in question were compared to profiles from cruises in the reference data set which were within a certain horizontal distance (in this case 2° latitude). The result of each comparison between two cruises is an offset, which for TA and DIC is additive. For DI346 and JC032, data from below 1500 m depth were compared and the profiles were based on density, whereas for DI332 the minimum depth was 1900 m and the profiles were based on depth (following Tanhua, 2010; Olsen et al., 2009). DI332 was compared with 14 cruises, DI346 with 9 cruises, JC032 with 2 cruises and JC031 with 2 cruises. This comparison with the CARINA data set, showed no offsets larger than $4 \, \mu mol \, kg^{-1}$ for DIC and $6 \, \mu mol \, kg^{-1}$ for TA (Wanninkhof et al., 2003). Therefore, the data that we submitted to BODC, CCHDO, and CDIAC contained no corrections, and contain the accuracies given in Table 5, defined for TA measurements as the standard deviation of CRMs per acid batch, and for DIC measurements as the standard deviation of all CRMs per cruise (Dickson et al., 2007; SOP #23).

The GLODAP initiative (Key et al., 2004), now in its version 2, compares newly added measurements since the CARINA initiative in the Atlantic, and contains the most comprehensive global data set available. This GLODAP 2 initiative has currently (April 2014) identified no recommendations for DIC and TA during DI332, DI346, and JC032, whilst it recommends to lower TA for JC031 by $10 \, \mu mol \, kg^{-1}$ (R. Key, personal communication, 2014).

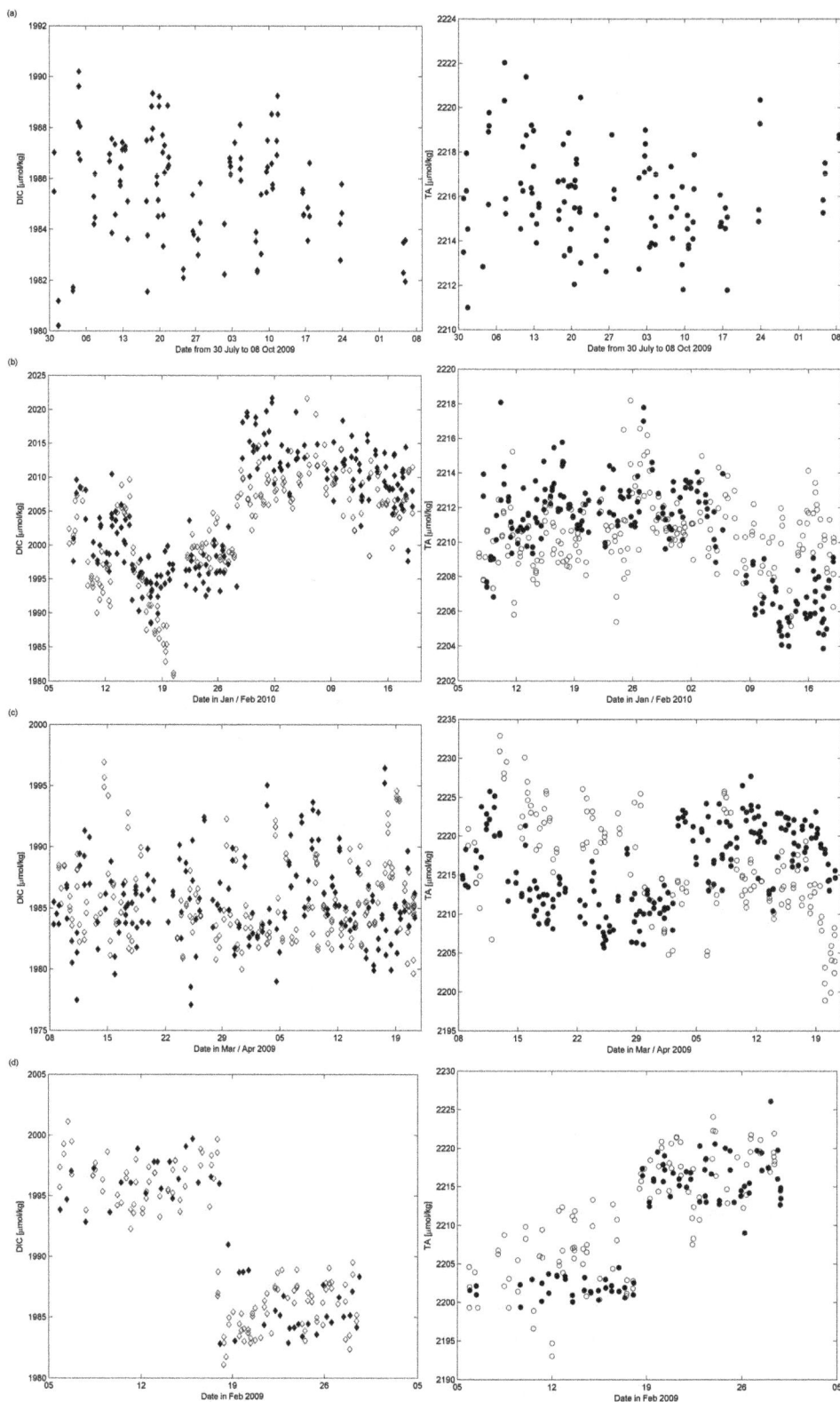

Figure 2. DIC and TA values of CRM analyses over time for (**a**) DI332, (**b**) DI346, (**c**) JC032, and (**d**) JC031. Solid symbols are DIC and TA values analysed on VINDTA #007, whilst open symbols are from VINDTA #004 during DI346 and from the stand-alone extractor during JC031 and JC032; only one instrument was used for samples from DI332. Please note that these values are plotted after outliers were identified and removed (see text in Sect. 4.1), but prior to re-calibration; offsets in these plots indicate for example a change in CRM (e.g. JC031) or a change in sample pipette volume (DI346), and different acid batches for TA.

Acknowledgements. We are very grateful to captains, officers, and crew of RRS *Discovery* and RRS *James Cook* for their support during the fieldwork and principal scientific officers S. Bacon (DI332), E. McDonagh (JC031), and B. King (JC032 and DI346) for support of the carbon research. We acknowledge funding by UK's NERC Strategic Ocean Funding Initiative (SOFI) NE/F01242X/1, supporting the UK Oceans 2025 programme (2007 to 2012), the EU CARBOCHANGE project (264879) and the UK Ocean Acidification Programme (NE/H017046/1) for supporting staff time for this work. We thank E. Madsen, E. Rathbone, I. Salter, J. Allen, R. Pidcock, J. Frommolet, K. Cox and S. Seeyave for taking samples for DIC and TA during D332. We also thank the editor, one anonymous reviewer and Are Olsen for careful comments, which have improved the manuscript.

U. Schuster was PI of carbon sampling and analyses for DI332, JC032, and DI346. A. J. Watson was PI of the SOFI grant. D. C. E. Bakker was PI of carbon sampling and analyses for JC031. A. M.de Boer was watch keeper during JC032. E. M. Jones was watch keeper during JC031. G. A. Lee was watch keeper during DI346. O. Legge was watch keeper during DI346 and aided quality control of data. A. Louwerse was watch keeper during DI346. J. Riley was watch keeper during JC031. S. Scally was watch keeper during JC032.

Edited by: D. Carlson

References

Bacon, S.: *RRS Discovery* cruise 332, 21 August to 25 September 2008. Arctic Gateway (WOCE AR7), National Oceanography Centre Southampton, Southampton, UK, National Oceanography Centre Southampton Cruise Report 53, 129 pp., 2010.

Broecker, W. S. and Peng, T. H.: Interhemispheric Transport of Carbon-Dioxide by Ocean Circulation, Nature, 356, 587–589, 1992.

Brown, P. J., Bakker, D. C. E., Schuster, U., and Watson, A. J.: Anthropogenic carbon accumulation in the subtropical North Atlantic, J. Geophys. Res., 115, C04016, doi:10.1029/2008JC005043, 2010.

Chipman, D. W., Takahashi, T., Breger, D., and Sutherland, S.: Carbon dioxide, hydrographic, and chemical data obtained during the *R/V Meteor* cruise 11/5 in the South Atlantic and northern Weddell Sea areas (WOCE sections A-12 and A-21), Oak Ridge National Laboratory (ORNL)/ Carbon Dioxide Information Analysis Center (CDIAC), Oak Ridge, Tennessee, USA, NDP 045, 61 pp., 1994.

Culberson, C. H.: Dissolved oxygen, in: WOCE operating manual, WHP Office Report WHPO 91-1, WOCE Report No. 68/91, available at: http://whpo.ucsd/edu/manuals/pdf/91-1/stal.pdf (last access: 10 March 2014), 15 pp., 1991.

Dickson, A. G. and Sabine, C. L.: Guide to best practice for ocean CO_2 measurments, PICES Special Publication, 3, 191 pp., 2007.

Gordon, L. I., Jennings, J. C. J., Ross, A. A., and Krest, J. M.: A suggested protocol for continuous flow automated analysis of seawater nutrients (phosphate, nitrate, nitrite and silicite acid) in the WOCE Hydrographic Program and the Joint Global Ocean Fluxes Study, in: WOCE operating manual, WHP Office Report WHPO 91-1, WOCE Report No. 68/91, available at: http://

whpo.ucsd/edu/manuals/pdf/91-1/stal.pdf (last access: 10 March 2014), 55 pp., 1993.

Hood, E. M., Sabine, C. L., and Sloyan, B. M.: The GO-SHIP Repeat Hydrography Manual: a Collection of Expert Reports and Guidelines, IOCCP Report Number 14, OCPO Publication Series Number 134; available at: http://www.go-ship.org/HydroMan.html. (last access: 10 March 2014), 2010.

Hydes, D. J., Aoyama, M., Aminot, A., Bakker, K., Becker, S., Coverly, S., Daniel, A., Dickson, A. G., Grosso, O., Kerouel, R., van Ooijen, J., Sato, K., Tanhua, T., Woodward, E. M. S., and Zhang, J. Z.: Determination of dissolved nutrients (N, P, Si) in seawater with high precision and inter-comparability using gas-segmented continuous flow analysers, in: The GO-SHIP Repeat Hydrography Manual: a Collection of Expert Reports and Guidelines, edited by: Hood, E. M., Sabine, C. L., and Sloyan, B. M., 87 pp., 2010.

Johnson, K. M., King, A. E., and Sieburth, J. M.: Coulometric TCO$_2$ analyses for marine studies; an introduction, Mar. Chem., 16, 61–82, 1985.

Johnson, K. M., Sieburth, J. M., Williams, P. J. l., and Braendstroem, L.: Coulometric total carbon dioxide analysis for marine studies: automation and calibration, Mar. Chem., 21, 117–133, 1987.

Johnson, K. M., Wills, K. D., Butler, D. B., Johnson, W. K., and Wong, C. S.: Coulometric Total Carbon-Dioxide Analysis for Marine Studies – Maximizing the Performance of an Automated Gas Extraction System and Coulometric Detector, Mar. Chem., 44, 167–187, 1993.

Joyce, T. and Corry, C.: Requirements for WOCE hydrographic program data reporting, WOCE Hydrographic Program Office, La Jolla, California, USA, WHPO Publication 90-1, review 2, 145 pp., 1994.

Kawano, T.: Method for Salinity (Conductivity Ratio) Measureme, in: The GO-SHIP Repeat Hydrography Manual: a Collection of Expert Reports and Guidelines, edited by: Hood, E. M., Sabine, C. L., and Sloyan, B. M., 2010.

Key, R. M., Kozyr, A., Sabine, C. L., Lee, K., Wanninkhof, R., Bullister, J. L., Feely, R. A., Millero, F. J., Mordy, C., and Peng, T. H.: A global ocean carbon climatology: Results from Global Data Analysis Project (GLODAP), Global Biogeochem. Cy., 18, GB4031, doi:10.1029/2004GB002247, 2004.

King, B. A. and Hamersley, D. R. C.: RRS James Cook cruise JC032, 07 March to 21 April 2009. Hydrographic sections across the Brazil Current and at 24° S in the Atlantic, National Oceanogrpahic Centre Southampton, Southampton, UK, 173 pp., 2010.

King, B. A. and Hamersley, D. R. C.: RRS Discovery Cruise 346; 05 January to 19 February 2010; the 2010 transatlantic hydrography section at 24.5° N, National Oceanography Centre, Southampton, Southampton, UK, 2012.

Langdon, C.: Determination of dissolved oxygen in seawater by Winkler titration using the amperometric technique, in: The GO-SHIP Repeat Hydrography Manual: a Collection of Expert Reports and Guidelines, edited by: Hood, E. M., Sabine, C. L., and Sloyan, B. M., 18 pp., 2010.

Macdonald, A. M., M.O., B., Wanninkhof, R., Lee, K., and Wallace, D. W. R.: A 1998–1992 comparison of inorganic carbon and its transport across 24.5° N in the Atlantic, Deep-Sea Res. Pt. II, 50, 3041–3064, 2003.

McDonagh, E. L. and Hamersley, D. R. C.: *RRS James Cook Cruise JC031*, 03 Feb to 03 Mar 2009. Hydrographic sections of Drake Passage, National Oceanography Centre, Southampton, Southampton, UKNOCS cruise report 39, 170 pp., 2009.

Mintrop, L., Perez, F. F., González-Dávila, M., Santana-Casiano, M. J., and Körtzinger, A.: Alkalinity determination by potentiometry: Intercalibration using three different methods, Ceinc. Mar., 26, 23–37, 2000.

Olsen, A., Key, R. M., Jeansson, E., Falck, E., Olafsson, J., van Heuven, S., Skjelvan, I., Omar, A. M., Olsson, K. A., Anderson, L. G., Jutterström, S., Rey, F., Johannessen, T., Bellerby, R. G. J., Blindheim, J., Bullister, J. L., Pfeil, B., Lin, X., Kozyr, A., Schirnick, C., Tanhua, T., and Wallace, D. W. R.: Overview of the Nordic Seas CARINA data and salinity measurements, Earth Syst. Sci. Data, 1, 25–34, doi:10.5194/essd-1-25-2009, 2009.

Peréz, F. F., Mercier, H., Vázquez-Rodriguez, M., Lherminier, P., Velo, A., Pardo, P. C., Rosón, G., and Ríos, A. F.: Atlantic Ocean CO_2 uptake reduced by weakening of the meridional overturning circulation, Nat. Geosci., 6, 146–152, doi:10.1038/NGEO1680, 2013.

Robinson, C. and Williams, P. J. L.: Development and assessment of an analytical system for the accurate and continual measurement of total dissolved inorganic carbon, Mar. Chem., 34, 157–175, 1991.

Rosón, G., Ríos, A. F., Pérez, F. F., Lavin, A., and Bryden, H. L.: Carbon distribution, fluxes, and budgets in the subtropical North Atlantic Ocean (24.5° N). J. Geophys. Res., 108, 3144, doi:10.1029/1999JC000047, 2003.

Sabine, C. L., Feely, R. A., Gruber, N., Key, R. M., Lee, K., Bullister, J. L., Wanninkhof, R., Wong, C. S., Wallace, D. W. R., Tilbrook, B., Millero, F. J., Peng, T. H., Kozyr, A., Ono, T., and Rios, A. F.: The oceanic sink for anthropogenic CO_2, Science, 305, 367–371, 2004.

Stalcup, M. C.: Salinity measurements, in: WOCE operating manual, WHP Office Report WHPO 91-1, WOCE Report No. 68/91, available at: http://whpo.ucsd/edu/manuals/pdf/91-1/stal.pdf (last access: 10 March 2014), 9 pp., 1991.

Tanhua, T.: Matlab Toolbox to Perform Secondary Quality Control (2nd QC) on Hydrographic Data, CDIAC, Oak Ridge, Tennessee, 2010.

Wanninkhof, R., Peng, T.-H., Huss, B., Sabine, C. L., and Lee, K.: Comparison of inorganic carbon system parameters measured in the Atlantic Ocean from 1990 to 1998 and recommended adjustments, Oak Ridge National Laboratory, Oak Ridge, Tennessee CDIC-140, 59 pp., 2003.

Watson, A. J., Nightingale, P. D., and Cooper, D. J.: Modelling atmosphere-ocean CO_2 transfer, Philos. T. R. Soc. Lon. B, 348, 125–132, 1995.

6

Global marine plankton functional type biomass distributions: *Phaeocystis* spp.

M. Vogt[1], **C. O'Brien**[1], **J. Peloquin**[1], **V. Schoemann**[2], **E. Breton**[3], **M. Estrada**[4], **J. Gibson**[5], **D. Karentz**[6],
M. A. Van Leeuwe[7], **J. Stefels**[7], **C. Widdicombe**[8], **and L. Peperzak**[2]

[1]Institute for Biogeochemistry and Pollutant Dynamics, Universitätsstrasse 16, 8092 Zürich, Switzerland

[2]Royal Netherlands Institute for Sea Research, P.O. Box 59, 1790 AB Den Burg (Texel), The Netherlands

[3]Université Lille Nord de France, ULCO, CNRS, LOG UMR8187, 32 Avenue Foch, 62930 Wimereux, France

[4]Institut de Ciències del MAR (CSIC), Passeig Maritim de la Barceloneta, 3749, 08003 Barcelona, Catalunya,
Spain

[5]Tasmanian Aquaculture and Fisheries Institute, University of Tasmania, Private Bag 50, Hobart Tasmania
7001, Australia

[6]University of San Francisco, College of Arts and Sciences, 2130 Fulton Street, San Francisco, CA 94117, USA

[7]University of Groningen, Centre for Ecological and Evolutionary Studies, Department of Plant Ecophysiology,
P.O. Box 14, 9750AA Haren, The Netherlands

[8]Plymouth Marine Laboratory, Prospect Place, The Hoe, Plymouth PL1 3DH, UK

Correspondence to: M. Vogt (meike.vogt@env.ethz.ch)

Abstract. The planktonic haptophyte *Phaeocystis* has been suggested to play a fundamental role in the global biogeochemical cycling of carbon and sulphur, but little is known about its global biomass distribution. We have collected global microscopy data of the genus *Phaeocystis* and converted abundance data to carbon biomass using species-specific carbon conversion factors. Microscopic counts of single-celled and colonial *Phaeocystis* were obtained both through the mining of online databases and by accepting direct submissions (both published and unpublished) from *Phaeocystis* specialists. We recorded abundance data from a total of 1595 depth-resolved stations sampled between 1955–2009. The quality-controlled dataset includes 5057 counts of individual *Phaeocystis* cells resolved to species level and information regarding life-stages from 3526 samples. 83 % of stations were located in the Northern Hemisphere while 17 % were located in the Southern Hemisphere. Most data were located in the latitude range of 50–70° N. While the seasonal distribution of Northern Hemisphere data was well-balanced, Southern Hemisphere data was biased towards summer months. Mean species- and form-specific cell diameters were determined from previously published studies. Cell diameters were used to calculate the cellular biovolume of *Phaeocystis* cells, assuming spherical geometry. Cell biomass was calculated using a carbon conversion factor for prymnesiophytes. For colonies, the number of cells per colony was derived from the colony volume. Cell numbers were then converted to carbon concentrations. An estimation of colonial mucus carbon was included a posteriori, assuming a mean colony size for each species. Carbon content per cell ranged from $9 \, \mathrm{pg \, C \, cell^{-1}}$ (single-celled *Phaeocystis antarctica*) to $29 \, \mathrm{pg \, C \, cell^{-1}}$ (colonial *Phaeocystis globosa*). Non-zero *Phaeocystis* cell biomasses (without mucus carbon) range from 2.9×10^{-5} to $5.4 \times 10^3 \, \mathrm{\mu g \, C \, l^{-1}}$, with a mean of $45.7 \, \mathrm{\mu g \, C \, l^{-1}}$ and a median of $3.0 \, \mathrm{\mu g \, C \, l^{-1}}$. The highest biomasses occur in the Southern Ocean below 70° S (up to $783.9 \, \mathrm{\mu g \, C \, l^{-1}}$) and in the North Atlantic around 50° N (up to $5.4 \times 10^3 \, \mathrm{\mu g \, C \, l^{-1}}$).

1 Introduction

Plankton functional types (PFTs; Le Quéré et al., 2005) and marine ecosystem composition are important for the biogeochemical cycling of many abundant elements on Earth, such as carbon, nitrogen, and sulphur (e.g. Weber and Deutsch, 2010). In recent decades, changes have been observed in marine plankton communities (Chavez et al., 2003; Reid et al., 2007; Hatun et al., 2009; Beaugrand and Reid, 2003), and these changes are likely to affect local and global biodiversity, fisheries and biogeochemical cycling. Marine ecosystem models based on PFTs (Dynamic Green Ocean Models; DGOMs) have been developed in order to study the lower trophic levels of marine ecosystems and the potential impact of changes in their structure and distribution (Le Quéré et al., 2005). DGOMs have been applied to a wide range of biological and biogeochemical questions (Aumont and Bopp, 2006; Hashioka and Yamanaka, 2007; Moore and Doney, 2007; Vogt et al., 2010; Weber and Deutsch, 2010). However, the validation of these models has proven difficult due to the scarcity of observational abundance and biomass data for individual PFTs.

The MARine Ecosystem DATa (MAREDAT) initiative is a community effort to provide marine ecosystem modellers with global biomass distributions for the major PFTs currently represented in marine ecosystem models (Buitenhuis et al., 2012; silicifiers, calcifiers, nitrogen fixers, DMS-producers, picophytoplankton, bacteria, microzooplankton, mesozooplankton and macrozooplankton). MAREDAT is part of the MARine Ecosystem Model Intercomparison Project (MAREMIP). All MAREDAT biomass fields are publicly available for use in model evaluation and development, and for other applications in biological oceanography.

The haptophyte *Phaeocystis* has been suggested to play a fundamental role in the global biogeochemical cycling of carbon and sulphur (Le Quéré et al., 2005). *Phaeocystis* is a globally distributed genus of marine phytoplankton with a polymorphic life cycle, alternating between flagellated, free-living cells of 3–9 µm in diameter and colonial stages which form colonies reaching several mm–cm (Rousseau et al., 1990; Peperzak et al., 2000; Peperzak and Gäbler-Schwarz, 2012; Chen et al., 2002; Schoemann et al., 2005). Three of the six recognised *Phaeocystis* species are known to form massive blooms of gelatinous colonies (Medlin and Zingone, 2007), which may contribute significantly to carbon export (Riebesell et al., 1995; DiTullio et al., 2000), although recent observations suggest that the contribution of *Phaeocystis* spp. to vertical flux of organic matter is small (Reigstad and Wassmann, 2007). In addition, *Phaeocystis* cells are important producers of dimethylsulphoniopropionate (DMSP), which is the marine precursor of the trace gas dimethylsulphide (DMS). DMS has been suggested to play an important role in cloud formation, and DMS production is the main recycling pathway of sulphur from the ocean to the land. Furthermore, *Phaeocystis* has been well documented as asso-

ciated with marked increases in seawater viscosity (Jenkinson and Biddanda, 1995; Seuront et al., 2007). In their review, Schoemann et al. (2005) conclude that it should be possible to derive a single unique parameterisation of *Phaeocystis* growth for global modelling. Hence, *Phaeocystis* has recently been included in a number of regional and global DGOMs (e.g. Wang and Moore, 2011).

Here, we present biomass data that were estimated from direct cell counts of colonial and single-celled *Phaeocystis*. We show the spatial and temporal distribution of *Phaeocystis* biomass, with a particular emphasis on the seasonal and vertical patterns. We discuss in detail our method for converting abundance to carbon biomass and note the uncertainties in the carbon conversions. Our biomass estimates are tailored to suit the needs of the modelling community for marine ecosystem model validation and model development, but they are also intended to aid biological oceanographers in the exploration of the relative abundances of different PFTs in the modern ocean and their respective biogeochemical roles, for the study of ecological niches in marine ecosystems and the assessment of marine biodiversity.

2 Data

2.1 Origin of data

Our data consists of abundance measurements from several databases (BODC, OBIS, OCB DMO, Pangaea, WOD09, US JGOFS[1]), and published and unpublished data from several contributing authors (E. Breton, M. Estrada, J. Gibson, D. Karentz, M. A. Van Leeuwe, J. Peloquin, L. Peperzak, V. Schoemann, J. Stefels, C. Widdicombe). Often, the online databases did not denote the method used for the quantitative analysis of *Phaeocystis* abundances. However, most known counts have been made using the common inverted microscopy and epifluorescence methods (Karlson et al., 2010). Both methods require the sampling of *Phaeocystis* colonies in Niskin bottles and the subsequent preservation of cells in Lugol's solution or another preservative. After storage of the sample prior to analysis, many scientists concentrate the sample through settling in counting chambers or filtration onto a polycarbonate filter.

Most conventional preservation agents cause the disintegration of the colonial matrix, such that colonial and single cells can no longer be distinguished. One preservation method based on a mixture of Lugol's, glutaraldehyde and iodine (Guiselin et al., 2009; Sherr and Sherr, 1993; Rousseau et al., 1990) is able to maintain colony structure (e.g. Karentz and Spero, 1995; Riebesell et al., 1995; Brown et al., 2008;

---

[1] BODC: British Oceanographic Data Centre; OBIS: The Ocean Biogeographic Information System, OCB DMO: Ocean Carbon and Biogeochemistry Coordination and Data Management Office, Pangaea: Data Publisher for Earth and Environmental Science, WOD: World Ocean Database Boyer et al. (2009), US JGOFS: US Joint Global Ocean Flux Study.

Wassmann et al., 2005), but this is not widely used. Due to these difficulties, only a few measurements resolve *Phaeocystis* life stages or morphotypes.

Table 1 summarizes the origin of all our data, sorted by database, principal investigator and the project during which measurements were taken. At present, the database contains 5057 individual data points from 3526 samples of 1595 depth-resolved stations.

2.2 Quality control

Given the low numbers of data points and the fact that *Phaeocystis* is a blooming species with a wide range of biomass concentrations, the identification and rejection of outliers in our dataset is challenging. We use Chauvenet's criterion to identify statistical outliers in the log-normalized biomass data (Glover et al., 2011; Buitenhuis et al., 2012). Based on the analysis, none of the stations was identified to yield biomasses with a probability of deviation from the mean greater than $1/2n$, with $n = 2547$ being the number of non-zero data summed up for all stations (two-sided z-score: $|zc| = 3.72$). In addition to the statistical testing of the biomass distribution, we also quality controlled the range of our cell abundances. We found that our maximum reported abundance of 19×10^7 cells l^{-1} is within the range of previously reported abundances: Schoemann et al. (2005) report maximum cell abundances of the order of ca. 10^7 cells l^{-1} in areas of colony occurrence (http://www.nioz.nl/projects/ironages). The largest bloom of *P. antarctica* was observed in Prydz Bay (http://www.nioz.nl/projects/ironages), with cell abundances measured up to 6×10^7 cells l^{-1}. Eilertsen et al. (1989) reported a maximum of 1.2×10^7 cells l^{-1} of *P. pouchetii* in the Konsfjord. For *P. globosa*, a maximal abundance of 20×10^7 cells l^{-1} has been observed, corresponding to a total biomass of ca. 10 mg C l^{-1} including mucus (Cadée and Hegeman, 1986; Schoemann et al., 2005). The latter biomass value is 20 times larger than the maximal biomass we report (5.4×10^3 µg C l^{-1}). Thus, based on statistical and observational evidence, none of the data were flagged.

2.3 Biomass conversion

We distinguish between single, colonial and unspecified *Phaeocystis* cells. While *Phaeocystis* is generally observed and counted under bloom conditions, a significant fraction of cells is non-colonial even during bloom conditions (V. Schoemann, auxillary data). Hence, in order to calculate the lower limit biomass, we have assumed unspecified cells to be single cells. To first order, this choice does not affect the order of magnitude of our cell biomass estimates, since cell carbon is of the same order of magnitude for both colonial and single cells (see below). We define total *Phaeocystis* biomass to consist of cell biomass and biomass contained in the mucus surrounding *Phaeocystis* colonies. For our calculation of total biomass, we chose unidentified cells to be in

the colonial stage. Hence, our cell biomass estimates represent a lower limit, and our total biomass estimates including colonial mucus represent an upper limit for global *Phaeocystis* biomass.

Biomass was determined from cell abundance using species- and form-specific conversion factors (Fig. 1). Similar conversion schemes have been previously described (e.g. Schoemann et al., 2005, and references therein). Total cell abundances were divided into single cells, colonial cells and undefined cell types. For each species, the mid-point of the range of reported cell diameters from the literature was used for single and colonial cells (Table 2; *P. globosa*: Rousseau et al., 2007; Schoemann et al., 2005; *P. antarctica*: Mathot et al., 2000; Rousseau et al., 2007; Schoemann et al., 2005; *P. pouchetii*: Wassmann et al., 2005; Rousseau et al., 2007).

Where the species was not specified, Southern Ocean cell counts were assumed to be *Phaeocystis antarctica*. For cell counts in other regions, the mid-point of the range of cell diameters for *P. pouchetii* and *P. globosa* was taken (Table 2; flagellates: 5.0 µm, colonial cells: 6.7 µm). From cell diameter we computed biovolume, assuming spherical geometry of all cell types. We then converted biovolume to carbon biomass using an empirical volume–carbon conversion formula for prymnesiophytes developed by Menden-Deuer and Lessard (2000, Table 2).

Most colonial cells were reported in the form of cell abundances. However, one dataset (*P. globosa*; number of data points: $n = 30$) provided colony counts only, but additionally reported the corresponding colony diameters. We used the reported colony diameter to calculate colony volume (assuming spherical colonies), and from this estimated the number of cells per colony using published conversion factors (Table 2; *P. globosa*: Rousseau et al., 1990; *P. antarctica*: Mathot et al., 2000; no colony-only cell counts reported for *P. pouchetii*). Total cell counts per colony were then converted to carbon biomass using the method described above.

We show biomass estimates based on cell carbon excluding colonial mucus as our lower limit for *Phaeocystis* biomass. The range of uncertainty for the lower limit biomass estimates is given by the uncertainty in cell diameters. Additional uncertainty is introduced where cell life form is not specified. The uncertainty introduced by this assumption is addressed by calculating a minimum cell biomass estimate treating all undefined cell types as single cells.

Estimates for colonial mucus are included to provide an upper limit for *Phaeocystis* biomass. Estimating mucus carbon from cell counts alone is problematic, as the ratio of mucus carbon to cell number increases with colony size. Colony size therefore needs to be known in order to calculate accurate estimates of mucus carbon. Only one of the datasets ($n = 30$) included information on colony size. Consequentially, we have used a standard colony diameter of 200 µm for all three species, based on a review of previously reported colony sizes: Verity et al. (2007) find most *P. pouchetii* colonies in their study range between 20–450 µm

Table 1. List of data contributors, in temporal order; Databases: BODC: British Oceanographic Data Centre, OBIS: Ocean Biogeographic Information System, US JGOFS: US Joint Global Ocean Flux Study, OCB: Ocean Carbon and Biogeochemistry, WOD09: World Ocean Database 2009; Institutes: AWI: Alfred-Wegener-Institute, Bremerhaven, Germany, IMARPE: Institut del Mar del Peru, Paita, Peru, IOS: Institute of Ocean Sciences, Sidney, Canada, MMBI: Murmansk Marine Biological Institute, Murmansk, Russia.

Entry No.	Database	Investigator/Institute	Project	Year(s)	Region	No. of data points	Reference(s)
1	BODC	D. Harbour	BOFS	1989–1991	North Atlantic	13	–
2	BODC	D. Harbour	JGOFS	1994	Arabian Sea	25	–
3	BODC	I. Joint	OMEX	1994–1995	North Atlantic	7	–
4	BODC	P. Tett	North Sea Project	1988–1989	North Sea	18	–
5	BODC	R. Uncles	LOIS	1994–1995	North Sea	19	–
6	BODC	P. Wassmann	OMEX	1994	North Atlantic	186	–
7	–	L. Peperzak		1992	Dutch coastal zone	64	Peperzak et al. (1998)
8	OBIS	P. Wassmann & T. Ratkova	ArcOD	1993–2003	Arctic	1815	–
9	OCB DMO	M. Silver	VERTIGO	2004	Hawaii	1	
10	Pangaea	P. Assmy	EIFEX	2004	Southern Ocean	28	Assmy (2007)
11	Schoemann et al. (2005)	G. Cadée	Marsdiep	1976–1985	Dutch coastal zone	2	Cadée and Hegeman (1986)
12	Schoemann et al. (2005)	G. Cadée	Marsdiep	1989–1992	Dutch coastal zone	3	Cadée (1991)
13	Schoemann et al. (2005)	G. Cadée	Marsdiep	1990	Dutch coastal zone	2	Cadée (1991)
14	Schoemann et al. (2005)	G. DiTullio		1996	Ross Sea, Antarctica	1	DiTullio et al. (2000)
15	Schoemann et al. (2005)	H. Fransz & G. Cadée	Marsdiep	1991	Dutch coastal zone	2	Fransz et al. (1992) Cadée and Hegeman (1993)
16	Schoemann et al. (2005)	B. Hansen		1988–1989	Barents Sea	6	Hansen et al. (1990)
17	Schoemann et al. (2005)	I. Jenkinson		1988	German Bight	12	Jenkinson and Biddanda (1995)
18	Schoemann et al. (2005)	S. Kang		1986	Weddell Sea, Antarctica	3	Kang and Fryxell (1993)
19	Schoemann et al. (2005)	B. Karlson		1993	Skagerrak Strait, North Sea	5	Karlson et al. (1996)
20	Schoemann et al. (2005)	K. Kennington		1996	Irish Sea	1	Kennington et al. (1999)
21	Schoemann et al. (2005)	A. Luchetta		1991	Barents Sea	1	Luchetta et al. (2000)
22	Schoemann et al. (2005)	S. Mathot		1994–1995	Ross Sea, Antarctica	35	Mathot et al. (2000)
23	Schoemann et al. (2005)	Palmisano		1984	McMurdo Sound, Antarctica	10	Palmisano et al. (1986)
24	Schoemann et al. (2005)	H. Pieters	Marsdiep	1978	Dutch coastal zone	1	Pieters et al. (1980)
25	Schoemann et al. (2005)	R. Riegman	Marsdiep	1991	Dutch coastal zone	4	Riegman et al. (1993)
26	Schoemann et al. (2005)	C. Robinson		1993	East Antarctica	1	Robinson et al. (1999)
27	Schoemann et al. (2005)	F. Scott		1992	East Antarctica	1	Scott et al. (2000)
28	Schoemann et al. (2005)	P. Tréguer		1988	Scotia Sea, Antarctica	1	Tréguer et al. (1991)
29	Schoemann et al. (2005)	F. Van Duyl	Marsdiep	1995	Dutch coastal zone	2	Van Duyl et al. (1998)
30	Schoemann et al. (2005)	E. Venrick	–	1994	REGION	1	Venrick (1997)
31	Schoemann et al. (2005)	S. Weaver	–	1994	REGION	1	Weaver (1979)
32	Schoemann et al. (2005)	T. Weisse		1975–1976	German Bight, North Sea	2	Weisse et al. (1986)
33	Schoemann et al. (2005)	G. Wolfe		1997	Labrador Sea	2	Wolfe et al. (2000)
34	WOD09	MMBI	–	1955–1997	Kola Bay (Barents Sea)	395	–
35	WOD09	IMARPE	–	1966–1977	Peruvian coastal zone	8	–
36	WOD09	IOS	–	1980	US coast (Oregon)	4	–
37	WOD09	University of Alaska	OCSEAP	1975–1977	Prince William Sound (Gulf of Alaska)	20	–
38	WOD09	AWI	IAPP	1991	Arctic	6	
39	–	C. Widdicombe	Western Channel Observatory	1992–2008	English Channel	1248	Widdicombe et al. (2010)
40	US JGOFS Data System	W. Smith, D. Caron & D. Lonsdale	AESOPS	1996–1997	Southern Ocean	184	–
41	–	D. Karentz	Icecolors	1986	Southern Ocean	74	Karentz and Spero (1995); Smith et al. (1992)
42	–	D. Karentz	GRINCHES	2004–2005	Ross Sea, Antarctica	14	
43	–	E. Breton	SOMLIT–MONITO	2006–2009	English Channel	216	E. Breton (unpublished data)
44	–	J. Gibson		1993–1995	East Antarctica	136	J. Gibson (unpublished data)
45	–	J. Peloquin	Ross Sea	2001–2005	Ross Sea, Antarctica	84	J. Peloquin (unpublished data)
46	–	M. Van Leeuwe & J. Stefels	Ant 16/3 R/V *Polarstern*	1999	Southern Ocean	33	Koeman (1999)
47	–	M. Estrada	Antarctic 85	1985	Weddell Sea, Antarctica	126	Estrada and Delgado (1990)
48	–	M. Estrada	Fronts	1985	Mediterranean Sea	156	Estrada (1991)
49	–	V. Schoemann	BGC of *Phaeocystis* colonies, EC-FP4	1994	Dutch coastal zone	80	Schoemann et al. (1998)

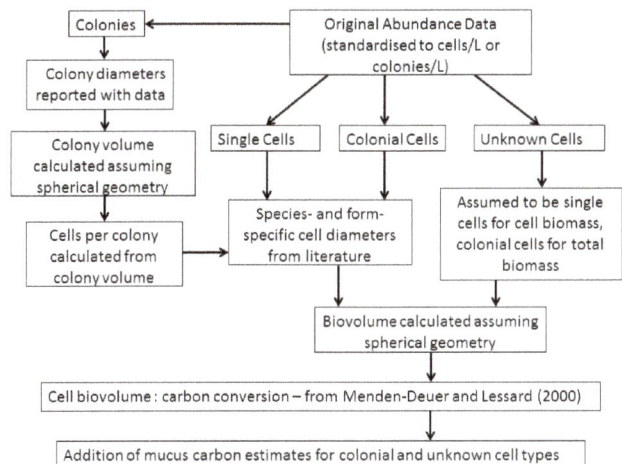

Figure 1. Flow diagram of methodology used to derive mean *Phaeocystis* biomass estimates from abundance data for single cells, colonial cells and unidentified cells. Abundance data was converted to biovolume, and a biovolume to carbon ratio was applied to derive biomass. Finally, an estimate of mucus carbon was added for colonial cell types.

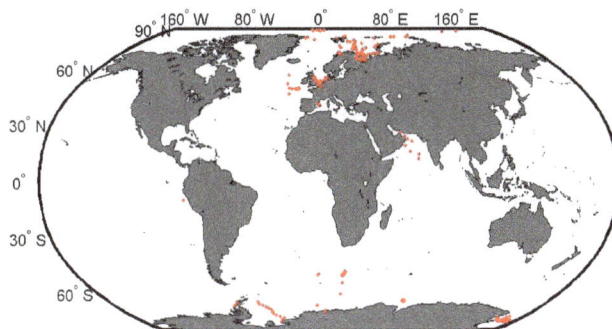

Figure 2. Global distribution of stations where *Phaeocystis* abundance counts were made available for this study. Most stations are located at temperate latitudes and in coastal areas.

in diameter; Reigstad and Wassmann (2007) observe most of their *P. pouchetii* colonies in a size range between 65–115 µm; Mathot et al. (2000) observe *P. antarctica* colonies to range from 9.3–560 µm; and Rousseau et al. (1990) report colony sizes of *P. globosa* to range from 10 µm–2 mm. In all references, larger colonies occured, but were rarer than the smaller colonies. In our data, *P. globosa* colonies range from 11–594 µm in diameter, with a mean diameter of 197 µm. Given that the samples of Verity et al. (2007), Mathot et al. (2000) and Rousseau et al. (1990) cover a similar range of sizes for all three species, and that the dataset that reports colony sizes confirms a mean colony size of ca. 200 µm, these findings suggest that the chosen standard diameter is a realistic value for a typical *Phaeocystis* bloom. Maximum sizes are reported in Schoemann et al. (2005) and Baumann et al. (1994), and range between 9 mm–3 cm for *P. globosa*, between 1.5–2 mm for *P. pouchetii*, and around 1.4–9 mm for *P. antarctica*. Given the lack of data on colony sizes, we are unable to quantify the impact of large colonies on average carbon biomass. However, huge colony sizes are likely to be geographically restricted to specific regions. We assess the uncertainty of our estimates by calculating mucus carbon for the minimum and maximum colony sizes reported for each species (Schoemann et al., 2005; Baumann et al., 1994). Estimates of minimal and maximal total carbon are included in our data base, but only mean total carbon including mucus will be discussed below.

Conversion factors have previously been published for estimating mucus biomass and number of cells from colony volume for *P. antarctica* (Mathot et al., 2000) and *P. globosa* (Rousseau et al., 1990). Using these estimates we calculated

the expected mucus biomass per cell (Table 2). Unspecified cell types were assumed to be colonial cells when calculating these upper estimates of *Phaeocystis* biomass.

For *P. pouchetii*, no direct mucus carbon conversion factor has been developed, but Verity et al. (2007) provides a conversion factor for colony volume to total colony biomass (Table 2; cells and mucus). Following the same procedure as for the other two species, we used this to calculate total biomass per cell. We then subtracted our cell biomass estimate for colonial cells to obtain an estimate of mucus carbon per cell for comparison with *P. globosa* and *P. antarctica* estimates.

Unspecified species outside of the Southern Ocean were given a total biomass per cell of 224 pg, which corresponds to the mean total biomass estimate for *P. globosa* and *P. pouchetii* (Table 2).

3 Results

3.1 Global distribution of abundance data

Of the 1595 stations contained in the database (Fig. 2), 83 % are located in the Northern Hemisphere (NH) and only 17 % in the Southern Hemisphere (SH; Fig. 3). Out of the 3526 samples, 2547 were reported as non-zero biomass, with 2054 non-zero abundances out of 2862 samples for the NH, and 493 non-zero abundances out of 664 samples for the SH (Table 3). Most measurements (53 %) were taken in the latitudinal band of 50–70° N (Fig. 3). When only data points with non-zero abundances are taken into account, we find that most non-zero data were collected between 60–80° N (64 %; Table 3), with relatively few non-zero abundances recorded between 50–60° N (11 %). Several latitudinal bands are undersampled. We could not collect data for the 40–20° S, 0–10° N and 30–40° N latitudinal bands. All in all, we have little non-zero data in tropical and sub-tropical latitudes from 40° S to 40° N, where sampling is targeted at other phytoplankton groups.

Table 2. Literature values for conversion factors from abundance to biomass. Cell diameters, biovolumes, carbon content and colony number conversions for *P. globosa*, *P. pouchetii*, and *P. antarctica*. Reported means with ranges given in parentheses.

| | *P. globosa* | | *P. pouchetii* | | *P. antarctica* | |
	Flagellate	Colonial	Flagellate	Colonial	Flagellate	Colonial
Diameter (μm)	5.5[1]	7.5[2]	5.0[3]	5.5[3]	4.8[4]	6.6[5]
	(3–8)	(4.5–10.4)	(2–8)	(3–8)	(2–7.5)	(3.2–10)
Biovolume[6] (μm^3)	87	217	65	87	56	151
	(14–268)	(48–589)	(4–268)	(14–268)	(4–221)	(17–524)
Carbon per cell (pg)	13[7]	29[8]	10[7]	13[9]	9[7]	21[10]
	(3–35)	(7–71)	(1–35)	(3–35)	(1–29)	(3–63)
Colony diameter (μm)		200		200		200
		(10–30 000)		(20–2000)		(25–9000)
Colony volume – cell number conversion[8,10] (V: [mm^{-3}])		$\log N_c = 0.51 \log V + 3.67$		$\log N_c = 0.537 \log V + 3.409$		$N_c = (\frac{V}{417})^{0.60}$
Colony volume – mucus carbon conversion		335 ng mm^{-3} [8]				213 ng mm^{-3} [10]
Colony volume – total carbon conversion			$\log C = 0.924 \cdot V + 3.947$ [9] C: [μg]; V: [mm^{-3}]			
Total carbon per cell including mucus (pg)		34		415		24
		(29–7768)		(29–6008)		(21–362)
Percent of total carbon associated with mucus contribution (pg)		14.6		96.9		14.6
		(0.2–99.6)		(1.4–94.3)		(55.8–99.8)

References for the cell diameters: [1] Rousseau et al. (2007); Schoemann et al. (2005); [2] Rousseau et al. (2007); [3] Wassmann et al. (2005); Rousseau et al. (2007); [4] Schoemann et al. (2005); Mathot et al. (2000); Rousseau et al. (2007); [5] Mathot et al. (2000); Rousseau et al. (2007). References for biovolume conversion, assuming spherical geometry of cells: [6] Hillebrand et al. (1999). Reference for the biovolume–carbon conversion: [7] Menden-Deuer and Lessard (2000). References for colony volume–cell number conversion and for colony volume–mucus biomass conversion: [8] Rousseau et al. (1990); [9] Verity et al. (2007) (colony volume–total biomass conversion); [10] Mathot et al. (2000).

While 60 % of measurements were taken in the upper 10 m of the water column, the mean sampling depth of our dataset is 27 m, and the median sampling depth is 10 m. Reported cell abundances were maximal at depths between 0–80 m. Observations and laboratory experiments suggest that *Phaeocystis* is well-adapted to low light conditions (Arrigo et al., 1999; Moore et al., 2007; Shields and Smith, 2009). In our database, the deepest occurrence of *Phaeocystis* was at 292 m at 65° N, 35° W (Barents Sea; OBIS dataset).

3.2 Temporal distribution of data

The data were collected from 1955–2009, with 79 % of measurements taken during the period of 1990–2009 (Fig. 4). 6 % (8 %) of (non-zero) measurements were taken in the 1950s, < 1 % (< 1 %) in the 1960s, < 1 % (1 %) in the 1970s, 14 % (10 %) in the 1980s, 55 % (60 %) in the 1990s, and 23 % (20 %) between 2000–2009.

Dividing the data into the four seasons for both hemispheres gives a first indication of the level of temporal bias (Table 4). In the Northern Hemisphere, 56 % (64 %) of all (non-zero) data were taken in spring, 29 % (31 %) in summer, 9 % (5 %) in autumn and 6 % (<1 %) in winter. For the Southern Hemisphere, 27 % (32 %) of data were collected in spring, 58 % (52 %) in summer, 13 % (16 %) in autumn and only 2 % (< 1 %) in winter. Hence, NH data is biased towards spring values, and SH data towards summer values.

3.3 *Phaeocystis* cell biomass distribution (mucus excluded)

Phaeocystis biomass estimates based on cell carbon only, without mucus carbon included, constitute a lower boundary for carbon biomass of this PFT in the global ocean. Since mucus carbon biomass is difficult to quantify based on *Phaeocystis* cell counts, many marine ecosystem models do not include a parameterisation of mucus carbon for this PFT. Thus, in the following section, our estimates of cell biomass represent a lower limit of carbon biomass for model validation. *Phaeocystis* biomasses span a wide range of concentrations, which is why we show log transformed biomass concentrations in all subsequent figures. However, we report only non log-transformed biomass concentrations in this manuscript for better comparability with the original data submission.

3.3.1 Global surface cell biomass characteristics

Phaeocystis biomass estimated from cell carbon alone is depicted in Fig. 5a for the surface layer of the ocean (0–5 m). The maximal biomass calculated from the reported cell abundances is 5449.3 μg C l^{-1}, located at 53° N at a depth of 0 m during the spring bloom (month of May). The maximal cell biomass in the Southern Hemisphere is 783.9 μg C l^{-1}, recorded in the Ross Sea in January (76.49° S, 171.97° E, depth 1 m). The mean of all reported non-zero cell biomass values is 45.7 μg C l^{-1}, and the median is 3.0 μg C l^{-1}. Of

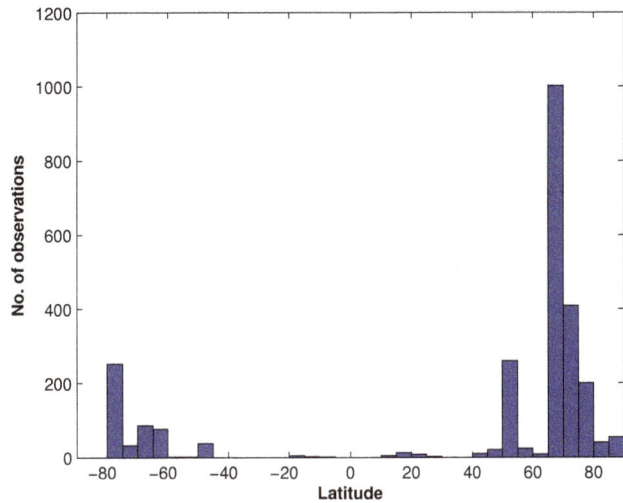

Figure 3. Number of *Phaeocystis* observations as a function of latitude for the period of 1950–2009. Most observations are located in the temperate and high latitudes of the Northern Hemisphere.

Table 3. Latitudinal distribution of abundance data in ten degree latitudinal bands (−90 to 90°). Number of data points for each latitudinal band. All: all measurements, non-zero: data with non-zero carbon biomass.

Latitudinal band	All data	Non-zero data
−90– −80°	0	0
−80– −70°	334	284
−70– −60°	283	162
−60– −50°	1	1
−50– −40°	37	37
−40– −30°	0	0
−30– −20°	0	0
−20– −10°	6	6
−10–0°	1	1
0–10°	0	0
10–20°	17	17
20–30°	10	10
30–40°	0	0
40–50°	152	30
50–60°	852	284
60–70°	1010	1010
70–80°	727	609
80–90°	94	94

all calculated cell biomasses, 40.1 % are in the range of 0–0.1 $\mu g \, C \, l^{-1}$, 55.6 % in the range of 0–1 $\mu g \, C \, l^{-1}$, and 67.5 % between 0 and 5 $\mu g \, C \, l^{-1}$. 94.8 % of all cell biomasses lie below 100 $\mu g \, C \, l^{-1}$.

Figure 5b shows the range of uncertainty for cell biomass in % resulting from the uncertainty in cell diameters reported for each species and life stage. Biomasses calculated using the higher estimates of cell diameter are 246 to 355 % higher than estimates calculated using mean cell dimensions. Biomasses calculated using the lower cell diameter estimates are between 4 and 26 % of the mean values. Uncertainties are highest when species or life form is not reported. Biomass estimates are highly sensitive to changes in cell size, and reduced uncertainty is only possible if cell measurements are available in addition to abundance data.

3.3.2 Latitudinal cell biomass distribution

Calculated cell biomasses do not follow a distinct latitudinal pattern (Fig. 6a). Highest cell biomasses occur at latitudes around 50° N and 80° S, lowest cell biomasses are calculated for latitudes around 20° S (Peruvian upwelling). Cell biomasses decrease from 50° N towards the pole in the Northern Hemisphere, but Southern Hemisphere concentrations increase polewards towards the Antarctic continent. Given that many of our data stem from coastal regions, we note that our latitudinal distributions are biased towards high coastal concentrations in some areas, as open ocean areas are still undersampled. However, cell biomass distributions confirm previous findings that *Phaeocystis* blooms occur in the temperate and high latitudes of both hemispheres, and that *Phaeocystis* is fairly ubiquitous, occurring in all major ocean basins.

3.3.3 Depth distribution of cell biomass

Figure 7 shows calculated cell biomass estimates for *Phaeocystis* in six different depth ranges (0–5 m, 5–25 m, 25–50 m, 50–75 m, 75–100 m and depths > 100 m). All depth bands have not been sampled at each station, and many datasets contain only surface measurements. Where depth profiles are available, cell biomass concentrations are generally highest in the surface layer and decrease with depth to 100 m (Fig. 6b). Cell biomasses are low between 100–300 m (mean non-zero biomass concentrations of 7.3 $\mu g \, C \, l^{-1}$), however, high *Phaeocystis* abundances are reported even at depths of close to 300 m in the Northern Hemisphere. The highest cell biomass reported below 100 m is 311.9 $\mu g \, C \, l^{-1}$ in the Arctic (66.42° N, 34.36° E) in late May, at a depth of 270 m. In the Southern Ocean, *Phaeocystis* cells are reported to a maximum depth of 200 m in the Weddell Sea during February and March, but biomass values below 100 m never exceed 0.01 $\mu g \, C \, l^{-1}$. Given the limited number of data points reported for this depth range, it is unclear how representative our data are of deep *Phaeocystis* cell biomasses in other sampling locations. This suggests that *Phaeocystis* should be sampled more regularly at depths between 100–300 m and below.

3.3.4 Seasonal distribution of cell biomass

Cell biomass distributions for the Northern and Southern Hemispheres show that the calculated *Phaeocystis* biomasses reflect those of a typical blooming species (Fig. 8a and b). In

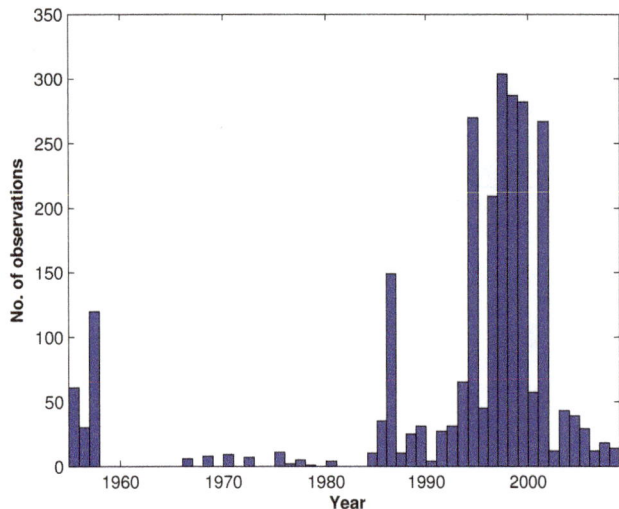

Figure 4. Number of observations for *Phaeocystis* species per year, for the years 1950–2009. Most counts were made after 1990.

Table 4. Seasonal distribution of abundance data for the Northern and Southern Hemispheres. Number of data points for each month. All: all data, non-zero: data with non-zero carbon biomass. 27 observations did not include the month when measurements were taken.

Month	Globe all	Globe non-zero	NH all	NH non-zero	SH all	SH non-zero
January	164	82	59	4	105	78
February	213	56	59	4	154	52
March	379	187	347	157	32	30
April	687	641	638	593	49	48
May	618	561	612	560	6	1
June	384	318	380	318	4	0
July	263	185	258	183	5	2
August	202	131	198	131	4	0
September	119	56	114	56	5	0
October	169	94	90	27	79	67
November	164	94	67	15	97	91
December	164	130	40	6	124	124
Spring	–	–	1597	1310	181	158
Summer	–	–	836	632	383	254
Autumn	–	–	271	98	87	79
Winter	–	–	158	14	13	2
Total	3526	2547	2862	2054	664	493

the NH, *Phaeocystis* blooms during the spring months, with the spread of the biomass distribution being a combination of the temporal development of a bloom, and different bloom starting times at different latitudes. In the SH, cell biomasses are highest in December and January. The temporal development mostly reflects Southern Ocean dynamics, as few samples were taken at latitudes below 40° S (compare Fig. 6b).

3.4   Total *Phaeocystis* biomass distribution (mucus included)

Biomass estimates including colonial mucus are given as an upper limit for our biomass estimates (Fig. 9a). Given that the ratio of mucus carbon to cell carbon is highly dependent on colony size, the addition of mucus carbon estimates introduces a high level of uncertainty to total biomass estimates where colony size data is unavailable. Calculating mucus carbon biomass based on the minimum and maximum reported colony sizes for each species (Schoemann et al., 2005) gives a huge range of values: percent colony carbon as mucus ranges from 0.2–99.6 % for *P. globosa*, 1.4–94.3 % for *P. antarctica* and 55.8–99.8 % for *P. pouchetii*. Using a standard colony diameter of 200 µm increases biomass estimates by a factor of 1.2 for colonial *P. globosa* and *P. antarctica* cells, but by 32.8 for *P. pouchetii* compared to estimates considering cell biomass alone. The contribution of (standard) mucus to total carbon per cell is 96.9 % for *P. globosa*, and 14.6 % for *P. pouchetii* and *P. antarctica* (Table 2) for this standard colony size. The difference between the three species leads to a larger contribution by the Northern Hemisphere species to total *Phaeocystis* biomass (Fig. 9a and b).

Total *Phaeocystis* biomass estimates including (standard) mucus range from $2.9 \times 10^{-5}\,\mu g\,C\,l^{-1}$ to $19\,823\,\mu g\,C\,l^{-1}$. The maximal total biomass ($19\,823\,\mu g\,C\,l^{-1}$) is 3.6 times higher

than the corresponding data point with the maximal cell biomass of $5449.3\,\mu g\,C\,l^{-1}$. This data point is associated with high cell numbers during a bloom of *P. pouchetii* off the coast of the Netherlands in the Wadden Sea. In contrast, the maximal total biomass in the Southern Hemisphere is only $918\,\mu g\,C\,l^{-1}$, and thus one order of magnitude lower than maximal total biomasses in the Northern Hemisphere (Fig. 9). The global mean of all reported non-zero total biomass values is $183.8\,\mu g\,C\,l^{-1}$, and the median is $11.3\,\mu g\,C\,l^{-1}$. While our publicly available dataset also contains an estimate of maximal and minimal total carbon biomass based on maximal and minimal reported colony sizes (and thus maximal and minimal mucus), we do not visualize these results here. Uncertainties in the mucus contribution to total biomass due to these uncertainties in colony size range from hundreds to thousands of percent, and total carbon biomass estimates are far from certain at this point in time.

4   Discussion

We have estimated the carbon biomass of the haptophyte *Phaeocystis* from microscopic determinations of cell abundances. This approach is associated with several uncertainties.

First, since the data included in this database are sparse, we may have biases that we cannot account for. Whether the biomass estimates truly represent global averages is unclear. Free-living cells of *Phaeocystis* are often ignored in experimental studies, while colonies are counted, despite the fact

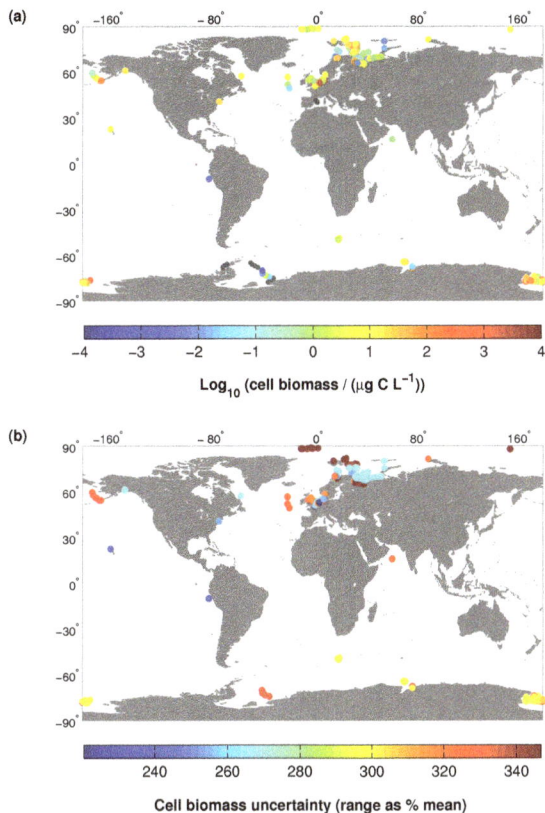

Figure 5. (a) Surface mean log-normalized *Phaeocystis* cell biomass concentrations in units of carbon (µg C l^{-1}) and **(b)** range of uncertainty in cell biomass in % of the mean, due to uncertainty in cell size. Black dots represent zero biomass values. Data has been log-transformed for a better visualization of the wide range of concentrations.

that there is always a background concentration of *Phaeocystis* cells when this genus is present in colonial form. Furthermore, even though *Phaeocystis* is ubiquitous (Schoemann et al., 2005), our data show a poor spatial resolution and data coverage outside the high-latitude coastal regions. Our biomass estimates for the coastal seas may not be representative of open ocean concentrations. Some areas such as the Pacific Ocean are clearly under-represented and we were not able to acquire any *Phaeocystis* measurements from the Northwest and West Pacific. Furthermore, there is a gap in our observations in the Arctic waters north of Siberia, and north of North America and in Greenland waters, despite published reports of high biomass off Greenland (Smith Jr., 1993). Our data is also seasonally biased in the Southern Hemisphere, with 58 % of the data acquired during the summer months. In addition, we note that *Phaeocystis* is only accurately counted at times when it is expected to form large blooms, when there is a strong likelihood that its abundance is high and when scientists are specifically looking for this group. Hence, low background concentrations of single-celled *Phaeocystis* will often be overlooked. Since the single-

celled life stages of *Phaeocystis* lack a clear morphological distinction, this gap in our current knowledge is unlikely to be resolved using microscopic methods, but will require genetic identification methods.

Second, there are methodological issues with the determination of abundance data that will influence our biomass calculations. Several data contributors do not report the life stage cells were in at the time of sampling, most likely due to the disruption of colony structure during cell fixation. This fact results in difficulties in distinguishing single and colonial cells. Hence, in order to obtain a lower limit on *Phaeocystis* cell biomass, we chose to assume undefined cells to be in the form of flagellates, which will bias the resulting biomass calculations. The ratio of free-living to colonial cells is highly variable, but a significant background concentration of free-living cells is present even during bloom conditions. Our assumption that all unspecified cells are flagellates is therefore likely to lead to an underestimation of *Phaeocystis* cell biomass.

Furthermore, non-blooming species such as *P. cordata*, *P. jahnii* or *P. scrobiculata* are not recorded explicitly in our abundance data, but may constitute a non-negligible fraction of total global *Phaeocystis* biomass in some oceanic regions.

Third, there are large uncertainties associated with the conversion of cell abundances to biomass. Cell measurements were only provided for very few datasets; for the majority of the database, biovolumes were calculated using mean published cell dimensions. Cell size is highly variable for all *Phaeocystis* species (Schoemann et al., 2005) and using a constant biovolume estimate for each species will underestimate the spatial and temporal variability that occurs in *Phaeocystis* biomass. Due to the differences in the reported size range, our estimates of cell carbon content are different from some previously reported figures. For example, our estimates of cell carbon content for *P. globosa* (Table 2; flagellates: 13 pg C cell^{-1}; colonial cells: 29 pg C cell^{-1}) are higher than estimates by Rousseau et al. (1990; flagellates: 11 pg C cell^{-1}; colonial cells: 14 pg C cell^{-1}), and our estimates for *P. antarctica* (Table 2; flagellates: 9 pg C cell^{-1}; colonial cells: 21 pg C cell^{-1}) are higher than those reported by Mathot (2000; flagellates: 3 pg C cell^{-1}; colonial cells: 14 pg C cell^{-1}) due to these differences in the reported mean cell diameters that were used to calculate the carbon estimates. Furthermore, literature values for the carbon conversion factor are only given for prymnesiophytes in general, but we lack information on the individual species of *Phaeocystis*, which may have a species-dependent, spatially and temporally varying cell carbon content.

Last, there is a large uncertainty associated with the addition of mucus carbon biomass due to the lack of data on cell forms, colony size and the amount of mucus per colonial cell. Greater use of preservation methods that maintain colony structure, along with routine colony size measurements, would allow for more reliable estimates of colonial mucus carbon. Further data on *Phaeocystis* colony sizes are

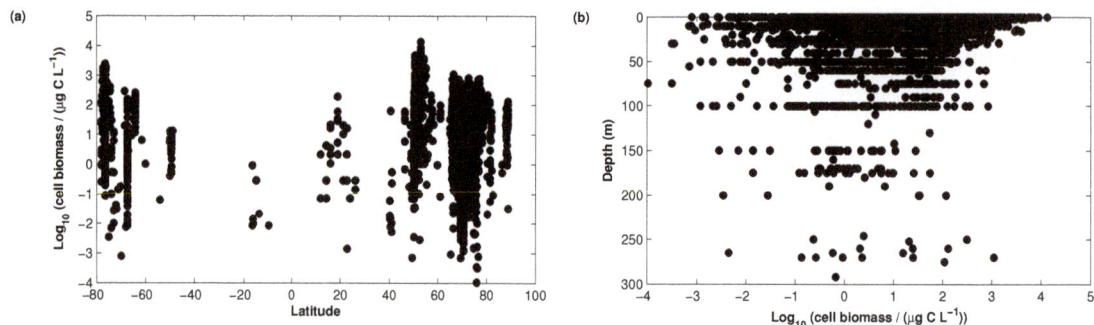

Figure 6. Distribution of non-zero log-normalized *Phaeocystis* cell biomass (μg C l^{-1}) (**a**) as a function of latitude and (**b**) as a function of depth.

clearly needed if mucus carbon is to be included in global biomass estimates and model validation. Moreover, there are uncertainties related to the structure of the mucilaginous carbon surrounding colonies. For example, an alternative method for estimating the total carbon biomass of *P. globosa* has been suggested by Van Rijssel et al. (1997) based on the observed hollow structure of the colonies. Van Rijssel et al. (1997) compute total biomass per cell based on a linear relationship between colony surface area and carbon content. A comparison of the estimated mean total carbon per *P. globosa* cell leads to significant differences. For our standard colonies of 200 μm diameter, we find total *P. globosa* carbon per cell to be 33.6 pg C cell^{-1} following Rousseau et al. (1990, Table 2); we compute an amount of 202.5 pg C cell^{-1} using Van Rijssel et al. (1997). The Rousseau relationship results in 9.6 ng C colony^{-1}, whereas the Van Rijssel relationship would lead to 58 ng C colony^{-1} for this species. Prior to the publication of Verity et al. (2007), the contribution of mucus carbon to total carbon per cell for *P. pouchetii* was done using the Rousseau et al. (1990) and Mathot et al. (2000) or the Van Rijssel et al. (1997) formulations (Reigstad and Wassmann, 2007). Using these relationships, Reigstad and Wassmann (2007) find a much lower contribution of mucus (10 %) to total carbon per cell than what we find using Verity et al. (2007, 96.9 %). Earlier estimates of *P. pouchetii* mucus carbon may thus not be compatible with our estimations. Clearly, future studies are needed to address this uncertainty in colony structure and mucus distribution, and the corresponding volume to biomass conversion factors.

5 Conclusions

This is the first attempt at creating a global *Phaeocystis* biomass database. At present, however, we are still far from being able to give a global estimate of *Phaeocystis* biomass concentration. Data are limited by lack of spatial and temporal resolution, and at most sampling sites we lack a seasonal cycle that would be necessary to determine reasonable estimates for annual mean biomass concentration. Annual and monthly mean biomasses are of particular interest for the

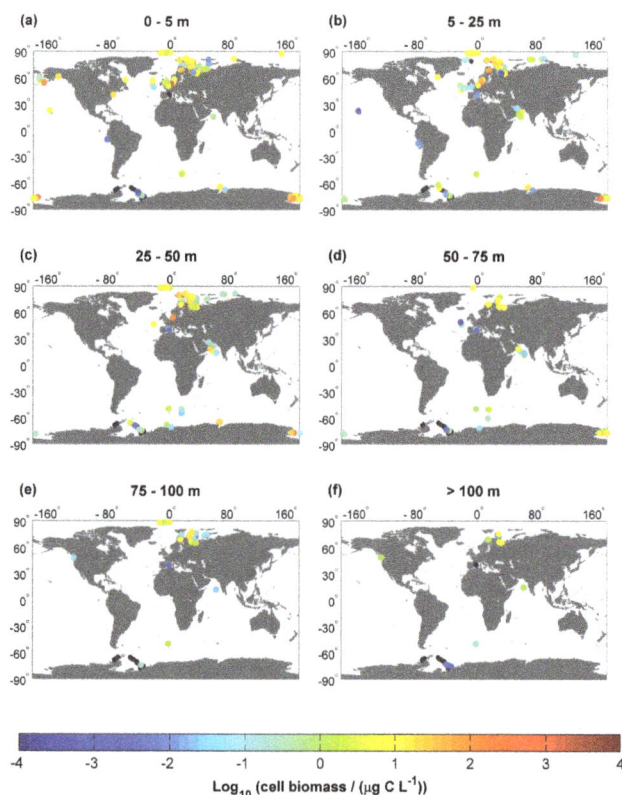

Figure 7. Log-normalized *Phaeocystis* cell biomass in units of carbon (μg C l^{-1}) at different depths (**a**) surface measurements (0–5 m) (**b**) measurements between 5–25 m (**c**) 25–50 m (**d**) 50–75 m (**e**) 75–100 m and (**f**) >100 m depth. Black dots represent zero biomass values.

modelling community, but these will only be meaningful if further microscopic data can be added to the database. Targeted explorations of marine ecosystems with the aim to determine phytoplankton biomass would be desirable, but such endeavours tend to be expensive and laborious. A marine census of species biomass would shed light on the relative importance of key marine plankton groups and their respective importance for global biogeochemical cycling.

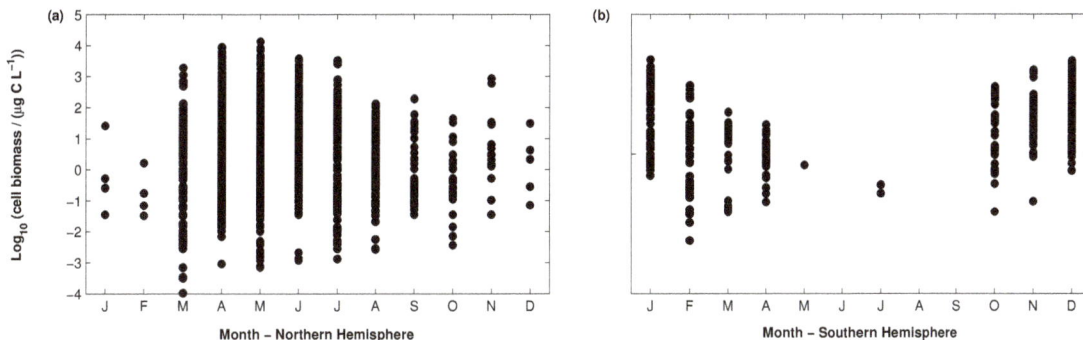

Figure 8. Seasonal distribution of log-normalized non-zero *Phaeocystis* cell biomass data for (**a**) the Northern and (**b**) the Southern Hemispheres.

Appendix A

A1 Data table

A full data table containing all biomass data points can be downloaded from the data archive PANGAEA, doi:10.1594/PANGAEA.779101. The data file contains longitude, latitude, depth, sampling time, abundance counts and biomass concentrations, as well as the full data references.

A2 Gridded netCDF biomass product

Monthly mean biomass data has been gridded onto a $360 \times 180°$ grid, with a vertical resolution of 33 depth levels (equivalent to World Ocean Atlas depths) and a temporal resolution of 12 months (climatological monthly means). Data has been converted to netCDF format for easy use in model evaluation exercises. The netCDF file can be downloaded from PANGAEA, doi:10.1594/PANGAEA.779101. This file contains total and non-zero abundances, cell biomasses and total biomass estimates. For all fields, the means, medians and standard deviations resulting from multiple observations in each of the 1° pixels are given. The ranges in cell and total biomasses due to uncertainties in cell size and life form are not included as variables in the netCDF product, but are given as ranges (minimum cell biomass, maximum cell biomass; minimum total biomass, maximum total biomass) in the data table.

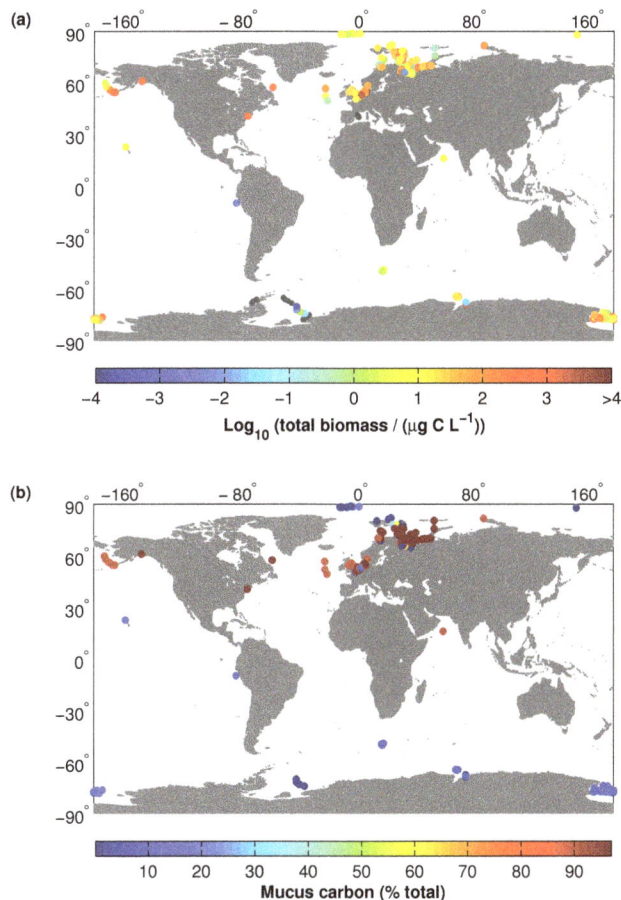

Figure 9. Estimates of (**a**) log-normalized total mean *Phaeocystis* biomass including colonial mucus for the surface layer (0–5 m) and (**b**) fraction of total mean surface biomass composed of mucus carbon. Zero values are not represented. The difference between the ratios of total carbon to cell carbon for the three species leads to a larger contribution by the Northern Hemisphere species to total *Phaeocystis* biomass.

Acknowledgements. We thank P. Assmy, G. C. Cadée, D. A. Caron, G. R. DiTullio, B. Hansen, I. R. Jenkinson, I. Joint, S.-H. Kang, B. Karlson, D. J. Lonsdale, S. Mathot, R. Riegman, M. W. Silver, W. O. Smith, P. Tett, P. Tréguer, R. Uncles, F. C. Van Duyl, E. L. Venrick, T. Weisse, G. V. Wolfe, and P. Wassmann for the permission to use and redistribute *Phaeocystis* data, and the BODC, JGOFS, OBIS OCB, PANGAEA and WOD databases for providing and archiving data. We also thank E. Buitenhuis for producing the gridded netCDF product, S. Doney for fruitful discussions on quality control, and S. Pésant for archiving the data. M. V. acknowledges funding from ETH Zürich. C. O'B.'s contribution to the research leading to these results has received funding from the European Community's Seventh Framework Programme (FP7 2007–2013) under grant agreement no [238366].

Edited by: S. Pesant

References

Aumont, O. and Bopp, L.: Globalizing results from ocean in situ fertilization studies, Global Biogeochem. Cy., 20, GB2017, doi:10.1029/2005GB002591, 2006.

Arrigo, K. R., Robinson, D. L., Worthen, R. B., Dunbar, R. B., DiTullio, G., R., VanWoert, M., and Lizotte, M. P.: Phytoplankton community structure and the drawdown of nutrients and CO_2 in the Southern Ocean, Science, 283, 365–367, doi:10.1126/science.283.5400.365, 1999.

Assmy, P.: Phytoplankton abundance measured on water bottle samples at station PS65/587-1, Alfred Wegener Institute for Polar and Marine Research, Bremerhaven, doi:10.1594/PANGAEA.603400, 2007.

Baumann, M. E. M., Lancelot, C., Brandini, F. P., Sakshaug, E., and John, D. M.: The taxonomic identity of the cosmopolitan prymnesiophyte Phaeocystis: a morphological and ecophysiological approach, J. Marine Syst., 5, 5–22, 1994.

Beaugrand, G. and Reid, P. C.: Long-term changes in phytoplankton, zooplankton and salmon related to climate, Glob. Change Biol., 9, 801–817, 2003.

Boyer, T. P., Antonov, J. I., Baranova, O. K., Garcia, H. E., Johnson, D. R., Locarnini, R. A., Mishonov, A. V., O'Brien, T. D., Seidov, D., Smolyar, I. V., and Zweng, M. M.: World Ocean Database 2009, edited by: Levitus, S., NOAA Atlas NESDIS 66, US Government Printing Office, Washington, D.C., 216 pp., DVDs, 2009.

Buitenhuis E. T., Vogt, M., Moriarty, R., Bednarsek, N., Doney, S. C., Leblanc, K., Le Quéré, C., Luo, Y., O'Brien, C., O'Brien, T., Peloquin, J. M., and Schiebel, R.: MAREDAT: Towards a World Ocean Atlas of MARine Ecosystem DATa, Earth Syst. Sci. Data Discuss., in preparation, 2012.

Brown, S. L., Landry, M. R., Yang, E. J., Rii, Y. M., and Bidigare, R. R.: Diatoms in the desert: Plankton community response to a mesoscal eddy in the subtropical North Pacific, Deep-Sea Res. Pt. II, 55, 1321–1333, 2008.

Cadée, G. C.: Phaeocystis colonies wintering in the water column?, Neth. J. Sea Res., 28, 227–230, 1991.

Cadée, G. C. and Hegeman, J.: Seasonal and annual variation in Phaeocystis pouchetii (Haptophyceae) in the westernmost inlet of the Wadden Sea during the 1973 to 1985 period, Neth. J. Sea Res., 20, 29–36, 1986.

Cadée, G. C. and Hegeman, J.: Persisting high levels of primary production at declining phosphate concentrations in the Dutch coastal area (Marsdiep), Neth. J. Sea Res., 31, 147–152, 1993.

Chavez, F. P., Ryan, J., Lluch-Cota, S. E., and Niquen, M.: From anchovies to sardines and back: Multidecadal change in the Pacific Ocean, Science, 299, 217–221, 2003.

Chen, Y. Q., Wang, N., Zhang, P., Zhou, H., and Qu, L. H.: Molecular evidence identifies bloom-forming Phaeocystis (Prymnesiophyta) from coastal waters of southeast China as Phaeocystis globosa, Biochem. Syst. Ecol., 30, 15–22, 2002.

DiTullio, G. R., Grebmeier, J. M., Arrigo, K. R., Lizotte, M. P., Robinson, D. H., Leventer, A., Barry, J. P., VanWoert, M. L., and Dunbar, R. B.: Rapid and early export of Phaeocystis antarctica blooms in the Ross Sea, Antarctica, Nature, 404, 595–598, 2000.

Eilertsen, H. C., Taasen, J. P., and Weslawski, J. M.: Phytoplankton studies in the fjords of West Spitzbergen: physical environment and production in spring and summer, J. Plankton Res., 11, 1245–1260, 1989.

Estrada, M.: Phytoplankton assemblages across a new Mediterranean front: changes from winter mixing to spring stratification, Oecologia Aquatica, 10, 157–185, 1991.

Estrada, M. and Delgado, M.: Summer phytoplankton distributions in the Weddell Sea, Polar Biol., 10, 441–449, 1990.

Fransz, H. G., Gonzalez, S. R., Cadée, G. C., and Hansen, F. C.: Long-term change of Temora longicornis (copepoda, Calanoida) abundance in a Dutch tidal inlet (Marsdiep) in relation to eutrophication, Neth. J. Sea Res., 30, 23–32, 1992.

Glover, D. M., Jenkins, W. J., and Doney, S. C: Modeling Methods for Marine Science, Cambridge University Press, Cambridge, UK, ISBN: 978-0-521-86783-2, 2011.

Guiselin, N., Courcot, L., Artigar, L. F., Le Jéloux, A., and Brylinski, J.-M.: An optimized protocol to prepare Phaeocystis globosa morphotypes for scanning electron microscopy observation, J. Microbiol. Meth., 77, 119–123, 2009.

Hansen, B., Berggreen, U. C., Tande, K. S., and Eilertsen, H. C.: Post-bloom grazing by Calanus glacialis, C. finmarchicus and C. hyperboreus in the region of the Polar Front, Barents Sea, Mar. Biol., 104, 5–14, 1990.

Hashioka, T. and Yamanaka, Y.: Ecosystem change in the western North Pacific associated with global warming using 3D-NEMURO, Ecol. Model., 202, 95–104, 2007.

Hatun, H., Payne, M. R., Beaugrand, G., Reid, P. C., Sando, A. B., Drange, H., Hansen, B., Jacobsen, J. A., and Bloch, D.: Large bio-geographical shifts in the north-eastern Atlantic Ocean: From the subpolar gyre, via plankton, to blue whiting and pilot whales, Prog. Oceanogr., 803, 149–162, 2009.

Hillebrand, H., Dürselen, C. D., Kirschtel, D., Pollingher, D., and Zohary, T.: Biovolume calculation for pelagic and benthic microalgae, J. Phycol., 35, 403–424, 1999.

Jenkinson, I. R. and Biddanda, B. A.: Bulk-phase viscoelastic properties of seawater: relationship with plankton components, J. Plankton Res., 17, 2251–2274, 1995.

Kang, S.-H. and Fryxell, G. A.: Phytoplankton in the Weddell Sea, Antarctica: Composition, abundance and distribution in water-column assemblages of the marginal ice-edge zone during austral autumn, Mar. Biol., 116, 335–348, 1993.

Karentz, D. and Spero, H. J.: Response of a natural Phaeocystis population to ambient fluctuations of UVB radiation caused by Antarctic ozone depletion, J. Plankton Res., 17, 1771–1789, doi:10.1093/plankt/17.9.1771, 1995.

Karlson, B., Edler, L., Granéli, W., Sahlsten, E., and Kuylenstierna, M.: Subsurface chlorophyll maxima in the Skagerrak-processes and plankton community structure, J. Sea Res., 35, 139–158, 1996.

Karlson, B., Cusack, C., and Bresnan, E. (Eds.): Microscopic and molecular methods for quantitative phytoplankton analysis, Paris, UNESCO, IOC Manuals and Guides, no. 55, 110 pp., 2010.

Kennington, K., Allen, J. R., Wither, A., Shammon, T. M., and Hartnoll, R. G.: Phytoplankton and nutrient dynamics in the northeast Irish Sea, Hydrobiologia, 393, 57–67, 1999.

Koeman, R. P. T.: Analyses van fytoplankton en microzooplankton van het Friese Front 1999, Rapportage van onderzoek in opdracht van het Rijksinstituut voor Kust en Zee (RIKZ), Haren, The Netherlands, 1999 (in Dutch).

Le Quéré, C., Harrison, S. P., Prentice, I. C., Buitenhuis, E. T., Aumont, O., Bopp, L., Claustre, H., Da Cunha, L. C., Geider, R., Giraud, X., Klaas, C., Kohfeld, K. E., Legendre, L., Manizza, M., Platt, T., Rivkin, R. B., Sathyendranath, S., Uitz, J., Watson, A. J., and Wolf-Gladrow, D.: Ecosystem dynamics based on plankton functional types for global ocean biogeochemistry models, Glob. Change Biol., 11, 2016–2040, doi:10.1111/j.1365-2486.2005.1004.x, 2005.

Luchetta, A., Lipizer, M., and Socal, G.: Temporal evolution of primary production in the central Barents Sea, J. Marine Syst., 27, 177–193, 2000.

Mathot, S., Smith Jr., W. O., Carlson, C. A., Garrison, D. L., Gowing, M. M., and Vickers, C. L.: Carbon partitioning within Phaeocystis antarctica (Prymnesiophyceae) colonies in the Ross Sea, Antarctica, J. Phycol., 36, 1049–1056, 2000.

Medlin, L. and Zingone, A.: A taxonomic review of the genus Phaeocystis, Biogeochemistry, 83, 3–18, doi:10.1007/s10533-007-9087-1, 2007.

Menden-Deuer, S. and Lessard, E. J.: Carbon to volume relationships for dinoflagellates, diatoms, and other protist plankton, Limnol. Oceanogr., 45, 569–579, 2000.

Moore, J. K. and Doney, S. C., Iron availability limits the ocean nitrogen inventory stabilizing feedbacks between marine denitrification and nitrogen fixation, Global Biogeochem. Cy., 21, GB2001, doi:10.1029/2006GB002762, 2007.

Moore, C. M., Hickman, A. E., Poulton, A. J., Seevaye, S., and Lucas, M. I.: Iron-light interactions during the Crozet Natural Iron Bloom Export Experiment (CROZEX): Part II – Taxonomic responses and elemental stoichiometry, Deep-Sea Res. Pt. II, 54, 2066–2084, doi:10.1016/j.dsr2.2007.06.015, 2007.

Palmisano, A. C., SooHoo, J. B., SooHoo, S. L., Kottmeier, S. T., Craft, L. L., and Sullivan, C. W.: Photoadaptation in Phaeocystis pouchetii advected beneath annual sea ice in McMurdo Sound, Antarctica, J. Plankton Res., 8, 891–906, 1986.

Peperzak, L. and Gäbler-Schwarz, S.: Current knowledge of the life cycles of Phaeocystis globosa and Phaeocystis antarctica (Prymnesiophyceae), J. Phycol., 48, 514–517, doi:10.1111/j.1529-8817.2012.01136.x, 2012.

Peperzak, L., Colijn, F., Gieskes, W.W.C. and Peeters, J.C.H.: Development of the diatom-Phaeocystis spring bloom in the Dutch coastal zone of the North Sea: the silicon depletion versus the daily irradiance threshold hypothesis. J. Plankton Res., 20(3), 517-537, doi:10.1093/plankt/20.3.517, 1998.

Peperzak, L., Colijn, F., and Peeters, J. C. H.: Observations of flagellates in colonies of Phaeocystis globosa (Prymnesiophyceae); a hypothesis for their position in the life cycle, J. Plankton Res., 22, 2191–2203, 2000.

Pieters, H., Kluytmans, J. H., Zandee, D. I., and Cadée, G. C.: Tissue composition and reproduction of Mytilus edulis in relation to food availability, Neth. J. Sea Res., 14, 349–361, 1980.

Reid, P. C., Johns, D. G., Edwards, M., Starr, M., Poulin, M., and Snoeijs, P.: A biological consequence of reducing Arctic ice cover: arrival of the Pacific diatom Neodenticula seminae in the North Atlantic for the first time in 800,000 years, Glob. Change Biol., 13, 1910–1921, doi:10.1111/j.1365-2486.2007.01413.x, 2007.

Reigstad, M. and Wassmann, P.: Does Phaeocystis spp. contribute significantly to vertical export of organic carbon?, Biogeochemistry, 83, 217–234, doi:10.1007/s10533-007-9093-3, 2007.

Riebesell, U., Reigstad, M., Wassmann, P., Noji, T., and Passow, U.: On the trophic fate of Phaeocystis pouchetii (haricot): VI. Significance of Phaeocystis-derived mucus for vertical flux, Neth. J. Sea Res., 33, 193–203, 1995.

Riegman, R., Rowe, A., Noordeloos, A. A., and Cadee, G. C.: Evidence for eutrophication induced Phaeocystis sp. blooms in the Marsdiep area (The Netherlands), in: Toxic Phytoplankton Blooms in the Sea, edited by: Smayda, T. J. and Shimizu, Y., Elsevier, 799–805, 1993.

Robinson, C., Archer, S. D., and le. B. Williams, P. J.: Microbial dynamics in coastal waters of East Antarctica: plankton production and respiration. Mar. Ecol.-Prog. Ser., 180, 23–36, 1999.

Rousseau, V., Mathot, S., and Lancelot, C.: Calculating carbon biomass of Phaeocystis sp. from microscopic observations, Mar. Biol., 107, 305–314, 1990.

Rousseau, V., Chrétiennot-Dinet, M.-J., Jacobsen, A., Verity, P., and Whipple, S.: The life cycle of Phaeocystis: state of knowledge and presumptive role in ecology, Biogeochemistry, 83, 29–47, doi:10.1007/s10533-007-9085-3, 2007.

Schoemann, V.: Effects of phytoplankton blooms on the cycling of manganese and iron in coastal waters, Limnol. Oceanogr., 43, 1427–1441, 1998.

Schoemann, V., Becquefort, S., Stefels, J., Rousseau, V., and Lancelot, C.: Phaeocystis blooms in the global ocean and their controlling mechanisms: A review, J. Sea Res., 53, 43–66, doi:10.1016/j.seares.2004.01.008, 2005.

Scott, F. J., Davidson, A. T., and Marchant, H. J.: Seasonal variation in plankton, submicrometre particles and size-fractionated dissolved organic carbon in Antarctic coastal waters, Polar Biol., 23, 635–643, 2000.

Seuront, L., Lacheze, C., Doubell, M. J., Seymour, J. R., Van Dongen-Vogels, V., Newton, K., Alderkamp, A. C., and Mitchell, J. G.: The influence of Phaeocystis globosa on microscale spatial patterns of chlorophyll a and bulk-phase seawater viscosity, Biogeochemistry, 83, 173–188, 2007.

Sherr, E. B. and Sherr, B. F.: Preservation and storage of samples for enumeration of heterotrophic protists, in: Handbook of Methods in Aquatic Microbial Ecology, edited by: Kemp, P. F., Sherr, B. F., Sherr, E. B., and Cole, J. J., Lewis Publishers, Boca Raton, 207–212, 1993.

Shields, A. R. and Smith, W. O.: Size-fractionated photosynthesis/irradiance relationships during Phaeocystis antarctica-dominated blooms in the Ross Sea, Antarctica, J. Plankton Res., 31, 701–712, doi:10.1093/plankt/fbp022, 2009.

Smith, R. C., Przelin, B. B., Baker, K. S., Bidigare, R. R., Boucher, N. P., Coley, T., Karentz, D., MacIntyre, S., Matlick, H. A., Menzies, D., Ondrusek, M., Wan, Z., and Waters, K. J.: Ozone depletion: ultraviolet radiation and phytoplankton biology in Antarctic waters, Science, 255, 952–959, 1992.

Smith Jr., W.: Nitrogen uptake and new production in the Greenland Sea: The spring Phaeocystis bloom, J. Geophys. Res., 98, 4681–4688, 1993.

Tréguer, P., Lindner, L., Leynaert, A., Panouse, M., and Jacques, G.: Production of biogenic silica in the Weddell-Scotia Seas measured with 32Si, Limnol. Oceanogr., 36, 1217–1227, 1991.

Van Duyl, F. C., Gieskes, W. W. C., Kop, A. J., and Lewis, W. E.: Biological control of short-term variations in the concentration of DMSP and DMS during a *Phaeocystis* spring bloom, J. Sea Res., 40, 221–231, 1998.

Van Rijssel, M., Hamm, C. E., and Gieskes, W. W. C.: *Phaeocystis globosa* (Prymnesiophyceae) colonies: hollow structures built with small amounts of polysaccharides, Eur. J. Phycol., 32, 185–192, 1997.

Venrick, E. L.: Comparison of the phytoplankton species composition and structure in the climax area (1973–1985) with that of station ALOHA (1994), Limnol. Oceanogr., 42, 1643–1648, 1997.

Verity, P. G., Whipple, S. J., Nejstgaard, J. C., and Alderkamp, A. C.: Colony size, cell number, carbon and nitrogen contents of *Phaeocystis pouchetii* from western Norway, J. Plankton Res., 24, 359–367, 2007.

Vogt, M., Vallina, S. M., Buitenhuis, E. T., Bopp, L., and Le Quéré, C.: Simulating dimethylsulphide seasonality with the Dynamic Green Ocean Model PlankTOM5, J. Geophys. Res.-Oceans, 115, C06021, doi:10.1029/2009JC005529, 2010.

Wang, S. and Moore, J. K.: Incorporating *Phaeocystis* into a South-ern Ocean ecosystem model, J. Geophys. Res., 116, C01019, doi:10.1029/2009JC005817, 2011.

Wassmann, P., Ratkova, T., and Reigstad, M.: The contribution of single and colonial cells of *Phaeocystis* pouchetii to spring and summer blooms in the north-eastern North Atlantic, Harmful Algae, 4, 823–840, 2005.

Weaver, S. S.: Ceratium in Fire Island Inlet, Long Island, New York (1971–1977), Limnol. Oceanogr., 24, 553–558, 1979.

Weber, T. S. and Deutsch, C.: Ocean nutrient ratios governed by plankton biogeography, Nature, 467, 550–554, doi:10.1038/nature09403, 2010.

Weisse, T., Grimm, N., Hickel, W., and Martens, P.: Dynamics of *Phaeocystis pouchetii* blooms in the Wadden Sea of Sylt (German Bight, North Sea), Estuar. Coast. Shelf Sci., 23, 171–182, 1986.

Widdicombe, C. E., Eloire, D., Harbour, D., Harris, R. P., and Somerfield, P. J.: Long term phytoplankton community dynamics in the Western English Channel, J. Plankton Res., 32, 643–656, 2010.

Wolfe, G. V., Levasseur, M., Cantin, G., and Michaud, S.: DMSP and DMS dynamics and microzooplankton grazing in the Labrador Sea: application of the dilution technique, Deep-Sea Res. Pt. I 47, 2243–2264, 2000.

Vertical distribution of chlorophyll a concentration and phytoplankton community composition from in situ fluorescence profiles: a first database for the global ocean

R. Sauzède[1,2], H. Lavigne[3], H. Claustre[1,2], J. Uitz[1,2], C. Schmechtig[1,2], F. D'Ortenzio[1,2], C. Guinet[4], and S. Pesant[5,6]

[1] Laboratoire d'Océanographie de Villefranche, CNRS, UMR7093, Villefranche-Sur-Mer, France

[2] Université Pierre et Marie Curie-Paris 6, UMR7093, Laboratoire d'océanographie de Villefranche, Villefranche-Sur-Mer, France

[3] Istituto Nazionale di Oceanografia e di Geofisica Sperimentale, Sgonico (OGS), Italy

[4] Centre d'Etudes Biologiques de Chizé, CNRS, Villiers en Bois, France

[5] MARUM, Center for Marine Environmental Sciences, Universität Bremen, Bremen, Germany

[6] PANGAEA, Data Publisher for Earth and Environmental Science, Bremen, Germany

Correspondence to: R. Sauzède (sauzede@obs-vlfr.fr)

Abstract. In vivo chlorophyll *a* fluorescence is a proxy of chlorophyll *a* concentration, and is one of the most frequently measured biogeochemical properties in the ocean. Thousands of profiles are available from historical databases and the integration of fluorescence sensors to autonomous platforms has led to a significant increase of chlorophyll fluorescence profile acquisition. To our knowledge, this important source of environmental data has not yet been included in global analyses. A total of 268 127 chlorophyll fluorescence profiles from several databases as well as published and unpublished individual sources were compiled. Following a robust quality control procedure detailed in the present paper, about 49 000 chlorophyll fluorescence profiles were converted into phytoplankton biomass (i.e., chlorophyll *a* concentration) and size-based community composition (i.e., microphytoplankton, nanophytoplankton and picophytoplankton), using a method specifically developed to harmonize fluorescence profiles from diverse sources. The data span over 5 decades from 1958 to 2015, including observations from all major oceanic basins and all seasons, and depths ranging from the surface to a median maximum sampling depth of around 700 m. Global maps of chlorophyll *a* concentration and phytoplankton community composition are presented here for the first time. Monthly climatologies were computed for three of Longhurst's ecological provinces in order to exemplify the potential use of the data product. Original data sets (raw fluorescence profiles) as well as calibrated profiles of phytoplankton biomass and community composition are available on open access at PANGAEA, Data Publisher for Earth and Environmental Science.

1   Introduction

Phytoplankton biomass is generally recognized to play a key role in the global carbon cycle, stressing the need for a better understanding of its spatio-temporal distribution and variability in the global ocean. Chlorophyll *a* concentration is widely used as a proxy to estimate phytoplankton biomass. The geographic and temporal distribution of this proxy is already well documented at a global scale thanks to synoptic remote sensing observations by ocean-color radiometry (OCR, McClain, 2009; Siegel et al., 2013). Nevertheless, OCR observations are restricted to the ocean surface layer, "sensing" only one-fifth of the so-called euphotic layer where phytoplankton photosynthesis is realized and which can sometimes extend to well below 100 m (Gordon and Mc-Cluney, 1975; Morel and Berthon, 1989). It is therefore essential to better resolve the global distribution of phytoplankton biomass in the vertical.

The vertical distribution of chlorophyll *a* can be estimated with greatest accuracy from the analysis of water samples by high-performance liquid chromatography (HPLC, Claustre et al., 2004; Peloquin et al., 2013). However, these in situ measurements are relatively scarce because their acquisition requires ship-based sampling and their analysis is costly. Moreover, because these measurements are made on water samples, the vertical resolution is generally weak (e.g., around one measurement every 10 m). The measurement of in vivo chlorophyll *a* fluorescence is widely used as a proxy for chlorophyll *a* concentration (Lorenzen, 1966). Besides dissolved oxygen concentration, fluorescence is the most measured biogeochemical property in the global ocean. The advantages of this method are as follows: (1) it can be easily measured in situ using reliable sensors; (2) the vertical resolution is high, yielding several values per meter; and (3) data are available in digital format immediately after their acquisition. The integration of fluorescence sensors on autonomous platforms (e.g., profiling floats, animals, gliders) has recently led to a sudden rise in the acquisition of in vivo chlorophyll *a* fluorescence data (Claustre et al., 2010a). However, the relationship between chlorophyll *a* fluorescence and phytoplankton biomass is highly variable and depends on several factors, including phytoplankton physiological state and community composition (Cunningham, 1996; Falkowski et al., 1985; Kiefer, 1973). The conversion of in situ chlorophyll *a* fluorescence measurements into phytoplankton biomass must therefore be done with great care.

FLAVOR (Fluorescence to Algal communities Vertical distribution in the Oceanic Realm) is a method developed to transform and combine large numbers of fluorescence profiles from various sampling sensors and platforms (Sauzède et al., 2015a). This neural network-based method generates vertical distributions of (1) chlorophyll *a* concentration and (2) phytoplankton community size indices (i.e., microphytoplankton, nanophytoplankton and picophytoplankton) based on the shape of in situ fluorescence profiles (i.e., normalized profiles) and the day and location of acquisition. In addition to chlorophyll *a* concentration, community composition is an essential variable that determines the possible impact of phytoplankton on oceanic carbon fluxes and climate change scenarios (e.g., Le Quere et al., 2005). Global data compilations of phytoplankton community composition from discrete water samples have recently been published in ESSD (Peloquin et al., 2013) but data remain rather sparse. It could be an invaluable source of information to have a database of phytoplankton community size indices with the same spatiotemporal resolution as the fluorescence data sets. It has now become possible using the FLAVOR method to transform and combine all available in situ fluorescence data into a single-reference database that comprises essential information on chlorophyll *a* concentration and phytoplankton community size indices vertical distributions.

Presently, the widely used climatology of the global vertical distribution of chlorophyll *a* concentration is published in the World Ocean Atlas 2001 (Conkright et al., 2002). The latter climatology is based on estimates from analyzed water samples available in the World Ocean Database (WOD, Levitus et al., 2013) and the World Data Center (WDC, http://gcmd.gsfc.nasa.gov/). This climatology, based on seven discrete depths (0-10-20-30-50-75-100 m), is mainly limited by the lack of in situ estimations of chlorophyll *a* concentration, which leads to a strong spatial interpolation of data. Moreover, the discrete depths used to compute the climatology fail to finely reproduce the vertical distribution of the phytoplankton biomass, especially in areas characterized by very deep (> 100 m) deep chlorophyll maxima (DCM) such as the core of subtropical oligotrophic gyres. Using FLAVOR, the potential of the high vertical (around one data point per meter) and spatial resolution of chlorophyll fluorescence measurements would improve the 3-D climatologies of chlorophyll *a* concentration significantly. Moreover, climatologies of phytoplankton community size indices could be created with a similar spatio-temporal resolution.

This paper presents a global compilation of chlorophyll fluorescence profiles obtained from online databases and from published and unpublished individual sources. These were converted into a global compilation of phytoplankton biomass (i.e., chlorophyll *a* concentration) and community composition using the FLAVOR method. Prior to the application of FLAVOR, a 10-step quality control procedure was specifically developed. The remaining profiles were then analyzed. As examples of application, we present the first maps of global mean chlorophyll *a* concentration for several oceanic layers as well as global maps of phytoplankton community size indices. To further assess the quality of the resulting database, the climatological chlorophyll *a* concentration computed here for the surface layer is compared to the climatological remotely sensed chlorophyll *a* concentration available from Modis Aqua. Moreover, monthly 3-D climatologies of chlorophyll *a* concentration and associated phytoplankton community size indices are analyzed for

Table 1. Summary of the contributions of the chlorophyll fluorescence profiles in the database presented in this study.

Data source/institute/investigator	Period	Number of fluorescence profiles	Percentage of data in the database	Website if available or contact for requests
National Oceanographic Data Center (NODC)	Jun 1958–Mar 2014	30 977	63.7 %	http://www.nodc.noaa.gov/
Oceanographic Autonomous Observations (OAO)	May 2008–Jan 2015	6092	12.5 %	http://www.oao.obs-vlfr.fr/
Laboratoire d'Océanographie de Villefranche (LOV) cruises	May 1991–Jan 2012	3320	6.8 %	claustre@obs-vlfr.fr, sauzede@obs-vlfr.fr
Japan Oceanographic Data Center (JODC)	Jan 1998–Jul 2004	2262	4.6 %	http://www.jodc.go.jp/
PANGAEA	Nov 1980–Apr 2009	2294	4.7 %	http://www.pangaea.de/
C. Guinet (data acquired by elephant seals, Guinet et al., 2013)	Dec 2007–Jan 2011	1908	3.9 %	christophe.guinet@cebc.cnrs.fr,
British Oceanographic Data Center (BODC)	Sep 1996–Nov 2008	1219	2.5 %	http://www.bodc.ac.uk/
Systèmes d'Informations Scientifiques pour la MER (SISMER)	Sep 1999–May 2008	237	0.5 %	http://www.ifremer.fr/sismer/
Australian Antarctic Data Center (AADC)	Jan 2001–Feb 2006	234	0.5 %	http://data.aad.gov.au/
Southern Ocean Iron RElease Experiment (SOIREE)	Feb 1999	57	0.1 %	http://www.uea.ac.uk/~e610/soiree/index.html

several ecological provinces defined by Longhurst (2010). Overall, the data set presented here can be readily exploited to deepen our understanding of the spatio-temporal distribution and variability of phytoplankton biomass and associated community composition in the global ocean. It is obviously a first step towards a database that will regularly be improved thanks to the ongoing intensification of chlorophyll a fluorescence profile acquisition by Bio-Argo profiling floats, gliders and mammals equipped with instruments.

2 Data and methods

2.1 Origins of in situ chlorophyll fluorescence measurements

The database presented in this study is available from PANGAEA, Data Publisher for Earth and Environmental Science in two formats: (1) the database containing all compiled raw fluorescence profiles (the raw database, http://doi.pangaea.de/10.1594/PANGAEA.844212, Sauzède et al., 2015b) and (2) the database containing the fluorescence profiles which

are calibrated into chlorophyll a concentration and associated phytoplankton community size indices (the calibrated database, http://doi.pangaea.de/10.1594/PANGAEA. 844485, Sauzède et al., 2015c). The data of in situ vertical fluorescence profiles compiled for creating the raw database were obtained from several available online databases as well as published and unpublished individual sources. The duplicates and single-surface values, which are not vertical profiles, were automatically removed (not integrated in the raw database). Finally, the raw database contains 268 127 fluorescence profiles. Following a robust quality control procedure detailed hereafter (Sect. 2.2), about 49 000 chlorophyll fluorescence profiles were converted into phytoplankton biomass (i.e., chlorophyll a concentration) and size-based community composition (i.e., microphytoplankton, nanophytoplankton and picophytoplankton). The origin of this calibrated database is summarized in Table 1. The majority of the data come from the National Oceanographic Data Center (NODC) and the fluorescence profiles acquired by Bio-Argo floats are available on the Oceanographic Autonomous Observations (OAO) web platform (63.7 and

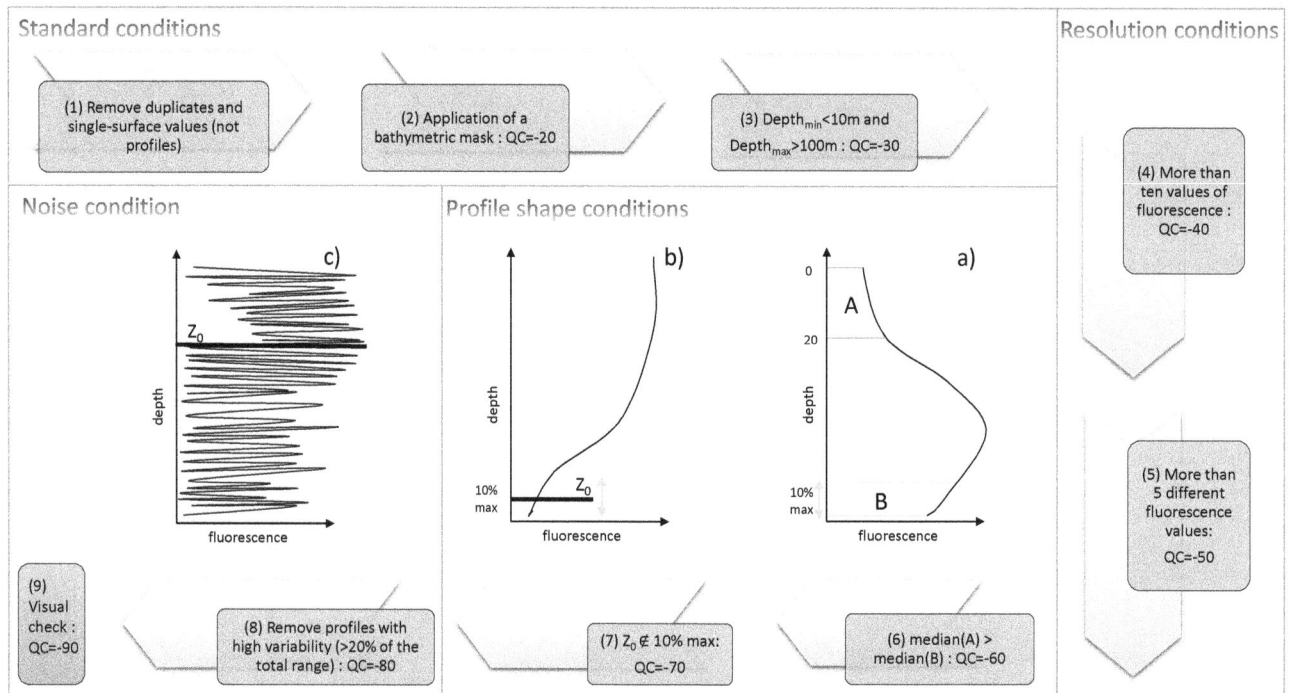

Figure 1. Schematic overview of the quality control procedure specifically developed for the database presented in this study. The fluorescence profiles represented in the **(a)**, **(b)** and **(c)** panels are examples of profiles which are rejected by the quality control steps (6), (7) and (8) respectively.

12.5 % respectively, see percentages of data in the database depending on their origin in Table 1).

Different modes of acquisition were used to collect the data presented in this study: (1) the CTD (conductivity, temperature and depth) profiles are acquired using a fluorometer mounted on a CTD rosette; (2) the OSD (Ocean Station Data) profiles are derived from water samples analyzed by fluorometry and are defined as "low" resolution profiles (Boyer et al., 2009); (3) the UOR (Undulating Oceanographic Recorder) profiles are acquired by a "fish" equipped with fluorometer and towed by a research vessel; (4) AP (Autonomous Platforms) profiles are acquired by Bio-Argo profiling floats or elephant seals equipped with a fluorometer (Claustre et al., 2010b; Guinet et al., 2013). Table 2 lists the number of profiles in the calibrated database according to these four modes of acquisition.

It is worth noting that the data acquired from gliders were not included in the database. Although glider data are extremely numerous, they are restricted to a very small spatiotemporal window. As a consequence, a database including glider data would likely be spatially and temporally biased, in contradiction with our first aim of building a global climatological database.

Table 2. Summary of the chlorophyll fluorescence profiles in the database presented in the study depending on the different modes of data acquisition.

Acquisition	Number of fluorescence profiles	Percentage of data in the database
CTD	27 433	56.4 %
OSD	10 831	22.3 %
UOR	2952	6 %
AP	7384	15.2 %
Total	48 600	

2.2 Quality control

In order to use the FLAVOR method (see details in Sect. 2.3), a specific and adapted data quality control procedure was developed and applied to each in situ chlorophyll fluorescence profile. This procedure was schematically implemented according to four main steps of data control (Fig. 1), each step being developed for discarding most, if not all, spurious fluorescence profiles that would deteriorate the quality of the database. Firstly, several basic tests were applied: (1) duplicates and single-surface values, which are not vertical profiles, were removed (these profiles were removed from the beginning of the process so they are not included in the socalled raw database); (2) coastal profiles were removed using

Table 3. Summary of the number of fluorescence profiles rejected at each step of quality control.

QC step number (see Fig. 1)	Number of fluorescence profiles deleted	% of data deleted
2	162 609	74 %
3	31 904	14.5 %
4	15 396	7 %
5	286	0.1 %
6	3569	1.6 %
7	2891	1.3 %
8	1597	0.7 %
9 – Visual check	244	0.1 %
Chauvenet's criterion and range criterion after calibration (see Sect. 2.3)	1031	0.5 %

a bathymetric mask of 500 m depth; (3) the uppermost measurement has to be located within the 0–10 m layer, while the deepest measurement has to be at or below 100 m. Secondly, tests on the profile vertical resolution are applied: (4) a minimum of 10 values per profile is required (i.e., condition on the vertical resolution acquisition); (5) a minimum of five non-equal values per profile are required (i.e., condition on the sensor resolution). Then, several tests are applied on the fluorescence profile shape. These conditions are based on the parameter used for the development of the FLAVOR method, Z_0, which is the depth at which the fluorescence profile returns to a constant background value (see details in Sect. 2.3 and examples in Fig. 1b and c). (6) The median of the fluorescence values from the surface down to 20 m has to be greater than the median of the values of the last 10 % of the deepest samples of the profile (see Fig. 1a); (7) the depth Z_0 has not to be within the last 10 % of the deepest samples of the profile (see Fig. 1b). Finally, a test on the noise of the profiles was developed and applied: (8) profiles with aberrant data caused by electronic noise are removed (i.e., variability greater than 20 % of the total profile range, see Fig. 1c). To finish, a visual check allowed all the remaining fluorescence profiles to be verified. The number of raw fluorescence profiles rejected at each step of the quality control procedure is presented in Table 3. Around 80 % of the raw fluorescence profiles were thus removed by this procedure. This step is an essential prerequisite for the development of a "clean" database of vertical distributions of phytoplankton biomass and community composition in the global ocean. The quality control procedure removed 77, 71, 28 and 25 % of the OSD, UOR, AP and CTD profiles, respectively, with profiles removed by the test on the bathymetry not taken into account.

2.3 Conversion of chlorophyll fluorescence into chlorophyll a concentration and phytoplankton community composition

In order to assess the vertical distribution of the total chlorophyll a concentration (hereafter, [TChl]) and the chlorophyll a concentration associated to each phytoplankton size index (hereafter, [microChl], [nanoChl] and [picoChl] for microphytoplankton, nanophytoplankton and picophytoplankton respectively), the FLAVOR method (Sauzède et al., 2015a) is applied to each chlorophyll fluorescence profile, satisfying the quality control procedure (see Sect. 2.2). In summary, FLAVOR is a neural network-based method which uses (1) the shape of the chlorophyll fluorescence profile (10 values from the normalized profile with values range between 0 and 1); (2) the depth Z_0, which is the depth at which the fluorescence profile returns to a constant background value (see examples of Z_0 depths represented by the horizontal red line for two profiles on Fig. 1b and c); and (3) the location (latitude and longitude) and the day of acquisition of the fluorescence profile as inputs. The outputs of FLAVOR are the vertical distributions of (1) [TChl] and (2) [microChl], [nanoChl] and [picoChl] with the same vertical resolution as the input raw fluorescence profile. FLAVOR is composed of two different neural networks: the first one was adapted to retrieve the vertical distribution of [TChl] and the second one to retrieve the vertical distributions of [microChl], [nanoChl] and [picoChl] simultaneously. Both neural networks were adapted and validated using a large database including 896 concomitant in situ vertical profiles of HPLC pigments and chlorophyll fluorescence. These profiles were collected as part of 22 oceanographic cruises representative of the global ocean in terms of trophic and oceanographic conditions, making the method applicable to most oceanic waters. The diagnostic pigment-based approach of Uitz et al. (2006), based on Claustre (1994) and Vidussi et al. (2001), was utilized to estimate the biomass associated with the three pigment-derived size classes for each profile. Finally, the data set of concurrent fluorescence profiles and HPLC-determined [TChl], [microChl], [nanoChl] and [picoChl] at discrete depths was used to establish the neural network-based relationships between the fluorescence profile shape and the vertical distributions of [TChl] and phytoplankton community. The schematic overview of the FLAVOR method is shown on Fig. 4 in Sauzède et al. (2015a). The global absolute errors of FLAVOR retrievals are 40, 46, 35 and 40 % for the [TChl], [microChl], [nanoChl] and [picoChl], respectively (Sauzède et al., 2015a).

Admittedly, the FLAVOR method has some limitations. The dependence of chlorophyll fluorescence on the light environment is probably intrinsically accounted for in the algorithm thanks to the geolocation and date of acquisition used as inputs for the training. However, one of the potential concerns with FLAVOR is that the impact of the daytime non-photochemical quenching (NPQ; see, e.g., Cullen

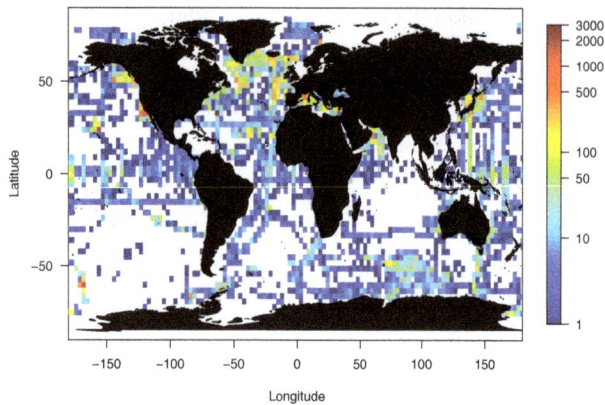

Figure 2. Geographic distribution of the 48 600 chlorophyll fluorescence profiles in the database that passed through all the steps of the quality control procedure. The color scale indicates the number of fluorescence profiles in boxes of 3° per 3°.

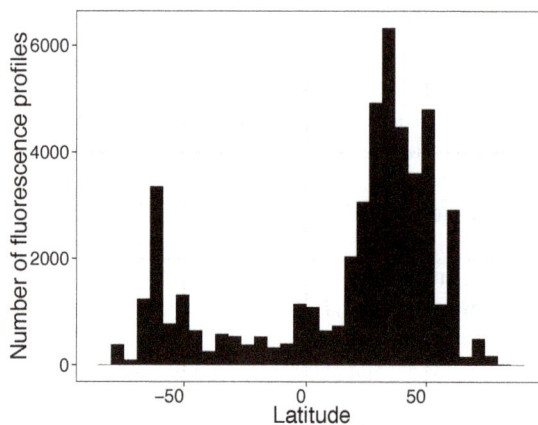

Figure 3. Frequency distribution of the 48 600 profiles of chlorophyll *a* concentration and associated phytoplankton community composition in the database as a function of latitude.

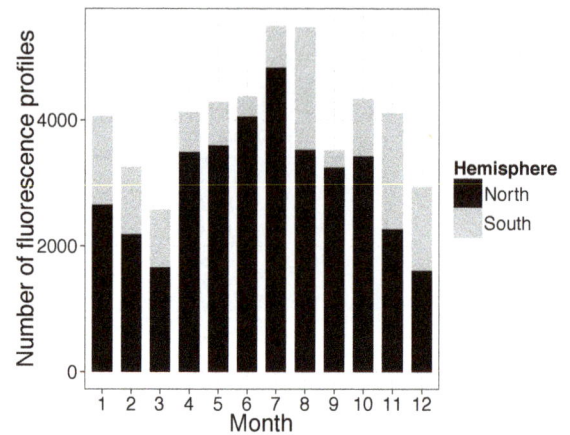

Figure 4. Temporal distribution of the 48 600 profiles of chlorophyll *a* concentration and associated phytoplankton community composition in the database as a function of months with black and gray colors, indicating the hemispheres of data acquisition.

and Lewis, 1995), responsible for a decrease in chlorophyll fluorescence values at high irradiance, is not accounted for by the method. The NPQ uncorrected fluorescence profile shape is indeed used to retrieve the vertical distribution of phytoplankton biomass (see details in Sauzède et al., 2015a). Note that, if density profiles are available together with fluorescence profiles, NPQ can be corrected using the method of Xing et al. (2012). This method involves substituting the fluorescence values acquired within the mixed layer by the maximum value within this layer.

It has been previously mentioned that FLAVOR is not adapted for the retrieval of chlorophyll *a* concentration on a fluorescence profile-by-profile basis (Sauzède et al., 2015a). Rather, FLAVOR and, hence, the resulting database, are relevant for large-scale investigations, e.g., development of climatologies of the vertical distribution of chlorophyll *a*, from which regional anomalies or temporal trends might be ev-

idenced. In fact, the method was validated using a global database and it is not excluded that the retrievals from FLAVOR might be regionally biased. For instance, Sauzède et al. (2015a) have shown that FLAVOR retrievals for the Southern, Arctic and Indian oceans are slightly less accurate than for the other basins. This is likely because the method is not constrained enough in these specific areas which are known for data scarcity. Additional details about the performance of the method for various oceanic basins are given in Sauzède et al. (2015a), in Figs. S3, S5–S7. Finally, it is worth recalling here that the relationships between the phytoplankton biomass or community composition profiles and the fluorescence profiles are assumed to be identical for profiles acquired before 1991 (not involved in the training data set because of lack of HPLC data) and after 1991 (only used for the training process). In the context of possible use of this database for supporting analysis in looking for trends or a shift in chlorophyll *a* time series, this assumption will have to be taken into consideration.

An additional step of quality control is further applied once the FLAVOR method has been operated. It is based on Chauvenet's criterion which is used to identify statistical outliers in the retrieved biomass data (Buitenhuis et al., 2013; Glover et al., 2011; O'Brien et al., 2013). The criterion was applied to the surface data of each profile (median of values from the surface down to 20 m). As Chauvenet's criterion is based on the assumption that the data follow a normal distribution, the analysis was performed on the log-normalized [TChl] surface values. Such a criterion removes aberrant data partially caused by the failure of the FLAVOR method (see number of profiles removed by Chauvenet's criterion in Table 3).

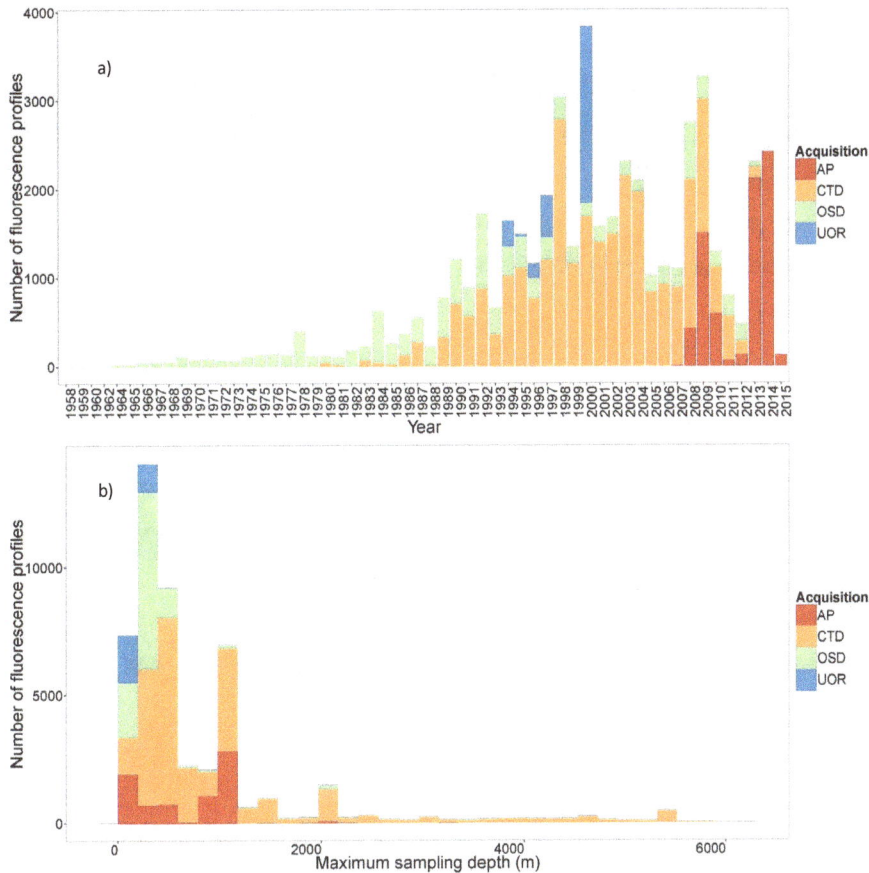

Figure 5. Frequency distribution of the 48 600 profiles of chlorophyll *a* concentration and associated phytoplankton community composition in the database as a function of: **(a)** years of acquisition and **(b)** the maximum depth of acquisition. Colors refer to the different modes of data acquisition.

3 Results and discussion

3.1 Spatial and temporal coverage of the database

The 48 600 chlorophyll fluorescence profiles which successfully passed all the steps of quality control were transformed into total chlorophyll *a* concentration and associated phytoplankton community size indices (i.e., microphytoplankton, nanophytoplankton and picophytoplankton) using FLAVOR (see details in Sect. 2.3). The resulting database covers all ocean basins with more profiles in the Northern Hemisphere (75 %) than in the Southern Hemisphere (25 %, see Figs. 2 and 3). However, the Southern Hemisphere remains relatively well represented with the profiles acquired by autonomous platforms and especially by elephant seals equipped with a fluorometer. Few data were acquired in the Indian Ocean and in the tropical South Atlantic and South Pacific (see Fig. 2). The highest numbers of fluorescence profiles are found at the BATS (the Bermuda Atlantic Time-series Study) and HOT (the Hawaii Ocean Time-series) time-series stations, which are located at 31.67° N–64.17° W and 22.75° N–158.00° W, respectively, and where data acquisi-

tion started in 1988. On the annual scale, the data acquisition appears evenly distributed, with a slight underrepresentation of autumn months (April to June) in the Southern Hemisphere (Fig. 4). The temporal distribution of fluorescence profiles in the database covers 56 years from 1958 to the present (Fig. 5a) and most of the observations were collected after the late 1980s. There are fewer observations from 2010 to 2012 because all data generally acquired by ship-based platforms have not been archived yet in the online databases. A significant increase in data density observed between 2013 and 2015 (in 2015, 124 profiles were acquired in half a month) mainly results from data acquired by Bio-Argo profiling floats. Around one-sixth of this global database has been sampled in only 2 years by the Bio-Argo platforms. This illustrates the potential of this new type of acquisition which is expected to dramatically increase the number of collected fluorescence profiles in the future.

Vertically, the database includes values of total chlorophyll *a* concentration and associated phytoplankton community composition from the surface down to a mean sampling

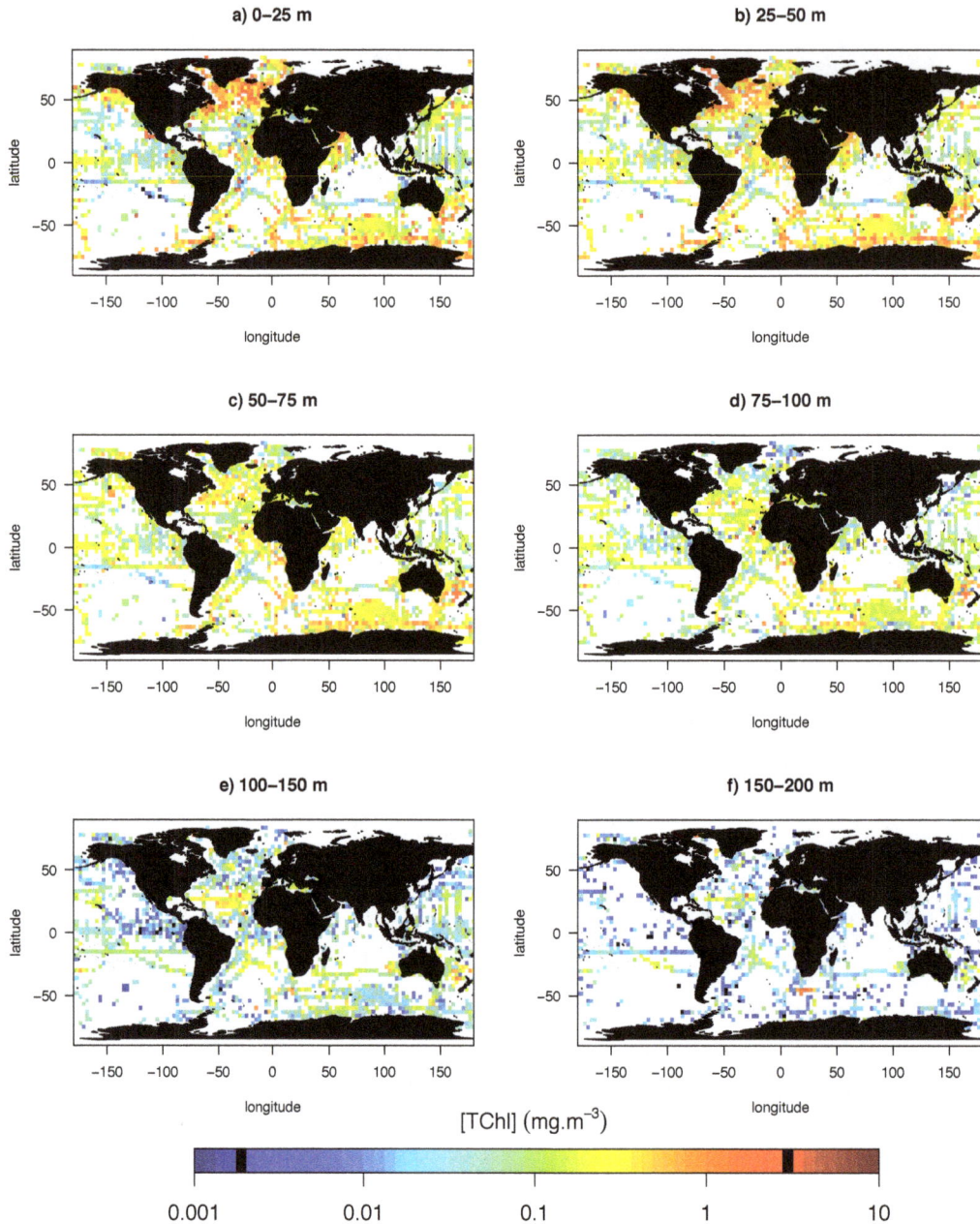

Figure 6. Median total chlorophyll *a* concentration (mg m^{-3}) scaled to a 3° spatial resolution for six vertical layers: (**a**) 0–25 m, (**b**) 25–50 m, (**c**) 50–75 m, (**d**) 75–100 m, (**e**) 100–150 m and (**f**) 150–200 m.

depth of 743 m (with a maximum sampling depth ranging from 100 to 6000 m; Fig. 5b).

3.2 Vertical distribution of the chlorophyll biomass

We present the database with respect to the vertical distribution of the total chlorophyll *a* concentration ([TChl]). Figure 6 displays the median [TChl] gridded within squares of 3° latitude by 3° longitude and over six vertical layers (0–25, 25–50, 50–75, 75–100, 100–150 and 150–200 m). In the sur-

face layer (0–25 m, see Fig. 6a), the [TChl] median is the highest in the North Atlantic and the lowest in the South Pacific subtropical gyre. The median [TChl] decreases with depth for all the data, except for data acquired in South Pacific and Atlantic subtropical gyres where the median [TChl] increases with depth. This increase is associated with the so-called deep chlorophyll maximum (DCM) that is typical of these oligotrophic regions (e.g., Cullen, 1982; Mignot et al., 2011, 2014).

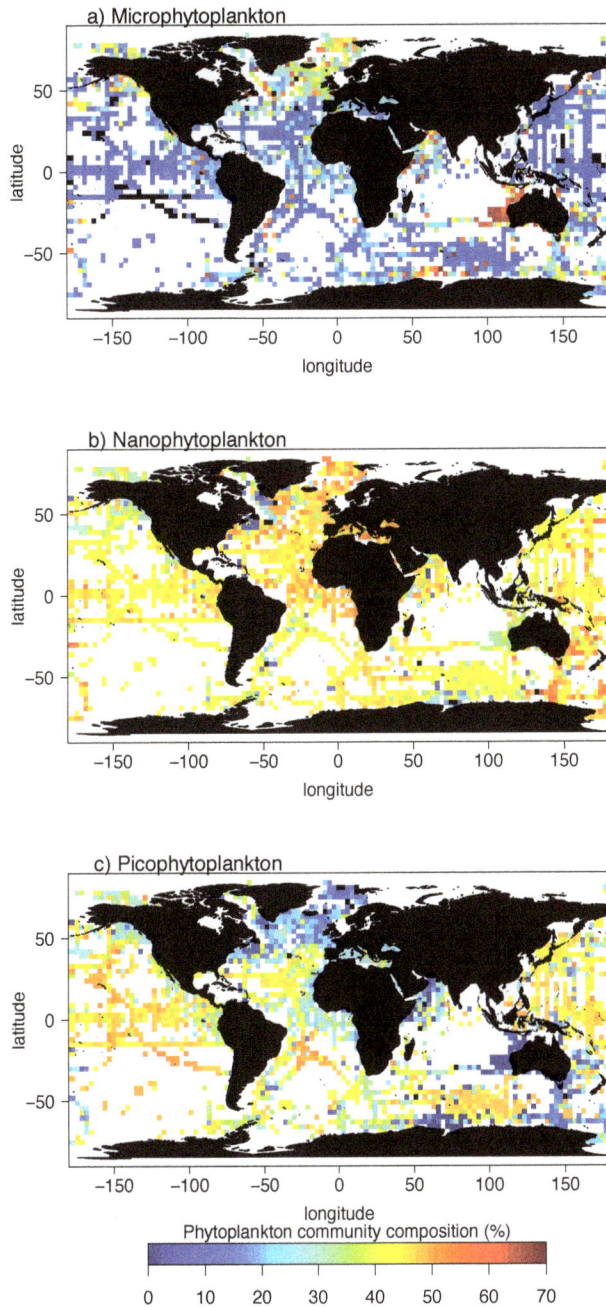

Figure 7. Mean relative contribution to the total chlorophyll *a* biomass (%) for the three phytoplankton size-based groups gridded and scaled to a 3° resolution within the 0–1.5 Z_e layer: (**a**) microphytoplankton, (**b**) nanophytoplankton and (**c**) picophytoplankton.

The global distribution of the phytoplankton community composition, given in terms of fraction of chlorophyll *a* concentration associated to micro-, nano- and picophytoplankton, is presented for the 0–1.5 Z_e layer (Fig. 7a, b and c respectively). Here Z_e, the euphotic depth is defined as the depth at which the irradiance is reduced to 1 % of its surface value. It was estimated according to the method of Morel and

Berthon (1989), using the [TChl] profiles derived from FLA-VOR. Figure 7 reveals general geographic patterns which are consistent with the knowledge about the ecological domains and biogeochemical provinces (Longhurst, 2010). On average microphytoplankton are dominant in the subarctic zone, with a relative contribution to the chlorophyll biomass reaching more than 70 % in these areas (Fig. 7a). Picophytoplankton are dominant in the subtropical gyres (South and North Pacific as well as South and North Atlantic), with a contribution reaching 45–55 % (Fig. 7c). Nanophytoplankton appear to be ubiquitous with a relatively stable contribution to biomass of 40–50 % (Fig. 7b).

To further assess the quality, range and representation of the FLAVOR-retrieved [TChl] database presented in this study, the retrieved surface [TChl] is compared to the remotely sensed [TChl]. In this context, the climatological [TChl] mean was extracted at a 9 km spatial resolution from NASA Modis Aqua archive for the time period covering 2002 to 2014. The extracted satellite [TChl] data were re-gridded to a $3° \times 3°$ spatial resolution. Similarly the FLAVOR-retrieved [TChl] values for the upper layer of the database (i.e., mean value calculated between the surface and 20 m) for the same period were re-gridded to $3° \times 3°$ squares. Figures 8 and 9 show that climatological averaged [TChl] from Modis Aqua and from the present database are generally consistent (Fig. 8a and b). The log-transformed ratio of the Modis Aqua to the database [TChl] estimates reveals a rather good agreement with a median value of −0.16 and a standard deviation of 0.58 (see histogram in Fig. 8c). Figure 9 displays the geographic distribution of the log-transformed ratio between the Modis Aqua and the database estimates of climatological surface [TChl]. The ratio shows no specific spatial bias. However, as it is mentioned in Sect. 2.3, FLA-VOR retrievals for the Southern, Arctic and Indian oceans are slightly less accurate than for the other basins; it is therefore possible that the estimation errors are greater in these areas. Moreover, this observation has to be nuanced considering the difficulties in retrieving accurate ocean color satellite [TChl] in these high-latitude environments (Gregg and Casey, 2004; Guinet et al., 2013; Johnson et al., 2013; Peloquin et al., 2013; Siegel et al., 2005; Szeto et al., 2011).

3.3 Example of application: climatological time series of the vertical distribution of chlorophyll *a* concentration and phytoplankton community composition

As an example of application, monthly climatologies were computed for three ecological provinces defined by Longhurst (2010) and well represented in the current data set (Fig. 10a): (1) the North Atlantic Subtropical Gyral Province West (NASW, Fig. 10b), (2) the Atlantic Subarctic Province (SARC, Fig. 10c) and (3) the North Pacific Subtropical Gyre Province (NPTG, Fig. 10d). Overall the time series of the vertical distribution in [TChl] are consistent

a) Modis Aqua mean

b) Database mean

[TChl] (mg.m^{-3})

c) Histogram

mean=−0.14
median=−0.13
sd=0.53

$\log_{10}(Chl_{surf}/Chl_{sat})$

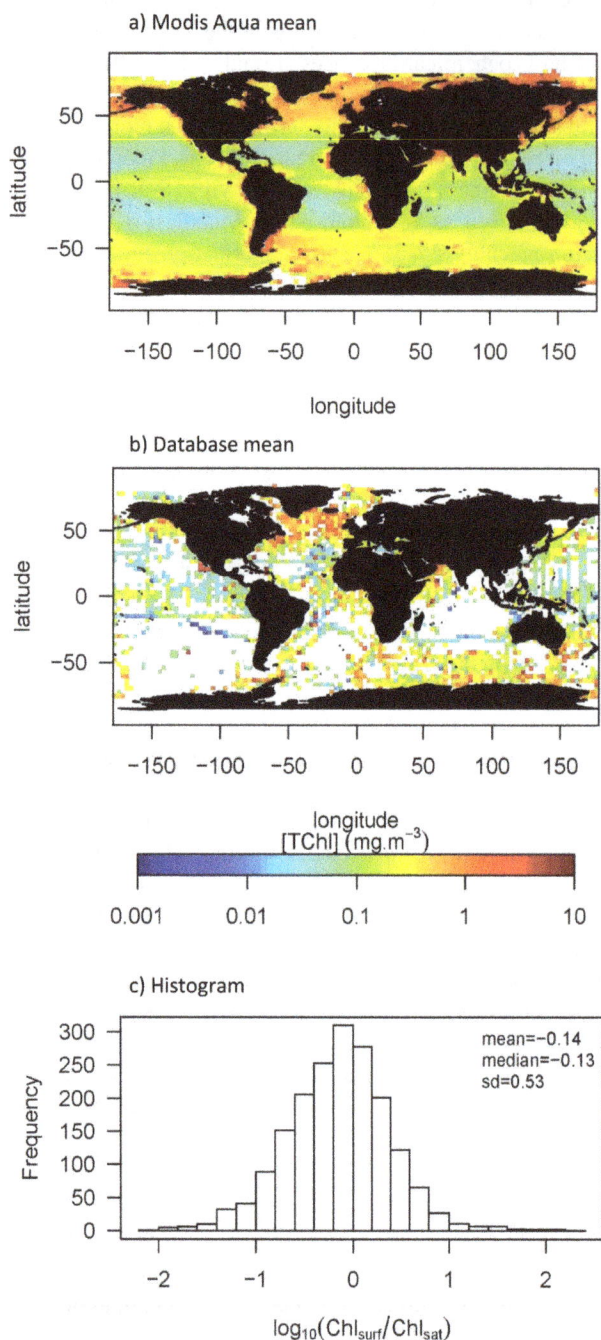

Figure 8. (a) Climatological mean (2002–2014) chlorophyll *a* concentration (mg m^{-3}) from Modis Aqua (scaled to a 3° resolution); (b) climatological mean (1958–2015) surface chlorophyll *a* concentration (mg m^{-3}) from the present database (averaged over the upper 20 m and scaled to a 3° resolution); (c) histogram of the log10 ratio of the chlorophyll *a* concentration from the database to the chlorophyll *a* concentration from Modis Aqua. The mean, median and standard deviation of the ratio are indicated in the figure. The color scale applies to panels (**a**) and (**b**).

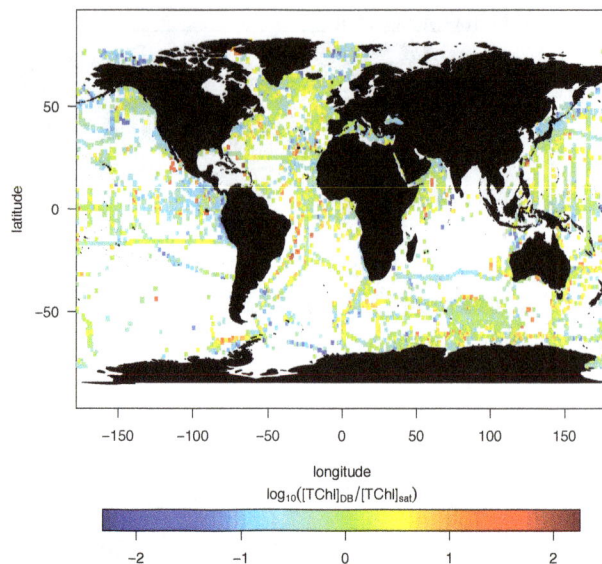

$\log_{10}([TChl]_{DB}/[TChl]_{sat})$

Figure 9. Geographic distribution of the log10-transformed ratio of the climatological mean surface [TChl] of the database over the upper 20 m of the water column ([TChl]$_{DB}$) and the climatological mean satellite [TChl] from Modis Aqua ([TChl]$_{sat}$). Both [TChl] data were scaled to a 3° resolution.

with expectations as detailed by Longhurst (2010). For the NASW province (Fig. 10b), [TChl] is relatively homogeneous from the surface to around 140 m from January to March; then the stratification of the water column leads to the establishment of a deep chlorophyll maximum (DCM) from April to November. Over the year, [TChl] varies in a restricted range of values (0.35–0.55 mg m^{-3}). The dominant phytoplankton groups are the nano- and the picophytoplankton with relative chlorophyll contribution reaching 40–45 % for both size-based groups. The contribution of microphytoplankton remains low throughout the year (10 %). For the SARC province, the phytoplankton bloom starts in May (as indicated by a significant increase in [TChl], Fig. 10c). The bloom continues for 4 to 5 months with [TChl] within the 1.5–2 mg m^{-3} range (with maximum values in July). The microphytoplankton contribution increases during the bloom and reaches a maximum (60 %) in August, whereas the nanophytoplankton relative contribution decreases from April to August. The contribution of picophytoplankton increases slightly all along the year to reach a maximum of about 40 % in December. For the NPTG province (Fig. 10d), a DCM (0.15–0.25 mg m^{-3}) is established at a depth of 100–125 m and persists all year long. This DCM deepens in summer, consistently with a deeper light penetration in the water column at this period. The [TChl] at DCM reaches a maximum value in June and July. The dominant phytoplankton groups are the nano- and the picophytoplankton with relative contribution reaching 45–50 % for both size-based groups and slight opposite temporal evolutions. The

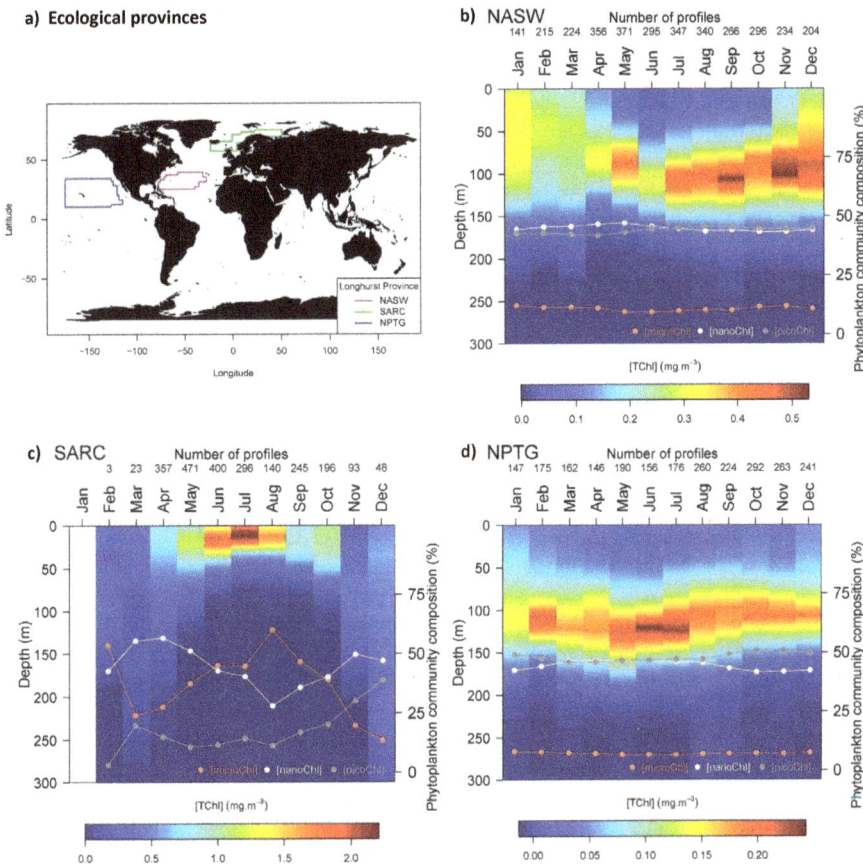

Figure 10. Monthly climatologies of the vertical distribution of the total chlorophyll *a* concentration and associated phytoplankton size-based groups for three ecological provinces defined by Longhurst (2010). (**a**) Geographic distribution of the considered provinces: North Atlantic Subtropical Gyral Province West (NASW), Atlantic Subarctic Province (SARC) and North Pacific Subtropical Gyre Province (NPTG). Climatologies obtained for the (**b**) NASW, (**c**) SARC and (**d**) NPTG. The color scale indicates the total chlorophyll *a* concentration (mg m^{-3}); the data points superimposed onto the colored monthly vertical profiles show the percentages of integrated chlorophyll *a* concentration associated with the micro-, nano- and picophytoplankton within the water column.

contribution of microphytoplankton remains low throughout the year ($< 10\%$).

4   Conclusions and recommendations for use

The phytoplankton biomass (i.e., chlorophyll *a* concentration) and phytoplankton community size indices were derived from chlorophyll fluorescence profiles using a dedicated calibration method (FLAVOR, Sauzède et al., 2015a). For the first time, in situ chlorophyll fluorescence profiles from various data centers have been collected and synthesized in a global data set to create unified and interoperable products related to chlorophyll *a* concentration and phytoplankton communities. This work can thus be considered as a first step towards the development of a 3-D climatological representation of chlorophyll *a* concentration and phytoplankton community composition. As mentioned before, we recall here that this database should not be used on a profile-by-profile basis. Instead, this database has rather to be used to derive climatologies from which regional or temporal trends might possibly be extracted. To date, and because of the lack of in situ vertical data, the identification of such trends has been based exclusively on surface remotely sensed data (Beaulieu et al., 2013; Boyce et al., 2010; Gregg, 2005; Gregg et al., 2002). Obviously, the present data set offers a potential refinement to improve open-ocean climatologies of chlorophyll *a* with respect to the vertical dimension.

Finally, this database has to be considered as a reference that has the potential to evolve. It is now clear that numerous fluorescence profiles will be acquired through robotic observations (e.g., Claustre et al., 2010b; Johnson et al., 2009). In fact, about one-sixth of the profiles of the present database have been sampled by Bio-Argo profiling floats in only 2 years. Therefore the database proposed here represents a first step towards a global single-reference database reconciling the oldest data sets of chlorophyll fluorescence with the future ones, mostly acquired remotely by autonomous platforms.

Acknowledgements. This paper is a contribution to the Remotely Sensed Biogeochemical Cycles in the Ocean (remOcean) project, funded by the European Research Council (grant agreement 246777), to the French Bio-Argo project funded by CNES-TOSCA and to the French "Equipement d'avenir" NAOS project (ANR J11R107-F). The French PROOF and CYBER programs are acknowledged for their support of cruises where in situ chlorophyll fluorescence profiles were acquired. We are grateful to all the project PIs who contributed data, as well as to the anonymous staff who took part in the data acquisition during the cruises.

Edited by: F. Schmitt

References

Beaulieu, C., Henson, S. A., Sarmiento, Jorge L., Dunne, J. P., Doney, S. C., Rykaczewski, R. R., and Bopp, L.: Factors challenging our ability to detect long-term trends in ocean chlorophyll, Biogeosciences, 10, 2711–2724, doi:10.5194/bg-10-2711-2013, 2013.

Boyce, D. G., Lewis, M. R., and Worm, B.: Global phytoplankton decline over the past century, Nature, 466, 591–596, doi:10.1038/nature09268, 2010.

Boyer, T. P., Antonov, J. I., Baranova, O. K., Garcia, H. E., Johnson, D. R., Locarnini, R. A., Mishonov, A. V, O'Brien, T. D., Seidov, D., Smolyar, I. V., and Zweng, M. M.: World Ocean Database 2009, edited by: Levitus, S., 2009.

Buitenhuis, E. T., Vogt, M., Moriarty, R., Bednaršek, N., Doney, S. C., Leblanc, K., Le Quéré, C., Luo, Y.-W., O'Brien, C., O'Brien, T., Peloquin, J., Schiebel, R., and Swan, C.: MAREDAT: towards a world atlas of MARine Ecosystem DATa, Earth Syst. Sci. Data, 5, 227–239, doi:10.5194/essd-5-227-2013, 2013.

Claustre, H.: The trophic status of various oceanic provinces as revealed by phytoplankton pigment signatures, Limnol. Oceanogr., 39, 1206–1210, 1994.

Claustre, H., Hooker, S. B., Van Heukelem, L., Berthon, J.-F., Barlow, R., Ras, J., Sessions, H., Targa, C., Thomas, C. S., van der Linde, D., and Marty, J.-C.: An intercomparison of HPLC phytoplankton pigment methods using in situ samples: application to remote sensing and database activities, Mar. Chem., 85, 41–61, doi:10.1016/j.marchem.2003.09.002, 2004.

Claustre, H., Antoine, D., Boehme, L., Boss, E., D'Ortenzio, F., D'Andon, O. F., Guinet, C., Gruber, N., Handegard, N. O., Hood, M., Johnson, K., Körtzinger, A., Lampitt, R., LeTraon, P. Y., Lequéré, C., Lewis, M., Perry, M. J., Platt, T., Roemmich, D., Testor, P., Sathyendranth, S., Send, U., and Yoder, J.: Guidelines towards an integrated ocean observation system for ecosystems and biogeochemical cycles, in: Proceedings of the OceanObs 09: Sustained Ocean Observations and Information for Society Conference (Vol.1), edited by: Hall, J., Harrison, D. E., and Stammer, D., ESA Publ., Venice, Italy, 2010a.

Claustre, H., Bishop, J., Boss, E., Bernard, S., Johnson, K., Lotiker, A., Ulloa, O., Perry, M. J., Uitz, J., Curie, M., Villefranche, D., Lazaret, C., Division, E. S., Berkeley, L., Road, O. C., Observation, E., Africa, S., Fermi, V., De Brest, C., Valley, O., and Jolla, L.: Bio-optical profiling floats as new observational tools for biogeochemical and ecosystem studies: potential synergies with ocean color remote sensing, in: Proceedings of the OceanObs 09: Sustained Ocean Observations and Information for Society Conference (Vol.2), edited by: Hall, J., Harrison, D. E., and Stammer, D., ESA Publ., Venice, Italy, 2010b.

Conkright, M. E., Locarnini, R. A., Garcia, H. E., O'Brien, T. D., Boyer, T. P., Stephens, C., and Antonov, J. I.: World Ocean Atlas 2001: Objective analyses, data statistics, and figures: CD-ROM documentation, US Department of Commerce, National Oceanic and Atmospheric Administration, National Oceanographic Data Center, Ocean Climate Laboratory, 2002.

Cullen, J. J.: The deep chlorophyll maximum: comparing vertical profiles of chlorophyll a, Can. J. Fish. Aquat. Sci., available at: http://agris.fao.org/agris-search/search.do?recordID=US201302185941 (last access: 23 February 2015), 1982.

Cullen, J. J. and Lewis, M. R.: Biological processes and optical measurements near the sea surface: Some issues relevant to remote sensing, J. Geophys. Res., 100, 13255, doi:10.1029/95JC00454, 1995.

Cunningham, A.: Variability of in-vivo chlorophyll fluorescence and its implication for instrument development in bio-optical oceanography, Sci. Mar., 60, 309–315, 1996.

Falkowski, P., Kiefer, D. A., Division, O. S., and Angeles, L.: Chlorophyll a fluorescence in phytoplankton: relationship to photosynthesis and biomass, J. Plankton Res., 7, 715–731, 1985.

Glover, D. M., Jenkins, W. J., and Doney, S. C.: Modeling Methods for Marine Science, Cambridge University Press, Cambridge, 2011.

Gordon, H. R. and McCluney, W. R.: Estimation of the depth of sunlight penetration in the sea for remote sensing, Appl. Optics, 14, 413–416, doi:10.1364/AO.14.000413, 1975.

Gregg, W. W.: Recent trends in global ocean chlorophyll, Geophys. Res. Lett., 32, L03606, doi:10.1029/2004GL021808, 2005.

Gregg, W. W. and Casey, N. W.: Global and regional evaluation of the SeaWiFS chlorophyll data set, Remote Sens. Environ., 93, 463–479, doi:10.1016/j.rse.2003.12.012, 2004.

Gregg, W. W., Conkright, M. E., Atlantic, N., Chlorophyll, B. O., Indian, S., Pacific, S., and Atlantic, S.: Decadal changes in global ocean chlorophyll, Geophys. Res. Lett., 29, 1–4, 2002.

Guinet, C., Xing, X., Walker, E., Monestiez, P., Marchand, S., Picard, B., Jaud, T., Authier, M., Cotté, C., Dragon, A. C., Diamond, E., Antoine, D., Lovell, P., Blain, S., D'Ortenzio, F., and Claustre, H.: Calibration procedures and first dataset of Southern Ocean chlorophyll a profiles collected by elephant seals equipped with a newly developed CTD-fluorescence tags, Earth Syst. Sci. Data, 5, 15–29, doi:10.5194/essd-5-15-2013, 2013.

Johnson, K., Berelson, W., Boss, E., Chase, Z., Claustre, H., Emerson, S., Gruber, N., Kortzinger, A., Perry, M., and Riser, S.: Observing Biogeochemical Cycles at Global Scales With Profiling Floats and Gliders Prospects for a Global Array, Oceanography, 22, 216–225, 2009.

Johnson, R., Strutton, P. G., Wright, S. W., McMinn, A., and Meiners, K. M.: Three improved satellite chlorophyll algorithms for the Southern Ocean, J. Geophys. Res.-Ocean., 118, 3694–3703, doi:10.1002/jgrc.20270, 2013.

Kiefer, D. A.: Chlorophyll a fluorescence in marine centric diatoms: Responses of chloroplasts to light and nutrient stress, Mar. Biol., 23, 39–46, doi:10.1007/BF00394110, 1973.

Le Quere, C., Harrison, S., Prentice, I., Buitenhuis, E., Aumont, O., Bopp, L., Claustre, H., de Cunha, L., Geider, R., Giraud, X., Klaas, C., Kohfield, K., Legendre, L., Manizza, M., Platt, T., Rivkin, R., Sathyendranath, S., Uitz, J., Watson, A., and Wolf-Gladrow, D.: Ecosystem dynamics based on plankton functional types for global ocean biogeochemistry models, Glob. Chang. Biol., available at: https://ueaeprints.uea.ac.uk/27956/ (last access: 19 March 2014), 2005.

Levitus, S., Antonov, J. I., Baranova, O. K., Boyer, T. P., Coleman, C. L., Garcia, H. E., Grodsky, A. I., Johnson, D. R., Locarnini, R. A., Mishonov, A. V., Reagan, J. R., Sazama, C. L., Seidov, D., Smolyar, I., Yarosh, E. S., and Zweng, M. M.: The World Ocean Database, Data Sci. J., 12, WDS229–WDS234, 2013.

Longhurst, A. R.: Ecological Geography of the Sea, available at: http://books.google.com/books?hl=fr&lr=&id=QdJZezzrCfQC&pgis=1 (last access: 22 January 2015), 2010.

Lorenzen, C. J.: A method for the continuous measurement of in vivo chlorophyll concentration, Deep Sea Res. Oceanogr. Abstr., 13, 223–227, doi:10.1016/0011-7471(66)91102-8, 1966.

McClain, C. R.: A decade of satellite ocean color observations, Ann. Rev. Mar. Sci., 1, 19–42, doi:10.1146/annurev.marine.010908.163650, 2009.

Mignot, A., Claustre, H., D'Ortenzio, F., Xing, X., Poteau, A., and Ras, J.: From the shape of the vertical profile of in vivo fluorescence to Chlorophyll-a concentration, Biogeosciences, 8, 2391–2406, doi:10.5194/bg-8-2391-2011, 2011.

Mignot, A., Claustre, H., Uitz, J., Poteau, A., D'Ortenzio, F., and Xing, X.: Understanding the seasonal dynamics of phytoplankton biomass and the deep chlorophyll maximum in oligotrophic environments: A Bio-Argo float investigation, Global Biogeochem. Cycles, 28, 856–876, doi:10.1002/2013GB004781, 2014.

Morel, A. and Berthon, J.-F.: Surface Pigments, Algal Biomass Profiles, and Potential Production of the Euphotic Layer: Relationships Reinvestigated in View of Remote-Sensing Applications, Limnol. Oceanogr., 34, 1545–1562, 1989.

O'Brien, C. J., Peloquin, J. A., Vogt, M., Heinle, M., Gruber, N., Ajani, P., Andruleit, H., Arístegui, J., Beaufort, L., Estrada, M., Karentz, D., Kopczynska, E., Lee, R., Poulton, A. J., Pritchard, T., and Widdicombe, C.: Global marine plankton functional type biomass distributions: coccolithophores, Earth Syst. Sci. Data, 5, 259–276, doi:10.5194/essd-5-259-2013, 2013.

Peloquin, J., Swan, C., Gruber, N., Vogt, M., Claustre, H., Ras, J., Uitz, J., Barlow, R., Behrenfeld, M., Bidigare, R., Dierssen, H., Ditullio, G., Fernandez, E., Gallienne, C., Gibb, S., Goericke, R., Harding, L., Head, E., Holligan, P., Hooker, S., Karl, D., Landry, M., Letelier, R., Llewellyn, C. A., Lomas, M., Lucas, M., Mannino, A., Marty, J.-C., Mitchell, B. G., Muller-Karger, F., Nelson, N., O'Brien, C., Prezelin, B., Repeta, D., Jr. Smith, W. O., Smythe-Wright, D., Stumpf, R., Subramaniam, A., Suzuki, K., Trees, C., Vernet, M., Wasmund, N., and Wright, S.: The MAREDAT global database of high performance liquid chromatography marine pigment measurements, Earth Syst. Sci. Data, 5, 109–123, doi:10.5194/essd-5-109-2013, 2013.

Sauzède, R., Claustre, H., Jamet, C., Uitz, J., Ras, J., Mignot, A., and D'Ortenzio, F.: Retrieving the vertical distribution of chlorophyll a concentration and phytoplankton community composition from in situ fluorescence profiles: A method based on a neural network with potential for global-scale applications, J. Geophys. Res.-Ocean., 120, 451–470, doi:10.1002/2014JC010355, 2015a.

Sauzède, R., Lavigne, H., Claustre, H., Uitz, J., Schmechtig, C., D'Ortenzio, F., Guinet, C., and Pesant, S.: Global compilation of chlorophyll fluorescence profiles from several data bases, and from published and unpublished individual sources, available at: http://doi.pangaea.de/10.1594/PANGAEA.844212, last access: 19 March 2015b.

Sauzède, R., Lavigne, H., Claustre, H., Uitz, J., Schmechtig, C., D'Ortenzio, F., Guinet, C., and Pesant, S.: Global data product of chlorophyll a concentration and phytoplankton community composition (microphytoplankton, nanophytoplankton and picophytoplankton) computed from in situ fluorescence profiles, available at: http://doi.pangaea.de/10.1594/PANGAEA.844485, last access: 19 March 2015c.

Siegel, D. A., Maritorena, S., Nelson, N. B., and Behrenfeld, M. J.: Independence and interdependencies among global ocean color properties: Reassessing the bio-optical assumption, J. Geophys. Res., 110, C07011, doi:10.1029/2004JC002527, 2005.

Siegel, D. A., Behrenfeld, M. J., Maritorena, S., McClain, C. R., Antoine, D., Bailey, S. W., Bontempi, P. S., Boss, E. S., Dierssen, H. M., Doney, S. C., Eplee, R. E., Evans, R. H., Feldman, G. C., Fields, E., Franz, B. A., Kuring, N. A., Mengelt, C., Nelson, N. B., Patt, F. S., Robinson, W. D., Sarmiento, J. L., Swan, C. M., Werdell, P. J., Westberry, T. K., Wilding, J. G., and Yoder, J. A.: Regional to global assessments of phytoplankton dynamics from the SeaWiFS mission, Remote Sens. Environ., 135, 77–91, doi:10.1016/j.rse.2013.03.025, 2013.

Szeto, M., Werdell, P. J., Moore, T. S., and Campbell, J. W.: Are the world's oceans optically different?, J. Geophys. Res., 116, C00H04, doi:10.1029/2011JC007230, 2011.

Uitz, J., Claustre, H., Morel, A., and Hooker, S. B.: Vertical distribution of phytoplankton communities in open ocean: An assessment based on surface chlorophyll, J. Geophys. Res., 111, C08005, doi:10.1029/2005JC003207, 2006.

Vidussi, F., Claustre, H., Manca, B. B., Luchetta, A., and Marty, J.-C.: Phytoplankton pigment distribution in relation to upper thermocline circulation in the eastern Mediterranean Sea during winter, J. Geophys. Res., 106, 19939, doi:10.1029/1999JC000308, 2001.

Xing, X., Claustre, H., Blain, S., D'Ortenzio, F., Antoine, D., Ras, J., and Guinet, C.: Quenching correction for in vivo chlorophyll fluorescence acquired by autonomous platforms: a case study with instrumented elephant seals in the Kerguelen region (Southern Ocean), Limnol. Oceanogr. Methods, 10, 483–495, doi:10.4319/lom.2012.10.483, 2012.

8

Data compilation of fluxes of sedimenting material from sediment traps in the Atlantic Ocean

S. Torres Valdés[1], S. C. Painter[1], A. P. Martin[1], R. Sanders[1], and J. Felden[2]

[1]Ocean Biogeochemistry and Ecosystems Research Group. National Oceanography Centre. European Way, Southampton, SO14 3ZH, UK
[2]Center for Marine Environmental Sciences. Universität Bremen. Leobener Strasse, POP 330 440, 28359 Bremen, Germany

Correspondence to: S. Torres Valdés (sinhue@noc.ac.uk)

Abstract. We provide a data set assemblage of directly observed and derived fluxes of sedimenting material (total mass, POC, PON, $bSiO_2$, $CaCO_3$, PIC and lithogenic/terrigenous fluxes) obtained using sediment traps. This data assemblage contains over 5900 data points distributed across the Atlantic, from the Arctic Ocean to the Southern Ocean. Data from the Mediterranean Sea are also included. Data were compiled from a variety of sources: data repositories (e.g. BCO-DMO, PANGAEA®), time-series sites (e.g. BATS, CARIACO), published scientific papers and data provided by the originating principal investigators (PIs). All sources are specified within the combined data set. Data from the World Ocean Atlas 2009 were extracted to coincide with flux data to provide additional environmental information where available. Specifically, contemporaneous data were extracted for temperature, salinity, oxygen (concentration, AOU and percentage saturation), nitrate, phosphate and silicate. Data show a broad range of flux estimates, with marked differences between ocean domains. Data also reveal important differences in the contribution that a given variable provides to the total mass flux, which is relevant towards understanding the factors that control the strength of the biological carbon pump.

1 Introduction

The export of particulate organic carbon (POC) from the sunlit upper layers to the ocean interior (deep waters and deep ocean sediments), known as the biological carbon pump (BCP), is an important component of the global carbon cycle. However, the BCP is not well constrained, with estimates of carbon transport ranging from 3.4–4.7 (Eppley and Peterson, 1979) to $\sim 20\,\mathrm{Gt\,C\,yr^{-1}}$ depending on the method used (e.g. Laws et al., 2000). In the North Atlantic alone, estimates of the BCP range fourfold from 0.55 to $1.94\,\mathrm{Gt\,C\,yr^{-1}}$, representing 10–20 % of global export fluxes (Sanders et al., 2014). The uncertainty in these estimates reflects the sparseness of observations and to a smaller extent the variety of methods employed.

Currently, several independently compiled data sets of ex-port flux exist but they reside in separate data repositories as individual data sets. Given the importance of the BCP for carbon storage and the need to further constrain its magnitude, we have attempted to bring together into a single compilation, all data currently available on POC fluxes (and related variables) obtained from sediment-trap deployments in the Atlantic Ocean, which includes the adjacent Arctic Ocean, the Atlantic sector of the Southern Ocean, and the Mediterranean Sea (as an adjacent sea). This data set assemblage was put together as part of the EU FP7 (Seventh Framework Programme) EURO-BASIN (Basin-scale Analysis, Synthesis and INtegration) programme (Work Package 2). Data were obtained from a variety of sources including data repositories (e.g. BCO-DMO, PANGAEA®), time-series sites (e.g. BATS, CARIACO), published scientific papers and directly from the originating PIs. All sources are specified within the data set.

2 Data

Observational studies of export often make use of sediment traps. These are deployed at depths of interest in order to collect particles sinking through the water column. Upon recovery, particles are then analysed primarily to determine their carbon content, but additionally the nitrogen content is also easily obtained. The quantities of ballasting material such as calcium carbonate and opal are also increasingly important variables. The flux rate of material captured by a trap can be calculated from knowledge of the duration a trap is deployed for and the aperture of the trap itself. By obtaining export fluxes at different depths, the efficiency of the carbon pump can be evaluated simply as the fraction of the flux leaving the surface that makes it to a given depth.

2.1 Data sources

The data assemblage presented here includes variables commonly measured in studies concerning export production. These include total mass (Tot_Mass) flux, POC flux, particulate organic nitrogen (PON) flux, biogenic silica (bSiO$_2$) flux, calcium carbonate (CaCO$_3$) flux, particulate inorganic carbon (PIC) flux, and terrigenous or lithogenic (Terr/Litho) material flux. Most data were obtained from data repositories and individual time-series websites. A total of 2679 data points (45 % of total) were derived from 32 smaller data sets obtained from the Data Publisher for Earth and Environmental Science (PANGAEA®) at http://www.pangaea.de; 111 data points (1.9 %) were obtained from the Biological and Chemical Oceanography Data Management Office (BCO-DMO) at http://bcodmo.org (Lee et al., 2009a, b); 1755 data points (29.5 %) were obtained from the Carbon Retention In A Colored Ocean Project Ocean Time Series (CARIACO) at http://www.imars.usf.edu/CAR (Montes et al., 2012); and 784 data points (13.2 %) from the Bermuda Atlantic Time Series (BATS) at http://bats.bios.edu. Data (428 data points, 7.2 %) were also obtained from published journal articles (e.g. Fischer et al., 2000; Bory et al., 2001; Hwang et al., 2009). Finally, a few data sets (190 data points, 3.2 %) were directly obtained from Principal Investigators (Bauerfeind et al., 2009; Lampitt et al., 2010). These combined data sets provide 5947 data points of variables related to export production. In this compilation, a "data point" is an independent measurement or calculation of a given variable. The full data set is organised in columns as indicated in Table 1. The names and acronyms of variables as presented in this table are the most commonly used and least ambiguous terms. Though not all individual data sets contain all the variables listed in Table 1. The total number of data points per variable are presented in Table 2. POC flux contains the largest number of observations (5206 data points) and PIC flux the lowest (1048 data points).

2.2 Methods commonly employed

Across data sets there is considerable inconsistency in the names of variables and reported units. For this compilation we have attempted, to the best of our knowledge, to identify the variable that was actually being dealt with. For instance "biogenic particulate silicon" vs. "biogenic particulate silica", with the first referring to the element silicon (Si) and the second referring to the mineral silica (SiO$_2$). Upon careful examination of such cases, when enough information was available to identify the correct variable, that variable was standardised so that it is internally consistent within this compilation (e.g. all Si related fluxes are consistently reported as SiO$_2$). In cases where available information was insufficient to clarify ambiguity, data sets were excluded from the compilation.

2.2.1 Sediment-trap types

The data include observations from a range of named sediment-trap designs: moored automatic Kiel sediment traps (Bauerfeind et al., 2009; Bauerfeind and Nöthig, 2011), cone-shaped SMT 230 Kiel and Mark VI/V traps (Wefer and Fischer, 1993), cone-shaped multi-sampling SMT 230 KMU traps (Romero et al., 2002; Fahl and Nöthig, 2007), conical particle interceptor traps (Antia et al., 1999), conical sediment McLane Mark-7 traps (Hwang et al., 2009), drifting Technicap PPS 5 sediment traps (Goutx et al., 2000), Kiel HDW traps (Jonkers et al., 2010), large-aperture time-series Kiel-type traps (Fischer et al., 2000, 2002; Iversen et al., 2010), Mark-VII automated sediment trap (CARI-ACO), McLane Mark 78G-21 (Jonkers et al., 2010), multisample moored conical traps (Bory et al., 2001), Parflux Mark 7G-13 time-series sediment trap (Honjo and Manganini, 1993; Jickells, 2003a, b, c, d; Lampitt et al., 2001, 2010), Aquatec Kiel-type sediment trap (Neuer et al., 1997, 2007), indented rotary sphere (IRS) settling velocity and time-series mode sediment traps (Peterson et al., 2005; Goutx et al., 2007; Lee et al., 2009a, b), SMT 234 Aquatec Meerestechnik Kiel trap (Helmke et al., 2005), surface-tethered particle interceptor traps (BATS), and PPS-5 traps (Jonkers et al., 2010). Information is sometimes insufficient to ascertain whether two models are identical. In many studies and within the various data sources, trap specifications are not described. Nevertheless, the specifications of the traps listed above are summarised briefly in Tables 3 and 4.

2.2.2 Sample collection procedure

Before deployment, the collecting cups of sediment traps are filled with ambient seawater. NaCl is typically added to increase the salinity to 40 (Antia et al., 1999;

Table 1. Data set column headers. All fluxes are reported as per day over the period of deployment. We note that only at BATS are samples collected in triplicate. Hence, other than BATS, data under the "replicate 1" and "average" headers contain the same information. Not applicable is denoted NA.

Data column	Label/Variable	Units	Description
1	ID	NA	Data point reference number.
2	Cruise/Project/Area/ STN/ Trap	NA	Cruise reference number or name, name of data, originating project, area where data was collected from, station, trap ID (depending on available information).
3	yyyymmdd1	NA	Date in the format year month day of sediment-trap deployment.
4	yyyymmdd2	NA	Date in the format year month day of sediment-trap recovery.
5	Duration	Days	Duration of sediment-trap deployment.
6	Lat	° N	Decimal latitude of sediment-trap deployment.
7	Long	° E	Decimal longitude of sediment-trap deployment.
8	Depth	m	Depth of sediment-trap deployment.
9	Samp_id1	NA	Sample/cup ID replicate 1.
10	Samp_id2	NA	Sample/cup ID replicate 2 (if available).
11	Samp_id3	NA	Sample/cup ID replicate 3 (if available).
12	Tot_Mass1	$mg\,m^{-2}\,d^{-1}$	Total mass flux replicate 1.
13	Tot_Mass2	$mg\,m^{-2}\,d^{-1}$	Total mass flux replicate 2 (if available).
14	Tot_Mass3	$mg\,m^{-2}\,d^{-1}$	Total mass flux replicate 3 (if available).
15	Tot_Mass_av	$mg\,m^{-2}\,d^{-1}$	Total mass flux average.
16	Tot_Mass_stdev	$mg\,m^{-2}\,d^{-1}$	Total mass standard deviation.
17	POC_1	$mg\,m^{-2}\,d^{-1}$	Particulate organic carbon flux replicate 1.
18	POC_2	$mg\,m^{-2}\,d^{-1}$	Particulate organic carbon flux replicate 2 (if available).
19	POC_3	$mg\,m^{-2}\,d^{-1}$	Particulate organic carbon flux replicate 3 (if available).
20	POC_Av	$mg\,m^{-2}\,d^{-1}$	Particulate organic carbon flux average.
21	POC_stdev	$mg\,m^{-2}\,d^{-1}$	Particulate organic carbon flux standard deviation.
22	PON_1	$mg\,m^{-2}\,d^{-1}$	Particulate organic nitrogen flux replicate 1.
23	PON_2	$mg\,m^{-2}\,d^{-1}$	Particulate organic nitrogen flux replicate 2 (if available).
24	PON_3	$mg\,m^{-2}\,d^{-1}$	Particulate organic nitrogen flux replicate 3 (if available).
25	PON_Av	$mg\,m^{-2}\,d^{-1}$	Particulate organic nitrogen flux average.
26	PON_stdev	$mg\,m^{-2}\,d^{-1}$	Particulate organic nitrogen flux standard deviation.
27	$bSiO_2$	$mg\,m^{-2}\,d^{-1}$	Biogenic silica flux average.
28	$bSiO_2$_stdev	$mg\,m^{-2}\,d^{-1}$	Biogenic silica flux standard deviation.
29	$bSiO_2$_Flag	NA	This flag indicates whether $bSiO_2$ was corrected or not for dissolution of the particular fraction in the sample collecting cups of traps (see Sect. 2.2.6).
30	Mol mass ratio silica to silicon	NA	Molecular mass ratio of silica to silicon used to scale Si to SiO_2 (see Sect. 2.2.6).
31	$CaCO_3$	$mg\,m^{-2}\,d^{-1}$	Calcium carbonate flux average.
32	$CaCO_3$_stdev	$mg\,m^{-2}\,d^{-1}$	Calcium carbonate flux standard deviation.
33	Terr/Litho	$mg\,m^{-2}\,d^{-1}$	Terrigenous or lithogenic (as reported) material flux average.
34	Terr_stdev	$mg\,m^{-2}\,d^{-1}$	Terrigenous or lithogenic material flux standard deviation.
35	PIC	$mg\,m^{-2}\,d^{-1}$	Particulate inorganic carbon flux average.
36	PIC_stdev	$mg\,m^{-2}\,d^{-1}$	Particulate inorganic carbon flux standard deviation.
37	Institution	NA	Affiliation institution of main author and/or data originator (when available).
38	Trap_Type	NA	Sediment-trap type and/or characteristics as described in the source study or data set.
39	Data source	NA	Link to data source and/or data source information.
40	doi	NA	Digital object identifier, as generated by PANGAEA®.
41	Event	NA	Event ID, as recorded by PANGAEA®.
42	Notes	NA	Notes.
43	WOA09_Temp	°C	Temperature from World Ocean Atlas 2009 (WOA09) Climatology.
44	WOA09_Sal	NA	Salinity from WOA09.
45	WOA09_DO	$\mu mol\,L^{-1}$	Dissolved oxygen concentration from WOA09.
46	WOA09_O_{2SAT}	%	Oxygen saturation from WOA09.
47	WOA09_AOU	$\mu mol\,L^{-1}$	Apparent oxygen utilisation from WOA09
48	WOA09_NO_3^-	$\mu mol\,L^{-1}$	Nitrate concentration from WOA09.
49	WOA09_$Si(OH)_4$	$\mu mol\,L^{-1}$	Silicate concentration from WOA09.
50	WOA09_PO_4^{3-}	$\mu mol\,L^{-1}$	Phosphate concentration from WOA09.

Table 2. Sinking material flux range, depth range distribution and number of data points available.

Variable	Number of data points	Range $\mathrm{mg\,m^{-2}\,d^{-1}}$		Depth range m	
		Min.	Max.	Min.	Max.
Total mass flux	4735	0.0	5584	15	5031
POC flux	5202	0.0	355.7	15	5031
PON flux	3996	0.0	57.9	20	5031
bSiO$_2$ flux	2895	0.0	590.5	117	5031
PIC flux	1048	0.04	81.4	117	4832
CaCO$_3$ flux	2631	0.0	2505.7	117	5031
Terr/Litho flux	2166	0.0	4528.8	117	5031

Bory and Newton, 2000; Fischer et al., 2002; Fahl and Nöthig, 2007; Neuer et al., 1997). Sufficient formalin to yield 2–3 % formaldehyde (wt/vol) or mercuric chloride (0.14 % final solution) is commonly added to poison the sample to preserve the content (Antia et al., 1999; Fischer et al., 2002; Fahl and Nöthig, 2007; Helmke et al., 2005; Bauerfeind et al., 2009). Following recovery of sediment traps, swimmers (i.e. zooplankton that feed on sedimenting material) are identified and removed from collecting cups (Antia et al., 1999; Bory et al., 2001; Lampitt et al., 2010). Sometimes the samples are sieved (1 mm mesh) to remove large swimmers (e.g. Fischer et al., 2000). Also, samples are sometimes centrifuged following the removal of swimmers and the supernatant is then analysed in order to take into account of any possible dissolution of the material collected (e.g. Waniek et al., 2005). Samples from trap cups are typically split to generate subsamples for the different types of analysis and filtered through preweighed filters which are rinsed with ammonium formate to remove salt and excess formalin (e.g. Bory et al., 2001). The reader is referred to the source references for details of a particular deployment.

2.2.3 Total mass flux

Total mass is obtained by weighing the dried matter collected on a filter. As such, it is sometimes referred to as "dry mass". Total mass flux (Tot_Mass$_{\mathrm{flux}}$, mg m^{-2} d^{-1}) is calculated as Tot_Mass$_{\mathrm{flux}} = \frac{M_w - F_w}{T \cdot A}$, where M_w is the mass dry weight (mg), F_w is the filter weight (mg), T is the deployment time (days), and A is the aperture trap area (m^2) (e.g. Bahr et al., 1997).

2.2.4 POC and PON fluxes

POC and PON are measured using an elemental CHN analyser (e.g. Fischer et al., 2000; Bahr et al., 1997). The fraction of C and N in a given sample is multiplied by Tot_Mass flux to yield POC and PON fluxes (mg m^{-2} d^{-1}). Aliquots destined for the determination of POC and PON are filtered onto combusted (6 h 400 °C GF/F) filters (e.g. Goutx et al.,

2000), or polycarbonate filters (25 or 47 mm) (e.g. Hwang et al., 2009). Before drying for CHN analysis, samples are rinsed with 1–6 N HCl to remove carbonate (Fischer et al., 2000; Goutx et al., 2000; Bory et al., 2001; Helmke et al., 2005; Iversen et al., 2010). Filters are then dried. Reported drying temperatures vary, but typically, filters are oven dried at 40 °C (e.g. Goutx et al., 2000), air-dried at 60 °C overnight (e.g. Hwang et al., 2009), or dried on a hot plate set at 80 °C (e.g. Wefer and Fischer, 1993; Helmke et al., 2005). Some authors use freeze drying instead (e.g. Fischer et al., 2002; Waniek et al., 2005).

As a term, POC is frequently used interchangeably with "organic carbon (C$_{\mathrm{org}}$)" (e.g. Fischer et al., 1996; Lampitt et al., 2001) or total organic carbon (Wefer and Fischer, 1991; Jonkers et al., 2010). POC, is sometimes estimated as C$_{\mathrm{org}}$ = C$_{\mathrm{total}}$ − C$_{\mathrm{CaCO_3}}$ (e.g. Romero et al., 2002), where C$_{\mathrm{total}}$ is the total carbon content of a sample and C$_{\mathrm{CaCO_3}}$ is the carbon content in calcium carbonate. Similarly, PON is sometimes referred to as total nitrogen (e.g. Lampitt and Antia, 1997; Jonkers et al., 2010). We suggest POC and PON are the most appropriate terms by reasons of method and most common usage in literature.

2.2.5 Calcium carbonate flux and particulate inorganic Carbon (PIC) flux

CaCO$_3$ and PIC fluxes are typically derived based on molar mass ratios. CaCO$_3$ has been estimated by multiplying PIC by 8.34 (Lampitt et al., 2010) or by 8.33 (Lampitt et al., 2001; Fischer et al., 2002; i.e. $\frac{\mathrm{CaCO_3}}{\mathrm{C}} \approx 8.33$ (though this is not usually explicitly stated). It has also been calculated as (C$_{\mathrm{total}}$ − C$_{\mathrm{org}}$) × 8.33 (e.g. Wefer and Fischer, 1991; Helmke et al., 2005) following CHN analysis; that is, PIC × $\frac{\mathrm{CaCO_3}}{\mathrm{C}}$. It has been also determined through mass loss following acidification and then weighing (e.g. Fahl and Nöthig, 2007). Hwang et al. (2009) refer to "biogenic CaCO$_3$", which they estimated by multiplying the "biogenic Ca" by 2.5; i.e. the molar mass ratio $\frac{\mathrm{CaCO_3}}{\mathrm{Ca}}$. In turn, they obtained "biogenic Ca" as the difference between total Ca and lithogenic Ca. The latter being 0.5 × Al, based on the ratio of Ca to Al of the average continental crust composition (e.g. Wedepohl, 1995; Rudnick and Gao, 2003). Total inorganic carbon is determined by coulometric titration (Hwang et al., 2009). CaCO$_3$ flux is sometimes corrected if organisms containing calcium carbonate, such as pteropods, are present in the sample (e.g. Bauerfeind et al., 2009).

PIC has been calculated as 12 % carbonate by weight (Antia et al., 1999; Bauerfeind et al., 2009), i.e. the C content in CaCO$_3$. PIC content has also been calculated from total Ca concentrations in samples as CaCO$_3$ (Bory et al., 2001). It is also estimated as C$_{\mathrm{total}}$ − C$_{\mathrm{org}}$; i.e. the difference between the C measured in filtered samples without removal of carbonate, and the C measured in samples treated with HCl. The term "inorganic carbon" as equivalent of PIC is sometimes used

Table 3. Sediment-trap specifications summary.

Trap type	Type of deployment	General specifications	Further information	Used by
Kiel type	Bottom tethered	Fiberglass-reinforced plastic baffle, funnel and frame. Baffle consists of 20 mm × 20 mm × 120 mm cells. 0.5 m² aperture, cone with a 34° angle. Bottom of funnel mounted into a PTFE (polytetrafluoroethylene) transfer cylinder. Rotary sampler with 21 collecting bottles.	Zeitzschel et al. (1978) Kremling et al. (1996)	Antia et al. (1999), Bauerfeind and Nöthig (2011) Bauerfeind et al. (2009) Fischer (2005, 2003a, b, c, d) Fischer et al. (2002, 2000) Iversen et al. (2010) Jonkers et al. (2010) Neuer et al. (1997, 2007) Peinert et al. (2001) Raab (2003) von Bodungen et al. (1995) Waniek et al. (2005) Žarić et al. (2005)
Parflux Mark V Parflux Mark VI Parflux Mark 78	Bottom tethered	1.15 m² aperture. 520 baffle 52 mm diameter cells with a 2.5 aspect ratio. 36° cone angle. Rotary sampler with 12, 13 or 25 sample cups. 0.5 m² aperture. 368 baffle 25 mm 42° cone angle. Rotary sampler with 13 sample cups. 0.5 m² aperture. 268 baffle 25 mm diameter cells with a 2.5 aspect ratio. 41° cone angle. Rotary sampler with 21 or 13 wide sample cups.	Honjo and Doherty (1988) Newton et al. (1994) www.mclanelabs.com	Bory et al. (2001) CARIACO time series Fischer (2005, 2003a, b, c, d) Fischer et al. (2002, 2000) Honjo and Manganini (1993) Hwang et al. (2009) Jickells et al. (1996) Jonkers et al. (2010) Lampitt and Antia (1997) Lampitt et al. (2001, 2010) Wefer and Fischer (1991) Žarić et al. (2005)
SMT 230 SMT 234	Bottom tethered	0.5 m² aperture. 41 sample cups. 0.5 m² aperture. 21 sample cups. Both with a 34° cone angle.	www.kum-kiel.de Helmke et al. (2010)	Fahl and Nöthig (2007) Helmke et al. (2005) Romero et al. (2002) Wefer and Fischer (1991)
Technicap PPS 3 Technicap PPS 5	Bottom Bottom and surface* tethered	0.125 m² aperture, cylindro-conical collector with a 2.5 aspect ratio (unbaffled). 1 m² aperture, 8 mm baffle cell diameter. Made of Fibreglass.	www.technicap.com Miquel et al. (2011)	Goutx et al. (2000)* Jonkers et al. (2010) DYFAMED time series
Indented Rotary Sphere; IRS	Surface tethered	It consists of a 15 cm diameter cylindrical particle interceptor. An indented rotating sphere valve about halfway down the 1.7 m total length of the trap, which leads to a skewed funnel delivering collected particles to a sample carrousel. The trap can be set to time series (TS) or settling velocities (SV) mode.	Peterson et al. (2005, 2009)	Lee et al. (2009b)

in the literature too (e.g. Lampitt and Antia, 1997; Lampitt et al., 2001).

2.2.6 Biogenic silica flux

Biogenic silica requires more care relative to other variables. This derives from the fact that multiple names and terms are used rather ambiguously. For this compilation we attempted, to the best of our knowledge, to identify the variable that was being dealt with in each instance. We based our evaluation on the methods used and in some cases by contacting originating PIs.

Biogenic silica is typically measured as dissolved silicon with colorimetric methods following extraction from particulate material. Several methods exists (e.g. Eggimann et al., 1980; DeMaster, 1981; Mortlock and Froelich, 1989; Müller and Schneider, 1993), but the methods most commonly used are based on an alkaline digestion method of Mortlock and Froelich (1989) (e.g. Antia et al., 1999; Bory et al., 2001; Salter et al., 2010) or the sequential leaching method of DeMaster (1981) as modified by Müller and Schneider (1993) (e.g. Wefer and Fischer, 1991, 1993; Romero et al., 2002; Helmke et al., 2005). The former is based on the extraction of opaline silicon into a 2 M solution of Na_2CO_3 at 85 °C for 5 h, after which the digested sample is measured with

Table 4. Sediment-trap specifications summary.

Trap type	Type of deployment	General specifications	Further information	Used by
Particle interceptor traps (PIT)	Surface tethered	Polycarbonate cylinder, 0.0039 m^2 aperture. Plastic baffling consists of circular openings of 1.2 cm diameter. The base holds a 90 mm Poretics polycarbonate membrane filter. The trap frame holds up 15 cylinders.	Bahr et al. (1997)	BATS
Surface tethered traps	Surface tethered	480 mm cylinders with a 125 mm diameter, and 200 mm-long baffle. Inserted in each cylinder are four 480 mm-long, 50 mm-diameter smaller cylinders.	Neuer et al. (2007)	Neuer et al. (2007)
OSU traps	Bottom tethered	2 : 1 height : diameter fiberglass plastic cone, 1 cm × 5 cm baffle, 10 cm aperture. Oregon State University (OSU)-made traps based on Soutar et al. (1977).	Dymond and Lyle (1994)	Dymond and Lyle (2003a, b)
Unnamed trap	Bottom tethered	Small trap, 118 mm diameter cylinder 490 mm working part and baffle. Lower part is conical and connected to a sample collecting flask.	Lisitsyn et al. (1995) Stein (1999)	Shevchenko (2000)
Unnamed drifters	Surface tethered	No information available		Irwin (2002a, b) Martin (2003a, b, c) NGOFS and Tande (2003) OMEX and Wassmann (2004a, b, c, d) OMEX and Wassmann (2004e, f, g, h)
Unnamed traps	Bottom tethered	No information available		Tett (2005) Thomsen and von Bodungen (2001a, b, c)

standard photometric methods using an autoanalyser (Mortlock and Froelich, 1989). The latter is an automated method designed to extract Si from a broader range of compounds. The extraction is carried out with a 1 M solution of NaOH also at 85 °C, but the digestion solution is cycled from and to a digestion vessel; a proportion runs through an autoanalyser and another fraction is circulated back to the digestion vessel until extraction is completed (Müller and Schneider, 1993). There is an issue however, associated with the use of either method. Since the end product of the extraction is Si, this then needs to be "scaled" back to silica (SiO_2). The method by Mortlock and Froelich (1989) uses a factor of 2.4 (the molar mass ratio of $\frac{SiO_2 \cdot 0.4H_2O}{Si}$), which accounts for the average water content of diatomaceous silica. The method by Müller and Schneider (1993) instead uses the molar mass ratio of $\frac{SiO_2}{Si} \approx 2.139$. Another method used is that of Koning et al. (2002) (e.g. Jonkers et al., 2010), which is based on the method by Müller and Schneider (1993). Given the molar mass ratios adopted, and given that some data sets report biogenic Si rather than biogenic SiO_2, which we converted using the molar mass ratio of 2.1, we include a column in the data assemblage where the ratio is indicated. We did this based on whether a study or source used either of the methods above, except when a "conversion factor" was explicitly

stated independently of the digestion method used. When information provided was "unclear" about the ratio used, this is pointed out.

In some cases, dissolved silica is first measured in the water used for the collecting cups. Dissolved silica is then measured again following the trap's recovery in order to correct for any opal dissolution (e.g. Jonkers et al., 2010). However, we note that not all the biogenic silica data in this compilation includes such a correction or its application to the data was not clear. We have flagged these data as follows: 0 when corrections were made, 1 when corrections were not made, and 2 when information was not available to ascertain either.

All calculations in the literature are related to the molar mass ratio of silica or its hydrated form (as above), to elemental silicon. However, the terminology used is rather inconsistent. Hwang et al. (2009) report opal as the result of multiplying "biogenic Si" by 2.4, where biogenic Si is the difference between the total Si and the lithogenic-Si (3.5 × Al, an approximation of the ratio of Si to Al in the continental crust). Bauerfeind et al. (2009) define "biogenic particulate silica (bPSi)" in their abstract, which suggests the compound "silicon dioxide" ($SiO_2 \cdot nH_2O$) is being dealt with. However, in the methods section, it is redefined as "biogenic particulate silicon (bPSi)"; hence it is the chemical element Si that seems to be dealt with. Further, Opal is defined as $2.1 \times bPSI$; i.e. the mass ratio ($\frac{SiO_2}{Si}$) multiplied by bPSi (Bauerfeind et al., 2009). The following terms are also commonly used in the literature: "$PSiO_2$" (von Bodungen et al., 1995; Peinert et al., 2001; Bauerfeind and Nöthig, 2011), "$PSiO_2$ and $BSiO_2$" (Fischer, 2003a), "PSi" (Fischer, 2005), "BSi" (Ragueneau et al., 2001), "$BSiO_2$" as the sum of $PSiO_2$ and $DSiO_2$ (e.g. Antia et al., 1999; Lampitt et al., 2001; Honjo and Manganini, 2003a, b, c), "$BSiO_2$" (Lampitt and Antia, 1997), "Opal" (Neuer et al., 1997, 2007), "bPSi or biogenic particulate silicon" (opal = $2.1 \times$ bPSi) (Bauerfeind et al., 2009), "opaline silica" (Lampitt et al., 2010) or "biogenic silica (opal)" (Antia et al., 1999; Fischer et al., 2002; Waniek et al., 2005).

Since the ballasting effect of the mineral SiO_2 (opal) is most germane to this database, and since the interest resides in identifying the opal related to particles of biological origin (hence the term "biogenic"), the data we report here is referred to as "biogenic silica" or "biogenic opal". When data was found to be reported as Si, these were converted to SiO_2 using the molar mass ratio $\frac{SiO_2}{Si}$. We suggest the use of lower-case "b" to refer to "biogenic" in combination with the chemical formula SiO_2 (i.e. $bSiO_2$), since upper-case B is the chemical symbol for the element Boron. We also suggest that whether samples of $bSiO_2$ are corrected for dissolution or not in the trap-collecting cups should clearly be stated in future studies.

2.2.7 Lithogenic and/or terrigenous material flux

Lithogenic fluxes are typically estimated as the difference between the total mass flux and what is termed either "biogenic flux", "biogenic matter" or "organic matter flux"; i.e. $CaCO_3$ + POC + $BSiO_2$ fluxes (e.g. Antia et al., 1999; Bauerfeind et al., 2009). For this purpose, in deriving "organic matter", POC is sometimes multiplied by 2 (Fischer et al., 2002; Bauerfeind et al., 2009) or 2.5 (e.g. Hwang et al., 2009), as this is considered to give a more representative flux of organic matter but, in the literature, this adjustment would benefit from a fuller explanation. Thus, biogenic flux (Bio_{flux}) is $Bio_{flux} = 2 \times POC_{flux} + CaCO_{3flux} + Opal_{flux}$. Lithogenic flux (Litho) is given by $Litho = Tot\_Mass_{flux} - Bio_{Flux}$. Some researchers estimate the lithogenic material from Al concentrations, under the assumption it contains 8.4 % Al (Bory et al., 2001) or by multiplying the Al concentration by 12.15 (e.g. Hwang et al., 2009). The latter is based on the assumption of a crustal Al composition of 8.2 % ($1/12.15 = 0.082$).

2.3 Data standardisation

This data assemblage contains fluxes from short-duration deployments (hour–days) to longer-duration deployments lasting from months to a year, or over a year. Hence, in order to standardise the data set, all values from long-term deployments, typically reported in grams per square metre per

year ($g\,m^{-2}\,yr^{-1}$) (e.g. Wefer and Fischer, 1991, 1993; Fischer et al., 2000; Fischer, 2005; Peinert et al., 2001), were converted to daily values, i.e. milligrams per square metre per day ($mg\,m^{-2}\,d^{-1}$), which is the unit most commonly reported. Long-term deployments, however, can be easily identified; a column is provided which specifies the duration of the deployment. A few daily values were reported in grams per square metre per day ($g\,m^{-2}\,d^{-1}$), and these were also converted to milligrams per square metre per day ($mg\,m^{-2}\,d^{-1}$) for consistency. In a few instances, POC, PIC, and $bSiO_2$ were reported in moles per square metre per year ($mol\,m^{-2}\,yr^{-1}$) (e.g. Antia et al., 1999; Dymond and Lyle, 2003a, b; Honjo and Manganini, 2003d, e, f; Fahl and Nöthig, 2007) or millimole per square metre per day ($mmol\,m^{-2}\,d^{-1}$) (Martin, 2003a, b, c). Again, for consistency, these were converted to milligrams per square metre per day ($mg\,m^{-2}\,d^{-1}$) using the appropriate molecular masses and or molecular mass ratios as required. A column of notes is included, and where unit conversions were done, these are pointed out.

3   Quality control

Given that the data compiled here derives from research already published, we assume that the originating authors have already undertaken steps necessary to assure data quality. We point out however, that attention should be paid to the fact that the use of slightly different "conversion factors" for a given variable inherently adds error to the data, with up to 20 % in the case of lithogenic flux when derived as the difference between total mass flux and "organic matter" flux, and where "organic matter" is calculated using conversion factors of 2 or 2.5 (Sect. 2.2.7). In the case of $bSiO_2$, the error generated is ~12 %, resulting from the use of 2.1, 2.139 or 2.4 when estimated from Si (Sect. 2.2.6). We did not attempt to "harmonise" the data by using a unique factor for a given variable, since this would involve modifying the data from that found in the original sources. However, in the light of the different deployment durations and traps used, different analytical methods employed, different calculation approaches and different units reported, here we have tried, as best as possible, to put the data together in a manner allowing users to trace original data sources for further scrutiny and so that users can decide how to handle the data further for the specific questions they may choose to tackle.

4   Ancillary data

Where possible, data from the World Ocean Atlas 2009 (WOA09) were extracted to coincide with flux data to provide additional environmental information (http://www.nodc.noaa.gov/OC5/WOA09/pr_woa09.html). Specifically, data were extracted for temperature, salinity, oxygen (concentration, AOU and percentage saturation), nitrate, phosphate and silicate. The extraction involves linear

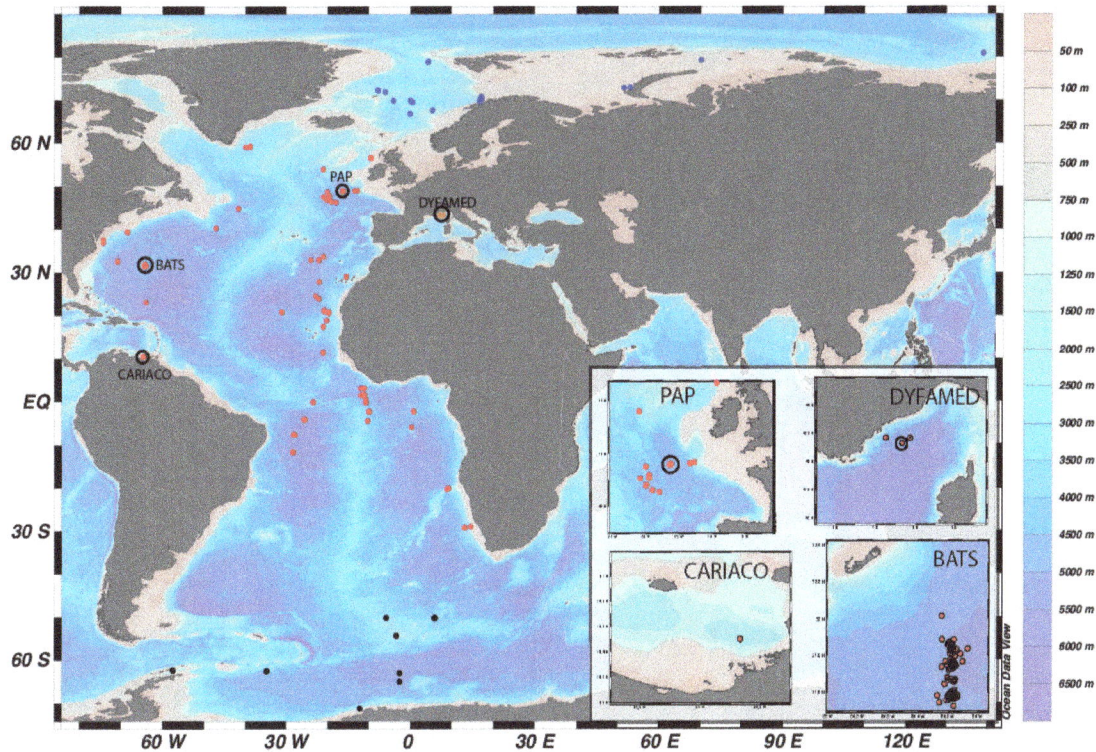

Figure 1. Map showing the location of sediment-trap deployments. Inset figures show expanded maps of regions where the CARIACO, BATS, PAP and DYFAMED time series are located. Locations are colour-coded per ocean domain: Atlantic (red dots), Mediterranean (orange dots), Arctic (blue dots), and Southern Ocean (black dots).

interpolation of WOA09 data to the latitude, longitude and depth of the flux data. Each environmental variable is a weighted average over the period of deployment. Note that as WOA09 is a climatology it cannot provide data for specific years. For example, if a mooring collected flux data from 1 November 1990 until 15 January 1991, the WOA data at the relevant point is averaged over the 76 days comprising the annual climatologies for November (for 30 days), December (for 31 days) and January (15 days). For temperature, salinity and oxygen variables, monthly climatologies are used above 1500 m and annual ones below. For nutrients, monthly climatologies are only available and used above 500 m. The distribution of WOA09 climatologies does not extend close to the coasts. Hence, given the proximity of the CARIACO time-series station to the mainland, ancillary data is not available for this site from WOA09.

5 Data distribution

Figure 1 shows the distribution of the sediment-trap deployments compiled in this data set. Data coverage spans from 1982 to 2011, with the largest amount of observations between 1990 and 2010 (Fig. 2). Figure 3 shows a map with the number of data points available on a $5° \times 5°$ grid. The most abundant contributions to this data set derive from es-

tablished time-series stations: BATS (31°40′ N, 64°10′ W) 784 data points, 13.2 %; CARIACO (10.5° N, 64.4° W) 1755 data points, 29.5 %; DYnamique des Flux Atmosphériques en MEDiterranée et leur évolution dans la colonne deau (DYFAMED) 43°25′ N, 07°52′ E, 401 data points, 6.7 % of total (Miquel et al., 2011); the Porcupine Abyssal plain (PAP), 49° N, 16°30′ W, 366 data points, 6.2 % of total (Lampitt et al., 2001, 2010); the North Atlantic Bloom Experiment (NABE), from 34° N, 21° W to 48° N, 21° W, 170 data points, 2.9 % of total (Honjo and Manganini, 1993; Martin, 2003a, b, c); and the European Station for Time series in the OCean (ESTOC), 29° N, 15.5° W, 124 data points, 2.1 % of total (Neuer et al., 1997). Missing in this compilation are data from the Progamme Océan Multidisciplinaire Méso Echelle (POMME, 30–60° N, 0–30° W), which are not yet publicly available. We are also aware of a few more recent data sets from the ESTOC, but these have copyright restrictions (e.g. Neuer et al., 2007).

Observation depths span from 15 down to 5031 m (Table 2). Tot_Mass, CaCO₃ and Terr/Litho exhibit the broadest range of export fluxes (0.0–5585 and 0.0–4529 mg m² d⁻¹, respectively), while PON and PIC exhibit the narrowest range (0.0–57.9 and 0.04–81.4 mg m² d⁻¹). The largest number of observations, with higher vertical resolution, have been made within the first 1000 m of the water column,

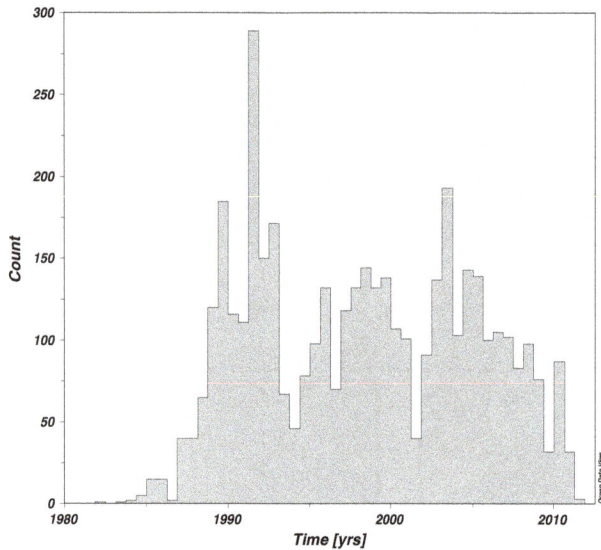

Figure 2. Data–time distribution histogram.

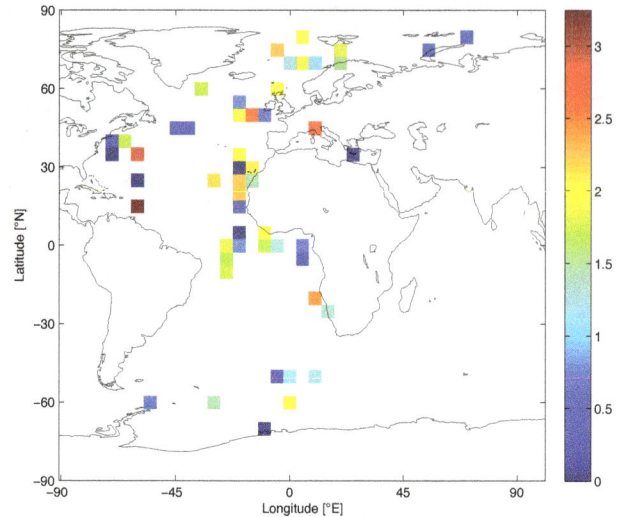

Figure 3. Map showing number of data points on a $5° \times 5°$ grid ($\log_{10}$-scale colour-code).

particularly in the upper 500 m (Fig. 4). At depths greater than 1000 m a preference for sampling at 3000 m is apparent in the data. With the exception of the PIC flux and lithogenic/terrigenous flux, which contain the lowest number of data points (Table 2), all other variables of particle export have been sampled at a similar vertical resolution (Fig. 4).

The overall pattern of particle export fluxes show the expected decrease from upper layers to depth as sinking particles decay and dissolve (Fig. 5). This is particularly clear in the Atlantic Ocean and Mediterranean Sea data (red and orange symbols in Fig. 5). POC, PON, $bSiO_2$ and $CaCO_3$ show a similar vertical structure, with the range of values at a given depth decreasing from surface to depth: from up to 550 mg m^2 d^{-1} POC and 58 mg m^2 d^{-1} PON at 30 m, up to 478 mg^2 d^{-1} $bSiO_2$ at 225 m, and 25 000 mg^2 d^{-1} $CaCO_3$ at 152 m, down to 0–6.5 mg^2 d^{-1} POC at ~5000 m, and up to 1.4 mg^2 d^{-1} PON, up to 27 mg^2 d^{-1} $bSiO_2$ and up to 15 mg^2 d^{-1} $CaCO_3$ at ~4700 m.

In the upper 1000 m, the largest fluxes of POC and PON occur in the Atlantic and the Arctic domains. Within the Arctic domain, the broad range of POC and PON fluxes in the upper 200 m derive from trap deployments off the northwest coast of Norway (~ 17° E, ~ 70° N). The largest fluxes of $bSiO_2$ in the upper 1000 m are found in the Atlantic and Southern Ocean domains. PIC fluxes are largest in the Atlantic and the Mediterranean Sea. In the Atlantic, PIC data show maximum values at 1400 m, which then decrease at greater depths (Fig. 5). These maximum values at 1400 m, though, derive from trap deployments north of Ireland and may result from the supply of PIC from the shelf or shelf-break front. $CaCO_3$ fluxes are rather comparable among ocean domains, though a larger range is found at about 150 m in the Atlantic. Terr/Litho fluxes show a broad

range of values within the upper 1250 m which reduce substantially at greater depths.

The proportion each export variable makes to total mass flux is shown in Fig. 6. The largest contributions of up to 80 % are provided by the Terr/Litho, $CaCO_3$ and $bSiO_2$ fractions but there is a broad range in the contribution each fraction makes to the total mass flux at all sampled depths and within the four ocean domains. In the case of $CaCO_3$ and Terr/Litho the range in the contribution from these fractions to total mass flux appears to narrow with depth which may reflect the attenuation of other variables with depth rather than any systematic change to Terr/Litho and $CaCO_3$ contributions. The largest contribution (up to 90 %) made by $bSiO_2$ to total mass flux derive from trap deployments in the Southern Ocean with apparent peaks at depths of 500 and 4500 m; elsewhere the $bSiO_2$ contribution is smaller but nevertheless a major component of the downward particle flux. The contribution made by POC to total mass flux is broadly similar within all four ocean domains and decreases from ~ 80 % in the upper 500 m to < 20 % at 5000 m revealing a marked attenuation with depth. Both PON and PIC typically contribute < 20 % to total mass flux, but, whilst the contribution from PIC remains fairly constant with depth, there is vertical attenuation of PON with depth such that at depths > 500 m the PON contribution is < 10 %.

6 Conclusions

We have assembled a data set of over 5900 data points of particle flux across the wider Atlantic Ocean and adjacent seas, which will be invaluable in determining seasonal and geographical variability in the biological carbon pump. Our initial examination of this data set already indicates important

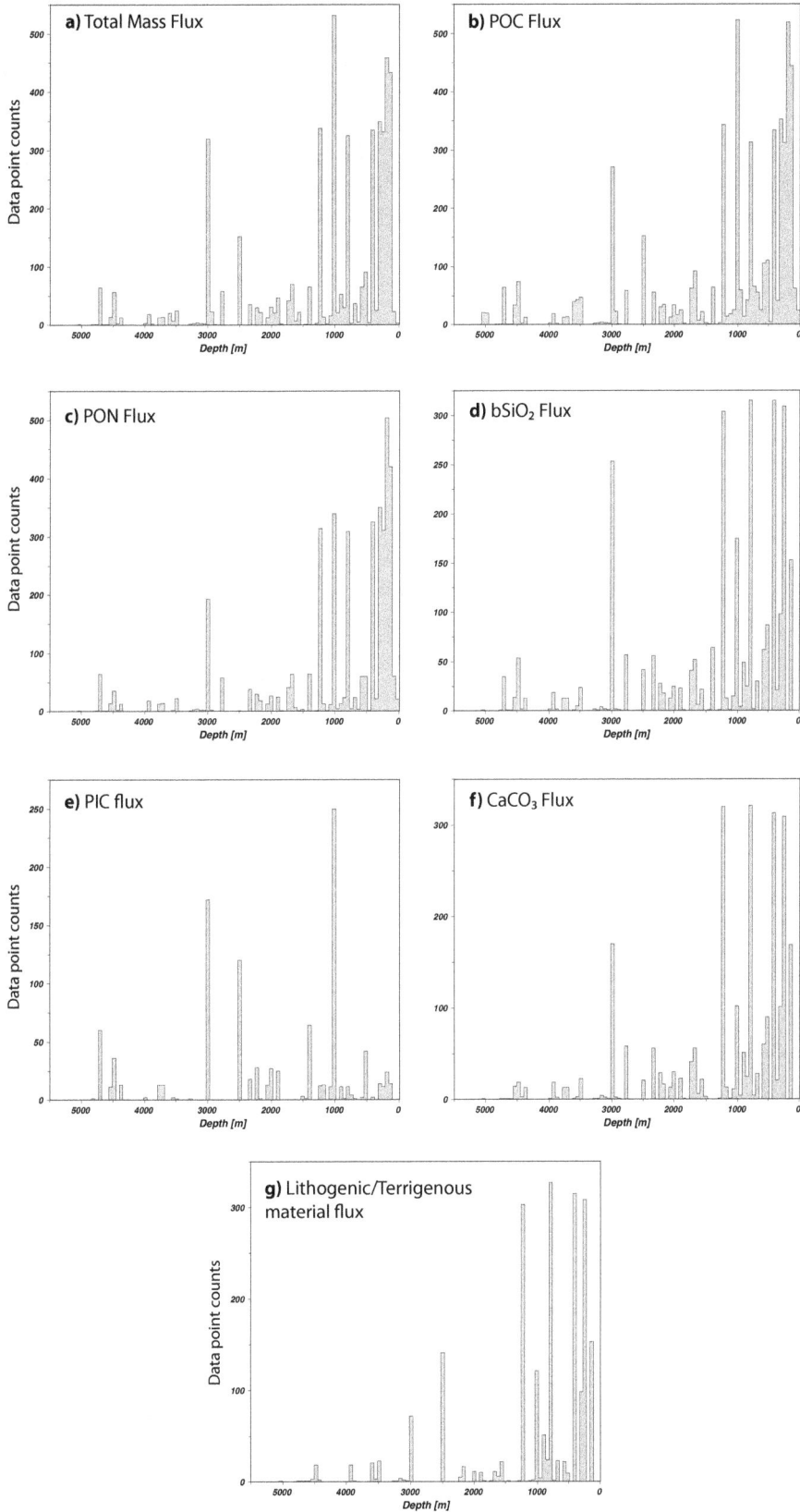

Figure 4. Data points available per depth; (**a**) total mass flux, (**b**) POC flux, (**c**) PON flux, (**d**) bSiO$_2$ flux, (**e**) PIC flux, (**f**) CaCO$_3$ flux, and (**g**) lithogenic/terrigenous material flux.

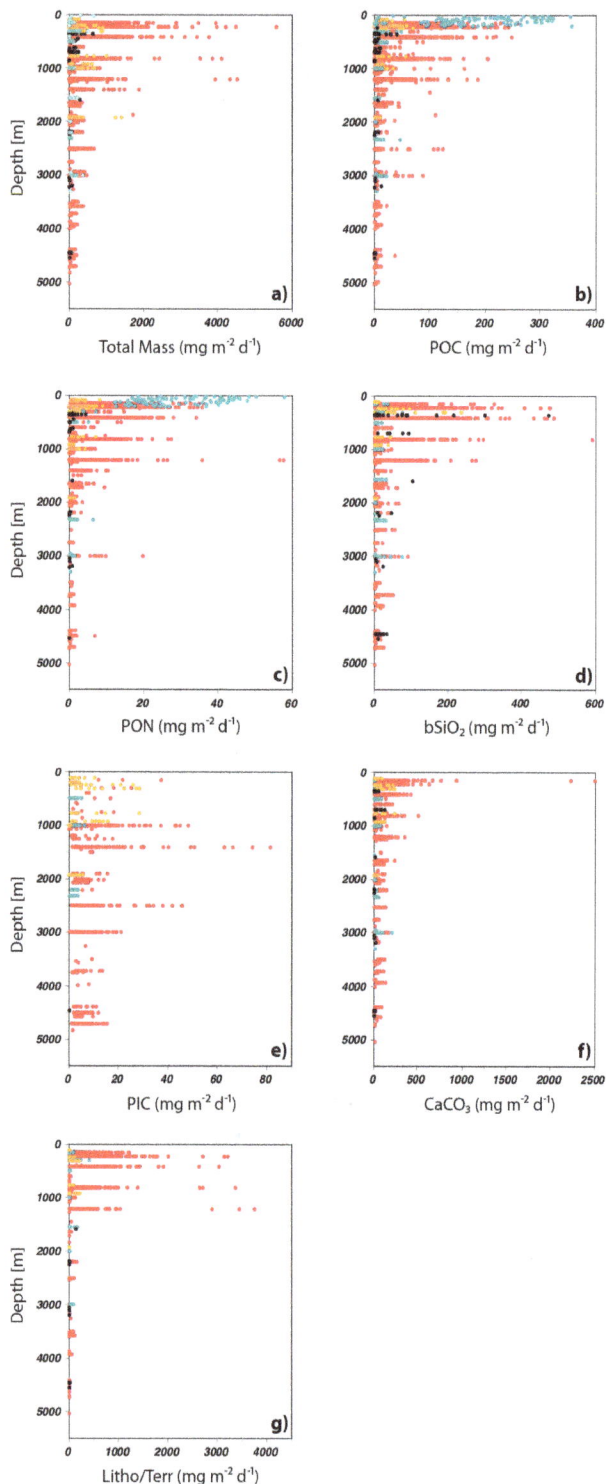

Figure 5. Downward fluxes plotted against depth; **(a)** total mass, **(b)** POC, **(c)** PON, **(d)** bSiO$_2$, **(e)** PIC, **(f)** CaCO$_3$, and **(g)** Litho/Terr. Data points are colour-coded per ocean domain: Atlantic (red dots), Mediterranean (orange dots), Arctic (blue dots), and Southern Ocean (black dots).

differences in the flux estimates between ocean domains and in the contribution particular flux variables make to the total mass flux, which may in turn indicate important differences in the strength of the BCP due to local environmental- and ecosystem-level forcing. Exploring the reasons for such differences remains a major scientific and societal problem particularly given projected changes to the future ocean and this data set will help in this endeavour. This data set has been submitted to the data repository PANGAEA® (http://www.pangaea.de), where it has been made available under doi:10.1594/PANGAEA.807946.

7 List of compiled data sets

Here we list all individual data sets. PANGAEA® digital object identifiers are also given.

- Antia, Avan N (2003): Particle fluxes of L2-B-92_trap.

- Antia, Avan N (2003): Particle fluxes of OMEX2_trap.

- Antia, Avan N (2003): Particle fluxes of OMEX3_trap.

- Antia, Avan N (2003): Particle fluxes of SEEP-7_trap.

- Antia, Avan N (2003): Particle Flux of SEEP-10_trap.

- Bahr, Fred; Bates, Nicolas R (2013): Total flux, particulate carbon and nitrogen from surface-tethered sediment traps at time series station BATS in 1988 and

- Bahr, Fred; Bates, Nicolas R (2013): Total flux, particulate carbon and nitrogen from surface-tethered sediment traps at time series station BATS in 1990.

- Bahr, Fred; Bates, Nicolas R (2013): Total flux, particulate carbon and nitrogen from surface-tethered sediment traps at time series station BATS in 1991.

- Bahr, Fred; Bates, Nicolas R (2013): Total flux, particulate carbon and nitrogen from surface-tethered sediment traps at time series station BATS in 1992.

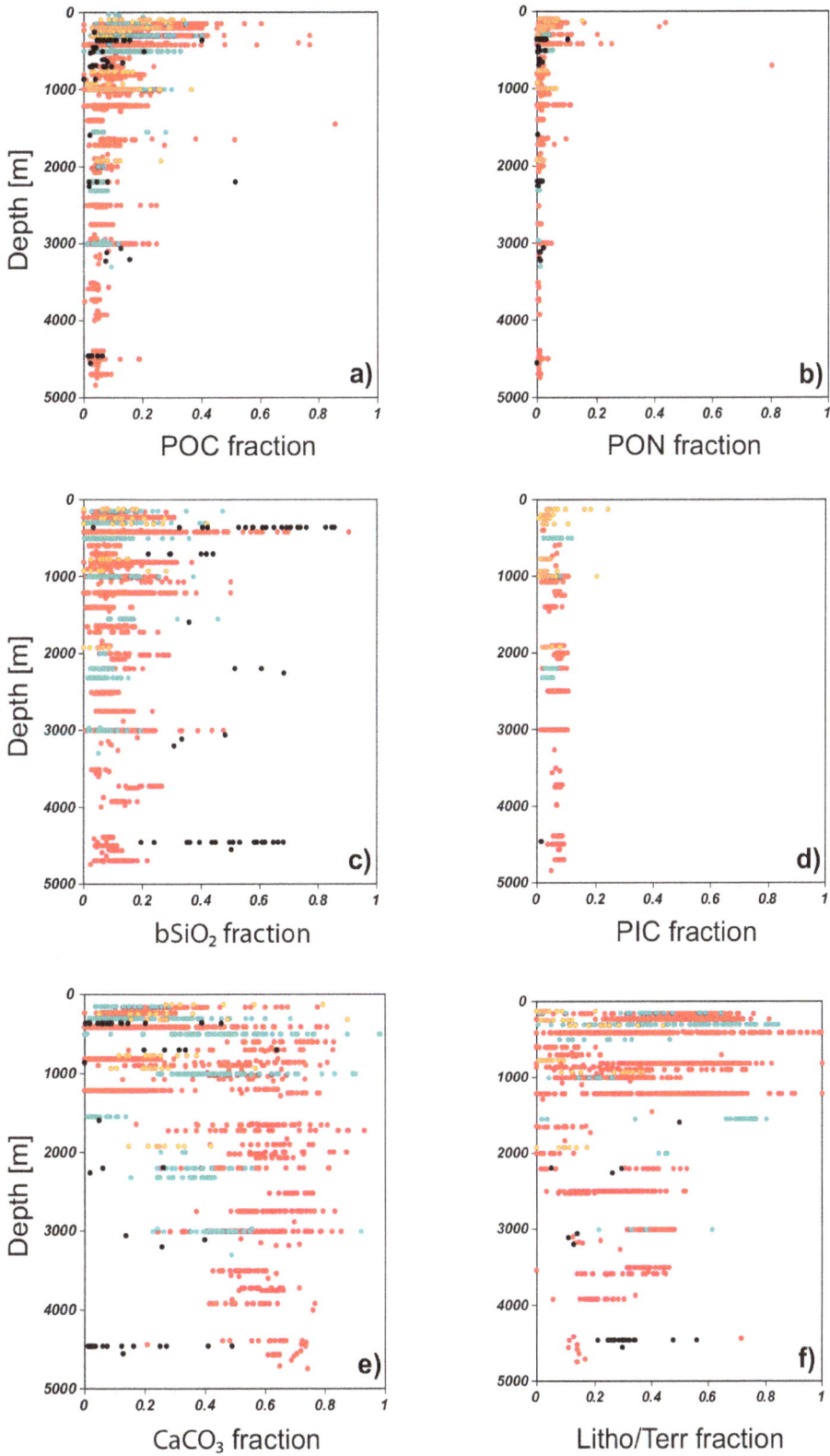

Figure 6. Fraction of POC, PON, and bSiO$_2$, PIC, CaCO$_3$ and Litho/Terr downward fluxes relative to Tot_Mass flux. Data points are colour-coded per ocean domain: Atlantic (red dots), Mediterranean (orange dots), Arctic (blue dots), and Southern Ocean (black dots). Few outliers (fraction > 1) were excluded from the graphs.

– Bahr, Fred; Bates, Nicolas R (2013): Total flux, particulate carbon and nitrogen from surface-tethered sediment traps at time series station BATS in 1993.

– Bahr, Fred; Bates, Nicolas R (2013): Total flux, particulate carbon and nitrogen from surface-tethered sediment traps at time series station BATS in 1994.

– Bahr, Fred; Bates, Nicolas R (2013): Total flux, particulate carbon and nitrogen from surface-tethered sediment traps at time series station BATS in 1995.

– Bahr, Fred; Bates, Nicolas R (2013): Total flux, particulate carbon and nitrogen from surface-tethered sediment traps at time series station BATS in 1996.

– Bahr, Fred; Bates, Nicolas R (2013): Total flux, particulate carbon and nitrogen from surface-tethered sediment traps at time series station BATS in 1997.

– Bahr, Fred; Bates, Nicolas R (2013): Total flux, particulate carbon and nitrogen from surface-tethered sediment traps at time series station BATS in 1998.

– Bahr, Fred; Bates, Nicolas R (2013): Total flux, particulate carbon and nitrogen from surface-tethered sediment traps at time series station BATS in 1999.

– Bahr, Fred; Bates, Nicolas R (2013): Total flux, particulate carbon and nitrogen from surface-tethered sediment traps at time series station BATS in 2000.

– Bahr, Fred; Bates, Nicolas R (2013): Total flux, particulate carbon and nitrogen from surface-tethered sediment traps at time series station BATS in 2001.

– Bahr, Fred; Bates, Nicolas R (2013): Total flux, particulate carbon and nitrogen from surface-tethered sediment traps at time series station BATS in 2002.

– Bahr, Fred; Bates, Nicolas R (2013): Total flux, particulate carbon and nitrogen from surface-tethered sediment traps at time series station BATS in 2003.

– Bahr, Fred; Bates, Nicolas R (2013): Total flux, particulate carbon and nitrogen from surface-tethered sediment traps at time series station BATS in 2004.

– Bahr, Fred; Bates, Nicolas R (2013): Total flux, particulate carbon and nitrogen from surface-tethered sediment traps at time series station BATS in 2005.

– Bahr, Fred; Bates, Nicolas R (2013): Total flux, particulate carbon and nitrogen from surface-tethered sediment traps at time series station BATS in 2006.

– Bahr, Fred; Bates, Nicolas R (2013): Total flux, particulate carbon and nitrogen from surface-tethered sediment traps at time series station BATS in 2007.

– Bahr, Fred; Bates, Nicolas R (2013): Total flux, particulate carbon and nitrogen from surface-tethered sediment traps at time series station BATS in 2008.

– Bahr, Fred; Bates, Nicolas R (2013): Total flux, particulate carbon and nitrogen from surface-tethered sediment traps at time series station BATS in 2009.

– Bahr, Fred; Bates, Nicolas R (2013): Total flux, particulate carbon and nitrogen from surface-tethered sediment traps at time series station BATS in 2010.

– Bahr, Fred; Bates, Nicolas R (2013): Total flux, particulate carbon and nitrogen from surface-tethered sediment traps at time series station BATS in 2011.

– Bauerfeind, Eduard; Nöthig, Eva-Maria (2011): Biogenic particle flux at AWI HAUSGARTEN from mooring FEVI3 at 2400 m. Alfred Wegener Institute, Helmholtz Center for Polar and Marine Research, Bremerhaven,

– Bauerfeind, Eduard; Nöthig, Eva-Maria; Beszczynska, Agnieszka; Fahl, Kirsten; Kaleschke, Lars; Kreker, Kathrin; Klages, Michael; Soltwedel, Thomas; Lorenzen, Christiane; Wegner, Jan (2009): Biogenic particle flux at AWI HAUSGARTEN from mooring FEVI1.

– Bauerfeind, Eduard; Nöthig, Eva-Maria; Beszczynska, Agnieszka; Fahl, Kirsten; Kaleschke, Lars; Kreker, Kathrin; Klages, Michael; Soltwedel, Thomas; Lorenzen, Christiane; Wegner, Jan (2009): Biogenic particle flux at AWI HAUSGARTEN from mooring FEVI2.

– Bauerfeind, Eduard; Nöthig, Eva-Maria; Beszczynska, Agnieszka; Fahl, Kirsten; Kaleschke, Lars; Kreker, Kathrin; Klages, Michael; Soltwedel, Thomas; Lorenzen, Christiane; Wegner, Jan (2009): Biogenic particle flux at AWI HAUSGARTEN from mooring FEVI3 at 300 m.

– Bauerfeind, Eduard; Nöthig, Eva-Maria; Beszczynska, Agnieszka; Fahl, Kirsten; Kaleschke, Lars; Kreker, Kathrin; Klages, Michael; Soltwedel, Thomas; Lorenzen, Christiane; Wegner, Jan (2009): Biogenic particle flux at AWI HAUSGARTEN from mooring FEVI4.

– Bauerfeind, Eduard; Nöthig, Eva-Maria; Beszczynska, Agnieszka; Fahl, Kirsten; Kaleschke, Lars; Kreker, Kathrin; Klages, Michael; Soltwedel, Thomas; Lorenzen, Christiane; Wegner, Jan (2009): Biogenic particle flux at AWI HAUSGARTEN from mooring FEVI7.

– Bory, A; Jeandel, Catherine; Leblond, Nathalie; Vangriesheim, Annick; Khripounoff, Alexis; Beaufort, Luc; Rabouille, Christophe; Nicolas, E; Tachikawa, Kazuyo; Etcheber, Henri; Buat-Menard, P (2012): Downward particle fluxes at the oligotrophic and mesotrophic site of the EUMELI program.

– Dymond, Jack R; Lyle, Mitchell W (2003): Particle fluxes of HAP-4_trap.

– Dymond, Jack R; Lyle, Mitchell W (2003): Particle fluxes of NAP_trap.

– Fahl, Kirsten; Nöthig, Eva-Maria (2007): Lithogenic and biogenic annual particle fluxes on the Lomonosov Ridge of trap LOMO-2.

– Fahl, Kirsten; Nöthig, Eva-Maria (2007): Lithogenic and biogenic particle fluxes on the Lomonosov Ridge of trap LOMO-2.

– Fischer, Gerhard (2003): Flux data of trap WR2.

– Fischer, Gerhard (2003): Flux data of trap WR3.

– Fischer, Gerhard (2003): Flux data of trap WR4.

– Fischer, Gerhard (2005): Particle fluxes for the sampling interval, various ratios and major nutrients at the Atlantic/Southern Ocean trapping sites.

– Fischer, Gerhard (2003): Particle fluxes of trap NU2.

– Fischer, Gerhard; Gersonde, Rainer; Wefer, Gerold (2002): (Table 4) Annual fluxes, percentages of total mass and most important elemental ratios at the PF and BO sites.

– Fischer, Gerhard; Ratmeyer, Volker; Wefer, Gerold (2012): Total organic carbon fluxes (<1 mm size fraction), and export fluxes to a depth of 1000 m in the Atlantic and Southern Ocean.

– Goutx, Madeleine; Momzikoff, André; Striby, L; Andersen, Valérie; Marty, Jean-Claude; Vescovali, Isabelle (2000): Fluxes of particulate organic carbon and nitrogen at DYNAPROC station.

- Helmke, Peer; Romero, Oscar E; Fischer, Gerhard (2005): Sampling intervals, fluxes, percentages of total flux, organic carbon and lithogen for sediment trap CB9.

- Honjo, Susumu; Manganini, Steven J (2003): Annual Particle fluxes of NABE-N34_trap.

- Honjo, Susumu; Manganini, Steven J (2003): Annual Particle fluxes of NABE-N48_trap.

- Honjo, Susumu; Manganini, Steven J (2003): Biogenic particle flux of trap NABE-N34.1.

- Honjo, Susumu; Manganini, Steven J (2003): Biogenic particle flux of trap NABE-N34.2.

- Honjo, Susumu; Manganini, Steven J (2003): Biogenic particle flux of trap NABE-N48.1.

- Honjo, Susumu; Manganini, Steven J (2003): Biogenic particle flux of trap NABE-N48.2.

- Hwang, Jeomshik; Manganini, Steven J; Montluon, D; Eglinton, Timothy I (2012): Biogeochemical properties of sinking particles intercepted at three depths on the NW Atlantic margin.

- Irwin, Brian (2002): POC, PON, chlorophyll, phaeophytin of BAF89/3_FTRAP1.

- Irwin, Brian (2002): POC, PON, chlorophyll, phaeophytin of BAF89/3_FTRAP2.

- Iversen, Morten Hvitfeldt; Nowald, Nicolas; Ploug, Helle; Jackson, George A; Fischer, Gerhard (2010): (Table 2) Total mass and organic carbon fluxes shown for each deployment.

- Jickells, Timothy D (2003): Particle fluxes of 19° N20° W_trap.

- Jickells, Timothy D (2003): Particle fluxes of 24° N23° W_trap.

- Jickells, Timothy D (2003): Particle fluxes of 28° N22° W_trap.

- Jickells, Timothy D (2003): Particle fluxes of Parflux7G_trap.

- Jonkers, Lukas; Brummer, Geert-Jan A; Peeters, Frank J C; van Aken, Hendrik M; de Jong, M Femke (2010): Shell flux and oxygen isotope data of North Atlantic foraminifera.

- Lampitt, Richard S; Antia, Avan N; Fischer, Gerhard (2006): Particle fluxes from sediment trap CB1.

- Lampitt, Richard S; Antia, Avan N; Fischer, Gerhard (2006): Particle fluxes from sediment trap CB2.

- Lampitt, Richard S; Antia, Avan N; Fischer, Gerhard (2006): Particle fluxes from sediment trap GBN3.

- Lampitt, Richard S; Antia, Avan N; Fischer, Gerhard (2006): Particle fluxes from sediment trap GBZ4.

- Lampitt, Richard S; Antia, Avan N; Fischer, Gerhard (2006): Particle fluxes from sediment trap WR1.

- Lampitt, Richard S; Antia, Avan N; Fischer, Gerhard (2006): Particle fluxes from sediment trap WR2.

- Lampitt, Richard S; Antia, Avan N; Fischer, Gerhard (2006): Particle fluxes from sediment trap WR3.

– Lampitt, Richard S; Antia, Avan N; Fischer, Gerhard (2006): Particle fluxes from sediment trap WR4.

– Lampitt, Richard S; Bett, Brian J; Kiriakoulakis, Kostas; Popova, E E; Ragueneau, Olivier; Vangriesheim, Annick; Wolff, George A (2001): Particle flux from sediment trap PAP-I.

– Lampitt, Richard S; Bett, Brian J; Kiriakoulakis, Kostas; Popova, E E; Ragueneau, Olivier; Vangriesheim, Annick; Wolff, George A (2001): Particle flux from sediment trap PAP-III.

– Lampitt, Richard S; Bett, Brian J; Kiriakoulakis, Kostas; Popova, E E; Ragueneau, Olivier; Vangriesheim, Annick; Wolff, George A (2001): Particle flux from sediment trap PAP-V.

– Lampitt, Richard S; Bett, Brian J; Kiriakoulakis, Kostas; Popova, E E; Ragueneau, Olivier; Vangriesheim, Annick; Wolff, George A (2001): Particle flux from sediment trap PAP-XIX.

– Lampitt, Richard S; Bett, Brian J; Kiriakoulakis, Kostas; Popova, E E; Ragueneau, Olivier; Vangriesheim, Annick; Wolff, George A (2001): Particle flux from sediment trap PAP-XV.

– Lampitt, Richard S; Bett, Brian J; Kiriakoulakis, Kostas; Popova, E E; Ragueneau, Olivier; Vangriesheim, Annick; Wolff, George A (2001): Particle flux from sediment trap PAP-XVIII.

– Lampitt, Richard S; Bett, Brian J; Kiriakoulakis, Kostas; Popova, E E; Ragueneau, Olivier; Vangriesheim, Annick; Wolff, George A (2001): Particle flux from sediment trap PAP-XX.

– Lampitt, Richard S; Bett, Brian J; Kiriakoulakis, Kostas; Popova, E E; Ragueneau, Olivier; Van-

griesheim, Annick; Wolff, George A (2001): Particle flux from sediment trap PAP-XXIIIa.

– Lampitt, Richard S; Bett, Brian J; Kiriakoulakis, Kostas; Popova, E E; Ragueneau, Olivier; Vangriesheim, Annick; Wolff, George A (2001): Particle flux from sediment trap PAP-XXV.

– Lampitt, Richard S; Salters, Vincent JM; de Cuevas, Beverly; Hartman, S; Larkin, Kate E; Pebody, C A (2010): Particle flux from sediment trap PAP-XXVI.

– Lampitt, Richard S; Salters, Vincent JM; de Cuevas, Beverly; Hartman, S; Larkin, Kate E; Pebody, C A (2010): Particle flux from sediment trap PAP-XXVII.

– Lampitt, Richard S; Salters, Vincent JM; de Cuevas, Beverly; Hartman, S; Larkin, Kate E; Pebody, C A (2010): Particle flux from sediment trap PAP-XXVIII.

– Lampitt, Richard S; Salters, Vincent JM; de Cuevas, Beverly; Hartman, S; Larkin, Kate E; Pebody, C A (2010): Particle flux from sediment trap PAP-XXXI.

– Lee, Cindy; Peterson, Michael L; Wakeham, Stuart G; Armstrong, Robert A; Cochran, J Kirk; Fukai, R; Fowler, Scott W; Hirschberg, David; Beck, Aaron; Xue, Jianhong (2013): Sediment Trap Data in the northwest Mediterranean Sea from the MedFlux project (2003 - 2005).

– Martin, John (2003): Particle interceptor data of sediment trap AT_II-119/4_Trap_A.

– Martin, John (2003): Particle interceptor data of sediment trap AT_II-119/4_Trap_B.

– Martin, John (2003): Particle interceptor data of sediment trap AT_II-119/5_Trap_C.

– Miquel, Juan-Carlos (2004): Particulate flux of carbon at DYNAPROC station.

– Miquel, Juan-Carlos; Marty, Jean-Claude (2004): Downward flux of particles and carbon at trap DYF1.

– Miquel, Juan-Carlos; Marty, Jean-Claude (2004): Downward flux of particles and carbon at trap DYF10.

– Miquel, Juan-Carlos; Marty, Jean-Claude (2004): Downward flux of particles and carbon at trap DYF11.

– Miquel, Juan-Carlos; Marty, Jean-Claude (2004): Downward flux of particles and carbon at trap DYF12.

– Miquel, Juan-Carlos; Marty, Jean-Claude (2004): Downward flux of particles and carbon at trap DYF13.

– Miquel, Juan-Carlos; Marty, Jean-Claude (2004): Downward flux of particles and carbon at trap DYF14.

– Miquel, Juan-Carlos; Marty, Jean-Claude (2004): Downward flux of particles and carbon at trap DYF16.

– Miquel, Juan-Carlos; Marty, Jean-Claude (2004): Downward flux of particles and carbon at trap DYF17.

– Miquel, Juan-Carlos; Marty, Jean-Claude (2004): Downward flux of particles and carbon at trap DYF18.

– Miquel, Juan-Carlos; Marty, Jean-Claude (2004): Downward flux of particles and carbon at trap DYF19.

– Miquel, Juan-Carlos; Marty, Jean-Claude (2004): Downward flux of particles and carbon at trap DYF20.

– Miquel, Juan-Carlos; Marty, Jean-Claude (2004): Downward flux of particles and carbon at trap DYF21.

– Miquel, Juan-Carlos; Marty, Jean-Claude (2004): Downward flux of particles and carbon at trap DYF23.

– Miquel, Juan-Carlos; Marty, Jean-Claude (2004): Downward flux of particles and carbon at trap DYF24.

– Miquel, Juan-Carlos; Marty, Jean-Claude (2004): Downward flux of particles and carbon at trap DYF25.

– Miquel, Juan-Carlos; Marty, Jean-Claude (2004): Downward flux of particles and carbon at trap DYF26.

– Miquel, Juan-Carlos; Marty, Jean-Claude (2004): Downward flux of particles and carbon at trap DYF27.

– Miquel, Juan-Carlos; Marty, Jean-Claude (2004): Downward flux of particles and carbon at trap DYF2-Calvi.

– Miquel, Juan-Carlos; Marty, Jean-Claude (2004): Downward flux of particles and carbon at trap DYF3.

– Miquel, Juan-Carlos; Marty, Jean-Claude (2004): Downward flux of particles and carbon at trap DYF3-Calvi.

– Miquel, Juan-Carlos; Marty, Jean-Claude (2004): Downward flux of particles and carbon at trap DYF4.

– Miquel, Juan-Carlos; Marty, Jean-Claude (2004): Downward flux of particles and carbon at trap DYF5.

– Miquel, Juan-Carlos; Marty, Jean-Claude (2004): Downward flux of particles and carbon at trap DYF5-Calvi.

– Miquel, Juan-Carlos; Marty, Jean-Claude (2004): Downward flux of particles and carbon at trap DYF6.

– Miquel, Juan-Carlos; Marty, Jean-Claude (2004): Downward flux of particles and carbon at trap DYF6-Calvi.

– Miquel, Juan-Carlos; Marty, Jean-Claude (2004): Downward flux of particles and carbon at trap DYF7.

– Miquel, Juan-Carlos; Marty, Jean-Claude (2004): Downward flux of particles and carbon at trap DYF7-Calvi.

– Miquel, Juan-Carlos; Marty, Jean-Claude (2004): Downward flux of particles and carbon at trap DYF8.

– Miquel, Juan-Carlos; Marty, Jean-Claude (2004): Downward flux of particles and carbon at trap DYF9.

– Neuer, Susanne; Ratmeyer, Volker; Davenport, Robert; Fischer, Gerhard; Wefer, Gerold (1997): Flux data from trap CI1.

– Neuer, Susanne; Ratmeyer, Volker; Davenport, Robert; Fischer, Gerhard; Wefer, Gerold (1997): Flux data from trap CI2.

– Neuer, Susanne; Ratmeyer, Volker; Davenport, Robert; Fischer, Gerhard; Wefer, Gerold (1997): Flux data from trap CI3.

– Neuer, Susanne; Ratmeyer, Volker; Davenport, Robert; Fischer, Gerhard; Wefer, Gerold (1997): Flux data from trap CI4.

– NGOFS; Tande, Kurt (2003): Particle flux of drift station JM9-1.

– OMEX Project Members; Wassmann, Paul (2004): Fluxes of trap JM10_DRIFT2.

– OMEX Project Members; Wassmann, Paul (2004): Fluxes of trap JM11_DRIFT1.

– OMEX Project Members; Wassmann, Paul (2004): Fluxes of trap JM3_DRIFT1.

– OMEX Project Members; Wassmann, Paul (2004): Fluxes of trap JM4_DRIFT1.

– OMEX Project Members; Wassmann, Paul (2004): Fluxes of trap JM5_DRIFT1.

– OMEX Project Members; Wassmann, Paul (2004): Fluxes of trap JM6_DRIFT1.

– OMEX Project Members; Wassmann, Paul (2004): Fluxes of trap JM7_DRIFT1.

– OMEX Project Members; Wassmann, Paul (2004): Fluxes of trap JM9_DRIFT1.

– Peinert, Rolf; Antia, Avan N; Bauerfeind, Eduard; von Bodungen, Bodo; Haupt, Olaf; Krumbholz, Marita; Peeked, Ilka; Ramseier, René O; Voß, Maren; Zeitzschel, Bernt (2005): Annual fluxes as measured by moored traps in different years for the Atlantic Province.

– Peinert, Rolf; Antia, Avan N; Bauerfeind, Eduard; von Bodungen, Bodo; Haupt, Olaf; Krumbholz, Marita; Peeken, Ilka; Ramseier, René O; Voß, Maren; Zeitzschel, Bernt (2005): Annual fluxes as measured by moored traps in different years for the Polar Province.

– Raab, Alexandra; von Bodungen, Bodo (2003): Monthly mean flux values at mooring station VP (Vring Plateau).

– Raab, Alexandra; von Bodungen, Bodo (2003): Monthly mean flux values at moring station NB (Lofoten Basin).

- Shevchenko, Vladimir P (2000): tab1+2.

- Tett, Paul (2005): Flux values of different LOIS-Trap Sites.

- Thomsen, C; von Bodungen, Bodo (2001): tab 3.1.1+2 Vertical particle flux and alkenones in mooring NB6.

- Thomsen, C; von Bodungen, Bodo (2001): tab 3.2.1+2 Vertical particle flux and alkenones in mooring OG4.

- Thunell, Robert C; Tappa, Eric (2013): Sediment Trap Data from the CARIACO Ocean Time Series (1995-2010).

- von Bodungen, Bodo; Antia, Avan N; Bauerfeind, Eduard; Haupt, Olaf; Koeve, Wolfgang; Machado, E; Peeken, Ilka; Peinert, Rolf; Reitmeier, Sven; Thomsen, C; Voß, Maren; Wunsch, M; Zeller, Ute; Zeitzschel, Bernt (1995): Flux data in the Greenland Basin from sediment trap OG4.

- von Bodungen, Bodo; Antia, Avan N; Bauerfeind, Eduard; Haupt, Olaf; Koeve, Wolfgang; Machado, E; Peeken, Ilka; Peinert, Rolf; Reitmeier, Sven; Thomsen, C; Voß, Maren; Wunsch, M; Zeller, Ute; Zeitzschel, Bernt (1995): Particle and nutrient flux data from mooring OG5 in the Greenland Basin.

- Waniek, Joanna J; Schulz-Bull, Detlef; Kuss, Joachim; Blanz, Thomas (2005): Deep ocean particle flux (>1000 m) and composition of particles collected by sediment traps at various sites in the northeast Atlantic (open ocean only) compiled from different sources.

- Waniek, Joanna J; Schulz-Bull, Detlef; Kuss, Joachim; Blanz, Thomas (2005): Interannual comparison of deep partical fluxes in the northeast Atlantic.

- Wefer, Gerold (2003): Particle fluxes of KG2_trap.

- Wefer, Gerold (2003): Particle fluxes of WS2_trap.

- Wefer, Gerold; Fischer, Gerhard (1991): (Table 2) Estimates of annual flux data in the South Atlantic.

- Wefer, Gerold; Fischer, Gerhard (1991): (Table 2a) Flux data of total mass and individual biogenic components for trap WS3.

- Wefer, Gerold; Fischer, Gerhard (1991): (Table 2b) Flux data of total mass and individual biogenic components for trap WS4.

- Wefer, Gerold; Fischer, Gerhard (1991): (Table 2c) Flux data of total mass and individual biogenic components for trap Polar_Front_1.

- Wefer, Gerold; Fischer, Gerhard (2003): Particle fluxes of CB1_trap.

- Wefer, Gerold; Fischer, Gerhard (2003): Particle fluxes of CB2_trap.

- Wefer, Gerold; Fischer, Gerhard (2003): Particle fluxes of CB4_trap.

- Wefer, Gerold; Fischer, Gerhard (2003): Particle fluxes of GBN6_trap.

- Wefer, Gerold; Fischer, Gerhard (2003): Particle fluxes of GBZ4_trap.

- Žarić, Snježana (2005): Planktic foraminiferal flux of sediment trap CB1_trap.

- Žarić, Snježana (2005): Planktic foraminiferal flux of sediment trap CB2_trap.

- Žarić, Snježana (2005): Planktic foraminiferal flux of sediment trap CB3_trap.

– Žarić, Snježana (2005): Planktic foraminiferal flux of sediment trap CB4_trap, lower.

– Žarić, Snježana (2005): Planktic foraminiferal flux of sediment trap CB4_trap, upper.

– Žarić, Snježana (2005): Planktic foraminiferal flux of sediment trap CB5_trap.

– Žarić, Snježana (2005): Planktic foraminiferal flux of sediment trap EA1_trap.

– Žarić, Snježana (2005): Planktic foraminiferal flux of sediment trap EA2_trap.

– Žarić, Snježana (2005): Planktic foraminiferal flux of sediment trap EA3_trap.

– Žarić, Snježana (2005): Planktic foraminiferal flux of sediment trap EA4_trap.

– Žarić, Snježana (2005): Planktic foraminiferal flux of sediment trap WA1_trap, 1232 m trap depth. Department of Geosciences, Bremen University,

– Žarić, Snježana (2005): Planktic foraminiferal flux of sediment trap WA1_trap, 4991 m trap depth. Department of Geosciences, Bremen University,

– Žarić, Snježana (2005): Planktic foraminiferal flux of sediment trap WA1_trap, 652 m trap depth. Department of Geosciences, Bremen University,

– Žarić, Snježana (2005): Planktic foraminiferal flux of sediment trap WA2_trap. Department of Geosciences, Bremen University,

– Žarić, Snježana (2005): Planktic foraminiferal flux of sediment trap WA3_trap, 5031 m trap depth. Department of Geosciences, Bremen University,

– Žarić, Snježana (2005): Planktic foraminiferal flux of sediment trap WAB1_trap, 4515 m trap depth. Department of Geosciences, Bremen University,

– Žarić, Snježana (2005): Planktic foraminiferal flux of sediment trap WAB1_trap, 727 m trap depth. Department of Geosciences, Bremen University,

– Žarić, Snježana (2005): Planktic foraminiferal flux of sediment trap WAB2_trap. Department of Geosciences, Bremen University,

– Žarić, Snježana (2005): Planktic foraminiferal flux of sediment trap WR2_trap.

– Žarić, Snježana (2005): Planktic foraminiferal flux of sediment trap WR3_trap.

– Žarić, Snježana (2005): Planktic foraminiferal flux of sediment trap WR4_trap.

– Žarić, Snježana (2005): Planktic foraminiferal flux of sediment trap WS1_trap.

– Žarić, Snježana (2005): Planktic foraminiferal flux of sediment trap WS2_trap.

– Žarić, Snježana (2005): Planktic foraminiferal flux of sediment trap WS3_trap.

– Žarić, Snježana (2005): Planktic foraminiferal flux of sediment trap WS4_trap.

Acknowledgements. Funding was provided by the European Commission FP7 EURO-BASIN (European Basin-Scale Analysis, Synthesis, and Integration; grant agreement 264 933EU). Additional support was provided by the National Environment Research Council (UK) national capability funds. The authors are grateful to Cindy Lee, Susanne Neuer, Richard Lampitt and Eduard Bauerfeind for the provision of additional data to that already publicly available. We are also grateful to Ian Salter for useful comments on an earlier version of this manuscript, and to two anonymous reviewers for their comments, suggestions and criticism, which helped improve this manuscript. Most figures were produced using the software package ODV.

Edited by: Y.-W. Luo

References

Antia, A. N., von Bodungen, B., and Peinert, R.: Particle flux across the mid-European continental margin, Deep-Sea Res. Pt. I, 46, 1999–2024, 1999.

Bahr, F., Kelly, R., Bates, N. R., Becker, S., Bell, S., Countway, P., Caporelli, E., Church, M. J., Close, A., Doyle, A., Gundersen, K., Hammer, M., Howse, F., Johnson, R., Goldthwait, S., Little, R., Morrison, R., Orcutt, K., Sanderson, M., Sherriff-Dow, R., Sorensen, J., Stone, S., Rathbun, C., and Waterhouse, T.: Bermuda Atlantic Time-Series Study: Methods, Tech. Rep., 1997.

Bauerfeind, E. and Nöthig, E.-M.: Biogenic particle flux at AWI HAUSGARTEN from mooring FEVI3 at 2400 m, Alfred Wegener Institute for Polar and Marine Research, Bremerhaven, Germany, 2011.

Bauerfeind, E., Noethig, E.-M., Beszczynska, A., Fahl, K., Kaleschke, L., Kreker, K., Klages, M., Soltwedel, T., Lorenzen, C., and Wegner, J.: Particle sedimentation patterns in the eastern Fram Strait during 2000–2005: Results from the Arctic long-term observatory HAUSGARTEN, Deep-Sea Res. Pt. I, 56, 1471–1487, 2009.

Bory, A. J. M. and Newton, P. P.: Transport of airborne lithogenic material down through the water column in two contrasting regions of the eastern subtropical North Atlantic Ocean, Global Biogeochem. Cy., 14, 297–315, 2000.

Bory, A., Jeandel, C., Leblond, N., Vangriesheim, A., Khripounoff, A., Beaufort, L., Rabouille, C., Nicolas, E., Tachikawa, K., and Etcheber, H.: Downward particle fluxes within different productivity regimes off the Mauritanian upwelling zone (EUMELI program), Deep-Sea Res. Pt. I, 48, 2251–2282, 2001.

DeMaster, D. J.: The Supply and Accumulation of Silica in the Marine-Environment, Geochim. Cosmochim. Ac., 45, 1715–1732, 1981.

Dymond, J. and Lyle, M.: Particle Fluxes in the Ocean and Implications for Sources and Preservation of Ocean Sediments, in: Material Fluxes on the Surface of the Earth, 125–142, The National Academies Press, 1994.

Dymond, J. and Lyle, M. W.: Particle fluxes of NAP_trap, 2003a.

Dymond, J. and Lyle, M. W.: Particle fluxes of HAP-4_trap, 2003b.

Eggimann, D. W., Manheim, F. T., and Betzer, P. R.: Dissolution and Analysis of Amorphous Silica in Marine Sediments, J. Sediment. Res., 50, 215–225, doi:10.1016/0967-0645(93)90034-K, 1980.

Eppley, R. W. and Peterson, B. J.: Particulate organic matter flux and planktonic new production in the deep ocean, Nature, 282, 677–680, 1979.

Fahl, K. and Nöthig, E.-M.: Lithogenic and biogenic particle fluxes on the Lomonosov Ridge (central Arctic Ocean) and their relevance for sediment accumulation: Vertical vs. lateral transport, Deep-Sea Res. Pt. I, 54, 1256–1272, 2007.

Fischer, G.: Particle fluxes of trap NU2, Geosciences, University of Bremen (GeoB), 2003a.

Fischer, G.: Flux data of trap WR2, Geosciences, University of Bremen (GeoB), 2003b.

Fischer, G.: Flux data of trap WR3, Geosciences, University of Bremen (GeoB), 2003c.

Fischer, G.: Flux data of trap WR4, Geosciences, University of Bremen (GeoB), 2003d.

Fischer, G.: Particle fluxes for the sampling interval, various ratios and major nutrients at the Atlantic/Southern Ocean trapping sites, Center for Marine Environmental Sciences (MARUM), 2005.

Fischer, G., Donner, B., Ratmeyer, V., Davenport, R., and Wefer, G.: Distinct year-to-year particle flux variations off Cape Blanc during 1988–1991: Relation to δ18O-deduced sea-surface temperatures and trade winds, J. Mar. Res., 54, 73–98, 1996.

Fischer, G., Ratmeyer, V., and Wefer, G.: Organic carbon fluxes in the Atlantic and the Southern Ocean: relationship to primary production compiled from satellite radiometer data, Deep-Sea Res. Pt. II, 47, 1961–1997, 2000.

Fischer, G., Gersonde, R., and Wefer, G.: Organic carbon, biogenic silica and diatom fluxes in the marginal winter sea-ice zone and in the Polar Front Region: interannual variations and differences in composition, Deep-Sea Res. Pt. II, 49, 1721–1745, 2002.

Goutx, M., Momzikoff, A., Striby, L., Andersen, V., Marty, J., and Vescovali, I.: High-frequency fluxes of labile compounds in the central Ligurian Sea, northwestern Mediterranean, Deep-Sea Res. Pt. I, 47, 533–556, 2000.

Goutx, M., Wakeham, S. G., Lee, C., Duflos, M., Guigue, C., Liu, Z., Moriceau, B., Sempere, R., Tedetti, M., and Xue, J.: Composition and degradation of marine particles with different settling velocities in the northwestern Mediterranean Sea, Limnol. Oceanogr., 52, 1645–1664, 2007.

Helmke, P., Romero, O., and Fischer, G.: Northwest African upwelling and its effect on offshore organic carbon export to the deep sea, Global Biogeochem. Cy., 19, GB4015, doi:10.1029/2004GB002265, 2005.

Helmke, P., Neuer, S., Lomas, M. W., Conte, M., and Freudenthal, T.: Cross-basin differences in particulate organic carbon export and flux attenuation in the subtropical North Atlantic gyre, Deep-Sea Res. Pt. I, 57, 213–227, 2010.

Honjo, S. and Doherty, K.: Large aperture time-series sediment traps; design objectives, construction and application, Deep-Sea Res., 35, 133–149, 1988.

Honjo, S. and Manganini, S.: Annual biogenic particle fluxes to the interior of the North Atlantic Ocean; studied at 34° N 21° W and 48° N 21° W, Deep-Sea Res. Pt. II, 40, 587–607, 1993.

Honjo, S. and Manganini, S. J.: Annual Particle fluxes of NABE-N34_trap, Woods Hole Oceanographic Institution, USA: U.S. JGOFS Data Management Office, 2003a.

Honjo, S. and Manganini, S. J.: Biogenic particle flux of trap NABE-N34.1, Woods Hole Oceanographic Institution, USA: U.S. JGOFS Data Management Office, 2003b.

Honjo, S. and Manganini, S. J.: Biogenic particle flux of trap NABE-N34.2, Woods Hole Oceanographic Institution, USA: U.S. JGOFS Data Management Office, 2003c.

Honjo, S. and Manganini, S. J.: Annual Particle fluxes of NABE-N48_trap, Woods Hole Oceanographic Institution, USA: U.S. JGOFS Data Management Office, 2003d.

Honjo, S. and Manganini, S. J.: Biogenic particle flux of trap NABE-N48.1, Woods Hole Oceanographic Institution, USA: U.S. JGOFS Data Management Office, 2003e.

Honjo, S. and Manganini, S. J.: Biogenic particle flux of trap NABE-N48.2, Woods Hole Oceanographic Institution, USA: U.S. JGOFS Data Management Office, 2003f.

Hwang, J., Manganini, S. J., Montlucon, D. B., and Eglinton, T. I.: Dynamics of particle export on the Northwest Atlantic margin, Deep-Sea Res. Pt. I, 56, 1792–1803, 2009.

Irwin, B.: POC, PON, chlorophyll, phaeophytin of BAF89/3_FTRAP1, Marine Environmental Data Service, Department of Fisheries and Oceans, Canada, 2002a.

Irwin, B.: POC, PON, chlorophyll, phaeophytin of BAF89/3_FTRAP2, Marine Environmental Data Service, Department of Fisheries and Oceans, Canada, 2002b.

Iversen, M. H., Nowald, N., Ploug, H., Jackson, G. A., and Fischer, G.: High resolution profiles of vertical particulate organic matter export off Cape Blanc, Mauritania: Degradation processes and ballasting effects, Deep-Sea Res. Pt. I, 57, 771–784, 2010.

Jickells, T., Newton, P., King, P., Lampitt, R., and Boutle, C.: A comparison of sediment trap records of particle fluxes from 19 to 48°N in the northeast Atlantic and their relation to surface water productivity, Deep-Sea Res. Pt. I, 43, 971–986, 1996.

Jickells, T. D.: Particle fluxes of 19° N20° W_trap, 2003a.

Jickells, T. D.: Particle fluxes of 24° N23° W_trap, 2003b.

Jickells, T. D.: Particle fluxes of 28° N22° W_trap, 2003c.

Jickells, T. D.: Particle fluxes of Parflux7G_trap, 2003d.

Jonkers, L., Brummer, G.-J. A., Peeters, F. J. C., van Aken, H. M., and De Jong, M. F.: Seasonal stratification, shell flux, and oxygen isotope dynamics of left-coiling N. pachydermaand T. quinquelobain the western subpolar North Atlantic, Paleoceanography, 25, PA2204, doi:10.1029/2009PA001849, 2010.

Koning, E., Epping, E., and Van Raaphorst, W.: Determining Biogenic Silica in Marine Samples by Tracking Silicate and Aluminium Concentrations in Alkaline Leaching Solutions, Aquat. Geochem., 8, 37–67, 2002.

Kremling, K., Lentz, U., Zeitzschel, B., Bull, D. E. S., and Duinker, J. C.: New type of timeseries sediment trap for the reliable collection of inorganic and organic trace chemical substances, Rev. Sci. Instrum., 67, 4360–4363, 1996.

Lampitt, R. and Antia, A.: Particle flux in deep seas: regional characteristics and temporal variability, Deep-Sea Res. Pt. I, 44, 1377–1403, 1997.

Lampitt, R. S., Bett, B. J., Kiriakoulakis, K., Popova, E. E., Ragueneau, O., Vangriesheim, A., and Wolff, G. A.: Material supply to the abyssal seafloor in the Northeast Atlantic, Prog. Oceanogr., 50, 27–63, 2001.

Lampitt, R. S., Salter, I., de Cuevas, B. A., Hartman, S., Larkin, K. E., and Pebody, C. A.: Long-term variability of downward particle flux in the deep northeast Atlantic: Causes and trends, Deep-Sea Res. Pt. II, 57, 1346–1361, 2010.

Laws, E. A., Falkowski, P. G., Smith Jr., W. O., Ducklow, H., and McCarthy, J. J.: Temperature effects on export product in the open ocean, Global Biogeochem. Cy., 14, 1231–1246, 2000.

Lee, C., Armstrong, R. A., Cochran, J. K., Engel, A., Fowler, S. W., Goutx, M., Masque, P., Miquel, J. C., Peterson, M., Tamburini, C., and Wakeham, S.: MedFlux: Investigations of particle flux in the Twilight Zone, Deep-Sea Res. Pt. II, 56, 1363–1368, 2009a.

Lee, C., Peterson, M. L., Wakeham, S. G., Armstrong, R. A., Cochran, J. K., Miquel, J. C., Fowler, S. W., Hirschberg, D., Beck, A., and Xue, J.: Particulate organic matter and ballast fluxes measured using time-series and settling velocity sediment traps in the northwestern Mediterranean Sea, Deep-Sea Res. Pt. II, 56, 1420–1436, 2009b.

Lisitsyn, A. P., Shevchenko, V. P., Vinogradov, M. E., Severina, O. V., Vavilova, V. V., and Mitskevich, I. N.: Particle fluxes in the Kara Sea dn Ob and Yenisey estuaries, Oceanology, 34, 683–693, 1995.

Martin, J.: Particle interceptor data of sediment trap AT_II-119/4_Trap_A, Woods Hole Oceanographic Institution, USA, 2003a.

Martin, J.: Particle interceptor data of sediment trap AT_II-119/4_Trap_B, Woods Hole Oceanographic Institution, USA, 2003b.

Martin, J.: Particle interceptor data of sediment trap AT_II-119/5_Trap_C, Woods Hole Oceanographic Institution, USA, 2003c.

Miquel, J. C., Martin, J., Gasser, B., Rodriguez-y Baena, A., Toubal, T., and Fowler, S. W.: Dynamics of particle flux and carbon export in the northwestern Mediterranean Sea: a two decade time-series study at the DYFAMED site, Prog. Oceanogr., 91, 461–481, 2011.

Montes, E., Müller-Karger, F., Thunell, R., Hollander, D., Astor, Y., Varela, R., Soto, I., and Lorenzoni, L.: Vertical fluxes of particulate biogenic material through the euphotic and twilight zones in the Cariaco Basin, Venezuela, Deep-Sea Res. Pt. I, 67, 73–84, 2012.

Mortlock, R. A. and Froelich, P. N.: A simple method for the rapid determination of biogenic opal in pelagic marine sediments, Mar. Chem., 36, 1415–1426, 1989.

Müller, P. J. and Schneider, R.: An automated leaching method for the determination of opal in sediments and particulate matter, Deep-Sea Res. Pt. I, 40, 425–444, 1993.

Neuer, S., Ratmeyer, V., Davenport, R., Fischer, G., and Wefer, G.: Deep water particle flux in the Canary Island region: seasonal trends in relation to long-term satellite derived pigment data and lateral sources, Deep-Sea Res. Pt. I, 44, 1451–1466, 1997.

Neuer, S., Cianca, A., Helmke, P., Freudenthal, T., Davenport, R., Meggers, H., Knoll, M., Santana-Casiano, J. M., Gonz alez Davila, M., Rueda, M.-J. E., and Llin as, O.: Biogeochemistry and hydrography in the eastern subtropical North Atlantic gyre. Results from the European time-series station ESTOC, Prog. Oceanogr., 72, 1–29, 2007.

Newton, P. P., Lampitt, R. S., Jickells, T. D., King, P., and Boutle, C.: Temporal and spatial variability of biogenic particles fluxes during the JGOFS northeast Atlantic process studies at 47° N, 20° W, Deep-Sea Res. Pt. I, 41, 1617–1642, 1994.

NGOFS and Tande, K.: Particle flux of drift station JM9-1, PAN-GAEA, 2003.

OMEX and Wassmann, P.: Fluxes of trap JM3_DRIFT1, PAN-GAEA, 2004a.

OMEX and Wassmann, P.: Fluxes of trap JM4_DRIFT1, PAN-GAEA, 2004b.

OMEX and Wassmann, P.: Fluxes of trap JM5_DRIFT1, PAN-GAEA, 2004c.

OMEX and Wassmann, P.: Fluxes of trap JM6_DRIFT1, PAN-GAEA, 2004d.

OMEX and Wassmann, P.: Fluxes of trap JM7_DRIFT1, PAN-GAEA, 2004e.

OMEX and Wassmann, P.: Fluxes of trap JM9_DRIFT1, PAN-GAEA, 2004f.

OMEX and Wassmann, P.: Fluxes of trap JM10_DRIFT2, PAN-GAEA, 2004g.

OMEX and Wassmann, P.: Fluxes of trap JM11_DRIFT1, PAN-GAEA, 2004h.

Peinert, R., Antia, A., Bauerfeind, E., von Bodungen, B., Haupt, O., Krumbholz, M., Peeken, I., Ramseier, R. O., Voß, M., and Zeitzschel, B.: Particle flux variability in the polar and Atlantic provinces of the Nordic Seas, in: The Northern North Atlantic: A Changing Environment, 53–68, Springer Verlag, 2001.

Peterson, M., Wakeham, S., Lee, C., Askea, M., and Miquel, J.: Novel techniques for collection of sinking particles in the ocean and determining their settling rates, Limnol. Oceanogr.-Meth., 3, 520–532, 2005.

Peterson, M. L., Fabres, J., Wakeham, S. G., Lee, C., Alonso, I. J., and Miquel, J. C.: Sampling the vertical particle flux in the upper water column using a large diameter free-drifting NetTrap adapted to an Indented Rotating Sphere sediment trap, Deep-Sea Res. Pt. II, 56, 1547–1557, 2009.

Raab, A.: Sedimente des Changeable-Sees, Oktoberrevolutions-Insel (Severnaja Zemlja), als Archive der Paläoumwelt Mittelsibiriens seit dem Frühweichsel = Changeable lake sediments, October Revolution Island (Severnaya Zemlya), as an archive for the environmental history in central Siberia since the early Weichselian, Berichte zur Polar- und Meeresforschung (Reports on Polar and Marine Research), Bremerhaven, Alfred Wegener Institute for Polar and Marine Research, 435, p. 115, 2003.

Ragueneau, O., Gallinari, M., Corrin, L., Grandel, S., Hall, P., Hauvespre, A., Lampitt, R., Rickert, D., Stahl, H., and Witbaard, A. T. R.: The benthic silica cycle in the Northeast Atlantic: annual mass balance, seasonality, and importance of non-steadystate processes for the early diagenesis of biogenic opal in deep-sea sediments, Prog. Oceanogr., 50, 171–200, 2001.

Romero, O., Boeckel, B., Lavik, G., Fischer, G., and Wefer, G.: Seasonal productivity dynamics in the pelagic central Benguela System inferred from the flux of carbonate and silicate organisms, J. Marine Syst., 37, 259–278, 2002.

Rudnick, R. L. and Gao, S.: Composition of the Continental Crust, in: Treatise on Geochemistry, 1–64, Elsevier, 2003.

Salter, I., Kemp, A. E. S., Lampitt, R. S., and Gledhill, M.: The association between biogenic and inorganic minerals and the amino acid composition of settling particles, Limnol. Oceanogr., 55, 2207–2218, 2010.

Sanders, R., Henson, S., Koski, M., De La Rocha, C., Painter, S. C., Poulton, A., Riley, J., Salihoglu, B., Visser, A., Yool, A., Bellerby, R., and Martin, A.: The Biological Carbon Pump in the North Atlantic, Prog. Oceanogr., accepted, 2014.

Shevchenko, V. P.: tab1+2, PANGAEA, 2000.

Soutar, A., Kling, S. A., Crill, P. A., Duffrin, E., and Bruland, K. W.: Monitoring the marine environment through sedimentation, Nature, 266, 136–139, 1977.

Stein, R.: Modern and late quaternary depositional environment of the St. Anna Trough Area, Northern Kara Sea, Berichte zur Polarforschung (Reports on Polar Research), Tech. Rep. 342, Alfred Wegener Institute for Polar and Marine Research, Bremerhaven, 1999.

Tett, P.: Flux values of different LOIS-Trap Sites, PANGAEA, 2005.

Thomsen, C. and von Bodungen, B.: tab 3.2.1+2 Vertical particle flux and alkenones in mooring OG4, PANGAEA, 2001a.

Thomsen, C. and von Bodungen, B.: tab 3.1.1+2 Vertical particle flux and alkenones in mooring NB6, 2001b.

Thomsen, C. and von Bodungen, B.: tab 3.3.1+2 Vertical particle flux and alkenones in mooring BI2, PANGAEA, 2001c.

von Bodungen, B., Antia, A., Bauerfeind, E., Haupt, O., Koeve, W., Machado, E., Peeken, I., Peinert, R., Reitmeier, S., Thomsen, C., Voss, M., Wunsch, M., Zeller, U., and Zeitzschel, B.: Pelagic processes and vertical flux of particles: an overview of a long-term comparative study in the Norwegian Sea and Greenland Sea, Geol. Rundsch., 84, 11–27, 1995.

Waniek, J., Schulz-Bull, D., Kuss, J., and Blanz, T.: Long time series of deep water particle flux in three biogeochemical provinces of the northeast Atlantic, J. Marine Syst., 56, 391–415, 2005.

Wedepohl, H. K.: The composition of the continental crust, Geochim. Cosmochim. Ac., 59, 1217–1232, 1995.

Wefer, G. and Fischer, G.: Annual primary production and export flux in the Southern Ocean from sediment trap data, Mar. Chem., 35, 597–613, 1991.

Wefer, G. and Fischer, G.: Seasonal patterns of vertical particle flux in equatorial and coastal upwelling areas of the eastern Atlantic, Deep-Sea Res. Pt. I, 40, 1613–1645, 1993.

Žarić, S., Donner, B., Fischer, G., Mulitza, S., and Wefer, G.: Sensitivity of planktic foraminifera to sea surface temperature and export production as derived from sediment trap data, Mar. Micropaleontol., 55, 75–105, 2005.

Zeitzschel, B., Diekmann, P., and Uhlmann, L.: A new multisample sediment trap, Mar. Biol., 45, 285–288, 1978.

9

CoastColour Round Robin data sets: a database to evaluate the performance of algorithms for the retrieval of water quality parameters in coastal waters

B. Nechad[1], K. Ruddick[1], T. Schroeder[2], K. Oubelkheir[2], D. Blondeau-Patissier[3], N. Cherukuru[4],
V. Brando[4], A. Dekker[4], L. Clementson[4], A. C. Banks[5,20], S. Maritorena[6], P. J. Werdell[7], C. Sá[8],
V. Brotas[8], I. Caballero de Frutos[9], Y.-H. Ahn[10], S. Salama[11], G. Tilstone[12], V. Martinez-Vicente[12],
D. Foley[13,†], M. McKibben[14], J. Nahorniak[14], T. Peterson[15], A. Siliò-Calzada[16], R. Röttgers[17], Z. Lee[18],
M. Peters[19], and C. Brockmann[19]

[1] Operational Directorate Natural Environment, Royal Belgian Institute for Natural Sciences (RBINS/ODNE),
100 Gulledelle Brussels, 1200, Belgium
[2] Commonwealth Scientific and Industrial Research Organisation (CSIRO), Land and Water, Environmental
Earth Observation Program, P.O. Box 2583, Brisbane, QLD 2001, Australia
[3] Charles Darwin University, 0815 Darwin, Australia
[4] Commonwealth Scientific and Industrial Research Organisation (CSIRO), P.O. Box 1666,
Canberra, ACT, Australia
[5] Hellenic Centre for Marine Research (HCMR), Institute of Oceanography, P.O. Box 2214,
Heraklion 71003, Crete, Greece
[6] Earth Research Institute (ERI), University of California, Santa Barbara, CA 93106-3060, USA
[7] NASA Goddard Space Flight Center, Greenbelt, MD 20771, USA
[8] Marine and Environmental Sciences Centre (MARE), Faculdade de Ciências da Universidade de Lisboa,
Campo Grande, 1749-016 Lisbon, Portugal
[9] Institute of Marine Sciences of Andalucia (ICMAN-CSIC) Puerto Real-Cádiz, 11519, Spain
[10] Korea Ocean Research & Development Institute (KORDI), Ansan, P.O. Box 29, 425–600, South Korea
[11] Faculty of Geo-information Science and Earth Observation (ITC), Department of Water Resource, University
of Twente, Hengelosestraat 99, 7500 AA Enschede, the Netherlands
[12] Plymouth Marine Laboratory, Prospect Place, The Hoe, Plymouth PL1 3DH, UK
[13] National Oceanic and Atmospheric Administration (NOAA), Southwest Fisheries Science Center,
110 Shaffer Road, Santa Cruz, CA 95060, USA
[14] College of Earth, Ocean and Atmospheric Sciences (CEOAS), Oregon State University, Corvallis, OR, USA
[15] Center for Coastal Margin Observation and Prediction and Institute of Environmental Health, Oregon Health
and Science University, 3181 SW Sam, Jackson Park Road, Portland, Oregon 97239, USA
[16] Environmental Hydraulics Institute of the University of Cantabria, Cantabria, Spain
[17] Institute of Coastal Research, Helmholtz-Zentrum Geesthacht, Centre for Materials and Coastal Research,
Max-Plank-Str. 1, 21502 Geesthacht, Germany
[18] School for the Environment, University of Massachusetts Boston, Boston, MA 02125, USA
[19] Brockmann Consult, Max-Planck-Str. 2, 21502 Geesthacht, Germany
[20] European Commission – Joint Research Centre (JRC), Institute for Environment and Sustainability,
Via Enrico Fermi 2749, Ispra (Va) 21027, Italy
†deceased

Correspondence to: B. Nechad (bnechad@naturalsciences.be)

Abstract. The use of in situ measurements is essential in the validation and evaluation of the algorithms that provide coastal water quality data products from ocean colour satellite remote sensing. Over the past decade, various types of ocean colour algorithms have been developed to deal with the optical complexity of coastal waters. Yet there is a lack of a comprehensive intercomparison due to the availability of quality checked in situ databases. The CoastColour Round Robin (CCRR) project, funded by the European Space Agency (ESA), was designed to bring together three reference data sets using these to test algorithms and to assess their accuracy for retrieving water quality parameters. This paper provides a detailed description of these reference data sets, which include the Medium Resolution Imaging Spectrometer (MERIS) level 2 match-ups, in situ reflectance measurements, and synthetic data generated by a radiative transfer model (HydroLight).

The data sets mainly consist of 6484 marine reflectance (either multispectral or hyperspectral) associated with various geometrical (sensor viewing and solar angles) and sky conditions and water constituents: total suspended matter (TSM) and chlorophyll a (CHL) concentrations, and the absorption of coloured dissolved organic matter (CDOM). Inherent optical properties are also provided in the simulated data sets (5000 simulations) and from 3054 match-up locations. The distributions of reflectance at selected MERIS bands and band ratios, CHL and TSM as a function of reflectance, from the three data sets are compared. Match-up and in situ sites where deviations occur are identified. The distributions of the three reflectance data sets are also compared to the simulated and in situ reflectances used previously by the International Ocean Colour Coordinating Group (IOCCG, 2006) for algorithm testing, showing a clear extension of the CCRR data which covers more turbid waters.

1 Introduction

Several studies on the intercomparison of ocean colour algorithms have been carried out to provide recommendations on appropriate methodologies and identify the domains of applicability and limitations or weaknesses of the algorithms, e.g. O'Reilly et al. (1998), Maritorena et al. (2006), Brewin et al. (2015), Odermatt et al. (2012), and Werdell et al. (2013). Except for the open ocean waters (or case 1 waters; Morel and Prieur, 1977), chlorophyll a algorithm studies, no substantial consensus was achieved regarding a convergence of approaches for the retrieval of in-water properties from satellite or in situ radiometric measurements in coastal waters.

The diversity of approaches is especially high in case 2 waters (Morel and Prieur, 1977) with higher complexity of the optical properties and larger ranges of in-water constituent concentrations. To understand how these elements can affect the performance of algorithms, the CoastColour Round Robin (CCRR) project was designed (Ruddick et al., 2010). The CCRR uses a variety of reference data sets to test algorithms and compare their accuracy for retrieving water quality (WQ) parameters. These WQ parameters include chlorophyll a (CHL) and total suspended matter (TSM) concentrations, inherent optical properties (IOPs), underwater light attenuation coefficients such as the diffuse attenuation of the downwelling irradiance (Kd) or the photosynthetically available radiation (PAR) with which a set of satellite data processing quality flags are associated.

Three types of data are being prepared for the CCRR: (a) match-ups, where in situ WQ is available simultaneously with a cloud-free Medium Resolution Imaging Spectrometer (MERIS) product; (b) in situ reflectances, where an in situ

water-leaving reflectance measurement (denoted by RLw, which is derived from the remote-sensing reflectance, Rrs, following RLw $= \pi$ Rrs) is available simultaneously with an in situ WQ; and (c) simulated RLw for specified sets of IOPs and geometrical conditions, using HydroLight. MERIS images are also provided for the selected regions where the remote-sensing WQ algorithms are tested.

The match-ups, the in situ reflectance and the simulated data sets are presented in Sect. 2, and the variability in WQ is characterized. The data from the three data sets are intercompared in Sect. 3. This study provides documentation for the publicly available data sets (as detailed in Sect. 4) which can be used as benchmarks for ocean colour algorithm testing in coastal waters in order to ultimately improve the remote-sensing algorithms.

2 Data

The in situ WQ parameters provided in the match-up data set and referred to hereafter as "match-up field measurements" are described in Sect. 2.1.1. The concurrent MERIS level 2 products, reported in Sect. 2.1.2, include the MERIS reflectances and WQ, denoted respectively as L2R and L2W, and level 2 flags.

The in situ reflectance data set, described in Sect. 2.2, consists of in situ TSM and CHL measurements collected simultaneously with reflectances that cover the spectral range 440–709 nm. Inclusion of the 709 nm band in these data sets is important because it allows testing of algorithms exploiting this MERIS band, which is unique amongst any other ocean colour mission spectral specifications, operational up to 2012, e.g. for the retrieval of CHL or fluorescence line

Table 1. Acronyms of in situ data sources, as well as associated websites where the original data and methodologies are available.

Acronym	Name
CEOAS/OSU (CEOAS)	College of Earth, Ocean and Atmospheric Sciences – Oregon State university (USA)
CSIC	Spanish Institute for Marine Sciences (Spain)
CSIR	Council for Scientific and Industrial Research (South Africa)
CSIRO	Commonwealth Scientific and Industrial Research Organisation (Australia)
EMECO	European Marine ECosystem Observatory http://www.emecodata.net
GKSS	Centre for Materials and Coastal Research, Helmholtz-Zentrum Geesthacht (Germany)
HCMR	Hellenic Centre for Marine Research (Greece)
Ifremer	French Research Institute for Exploration of the Sea (France) http://wwz.ifremer.fr/lerpc/Activites-et-Missions/Surveillance/REPHY
ITC	International Institute for Geo-Information Science and Earth Observation (Netherlands)
KORDI	Korea Ocean Research and Development Institute (South Korea)
MII	Marine Institute of Ireland (Ireland) http://data.marine.ie
MSU	Mississippi State University (USA)
NOAA	National Oceanic and Atmospheric Administration (USA)
NOMAD	NASA bio-Optical Marine Algorithm Dataset, http://seabass.gsfc.nasa.gov
PML	Plymouth Marine Laboratory (UK)
RBINS	Royal Belgian Institute for Natural Sciences (Belgium)
UCSB	University of California at Santa Barbara, Earth Research Institute (USA)
UNICAN	Environmental Hydraulics Institute of the University of Cantabria (Spain)

height using reflectance at band 709 nm combined with other bands in or around the phytoplankton absorption peak.

The artificial data set, based on radiative transfer simulations, is presented in Sect. 2.3. The match-up, in situ reflectance and simulated data sets come from 18 research institutes or databases (Table 1).

2.1 Match-up data set

The measurements in the match-up data set cover various water types from ocean and coastal regions called CoastColour sites, and consist of a collection of biogeochemical and optical measurements (inherent and apparent optical properties, hereafter referred to as IOPs and AOPs) along with the associated metadata. Only the WQ parameters for which remote-sensing algorithms are tested within the CCRR, such as CHL and TSM (see Table 2), are described in this paper, although supplementary oceanographic parameters are also included in the match-up database.

The match-up field measurements were collected at 17 CoastColour sites, selected in the framework of the CCRR (Fig. 1), where in situ WQ parameters from 2005 to 2010 were available, and measured above 5 m depth. MERIS L2R and L2W products from 2005 to 2010, derived at match-up locations, are included in the match-up data set, but only those of MERIS L2R are described in this paper.

The temporal availability of these data displayed in Fig. 2 shows unbalanced distributions over the CoastColour sites. The seasonal distribution of the match-up field measurements varies from one site to another (Fig. 3). For example, for chlorophyll a measurements, 52 % of the Acadia data were collected during the period June–August, 67 %

of Chesapeake Bay data during September–November, and 100 % of Benguela data during March–May; the seasonal distribution may also vary within each site between the different WQ parameters. From all the sites, the ensemble of temperature, salinity, chlorophyll a, particulate organic matter (PIM), and particulate inorganic matter (POM) measurements is evenly balanced throughout the seasons. During December–February, fewer TSM, turbidity, a, a_p, a_{phy}, a_d, and a_g measurements are available than during the other periods (about 13 to 18 % of the data), while the quantity of AOP data is significantly lower (2 to 9 % of the data).

2.1.1 Match-up field measurements

The number of stations where metadata and biogeochemical, IOP, and AOP data were collected over the CoastColour sites are reported in Table 3a and b. The availability of measurements throughout the sites varies from one parameter to another; for example, chlorophyll a concentration measurements are available from 16 sites, while the scattering coefficient spectra are provided at 2 sites.

Metadata, including depth, temperature, and salinity, exceed 20 000 for each parameter, whereas the numbers of bio-geochemical data, IOPs, and AOPs are much lower: 11 208 chlorophyll a concentration measurements, 538 TSM measurements, 957 reflectance spectra (the other AOP data do not reach 200 data each), and fewer than 700 IOP data (for each parameter) except for turbidity ($N = 2187$).

The number of CHL and turbidity measurements collected at the North Sea site constitute 77.0 and 99.8 % of the measurements respectively, while smaller numbers of TSM and RLw data are provided from the North Sea site: 39.4 and

Table 2. Metadata, IOPs, and AOPs given at wavelength λ, and biogeochemical in situ measurements available for the CoastColour sites. The two notations Chl *a* and TChl *a* refer to chlorophyll *a* concentration measured by high-performance liquid chromatography (HPLC) and by fluorometry respectively.

Metadata	Notation	Units	Concentrations	Notation	Units
Date, time		–	Chlorophyll *a* (fluorometry)	Chl *a*	$\mathrm{mg\,m^{-3}}$
Station, cruise		–	Total chlorophyll *a* (HPLC)	TChl *a*	$\mathrm{mg\,m^{-3}}$
File name, File_id (station)		–	TSM	TSM	$\mathrm{g\,m^{-3}}$
Latitude, longitude		degrees	Non algal particulate matter	NAP	$\mathrm{g\,m^{-3}}$
Wind speed		$\mathrm{m\,s^{-1}}$	Particulate inorganic matter	PIM	$\mathrm{g\,m^{-3}}$
Cloud cover		–	Particulate organic matter	POM	$\mathrm{g\,m^{-3}}$
Measurement depth		m	CDOM fluorescence	CDOMf	Qse
Secchi depth		m			
Water depth		m	Flags	Notation	Units
Photic depth	$Z_{p\%}$	m	General flag	Flag	–
Mixed layer depth	MLD	m	Location flag	Location_flag	–
Temperature		°C	Time flag	Time_flag	–
Salinity		psu	Chlorophyll *a* method	Chla_flag	–
Provider		–	CoastColour product	CCP_flag	–

IOPs	Notation	Units	AOPs	Notation	Units
Total absorption coefficient	$a(\lambda)$	$\mathrm{m^{-1}}$	Remote-sensing reflectance	Rrs (λ)	$\mathrm{sr^{-1}}$
Particles absorption coefficient	$a_\mathrm{p}(\lambda)$	$\mathrm{m^{-1}}$	Water-leaving reflectance	RLw (λ)	–
NAP absorption coefficient	$a_\mathrm{NAP}(\lambda)$	$\mathrm{m^{-1}}$	Water-leaving radiance (or	Lw (λ)	$\mathrm{mW\,cm^{-2}}$
Absorption by phytoplankton	$a_\mathrm{ph}(\lambda)$	$\mathrm{m^{-1}}$	above-water upwelling		$\mathrm{\mu m^{-1}\,sr^{-1}}$
Absorption by detritus	$a_\mathrm{d}(\lambda)$	$\mathrm{m^{-1}}$	radiance)		
CDOM absorption coefficient	$a_g(\lambda)$	$\mathrm{m^{-1}}$	Above-water downwelling	Es (λ)	$\mathrm{mW\,cm^{-2}}$
Total (back)scattering coefficient	$b_\mathrm{(b)}(\lambda)$	$\mathrm{m^{-1}}$	irradiance (or incident		$\mathrm{\mu m^{-1}}$
NAP scattering coefficient	$b_\mathrm{NAP}(\lambda)$	$\mathrm{m^{-1}}$	irradiance)		
NAP backscattering coefficient	$b_\mathrm{bNAP}(\lambda)$	$\mathrm{m^{-1}}$	Downwelling irradiance	Ed (λ)	$\mathrm{mW\,cm^{-2}}$
Backscattering ratio	$b_\mathrm{bp}(\lambda)/b_\mathrm{p}(\lambda)$	–			$\mathrm{\mu m^{-1}}$
Total beam attenuation coefficient	$c(\lambda)$	$\mathrm{m^{-1}}$	Diffuse attenuation of Ed	Kd (λ)	$\mathrm{m^{-1}}$
Particles beam attenuation coefficient	$c_\mathrm{p}(\lambda)$	$\mathrm{m^{-1}}$	Diffuse attenuation of PAR	K_par	$\mathrm{m^{-1}}$
Turbidity		FNU, FTU			

5.6 % of the total CCRR match-up field TSM and reflectance data respectively. When excluding the turbidity data, 91.6 % of the IOP measurements are contributed from the southern California (38.7 %), North Sea (22.9 %), Florida (7.6 %), and Great Barrier Reef region (7.0 %) sites.

The methods of chlorophyll *a*, TSM, IOPs, and Rrs measurements performed by each data contributor are briefly described below. Chlorophyll *a* measurement methods by the different laboratories are summarized in Table 4.

Chlorophyll *a* and TSM

Chlorophyll *a* concentrations were measured by either high-performance liquid chromatography (HPLC), fluorometry, or spectrophotometry. In the following, TChl *a* refers to chlorophyll *a* measurements determined by HPLC and Chl *a* denotes chlorophyll *a* obtained by fluorometry or spectrophotometry. TSM concentrations were collected at nine sites: the eastern Mediterranean Sea (hereafter E. Md. Sea), the

Baltic Sea and E. Md. Sea, the Great Barrier Reef region (referred to hereafter as the GBR region), the Indonesian waters, Morocco and the western Mediterranean Sea (hereafter Morocco-W. Md. Sea), the North Sea, the Red Sea, and Tasmania coastal waters.

In the CEOAS data set, 422 TChl *a* data were measured from 2006 to 2009 at the Oregon–Washington site and 2 at the central California site. Samples were stored at −80 °C until HPLC analysis. The distribution of TChl *a* measurements from Oregon–Washington is seasonally unbalanced with 8 % of the measurements collected during the period of December–February, 38 % in March–May, 50 % in June–August, and 50 % in September–November.

The CSIC data set contains 736 Chl *a* and 667 POM measurements collected in the Gulf of Cádiz (southwest Iberian Peninsula) within the Morocco-W. Md. Sea site. The measurements were taken in the nearshore area (< 30 km) of the Guadalquivir estuary from 2005 to 2007, and offshore dur-

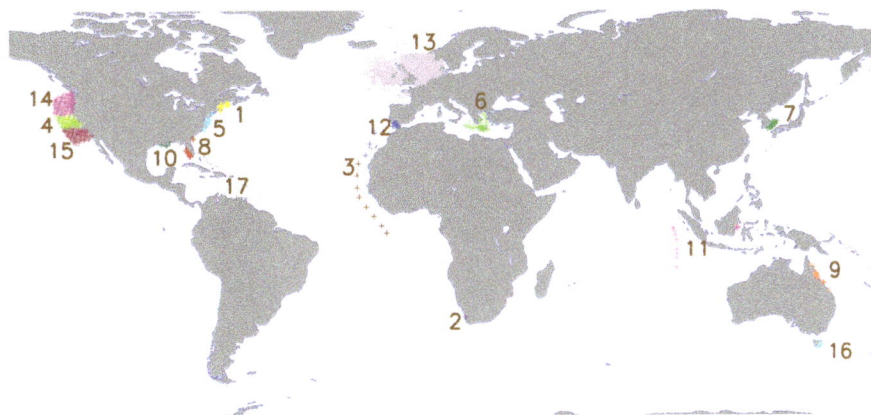

Figure 1. The distribution of the in situ data within the 17 CoastColour sites which are, numbered alphabetically, the coastal waters off (1) Acadia; (2) Benguela; (3) Cape Verde; (4) central California; (5) Chesapeake Bay; (6) the eastern Mediterranean Sea (referred to hereafter as E. Md. Sea); (7) the East China Sea; (8) Florida; (9) the Great Barrier Reef region (hereafter GBR region); (10) Gulf of Mexico; (11) Indonesia; (12) Morocco and western Mediterranean Sea (hereafter Morocco-W. Md. Sea); (13) the North Sea region extending to the English Channel, the Celtic and Irish seas, the Bay of Biscay, and southern Brittany (all referred to as the North Sea); (14) Oregon–Washington; (15) southern California; (16) Tasmania; and (17) Trinidad and Tobago.

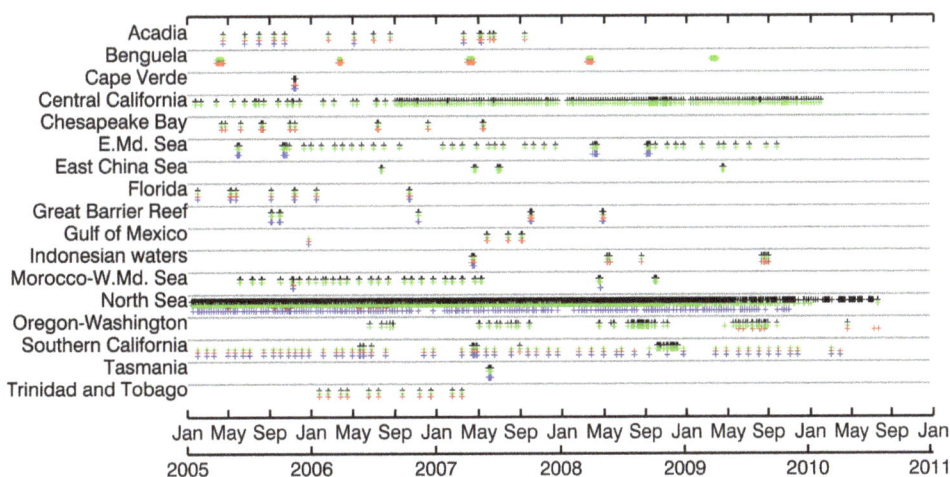

Figure 2. Time availability of at least one parameter available from the CoastColour sites within the match-up field measurements: metadata (black) (excluding the date, time, geographical coordinates, and data provider), biogeochemical data (green), AOPs (red), and IOPs (blue).

ing 2008 with slightly fewer measurements during the periods June–August (19 % of the data). Chlorophyll analysis was conducted by filtering samples of 500 mL through Whatman GF/F glass fibre filters (0.7 μm pore size), extracting in 90 % acetone, and measuring chlorophyll a by standard fluorometric methods using a Turner Designs model 10 fluorometer following JGOFS protocols (IOC/UNESCO, 1994). TSM concentrations were measured gravimetrically on pre-weighted Whatman GF/F (0.7 μm pore size) after rinsing with distilled water, following JGOFS protocols (IOC/UNESCO, 1994). Organic matter lost on ignition was determined by reweighting the filters after 3 h in the oven at 500 °C, giving the concentrations of PIM and POM (by subtraction). TSM and PIM measurements, contaminated by salt

(filters not correctly rinsed), show low variability in TSM and PIM, with 90 % of TSM measurements comprised between 31.1 and 48.3 g m^{-3}. Therefore, only Chl a and POM measurements are retained from the initial CSIC data set.

The CSIR chlorophyll data were collected from the Benguela coastal surface waters and measured using the standard fluorometric method of Parsons et al. (1984) with a Turner Designs 10AU fluorometer. A total of 131 Chl a measurements are available from March to April for years 2005–2009.

The CSIRO data set consists of data collected at 63 stations in the GBR region from 2005 to 2008 (where 25, 19, and 55 % are available from March to May, June to August, and September to November respectively) and at 21 stations

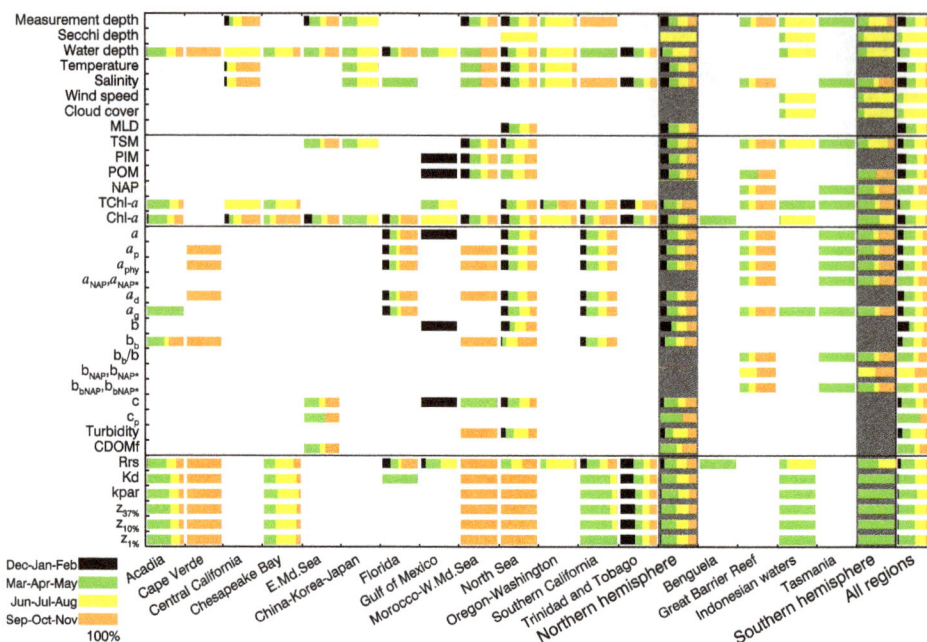

Figure 3. Seasonal availability of the metadata, biogeochemical, IOP, and AOP measurements from the CoastColour sites within the CCRR match-up data set.

in the Tasmanian waters in May 2007. Water samples were filtered through Whatman GF/F glass fibre filters with 0.7 μm nominal pore size and stored in liquid nitrogen until analysis by HPLC. The analyses conducted on the data set collected before July 2004 followed the method of Wright et al. (1991), while the method of Van Heukelem and Thomas (2001) was used for the subsequent campaigns (Oubelkheir et al., 2006; Blondeau-Patissier et al., 2009). For TSM analysis, the filters were pre-ashed at 450 °C, pre-washed in 100 mL of Milli-Q water, dried and pre-weighted. The samples were rinsed with 50 mL of distilled water and stored in Petri slides at 4 °C. The filters were dried at 60 °C (van der Linde, 1998).

The EMECO data set is provided by the International Council for the Exploration of the seas (ICES) and Smart-buoys data by the Centre for Environment, Fisheries and Aquaculture Science (Cefas), totaling 6274 stations with Chl a measurements, calibrated by HPLC. The distribution of these measurements is slightly unbalanced between the seasons (29 % of data available during March–May and 19 % in September–November).

The GKSS TSM and TChl a measurements were collected at 48 stations in the North Sea. TChl a and TSM measurements follow the protocol described in Doerffer and Schönfeld (2009). The sampling is equally distributed between the periods April–May, June–July, and September–October of years 2005–2006, with no measurements during December–February.

The HCMR data were collected at transect stations, where samples were taken in Niskin bottles from HCMR RV *Aegaeo*, in the E. Md. Sea site. For Chl a measurements, the

filtrations were performed using 47 mm diameter nucleopore filters consisting of Millipore® polycarbonate membrane filters, with 0.2 μm nominal pore size; Chl a was measured using Turner 00-AU-10 and Turner TD700 fluorometers using EPA Method 445 (Holm-Hansen et al., 1965) adapted by Arar and Collins (1992). For TSM measurements, the samples were filtered through 47 mm diameter, Isopore™ 0.45 μm polycarbonate membrane filters (Millipore®). After filtration of water samples, the filters were rinsed with Milli-Q water to remove salt. The filters were dried in the oven at 60 °C. In total, 294 Chl a measurements were collected from 2005 to 2009. Unbalanced percentages of Chl a data of 18 and 32 % are available from the periods June–August and September–November respectively. TSM measurements are available at 45 stations, sampled during years 2005 and 2008, with 47, 13, and 40 % of the data taken during the periods March–May, June–August, and September–November respectively.

The Ifremer data set consists of 975 Chl a measurements collected at 30 different locations within the Armorican Shelf (northwest of France), from 2005 to 2009. Data are available from the French phytoplankton surveillance network (REseau PHYtoplankon, REPHY; Gohin, 2011). Fluorometric measurements of Chl a were performed mostly in laboratory using a Turner C7 and C3. Over the four periods (seasons) from December–February to September–November, there are 18, 27, 32, and 23 % of the total number of Chl a measurements respectively.

The ITC measurements of Chl a and TSM were carried out in the Mahakam Delta waters from the upstream turbid

Table 3. Number of matchup-up field measurements provided by parameter (lines) and by site (columns).

(a) Number of metadata and biogeochemical match-up field measurements.

WQ	Acadia	Benguela	Cape Verde	Central California	Chesapeake Bay	E. Md. Sea	East China Sea	Florida	GBR region	Gulf of Mexico	Indonesian waters	Morocco–W. Md. Sea	North Sea	Oregon–Washington	Southern California	Tasmania	Trinidad & Tobago	Total
Measurement depth				650		433	78		78		119	738	27 837	566	126	21		30 646
Secchi depth											119		28					147
Water depth	76		8	2	81	139	78	85		41	110	63	245	381	7		11	1327
Temperature				223			77					63	25 530	429				26 322
Salinity				223			77	4	63			63	24 704	427	122	20	11	25 714
Wind speed											119							119
Cloud cover											113							134
MLD													124					124
TSM						45	78		63		119		212			21		538
PIM										6		667	48					721
POM								32		6		667	48					753
NAP									63							21		84
TChl a	40			2	69				63	41	4		239		247	21	5	1153
Chl a	25	131		606	12	294	47	84		6	96	736	7468	136	403		11	10 055

(b) Number of IOP and AOP match-up field measurements.

WQ	Acadia	Benguela	Cape Verde	Central California	Chesapeake Bay	E. Md. Sea	East China Sea	Florida	GBR region	Gulf of Mexico	Indonesian waters	Morocco–W. Md. Sea	North Sea	Oregon–Washington	Southern California	Tasmania	Trinidad & Tobago	Total
a								63	63	6			117		342	19		610
a_p			7					66	63			3	188		346	21		694
a_{phy}			7					66	62			3	176		346	21		681
a_{NAP}, a_{NAP*}									63							21		84
a_d			7					66				3	188		347			611
a_g	4							65	63		4		129		342	19		626
b										6			54					60
b_b	23		7									3	28		269			330
b_b/b									63							21		84
b_{NAP}, b_{NAP*}									25									25
b_{bNAP}, b_{bNAP*}									63							21		84
c						139				6		6	116					267
c_p						34												34
Turbidity												30	2157					2187
CDOMf						132												132
K_d	42		8	69				4			8	3	6		16		11	167
R_{Lw}	76	84	8	81				85	15	47	127	3	54	47	319		11	957
k_{par}	38		5	35							8	3	4		15		10	118
$z_{37\%}$	42		8	69							8	3	5		16		10	161
$z_{10\%}$	42		8	66							8	3	6		15		10	158
$z_{1\%}$	41		8	61							8	3	6		11		10	148

Table 4. Instrument and methods of chlorophyll *a* measurement in the CCRR match-up data set.

Data provider	Instrument	Filters, diameter (mm), nominal pore size (μm)	Tchl *a* measurement method (HPLC)	Chl *a* measurement method
CEOAS	–	Whatman GF/F, N/A, 0.7	HPLC	–
CSIC	Turner Model 10	Whatman GF/F, N/A, 0.7	–	JGOFS protocols; IOC/UNESCO (1994)
CSIR	Turner 10AU		–	Parsons et al. (1984)
CSIRO		GF/F, 47, 0.7	Wright et al. (1991), Van Heukelem and Thomas (2001)	–
EMECO	5LEDs (Ferrybox)		–	In vivo fluorometry
GKSS	–	–	Doerffer and Schönfeld (2009)	–
HCMR	Turner 10AU Turner TD700	Millipore polycarbonate membrane filters, membrane polycarbonate, 47, 0.2	–	EPA Method 445; Holm-Hansen et al. (1965), adapted by Arar and Collins (1992)
Ifremer	Turner C7, C3	–	–	Fluorometry
IOW	–	–	–	Fluorometry
ITC		Membrane filter, 47, 0.45	–	Spectrophotometry; Clesceri et al. (1998)
NOAA	–	–	–	Fluorometry
NOMAD	Various (see references)		Hooker et al. (2005)	Werdell and Bailey (2005), Pegau et al. (2003)
PML	Hypersil 3 mm C8 Thermo Separations and Agilent		Barlow et al. (1997); Llewellyn et al. (2005)	–
UCSB	Turner 10AU		Van Heukelem and Thomas (2001)	Strickland and Parsons (1972)
UNICAN	Hach Lange DR-5000		–	Spectrophotometry; Clesceri et al. (1998)

Mahakam River down to the clear water situated in the seaward area influenced by the Makassar Strait. From each station, two 1 L bottles of surface water samples were taken and then stored onboard in cool and dark conditions until their processing in the laboratory. TSM concentrations were determined using the gravimetric method. Water samples were filtered through previously weighted 47 mm diameter filters (Whatman GF/F filters, pore size of 0.45 μm). The filters were dried and reweighed (Clesceri et al., 1998). Chl *a* concentrations were measured using a spectrophotometer after the water samples had been filtered through 47 mm diameter filters (membrane filter, pore size of 0.45 μm) (Clesceri et al., 1998). The Chl *a* and TSM measurements cover the wet (May) and dry (August) seasons in 2008 and the dry season in August 2009, with a total of 119 stations.

The KORDI data set includes 47 Chl *a* and 78 TSM measurements collected at the East China Sea site. Samples were filtered through a 25 mm diameter GF/F glass fibre filter. Chl *a* measurements were performed through the methanol-extraction method using a PerkinElmer Lambda 19 dual-beam spectrophotometer. TSM and Chl *a* data are available

from cruises carried out during April and June 2007 and April 2009, and 31 % of TSM data are available from measurements made in July 2006. During the periods of April–May and June–July, respectively 41 and 59 % of TSM measurements are available, while 68 and 32 % of the Chl *a* data are provided for these periods.

The NOAA Chl *a* measurements were performed based on in vitro fluorescence measurements following 24 h dark period extractions in acetone. A total of 136 measurements are available from the Oregon–Washington site sampled from July to September 2008; 122 Chl *a* data from southern California acquired during the period September–November in 2008; and 606 Chl *a* data from the central California site, measured from 2005 to 2010. From the periods of September–November and June–August, respectively 52 and 30 % of the NOAA Chl *a* collection are available.

The NASA bio-Optical Marine Algorithm Dataset (NOMAD) presents a large collection of bio-optical data in ocean and coastal waters (Werdell and Bailey, 2005). The NASA SeaWiFS Bio-optical Archive and Storage System (SeaBass; Werdell et al., 2003), the source of the NOMAD data set, includes both the HPLC and fluorometric methods. HPLC methods may have differed between laboratories in order to separate different types of pigments, which may depend on the predominant component of chlorophyll (Hooker et al., 2005). HPLC-derived TChl a measurements in the NOMAD data set are the sum of monovinyl and divinyl chlorophyll a, plus chlorophyllide a, allomers, and epimers (Werdell and Bailey, 2005). The NOMAD TChl a data set constitutes 24 % of the total TChl a measurements gathered within the Coast-Colour match-up data set. From 2005 to 2007, 175 TChl a data were collected from the six CoastColour sites – Acadia (40), Chesapeake Bay (69), Gulf of Mexico (41), Indonesian waters (4), southern California (16), and Trinidad and Tobago (5) – and 142 Chl a measurements from Acadia (25 data), Chesapeake Bay (12), Florida (84), Gulf of Mexico (6), Indonesian waters (4), and Trinidad and Tobago (11).

The PML data set was collected during RV *Aegaeo* and RV *James Clark Ross* cruises in the MOS-2 and L4 areas respectively. The extraction of chlorophyll was performed in acetone including apo-carotenoate, and the separation used reversed-phase HPLC with 30 s of sonification and 5 min of centrifugation (4000 rpm) (Barlow et al., 1997). In the PML data set, divinyl-chlorophyll a, chlorophyllide-a and chlorophyll a isomers and epimers are added to chlorophyll a (Barlow et al., 1997). For TSM measurements, 2 to 4 L seawater samples were filtered in triplicates and washed with Milli-Q water. Filters were pre-ashed at 450 °C for 4 h, pre-washed in 500 mL of Milli-Q water, oven-dried at 75 °C for 24 h, and pre-weighted (van der Linde, 1998). A total of 191 pairs of Chl a and TChl a and 136 TSM measurements were collected by PML between 2005 and 2009. The distributions of Chl a, TChl a, and TSM measurements are overall well balanced across seasons.

The UNICAN data set includes 28 TSM and Chl a measurements collected in the North Sea region (the Bay of Biscay) in July 2010. Chl a was measured through a Hach Lange DR-5000 with Whatman GF/F filter following the spectrophotometric method described by Clesceri et al. (1998) (trichrometric method), using a white reference to control the quality of the measurements. TSM was estimated using a gravimetric method after filtration through GF/C glass fibre filters.

Inherent optical properties

IOP measurements were collected at 11 sites (blue symbols in Fig. 2). The measurement methods for the total absorption coefficient, a; absorption by CDOM, a_g; absorption by particles, a_p; absorption by detritus, a_d; absorption by phytoplankton pigments, a_{phy}; scattering b and backscattering coefficients b_b; total beam attenuation coefficient, c; and particle beam attenuation, c_p, are briefly described below.

For the CSIRO measurements of a, a_p, and a_{phy}, carried in the GBR region and Tasmania coastal waters, samples were filtered using a 25 mm Whatman GF/F filter with 0.7 μm nominal pore size and then stored in liquid nitrogen (Oubelkheir et al., 2006; Blondeau-Patissier et al., 2009). CDOM absorption was determined after filtration through polycarbonate filters (Millipore) of 0.22 μm nominal pore size, and water samples were filtered immediately after collection and stored in cool and dark conditions until analysis (Tilstone et al., 2003). The backscattering coefficients were measured using HOBI Labs HydroScat-6. The spectral dependency of the scattering coefficient was modelled as a hyperbolic function of wavelengths, using bands 412, 488, 510, 532, 555, and 650 nm (Oubelkheir et al., 2006; Blondeau-Patissier et al., 2009).

In the HCMR data set collected in the E. Md. Sea, 139 measurements of c_p are provided at 470, 660, and 670 nm (available at least at one of these wavelengths), and 34 measurements of c_p are given at 670 nm. The beam attenuation coefficients were measured using a 0.25 m path length transmissometer Chelsea Technologies Group Ltd Alpha Tracka II, emitting at 470 nm. The instrument was mounted on RV *Aegaeo*'s permanent CTD rosette frame for casts through the water column. The data were quality-controlled, filtered, and binned at 1 m intervals (Karageorgis et al., 2012).

MSU IOP data consist of six measurements of a, b and c coefficients collected at the Gulf of Mexico site.

The NOMAD absorption coefficients a_p, and a_g, and absorption by detritus a_d, were derived by spectroscopy at six CoastColour sites (Acadia, Cape Verde, Florida, Indonesian waters, Morocco-W. Md. Sea, and southern California). Note that for the Indonesian waters, only a_g is provided. These data have been quality-controlled, removing unreasonable data and instrument artifacts (Werdell, 2005). The spectral backscattering coefficient provided in NOMAD data set was obtained using HOBI Labs HydroScat-2 and HydroScat-6 sensors, WET Labs ECObb and ECOVSF sensors, and Wyatt Technology Corporation DAWN photometers. The details on b_b data processing are given in Werdell (2005).

Absorption coefficient spectra were measured by PML at 5 m depth in the North Sea, using the WET Labs ac9+. As reported in Martinez-Vicente et al. (2010), the measurements were corrected to account for temperature, salinity, and scattering effects. The samples were filtered through 47 mm diameter Whatman Anopore membranes (0.2 μm pore size), using pre-ashed glassware. Absorption coefficients were determined on the spectrophotometer and a 10 cm quartz cuvette from 350 to 750 nm, relative to a bi-distilled Milli-Q reference blank. a_g was calculated from the optical density and the cuvette pathlength, then the baseline offset was subtracted from a_g (Groom et al., 2009). The measurement of a_{phy} followed the method of Tassan and Ferrari (1995). The coefficients a_p and a_{phy} were measured using a PerkinElmer Lambda 2 spectrophotometer, and 25 mm GF/F filters (Tilstone et al., 2012). a_p were determined before and after pigment extraction using NaClO 1 % active chloride from 350 to

Table 5. Number and period(s) of match-up field RLw measurements in each CoastColour site and by data provider.

CoastColour site	Data provider	Number	Period
Acadia	NOMAD	76	Apr 2005 to Sep 2007
Benguela	CSIR	84	Mar 2005 to Mar 2008
Cape Verde	NOMAD	8	Oct 2005, Nov 2005
Chesapeake Bay	NOMAD	81	Mar 2005, May 2007
Florida	NOMAD	85	Jan 2005, Oct 2006
Great Barrier Reef	CSIRO	15	Sep 2007, Apr 2008
Gulf of Mexico	MSU(6) NOMAD(41)	47	Dec 2005 May 2007 to Jul 2007
Indonesian waters	ITC(119) NOMAD(8)	127	May 2008, Aug 2009 Apr 2007
Morocco-W. Md. Sea	NOMAD	3	Oct 2005
North Sea	GKSS(48) NOMAD(6)	54	Apr 2005 to Jul 2006 Oct 2005
Oregon–Washington	CEOAS	47	May 2009 to Jul 2010
Southern California	UCSB(303) NOMAD(16)	319	Jan 2005 to Mar 2010 May 2006 to Aug 2007
Trinidad and Tobago	NOMAD	11	Jan 2006 to Mar 2007
All		957	Jan 2005 to Jul 2010

750 nm. The scattering measurements were performed using an ECO VSF-3 sensor (Martinez-Vicente et al., 2010).

Backscattering coefficients provided by UCSB were estimated from profiled measurements of the total volume scattering function β at 140°, using a HobiLabs HydroScat-6, collected at the southern California site. These measurements were corrected for light attenuation along the photon path to the instrument detector (σ correction of Maffione and Dana, 1997) using concurrent absorption spectra (Kostadinov et al., 2007) for measurements up to 2005, and concurrent beam attenuation and absorption modelled from the diffuse attenuation coefficient for downwelling irradiance and the irradiance reflectance (see Antoine et al., 2011, for details). A total of 269 backscattering spectra initially measured at 442, 470, 510, 589, and 671 nm were interpolated at 412, 470, 510, and 589 nm assuming a λ^{-1} spectral dependency of the backscattering coefficient. UCSB absorption spectra up to 2005 were obtained using vertical profiles of WET Labs ac-9 measurements, after application of pure water calibration, as well as standard temperature, salinity, and scattering corrections (WET Labs ac-9 Protocol, 2003). Surface absorption values were derived from the upper 15 m absorption spectra, after filtering incomplete, negative, or extreme values; spectra were linearly interpolated at 412, 443, 490, 510, 530, 555, 620, and 665 nm (Kostadinov et al., 2007). Measurements of a_{phy}, a_g, and a_d spectra were obtained using a

Shimadzu UV2401-PC spectrophotometer. CDOM samples were filtered on 0–2 μm Poretics membranes, while GF/F filters were used to retain total particulate matter for a_p measurement, corrected for pathlength effects following Guillocheau (2003). Pigment extraction was performed in 100 % methanol.

Apparent optical properties: water-leaving reflectance

A total of 957 match-up field RLw spectra were collected at 13 CoastColour sites and provided from eight data providers, covering a variety of time periods as listed in Table 5. About 33 % of these data are provided for the southern California region, 13 % for the Indonesian coastal waters site, 9 % for the Benguela and Florida sites, and 8 % from the Acadia and Chesapeake Bay sites. Less than 19 % of the data set is provided from the rest of the CoastColour sites. Hyperspectral RLw measurements are available from the GBR region, the North Sea, and the Indonesian waters.

The instruments and methods of RLw measurements are summarized in Table 6 and briefly described below.

The CEOAS radiometric measurements in the Oregon–Washington site were performed using a Satlantic Hyper-Pro II instrument, equipped with two hyperspectral sensors to vertically profile the upwelling radiance, Lu, and downwelling irradiance, Ed, in the water column, plus a separate

Table 6. Instruments and methods of measurement of RLw in the CCRR match-up data set. θ_v and $\Delta\varphi$ denote respectively the sensor zenith angle and its azimuth angle relative to the sun. Ed, Lu, and Lsky denote respectively the downwelling irradiance, the upwelling radiance, and the sky radiance measured along the viewing angle θ_v. The indices + and − refer to measurements just above and below the water surface respectively.

Data provider	Instruments	Method	Reference
CEOAS	3 Satlantic HyperPro	Underwater profiling of Lu−, Ed−, and above water Ed+	http://satlantic.com/sites/default/files/documents/ProSoft-7.7-Manual.pdf
CSIR	2 TriOS RAMSES	Floating buoy attached to ship, measuring Lu−, Ed+	N/A
CSIRO	1 TriOS RAMSES	Above water Lu+, Lsky, Ed+; viewing $\theta_v = 45°$, $\Delta\varphi \sim 135°$	Tilstone et al. (2003)
GKSS	3 TriOS RAMSES	Lu+, Lsky, Ed+; viewing $\theta_v = 45°$, $\Delta\varphi \sim 135°$	N/A
ITC	2 TriOS RAMSES	Lu+, Lsky, Ed+, $\theta_v = 40°$, $\Delta\varphi = 135°$	N/A
MSU	N/A	N/A	N/A
NOMAD	Various	In-water profiling, or above-water instruments	Werdell and Bailey (2005)
UCSB	ASD spectrometer, Biospherical PRR-600	Merging RLw from in-water profiling and above-water ASD reflectance	Toole et al. (2000)

surface sensor mounted high on the ship deck that measures the above-water downwelling irradiance, Es. Processing of the collected data was performed using Satlantic ProSoft software version 8.1.3_1 (see http://satlantic.com/sites/default/files/documents/ProSoft-7.7-Manual.pdf for equations). In summary, the above-water radiance, Lw, is calculated by extrapolating the profiled Lu measurements to the subsurface (Lu(0^-)) and then accounting for the air–sea interface: $Lw = Lu(0^-)(1 − \rho)/n_w^2$, where ρ is the Fresnel reflectance of the air–sea interface (set to 0.021) and $n_w = 1.345$ is the refractive index of seawater. The surface irradiance reflectance is then obtained by RLw = π Lw / Es. Of the 137 wavelengths measured by the HyperPro II, this study presents data from 21 wavelengths covering 412 to 780 nm for RLw.

In the Benguela site, the CSIR used a Satlantic radiometer mounted on a floating buoy attached to the ship in order to measure the upwelling radiance Lu and the downwelling irradiance Ed at 0.66 m below the water surface. Lu was extrapolated to Lw by means of the upwelling diffuse attenuation coefficient, Ku, as described by Albert and Mobley (2003). RLw was estimated from Lw and Ed using a reflectance inversion algorithm optimized for local conditions.

The CSIRO RLw measurements in the GBR region were conducted under stable clear-sky conditions using one TriOS RAMSES instrument. Subsequent water-leaving radiance, Lw; sky radiance, Lsky; and Spectralon upwelling radiance, Lspec, were measured. Irradiance was calculated from Spectralon measurements according to Ed = π Lspec C, where C is the reflectance correction factor accounting for non-perfect

Lambertian panel properties. Water-leaving reflectance was calculated according to the REVAMP protocol (Tilstone et al., 2003) by applying a sky correction factor.

The GKSS radiometric measurements were conducted onboard ferry cruises in the North Sea region, using three TriOS RAMSES radiometers that simultaneously measure Lu at 45° viewing angle, Es, and Lsky, with an azimuth angle between 130 and 140° relative to the sun. The water-leaving reflectance RLw was computed according to RLw = π(Lu − ρ_{sky}Lsky)/Ed, where the specular reflectance ρ_{sky} is computed using the Fresnel law, as a function of the refractive index for the mean salinity along the transect.

The ITC measurements carried out in the Indonesian waters used two TriOS RAMSES spectroradiometers. The surface water upwelling and sky downwelling radiance measurements, Lu and Lsky, were measured sequentially at 40° zenith angle and at 40° nadir angle respectively. The irradiance sensor was mounted on an aluminium pole on top of the boat, pointing upward. The boat was positioned on a station to point the radiance sensor at a relative azimuth angle of 135° away from the sun. The sensors measured over the wavelength range 350–950 nm with a sampling interval of approximately 3.3 nm. The measurements were conducted under different cloud conditions. The sky radiance reflected by the water surface, ρ_{sky}, was estimated by assuming very small (but not zero) water-leaving reflectance in the near infrared and that ρ_{sky} values were less than 0.07, which is the highest value of scattered cumulus clouds by Mobley (1999). The result of ρ_{sky} values were relatively similar with ρ_{sky} values given by Mobley (1999) for each cloud type condition. The water-leaving reflectance was obtained following the equation $RLw = \pi(Lu - \rho_{sky}Lsky)/Ed$.

The MSU radiometric measurements are provided for the Gulf of Mexico in the Mississippi Sound area (around Gulf-port). The reflectance spectra were measured at 14 wavelengths in the spectral range 380–780 nm.

From the NOMAD database, Lw and Es measurements were extracted for the match-up locations between 2005 and 2010 and converted to RLw spectra. Various instruments were used for the measurements of the remote-sensing reflectance, Rrs, in the NOMAD data set (Werdell and Bailey, 2005), including in-water profiling or above-water measurements. All in- and above-water data from various instruments and data providers were consistently processed to Rrs, with the methods described in Werdell and Bailey (2005).

The UCSB RLw measurements in the southern California region were obtained using above-water radiometric measurements of one Dual FieldSpec spectrometer (ASD) instrument and underwater measurements of a Biospherical Instruments (San Diego, California) profiling reflectance radiometer (PRR-600), as described by Toole et al. (2000). Sea surface radiance, Ls, at viewing zenith angle of 45°; sky radiance (which would be reflected into Ls), Lsky; and Spectralon upwelling radiance, Lspec, were measured by the FieldSpec spectrometer. The above water reflectance was estimated following Toole et al. (2000): the above-water irradiance was calculated from Spectralon measurements according to $Ed = \pi Lspec/\rho_{spec}$, where ρ_{spec} is the reflectance of the plaque; the water-leaving reflectance was calculated as $RLw = \pi(Ls - \rho Lsky)/Ed - residual(750)$, where residual(750) corrects for any residual reflected sky radiance, assuming zero water-leaving radiance at 750 nm. Underwater downwelling irradiance, Ed^-, and upwelling radiance, Lu^-, were measured along vertical profiles using the Biospherical PRR-600 and then interpolated to above-water radiance and irradiance respectively, leading to a new estimate of RLw spectra, which were merged with FieldSpec reflectances (see Toole et al., 2000, for details).

2.1.2 MERIS data

MERIS CoastColour processing (see flow chart in Fig. 4) is applied to MERIS Level 1 Full Resolution Full Swath (FRS) to produce MERIS level 2 match-up data sets, namely MERIS water-leaving reflectance (L2R) and MERIS water quality products (L2W), over the CoastColour sites. Here, a brief description of MERIS CoastColour processing is given.

MERIS FRS products, including auxiliary data such as surface pressure, ozone, geographical location (used to identify products having an overlap with one of the test sites), viewing and sun angles, and solar flux, are processed with the Accurate MERIS Ortho-Rectified Geolocation Operational Software (AMORGOS processor, developed by ACRI-ST within ESA GlobCover project), yielding geometrically corrected MERIS child products (FSG). The L1P processor

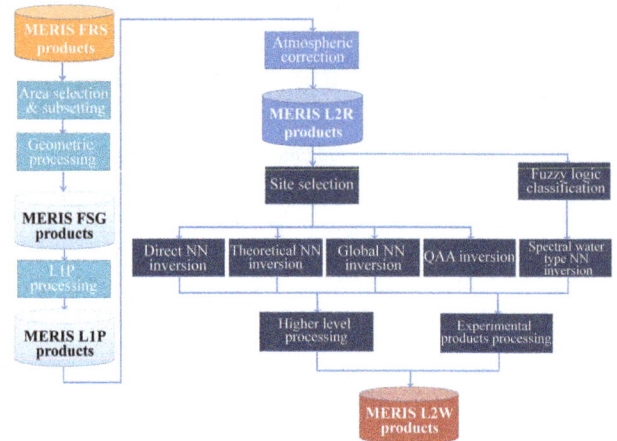

Figure 4. MERIS CoastColour processing.

subscenes the FSG data; applies the radiometric and smile corrections; and performs equalization following Bouvet and Ramoino (2010) and pixel classification, screening cloud pixels.

The L1P product, which contains the top of atmosphere radiance reflectance (TOA), is then atmospherically corrected to determine the water-leaving radiance reflectance, following the steps described in Doerffer (2011), which yields the L2R products. Furthermore, water pixels are classified according to their TOA reflectances and available geographical information, and L2W products are generated using various ocean colour algorithms. A complete list of the parameters contained in L2R and L2W products is given in Table 7.

Boxes of 5×5 pixels are extracted from L1P, L2R, and related L2W products at all match-up locations present for a given test site and are stored in three files associated with the site. Further processing is performed to average MERIS L2R spectra in each 5×5 box, discarding low-quality pixels (see the list of critical flags in Table 7) and yielding the mean reflectance, referred to hereafter as MERIS RLw, and its standard deviation. Other L2W and atmospheric products are also averaged over the 5×5 box (see the list in Table 8).

Finally, around each match-up location MERIS L1P, L2R, and L2W subscenes are provided in BEAM-DIMAP (".dim") format, and are associated with a KMZ file for quick visualization of area location via Google Earth.

With respect to the match-up field RLw data set, the MERIS RLw data set includes supplementary data from the following regions: the central California, E. Md. Sea and East China Sea, and Tasmania coastal waters, and extended data from Morocco-W. Md. Sea and the North Sea concurrent with extra match-up field WQ measurements (IOPs and/or biogeochemical data sets). The MERIS RLw data set is not available for all the locations of the match-up field RLw measurements (e.g. Benguela, Indonesian waters, GBR region), either because no MERIS image is available within 1 h of the match-up field measurement or because MERIS pixels

Table 7. The Level 2 products provided the MERIS match-up data set, as a 5 × 5 box around the locations of the match-up field measurements. The "critical" flags listed in italic font are associated with pixels being rejected (if the flags are raised) in the post-processed MERIS match-up data set.

Navigation	Description	Units	L2R, L2W	Description	Units
ProdID		–	reflec_x	RLw at λ (nm)	–
CoordID	ID of location	–	b_tsm	Scattering coefficient at 443 nm	m^{-1}
Name	Match-up name	–	a_tot	Total absorption coefficient (443 nm)	m^{-1}
Latitude, longitude	Geographical	degrees			
	coordinates		**Atmosphere**	**Description**	**Units**
Date, time			tau_nnn	Aerosol optical thickness at λ = nnn (nm)	
lat_corr, lon_corr	Ortho-corrected latitude/longitude	degrees	ang_443_865	Aerosol Ångström coefficient between 443 and 865 nm	–
dem_alt	DEM[a] model altitude				
dem_rough	Roughness at sight with		**Ancillary**	**Description**	**Units**
	intersection of line of	degrees	zonal_wind	ECMWF[c] zonal wind	$m\,s^{-1}$
	WGS84[b] ellipsoid taken		merid_wind	ECMWF meridional wind	$m\,s^{-1}$
	from DEM		glint_ratio	Glint ratio	–
sun_, view_zenith	Sun, view zenith angle	degrees	atm_press	ECMWF atmospheric pressure at mean sea level	hPa
sun_, view_azimuth	Sun, view azimuth angle	degrees			
pins		–	ozone	ECMWF ozone concentration	DU
ground_control_points		–	rel_hum	ECMWF relative humidity at 850 hPa	%
detector_index	Index of MERIS pixel	–			

Flags	Description	Flags	Description
land	Land pixel	coastline	Pixel is part of a coastline
water	Water pixel	cosmetic	Cosmetic flag
cloud_ice	Very high Rtoa indicating cloud, ice, or snow pixel	duplicated	Pixel has been duplicated (filled in)
bright	Bright pixel	f_meglint	Pixel corrected for glint
sunglint	Pixel affected by sun glint	f_loinld	Low inland water flag
glint_risk	Glint correction not reliable on the pixel	f_island	Island flag
		f_landcons	Land product available
suspect	Suspect flag (from L1[d])	f_ice	Ice pixel
invalid	Pixel is invalid	f_cloud	IDEPIX[f] final cloud flag
solzen	High sun zenith angle	f_bright	IDEPIX bright pixel
ancil	Unreasonable data for ozone or pressure	f_bright_rc	IDEPIX old bright pixel
		f_low_p_pscatt	IDEPIX test on apparent scattering
has_flint	If the atmospheric correction used the flint processor	f_low_p_p1	IDEPIX test on P1
		f_slope_1	IDEPIX spectral slope test 1 flag
l1_flags	Level 1 classification and quality flag	f_slope_2	IDEPIX spectral slope test 2 flag
		f_bright_toa	IDEPIX second bright pixel test
l1p_flags	Pixel classification flag (e.g. cloud screening, land, water)	f_high_mdsi	IDEPIX MDSI[g] above threshold
		f_snow_ice	IDEPIX snow/ice flag
atc_oor	If RLw is out of the expected range (as set in the NN[e])	agc_flags	Flag specific to the atmospheric and flint correction
toa_oor	Input Rtoa is out of the NN training range	agc_land	Land pixel
tosa_oor	Input Rtosa is out of the NN training range	agc_invalid	Pixel not considered for processing

[a] DEM refers to the digital elevation model of altitude; [b] WGS84 refers to the World Geodetic Standard 1984; [c] ECMWF is the European Centre for Medium Range Weather Forecast; [d] L1 is MERIS level 1 product; [e] NN is the atmosphere neural network algorithm; [f] IDEPIX is a generic pixel classification algorithm for optical Earth observation sensors; [g] MDSI is the MERIS differential snow index.

Table 8. The 5×5 box averaged L2R, L2W, and atmospheric parameters derived from the MERIS match-up data set.

Navigation	Description	Units	L2R, L2W	Description	Units
Fid	Match-up name	–	RLw_xxx[a]	RLw at λ (nm)	–
Latitude, longitude	Geographical coordinates	degrees	b_tsm[a]	Scattering coefficient at 443 nm	m^{-1}
			a_tot[a]	Total absorption coefficient (443nm)	m^{-1}
Date, time					
sun_, view_zenith	Sun, view zenith angle	degrees	Atmosphere	Description	Units
sun_, view_azimuth	Sun, view azimuth angle	degrees	tau_nnn[a]	Aerosol optical thickness at $\lambda = nnn$ (nm)	
			ang_443_865[a]	Aerosol Ångström coefficient between 443 and and 865 nm	–

Box-averaging information	Description	Units	Ancillary	Description	Units
N(var[b])	Number of pixels within the 5×5 box where valid var was retrieved	–	zonal_wind	ECMWF[c] zonal wind	$m\,s^{-1}$
			merid_wind	ECMWF meridional wind	$m\,s^{-1}$
			glint_ratio	Glint ratio	–
std(var[b])	Standard deviation of var over the N valid pixels in the 5×5 match-up box	var unit	atm_press	ECMWF atmospheric pressure at mean sea level	hPa
			ozone	ECMWF ozone concentration	DU
			rel_hum	ECMWF relative humidity at 850 hPa	%

[a] Averaged over N valid pixels in the 5×5 box around the match-up location. [b] The variable var refers to one of the MERIS L2 products listed under L2R, L2W, and atmosphere data types. [c] ECMWF is the European Centre for Medium Range Weather.

are flagged as cloud, land, suspect, sunglint, or invalid. After rejection of the flagged pixels, 457 MERIS RLw spectra remain from the CoastColour sites. About 80 % of these spectra are available from the North Sea region and match in situ measurements of temperature, salinity, and/or turbidity.

2.2 In situ reflectance data set

The in situ reflectance data set comprises a set of 336 RLw spectra sampled at nine MERIS bands from 412 to 709 nm, and collected simultaneously with CHL and/or TSM measurements at five CoastColour sites, from August 2002 to August 2009. The number of RLw data per site and per data provider, and their periods of measurement, are presented in Table 9. Part of these spectra, measured in Benguela, Indonesian waters, and the North Sea (the GKSS data set), are derived from the match-up field hyperspectral RLw data.

With respect to the match-up field data set, the in situ reflectance data set includes 266 spectra already given in the match-up field data set from the Benguela, Indonesian waters, North Sea (provided by the GKSS), and Oregon–Washington sites, plus supplementary data from the Mediterranean Sea and the North Sea (provided by RBINS; see details of measurement method hereafter) and data from Benguela covering year 2002. It excludes the entire RLw data from the Acadia, Cape Verde, Chesapeake Bay, Florida, GBR region, Gulf of Mexico, Morocco-W. Md. Sea, southern California, and Trinidad and Tobago sites, and the NOMAD RLw measurements subset collected at the North Sea and the Indonesian waters, because no CHL and/or TSM and/or RLw spectra up to 709 nm are available. The total number of RLw spectra available within the match-up field and in situ reflectance data sets is $N = 1027$ (with no overlapping data).

The RBINS radiometric measurements were acquired in the North Sea and Mediterranean Sea using three TriOS RAMSES radiometers that simultaneously measure Es and the radiances Lw and Lsky at 40 and 140° viewing angles respectively with 135° azimuth angle relative to the sun (Ruddick et al., 2006).

The CHL data were measured by HPLC in all the sites except for measurements taken in Benguela after year 2002 (fluorometry) and in the Indonesian waters (spectrophotometry) (Table 9). The total numbers of in situ CHL and TSM data are 294 and 186 respectively.

2.3 Simulated data set

Radiative transfer simulations were performed with Hydro-Light version 5.0 (Mobley and Sundman, 2008), using the atmospheric, air–sea interface, and sun and viewing angle characteristics as presented in Table 10, and the specific IOPs (SIOP) for mineral particles (denoted by MP), phytoplankton, and $a_g(443)$ as given in Table 11. The SIOPs include the specific absorption coefficients for phytoplankton, a_p^*, and for MP, a_{MP}^*; the spectral slope of a_{MP}^*, denoted by S_{MP}; the specific scattering coefficient for MP, b_{MP}^*; the spectral variation in the beam attenuation coefficient for phytoplankton,

Table 9. The number and period(s) of measurement of in situ RLw and TSM and/or CHL concentrations, collected at each CoastColour site within the in situ reflectance data set. The methods for CHL and TSM measurements are also provided.

CoastColour site (data provider)	Number of RLw spectra CHL, TSM	Period	CHL method	TSM method
Benguela (CSIR)	135, 135, 0	Aug 2002 to Mar 2008	year 2002: HPLC; other years: fluorometric	–
Indonesian waters (ITC)	119, 92, 119	May 2008, Aug 2009	Spectrophotometry	Gravimetric, GF/F
Mediterranean Sea (RBINS)	7, 7, 7	Mar 2009	HPLC, 90 % acetone, cell homogenizer	Gravimetric, GF/F
North Sea (GKSS)	48, 48, 48	Apr 2005 to Jul 2006	HPLC	Gravimetric, GF/F
North Sea (RBINS)	12, 12, 12	Apr 2006 to Jun 2009	HPLC, 90 % acetone, cell homogenizer	Gravimetric, GF/F
Oregon–Washington (CEOAS)	15, 15, 0	May 2009 to Aug 2009	HPLC	–
Total	321	Aug 2002 to Aug 2009	–	–

Table 10. Atmospheric, air–sea interface, and solar and viewing geometry specifications in CCRRv1.

Parameter	Values
Sun angles	Zenith: 0, 40, and 60°; azimuth: 0°
Viewing angles	Zenith: 0°; azimuth: 90°
Surface wind speed	$5\,\mathrm{m\,s}^{-1}$
Cloud fraction	0
Sky radiance distribution	Semi-empirical sky model; Harrison and Coombes (1988)
Direct and diffuse sky irradiances	Semi-empirical sky model RADTRAN

γ_{CHL}, and for MP, γ_{cMP}; and the spectral slope of CDOM absorption, S_{CDOM}.

This simulated data set is denoted as "CCRRv1" to facilitate comparison with future versions, e.g. with variability in the specific inherent optical properties.

A total of 5000 triplets of CHL and MP concentrations and $a_g(443)$ were generated according to the following:

- A random number function modelling a log-normal probability density function was used for CHL.

- The associated MP and $a_g(443)$ values were also generated by a random number function but constrained to yield reasonable covariation of the triad, comparable to that reported by Babin et al. (2003b) from in situ measurements in coastal European waters.

Figure 5a and b show the distributions of the simulated MP and $a_g(443)$ vs. CHL concentrations and their co-variations.

Based on these concentrations and SIOP models, a set of hyperspectral (2.5 nm resolution) data were generated, including the total absorption a, scattering b, and backscattering b_b coefficients; the phytoplankton absorption coefficient, a_{phy}; and the ratio $b_b/a + b_b$. For each in-water content (5000 cases) and sun angle (3 cases), HydroLight computed RLw and the diffuse downwelling irradiance attenuation spectra, Kd, as well as the photosynthetically available radiation, PAR. The spectra were further spectrally subsampled to (a) MERIS band central wavelengths (412.5, 442.5, 490, 510, 560, 620, 665, 681.25, 708.75, 753.75, 761.875, 865, 885, and 900 nm), (b) MODIS bands (412, 443, 469, 488, 531, 547, 645, 667, 678, 748, 859, and 869 nm) and (c) SeaWiFS bands (412, 443, 490, 510, 555, 670, 765, and 865 nm). In the following, only spectra generated at MERIS bands are presented.

3 Results and discussion

The distributions of water depth, temperature and salinity, CHL and TSM concentrations, IOPs, and AOPs are presented in Sects. 3.1–3.6, followed by the analysis of the covariation between CHL and TSM and bio-optical relationships existing in the CCRR data sets (Sect. 3.7).

The distributions of CHL, TSM, and IOPs in the match-up field data set and the in situ reflectance data set are related to the AOPs measured throughout the CoastColour sites. The similarities/differences in these relationships characteristic

Table 11. The inherent optical properties as established in CCRRv1.

Parameter and value	Description	Reference
$c_{phy}(660\,nm) = 0.407\,CHL^{0.795}$	Phytoplankton beam attenuation coefficient at 660 nm	Loisel and Morel (1998)
$\gamma_{CHL>2} = 0$ $\gamma_{CHL\leq2} = 0.5\log_{10}(CHL) - 0.3$	Spectral variation in c_{phy} (power law exponent)	Morel et al. (2002)
$\beta_{phy}(\lambda)$: Fournier–Forand	Phytoplankton scattering phase function with $b_{bphy}/b_{phy} = 0.006$	Similar to Morel et al. (2002)
$a_p^*(\lambda) = A(\lambda)CHL^{B(\lambda)}$	Phytoplankton specific absorption coefficient	Bricaud et al. (1998)
$b_{MP}^*(555\,nm) = 0.51\,m^2g^{-1}$	Specific scattering coefficient for MP	Babin et al. (2003a)
$\beta_{MP}(\lambda)$: Petzold	MP scattering phase function	Mobley (1994)
$a_{MP}^*(443\,nm) = 0.04\,m^2g^{-1}$	Specific absorption coefficient for MP	Babin et al. (2003b)
$S_{MP} = -0.0123\,nm^{-1}$	Spectral slope of a_{MP}^* (exponential)	Babin et al. (2003b)
$\gamma_{cMP} = -0.3749$	Spectral variation in the beam attenuation coefficient for MP (power law), giving $b_p^{715}/b_p^{555} = 0.925$	In agreement with Babin et al. (2003a)
$S_{CDOM} = -0.0176\,nm^{-1}$	Spectral slope of a_g (exponential)	Babin et al. (2003b)

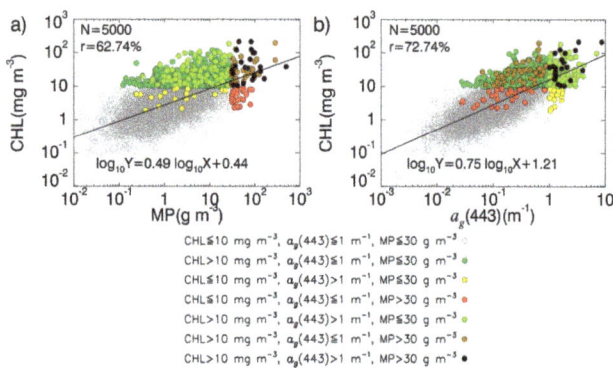

Figure 5. The simulated (**a**) MP and (**b**) $a_g(443)$ vs. the simulated CHL concentrations, in the CCRRv1. The colours represent the ranges of MP, CHL, and $a_g(443)$ as reported in the key above.

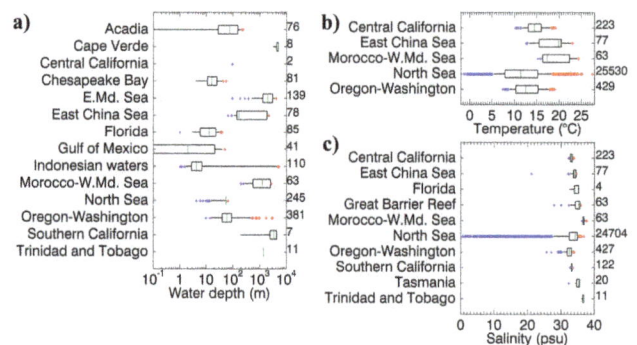

Figure 6. The distribution of (**a**) water depth (m), (**b**) temperature (°C), and (**c**) salinity (psu) as given in the in situ data set at all available depths. The black boxes delimit the 25th and 75th percentiles of the data and the black horizontal lines show the extension of up to the 5th and 95th percentiles. The green line represents the median value and the blue (red) "+" the minimum (maximum) plot values below (above) the 5th (95th) percentile. The number of measurements taken at each test site is reported on the right axis of the graph. The scale is logarithmic for the water depth.

of these sites may shed light on the common (universal) bio-optical relationships and/or emphasize some more regional features, which is of interest for remote-sensing algorithm development and validation. The bio-optical relationships within the match-up field and in situ data sets are also compared to the models, as well as to the ranges of TSM, CHL, and CDOM concentrations assumed in the simulated CCRRv1.

3.1 Water depth, temperature, and salinity

The CoastColour sites are characterized by different distributions of water depth, temperature, and salinity (Fig. 6). The median water depth varies from 2 m in the Gulf of Mexico to more than 1000 m in the Morocco-W. Md. Sea, Trinidad and Tobago, E. Md. Sea, southern California, and Cape Verde sites (Fig. 6a). The sea surface temperature in the North Sea ranges from −0.6 to 26 °C, encompassing the ranges of tem-

perature reported at the four other sites (Fig. 6b), probably due to the quasi-continuous sampling in the North Sea throughout the cold and warm seasons (Fig. 2). The frequent sampling of salinity in the North Sea across seasons is exhibited in the large range of this measurement (0.5–37 psu). About 82 % of salinity data measured in the CoastColour sites exceed 32 psu (Fig. 6c).

Note, however, that these distributions may not represent all the conditions within which the entire in situ measurements were collected, since the time windows of the metadata (excluding the date, time, and geographic coordinates) do not always cover those of the measurement of the biogeochemical data, IOPs, and AOPs (Figs. 2 and 3).

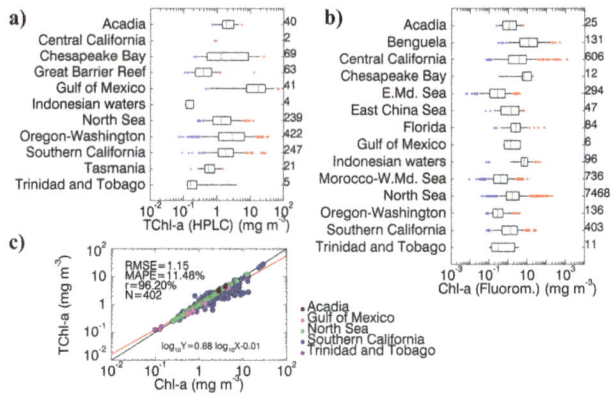

Figure 7. The distribution of (**a**) TChl a and (**b**) Chl a concentrations (in $mg\,m^{-3}$) as given in the in situ data set at all measurement depths, and (**c**) Chl a vs. TChl a. The number of measurements taken at each test site is reported on the right axis of the graph. The graphical convention in panels (**a**) and (**b**) is identical to Fig. 6. In panel (**c**) the solid line represents the 1 : 1 ratio, the dashed lines ±30 %, and the red line the linear regression fitting the log-transformed TChl a and Chl a measurements.

3.2   Chlorophyll a concentration

TChl a (HPLC method) and Chl a (fluorometric method) span from 0 to extremely high values ($> 1000\,mg\,m^{-3}$) in the central California site (Fig. 7a, b). TChl a values vary by about 2 orders of magnitude in most of the sites. The low number of TChl a measurements (≤ 5) in the data from the Indonesian waters and Trinidad and Tobago sites may explain the reduced variability observed there. With the higher number (temporal and spatial coverage) of Chl a measurements, larger ranges of variability are found in the measurements from the Indonesian waters and about 7 orders of magnitude from the measurements taken in the central California site. For most of the sites, Chl a varies at least 3 orders of magnitude.

Chlorophyll a concentrations (either Chl a or TChl a) exhibit median values less than $1\,mg\,m^{-3}$ from the E. Md. Sea, GBR region, Morocco-W. Md. Sea, Tasmania, and Trinidad and Tobago sites. Some of these sites have been extensively studied and characterized as ultra- to oligotrophic (CHL $\leq 1\,mg\,m^{-3}$) or mesotrophic to eutrophic waters:

- The eastern Mediterranean Sea is oligotrophic due to nutrient limitations. CHL ranges from $\sim 0.02\,mg\,m^{-3}$ in the Cyprus eddy to $0.3\,mg\,m^{-3}$ during the winter bloom (Groom et al., 2005). Similar ranges of CHL were reported in the ultra-oligotrophic eddies of the western Mediterranean Sea (Loisel et al., 2011).

- In the GBR region the water composition is largely influenced by the land use in the adjacent catchments (Schaffelke et al., 2012). Chlorophyll a concentrations are generally low, with median values ranging from

$0.1\,mg\,m^{-3}$ inshore to $0.25\,mg\,m^{-3}$ offshore along a cross-shelf gradient (Brodie et al., 2007).

- The eastern Atlantic off the Morocco coast is characterized by nutrient-rich waters (Freudenthal et al., 2002) and by the upwelling regime from April to September. Based on a single vertical profile in the chlorophyll maximum layer off the Moroccan coast in September 1999, the average CHL was estimated at about $1.4\,mg\,m^{-3}$ (Dolan et al., 2002), while Oubelkheir et al. (2005) found that surface CHL ranged from 0.01 to $3.75\,mg\,m^{-3}$ during the same cruise; these reported maxima values lie at the upper end (between the 75th and 95th percentiles) of the CCRR match-up field data range collected at 5 m depth.

- In the data from the central California site, the variations in Chl a are primarily determined by sea surface temperature and wind-driven coastal upwelling loading nutrient-rich waters (Chavez et al., 2002). This site exhibits the widest range of CHL variability (> 6 orders of magnitude).

In the data from the Acadia, East China Sea, Florida, North Sea, Indonesian waters, Oregon–Washington, and southern California sites, the median CHL ranges from 1 to $10\,mg\,m^{-3}$. For the Benguela, Chesapeake Bay, and Gulf of Mexico sites, the concentration of chlorophyll a exceeds $10\,mg\,m^{-3}$. It may exceed $50\,mg\,m^{-3}$ during algal blooms in the Benguela upwelling system (Probyn, 1985) and reach very high values (CHL $> 500\,mg\,m^{-3}$) during a dinoflagellate bloom of *Ceratium balechii* (Pitcher and Probyn, 2011).

The data from Oregon–Washington encompass a wide range of temporal and spatial variability. TChl a, collected between April and September during years 2006 to 2010, varies over 3 orders of magnitude, up to $33\,mg\,m^{-3}$ with a median value of $2.9\,mg\,m^{-3}$, while Chl a spans from 0.07 to $4\,mg\,m^{-3}$ during the period July–September 2008 with a median value of $0.3\,mg\,m^{-3}$. This is due to the productive upwelling season and the low-productivity downwelling season, more productive areas onshore, and less productivity near Oregon than to the north, close to Washington and in the Columbia River plume. It is also possible that variability in the data set is due to slight differences in sampling protocols between the laboratory groups although this would likely be minimal.

In Chesapeake Bay, a distribution similar to the match-up data was described in Tzortziou et al. (2007) based on measurements performed in 2001 where the mean CHL value was about $15\,mg\,m^{-3}$, and higher CHL values up to $74\,mg\,m^{-3}$ occurred during spring and summer periods.

Overall, the chlorophyll a match-up data set collected for the CCRR exercise are representative of the distributions reported in the literature. Moreover, the measured Chl a and TChl a in the CoastColour sites show a high correlation ($r = 96.2\%$, $N = 402$) with mean absolute percentage error

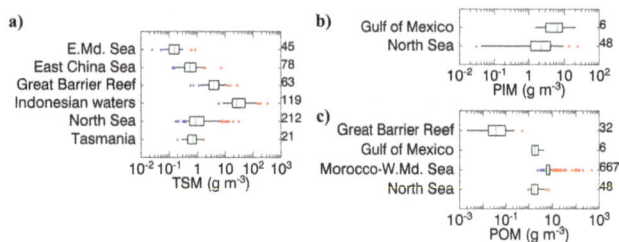

Figure 8. The distribution of (**a**) TSM ($g\,m^{-3}$), (**b**) PIM ($g\,m^{-3}$) and (**c**) POM ($g\,m^{-3}$) as given in the in situ data set at all measurement depths. The number of measurements taken at each test site is reported on the right axis of the graphs. The graphical convention is identical to Fig. 6.

(MAPE) equal to 11.5 % (Fig. 7c). Most of the discrepancies between TChl a and Chl a are noticed in measurements from the southern California site. When this site is excluded, a significantly lower MAPE is obtained for the seven sites (MAPE = 3.6 % with correlation $r = 99.8$ %).

3.3 TSM, turbidity, Kd, and Kpar

The distributions of TSM are reported in Fig. 8a. PIM and POM concentrations, measured over two and four Coast-Colour sites respectively, and their distributions are indicated in Fig. 8b and c.

The measurements from the E. Md. Sea show the lowest TSM concentrations (TSM < 1 $g\,m^{-3}$, $N = 45$), whereas the region of Indonesian waters exhibits the highest values (median TSM > 20 $g\,m^{-3}$, $N = 119$). In Tasmania, TSM varies between 0.1 and 2 $g\,m^{-3}$ (Cherukuru et al., 2014). The median TSM concentrations observed from the E. Md. Sea, East China Sea, Tasmania, North Sea, GBR region, and Indonesian waters sites are 0.2, 0.6, 0.7, 0.9, 3.8, and 26 $g\,m^{-3}$ respectively.

Turbidity measurements are provided at two sites (see Fig. 9a). The distribution of turbidity matches that of TSM over the North Sea – likely due to significantly overlapping periods where TSM and turbidity measurements were collected (see the green and red colours in Fig. 2).

The ranges of Kd (443) and Kpar measurements (Fig. 9c, d) show similar differences amongst the Acadia, Cape Verde, Chesapeake Bay, Indonesian waters, Morocco-W. Md. Sea, North Sea, southern California, and Trinidad and Tobago sites: the highest mean values are observed in Acadia and Chesapeake Bay (corresponding to the lowest mean values of $Z_{1\%} < 20$ m; see Fig. 9b), and the lowest in Cape Verde, Morocco-W. Md. Sea, and the Indonesian waters, which correspond to the highest mean values of $Z_{1\%} > 60$ m found at these three sites.

Kd (and Kpar) values are lower in the Indonesian waters than in the North Sea site. However, the Secchi disk data sets for these two sites, larger than the Kd (and Kpar) data set,

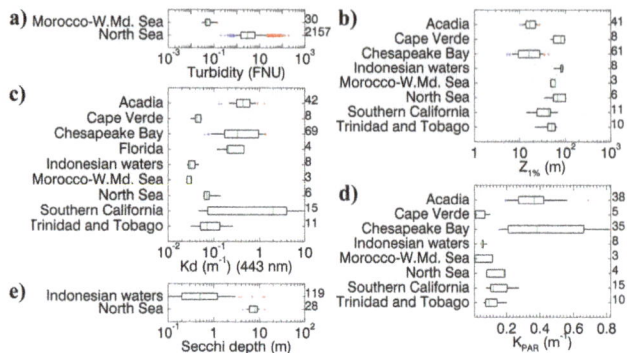

Figure 9. The distribution of (**a**) turbidity (FNU for the North Sea and FTU for Morocco-W. Md. Sea), (**b**) the photic depth $Z_{1\%}$ (m), (**c**) Kd at 443 nm (m^{-1}), (**d**) Kpar (m^{-1}), and (**e**) Secchi depth (m). The scale is logarithmic for turbidity and Kd, and linear elsewhere. The number of measurements taken at each test site is reported on the right axis of the graph. The graphical convention is identical to Fig. 6.

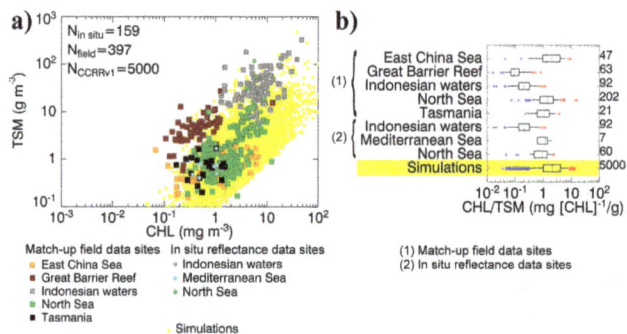

Figure 10. The distribution of (**a**) CHL concentrations ($mg\,m^{-3}$) vs. TSM concentrations ($g\,m^{-3}$) from the in situ reflectance data set (in the three sites as indicated in the key) plotted as filled circles, and from the match-up data set, including Chl a and TChl a (where available), and the associated match-up field TSM concentrations (in the six sites indicated in the key) and plotted as filled squares, both superimposed on the simulated data (yellow circles). (**b**) CHL / TSM ratio ($mg\,[CHL]^{-1}\,g^{-1}$) from the match-up, in situ reflectance and simulated data sets. The graphical convention in panel (**b**) is identical to Fig. 6; the yellow colour distinguishes the simulated data set from the in situ measurements.

suggest a higher water clarity in the North Sea than in the Indonesian waters (Fig. 9e).

3.4 CHL vs. TSM

The co-variation of CHL with TSM from the in situ reflectance data set at 159 locations (where both CHL and TSM are available) is compared to the co-variation of CHL with TSM from the match-up field data set at 1062 locations. Both co-variations can be visually compared to that of CHL vs. TSM from the simulated data set (Fig. 10a). The distribution

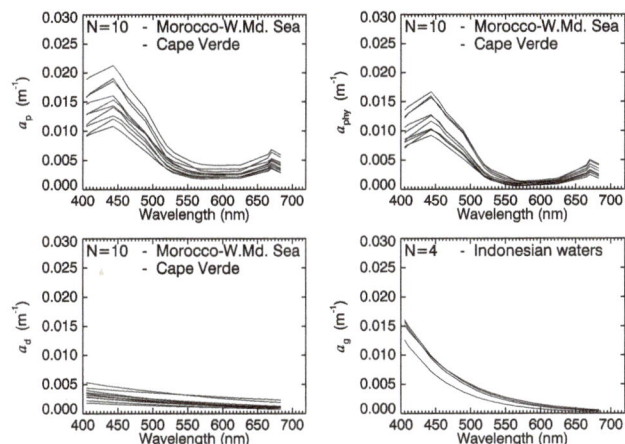

Figure 11. The match-up field absorption spectra provided from Morocco-W. Md. Sea, Cape Verde, and Indonesian waters.

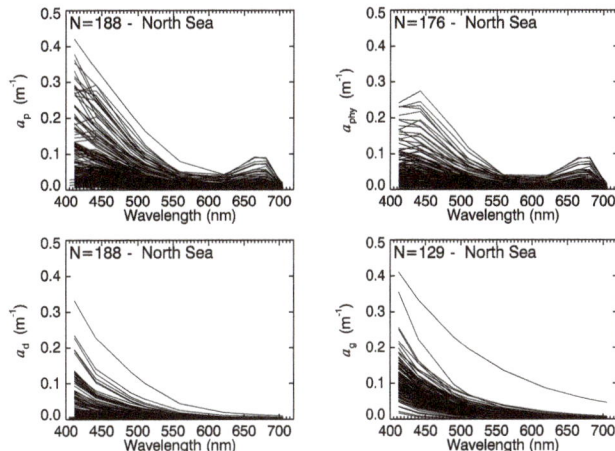

Figure 12. The match-up field absorption spectra provided from the North Sea site.

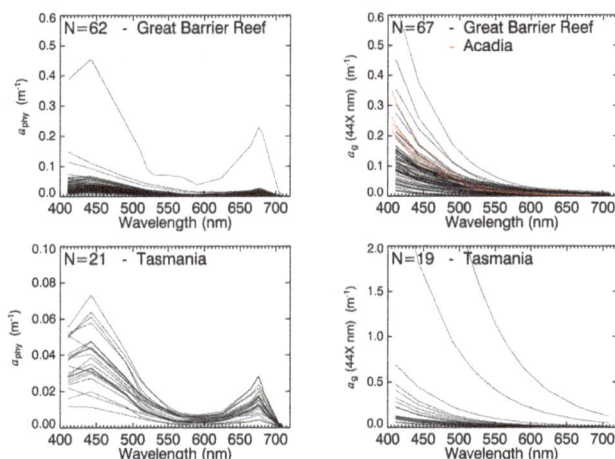

Figure 13. The match-up field absorption spectra provided from the GBR region and Acadia (upper panel) and Tasmania (bottom) sites.

of the ratio CHL / TSM is shown for the match-up field measurements (Fig. 10b).

The co-variations of CHL and TSM are generally consistent for the majority of in situ and match-up test sites, showing a general tendency of CHL increasing with TSM, as reported in Babin et al. (2003b, their Fig. 2). The simulated data fit better the distributions of CHL and TSM collected in the North Sea, since their models adopted the distributions documented in Babin et al. (2003b), based on measurements taken in European coastal waters including the North Sea.

As previously reported in Sects. 3.2 and 3.3, various CHL and TSM ranges are observed in the match-up field measurements throughout the CoastColour sites (the GBR region, the North Sea, and Tasmania coastal waters). The in situ reflectance data set showed differences in CHL and TSM ranges between the Indonesian waters, the North Sea, and the Mediterranean Sea (Fig. 10).

The simulated data encompass all the ranges covered by the in situ CHL and TSM (from the in situ reflectance data set), and partially the ranges of the match-up field data: excluding few measurements collected in the Indonesian waters and GBR region sites associated with very low CHL / TSM ratio (see Fig. 10b).

Large variability in the ratio CHL / TSM from the match-up field measurements is noticeable amongst the six Coast-Colour sites, spanning over 3 orders of magnitude (see Fig. 10b). The GBR region and Indonesian coastal waters sites present the lowest median value of the ratio CHL / TSM (from 0.1 to $0.2 \, \mathrm{mg} \, [\mathrm{CHL}]^{-1} \, \mathrm{g}^{-1}$), being approximately 10 times lower than the median magnitudes measured in the North Sea and Tasmania (around 1.4 and $1.1 \, \mathrm{mg} \, [\mathrm{CHL}]^{-1} \, \mathrm{g}^{-1}$ respectively). The East China Sea site exhibits the highest median value of CHL / TSM of $2 \, \mathrm{mg} \, [\mathrm{CHL}]^{-1} \, \mathrm{g}^{-1}$.

Identical CHL and TSM data (92 pairs) from the Indonesian waters are available both in the in situ reflectance

data set and the match-up field measurements, giving identical distributions of CHL vs. TSM and CHL / TSM ratio. From the Mediterranean Sea site, the in situ reflectance data set collected during March 2009 shows a median value of CHL / TSM of $1.2 \, \mathrm{mg} \, [\mathrm{CHL}]^{-1} \, \mathrm{g}^{-1}$ (only seven data). From the North Sea site, the distribution of CHL / TSM ratios in the in situ reflectance data set (60 data) is slightly shifted towards lower values relative to the ratios estimated from the match-up field measurements (202 data).

3.5 Inherent optical properties

The match-up field absorption coefficient spectra can be classified into four groups, starting from the sites where the lowest amplitudes around 443 nm are observed, to the highest amplitudes: (a) Morocco-W. Md. Sea, Cape Verde, and the Indonesian waters sites (Fig. 11); (b) the North Sea (Fig. 12);

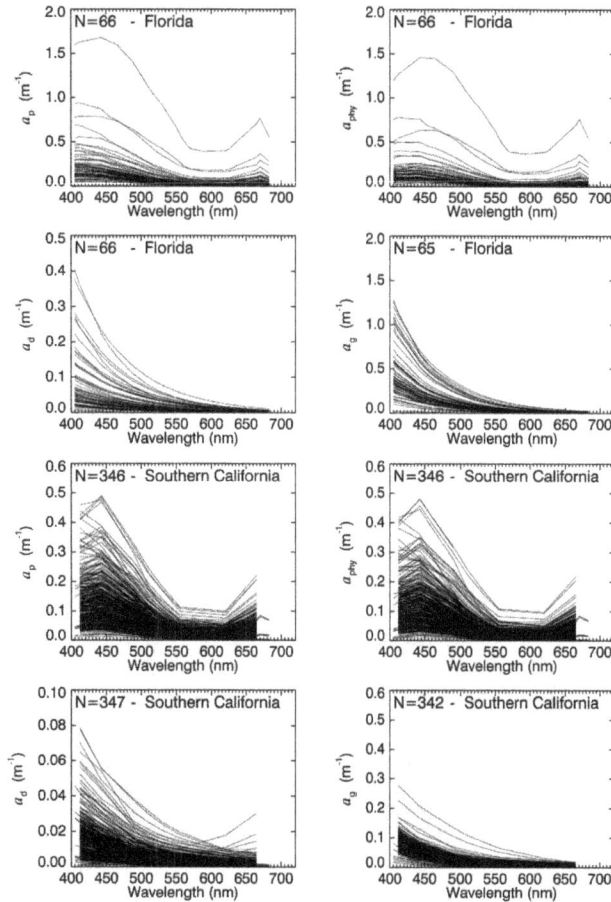

Figure 14. The match-up field absorption spectra provided from the Florida (upper panel) and southern California (bottom) sites.

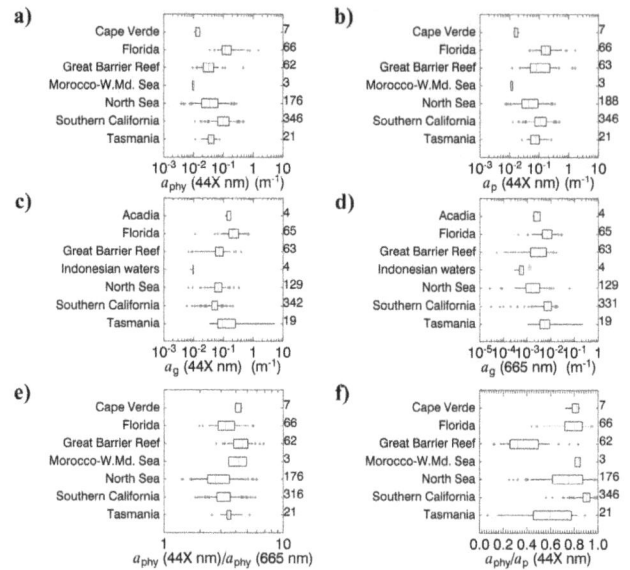

Figure 15. The distributions of (**a**) $a_{phy}(44X)$, (**b**) $a_p(44X)$, (**c**) $a_g(44X)$, (**d**) $a_g(665)$, (**e**) the ratios $a_{phy}(44X)/a_{phy}(665)$, and (**f**) $a_{phy}/a_p(443)$ measured at the CoastColour sites. When coefficients at wavelength 443 nm are missing, they are replaced by data at 440 or 442 nm. The graphical convention is identical to Fig. 6.

(Fig. 3) may partly explain this general discrepancy between the $a^*_{phy}(44X)$ data, which can be highly impacted during algal bloom events.

For $a_p(44X)$ data, the median values are 0.012 and 0.016 m^{-1} in Morocco-W. Md. Sea and Cape Verde sites respectively (Fig. 15b). Noticeably higher median values are observed from the North Sea data (0.044 m^{-1}), Tasmania (0.078 m^{-1}), and the GBR region (0.083 m^{-1}), and exceed 0.1 m^{-1} in the Florida and southern California sites.

The coefficients a_g taken around 443 nm and at 665 nm span over 3 and 4 orders of magnitude respectively throughout the CoastColour sites (Fig. 15c, d). The Florida measurements exhibit the highest median values around 443 nm, exceeding 0.2 m^{-1} with a high median value of 0.006 m^{-1} at 665 nm. Conversely, the southern California measurements show the highest median value of a_g around 665 nm exceeding 0.007 m^{-1}, with a significantly low median value of 0.05 m^{-1} at 443 nm. Note the overall similar distributions of $a_g(443)$ from the GBR region and the North Sea sites (median values around 0.07 m^{-1}), while $a_g(665)$ shows a significant shift towards higher values in the data from the GBR region. The Tasmania data contain two extreme spectra of a_g (see Fig. 13), which slightly increases the median value for this site up to 0.09 m^{-1}, above the value observed in the GBR region.

The ratio $a_{phy}(44X)/a_{phy}(665)$ shows the lowest median value of 2.7 in the North Sea and the highest value of 4.6 in the GBR region (Fig. 15e), which is inversely related to the distribution of CHL/TSM: the highest median

(c) the GBR region, Acadia, and Tasmania sites (Fig. 13); and (d) the southern California and Florida sites (Fig. 14). All the absorption coefficient spectra exhibit a large variability at shorter wavelengths (around 443 nm, denoted by 44X to refer to 440, 442, or 443 nm) and for a_p and a_{phy} around the phytoplankton absorption peak at 665 nm.

The median values for the available $a_{phy}(44X)$ data span between 0.01 m^{-1} in the Morocco-W. Md. Sea and Cape Verde sites and 0.1 m^{-1} in the Florida site (Fig. 15a). The median values encountered in the GBR region, North Sea, and Tasmania are between 0.031 and 0.039 m^{-1}. Note that the median concentrations of chlorophyll a from the GBR region and Tasmania sites are between 0.4 and 0.6 mg m^{-3}, which is 2 to 3 times lower than in the North Sea (1.3 mg m^{-3}). This indicates that, on average (for the available measurements sampled), the chlorophyll-specific absorption coefficients around 443 nm from the North Sea are lower than from the other two sites (as a comparison see spectra in Tilstone et al., 2012, for the North Sea and Blondeau-Patissier et al., 2009, for Australia). The different periods of sampling throughout the seasons for each site

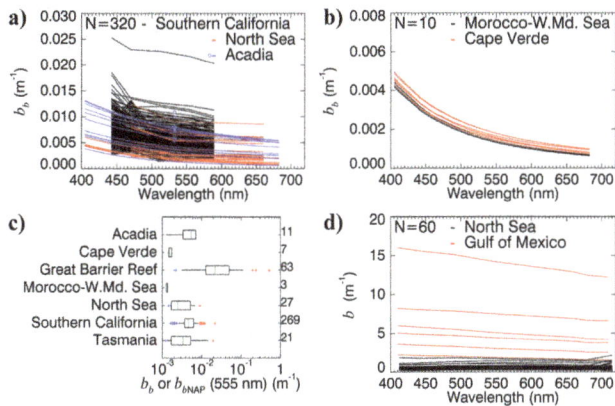

Figure 16. The spectra of (**a**) b (m^{-1}) measured in the Gulf of Mexico and North Sea sites and (**b**) b_b (m^{-1}) measured in five Coast-Colour sites (note that the coefficients from the southern California were limited to the spectral range 442–589 nm).

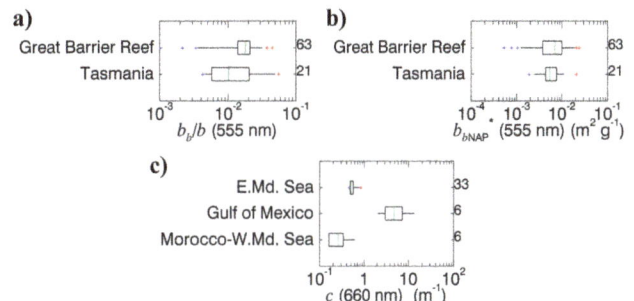

Figure 17. The distributions of (**a**) b_b/b, (**b**) b_{bNAP}^* (m^2 g^{-1}) and (**c**) c (660) (m^{-1}). Note the different scaling used for these plots. The graphical convention in Fig. 6 is used.

value of CHL / TSM of 1.5 (mg [CHL]$^{-1}$ g^{-1}) is observed in the North Sea (Fig. 10b), while a lower value of ~ 0.1 (mg [CHL]$^{-1}$ g^{-1}) is found in the GBR region.

The distributions of $a_{phy}/a_p(440)$ in the GBR region, North Sea, and Tasmania sites (see Fig. 15f) nearly follow the distributions of the associated CHL / TSM, marked by the lowest median values in the GBR region, Tasmania, and the North Sea sites (0.32, 0.60 and 0.75 respectively), and higher values (> 0.80) in the Cape Verde, Florida, Morocco-W. Md. Sea, and southern California sites. The large variability in CHL / TSM in the North Sea and GBR region (spanning over 3 and 2 orders of magnitude; see Fig. 8) can be related to the high variability in $a_{phy}/a_p(440)$ (about 10-fold magnitudes). From the Morocco-W. Md. Sea site, the number of $a_{phy}/a_p(440)$ data is too low ($N = 3$) compared to the number of CHL / TSM measurements ($N = 665$), yielding a mismatch between both distributions.

The southern California measurements of b_b show the highest variability in shapes and amplitudes, with values at 555 nm spanning from 0.0016 to 0.0216 m^{-1}, encompassing the ranges of b_b measurements from the Acadia and North Sea sites (Fig. 16a). From the Cape Verde and Morocco-W. Md. Sea sites, only 10 b_b spectra are available, lying at the bottom limit of b_b measurements from the three previous sites (Fig. 16b). The noticeable shift between the ranges of b_b measured in the Acadia and Cape Verde sites may partly explain the shift between Kd (or Kpar) in Acadia and Cape Verde (Fig. 9a).

The distributions of the total backscattering coefficients from Acadia, Cape Verde, Morocco-W. Md. Sea, North Sea, and southern California, and of non-algal particles' backscattering coefficients collected at the GBR region and Tasmania sites at 555 nm, are presented in Fig. 16c. Quite similar median values of $b_b(555)$ are observed in Acadia, southern

California, and Tasmania coastal waters, being respectively 0.0041, 0.0040, and 0.0034 m^{-1} (Fig. 16c).

In the GBR region, the coefficient $b_{bNAP}(555)$ spans over 3 orders of magnitude around the highest median value 0.021 m^{-1}. The distributions of $b_{bNAP}(555)$ coefficients and TSM (see Fig. 8) differ notably between Tasmania and GBR region.

Lower b_b values are found in the Cape Verde and Morocco-W. Md. Sea sites, where only a few backscattering measurements (< 10) are available, showing a limited variability.

The total scattering coefficients provided in the North Sea and Gulf of Mexico exhibit high relative variability in the two sites, with the highest amplitudes measured in the Gulf of Mexico (Fig. 16d).

The scattering to backscattering ratio b_b/b at 555 nm and the mass-specific non-algal particulate backscattering, b_{bNAP}^*, at 555 nm are available exclusively from the Tasmanian and GBR coastal waters (Fig. 17a, b). Most of $b_b/b(555)$ values from the GBR region lie above the 75th percentile of b_b/b measurements in Tasmania, their respective median values being 0.02 and 0.01 (Fig. 17a). Although different distributions are described by $b_{bNAP}(555)$ coefficients and TSM from the GBR region and Tasmania sites, the range of $b_{bNAP}^*(555)$ observed from the Tasmania site is within that observed from the GBR region (Fig. 17b), which spans from 5×10^{-4} to 5×10^{-2} m^2 g^{-1}. Generally similar median values are found in the GBR region and Tasmania site: 0.0053 and 0.0075 m^2 g^{-1} respectively.

The beam attenuation coefficients measured in the Gulf of Mexico cover the spectral range 410 to 710 nm; those measured in the E. Md. Sea are given at 470, 660, and 670 nm; and those in Morocco-W. Md. Sea are at 660 nm (only coefficients at 660 nm are reported here, Fig. 17c). The coefficients $c(660)$ span over 3 orders of magnitude, ranging from 0.04 to 0.9 m^{-1} in Morocco-W. Md. Sea and the E. Md. Sea (with median values being respectively 0.12 and 0.46 m^{-1}) and from 2 to 13 m^{-1} in the Gulf of Mexico (with a median value equal to 4.91 m^{-1}).

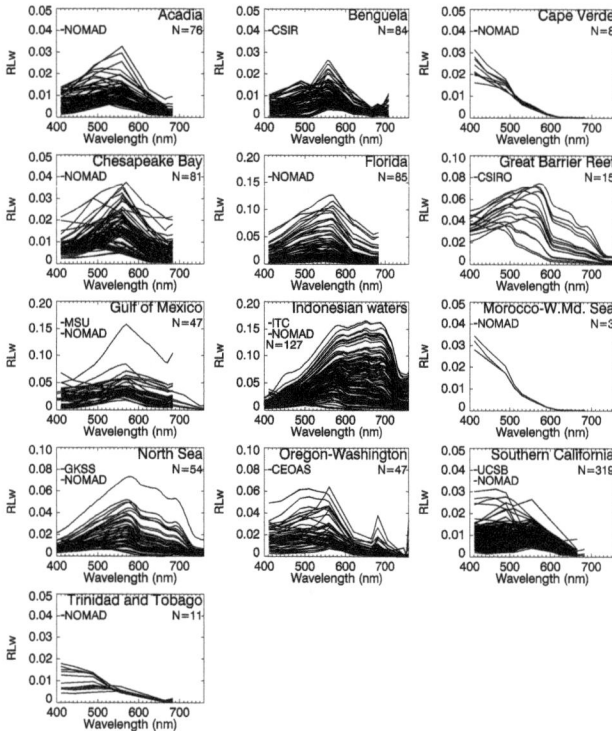

Figure 18. Match-up field RLw provided from 13 CoastColour sites by the eight data providers indicated in the figures. Note the different scales used for the sites.

Concurrent measurements of $a_g(443)$, TSM, and CHL available at the GBR region, Tasmania, and North Sea sites, and of $a_g(443)$, $a_{phy}(443)$, and $a_d(443)$ measured at the GBR region, Tasmania, southern California, and Florida sites, show large variability in water optical properties and biogeochemical parameters, covering the case 1 and case 2 waters (see ternary plots in the Supplement). Note, however, that these plots do not provide an accurate overview of the water masses sampled for this study.

3.6 Water-leaving radiance reflectance

The match-up field reflectance measurements, the MERIS RLw (both in the CCRR match-up data set), and the in situ reflectance spectra (in the CCRR in situ reflectance data set) are presented successively in Figs. 18–20. Note that the percentage of RLw data per site available from the in situ measurements (i.e. the match-up field and the in situ reflectance data sets) is different from that of the MERIS RLw data set: 31, 13, and 12 % of the in situ RLw data are provided from the southern California, Benguela, and Indonesian waters respectively; about 8 % from the Chesapeake Bay and Florida sites; and less than 7 % of RLw data from the North Sea site, while 80 % of MERIS RLw measurements are provided for the North Sea.

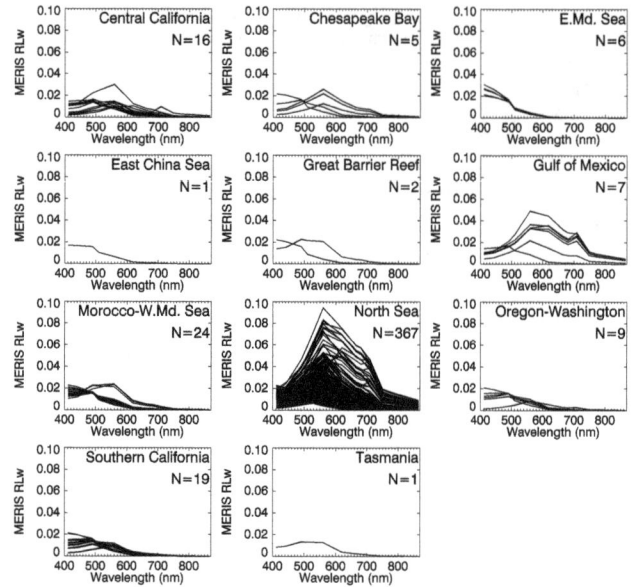

Figure 19. MERIS RLw provided in the CCRR match-up data set for the 11 CoastColour sites.

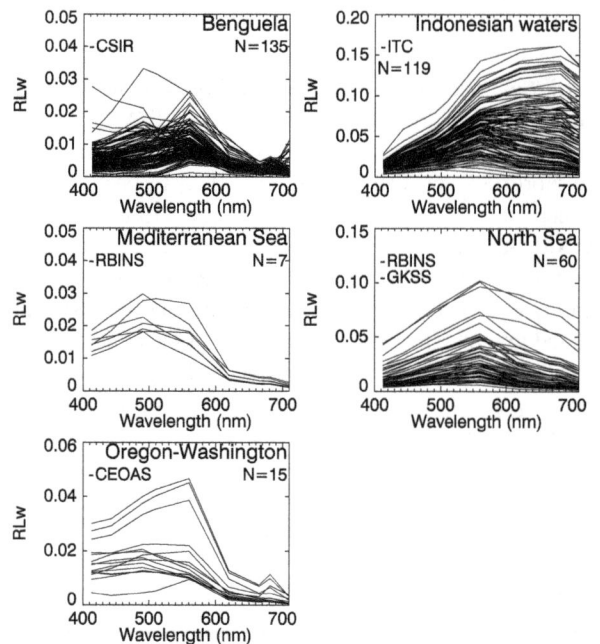

Figure 20. The in situ reflectance spectra provided from five Coast-Colour sites. Note the different scales used for the sites. The spectra from the North Sea site coloured in red are provided by GKSS.

The match-up field RLw measurements from the southern California, Morocco-W. Md. Sea, and Benguela sites present the generally lowest amplitudes amongst the CoastColour sites, where more than 75 % of the RLw values at 555 nm, RLw(555), are less than 0.01, and only 6 % of the collected spectra have RLw(555) > 0.02. For the southern California site, this is in agreement with the extremely high absorption coefficients reported earlier (see Fig. 14); from the Morocco-W. Md. Sea site, the three low match-up field reflectance spectra observed during October 2005 can be associated with the three relatively low absorption spectra measured during the same period (Fig. 11); and for the eutrophic waters of the Benguela site, these low reflectances can be explained by the high phytoplankton absorption and possibly high detrital and/or CDOM absorption. Note that MERIS RLw spectra for the southern California and Morocco-W. Md. Sea sites show ranges of RLw amplitudes comparable to those of the match-up field spectra.

The MERIS RLw measurements from the E. Md. Sea site show low RLw(555) < 0.01 and higher values at 412 nm ranging from 0.015 to 0.03 and inversely the in situ RLw measurements are slightly higher, up to 0.027, with relatively lower RLw at 412 nm < 0.019. These generally low reflectance values observed by MERIS and in situ are related to the clear oligotrophic waters of the Mediterranean Sea (95 % of Chl a data are less than $2\,\mathrm{mg\,m^{-3}}$, with a median value of $0.3\,\mathrm{mg\,m^{-3}}$; Fig. 7). The difference in the spectral shapes between MERIS and in situ measurements can be explained partly by the different periods of observations which were conducted in March 2009 for the in situ data: while only two measurements were available from MERIS in March 2008, all the other measurements were collected during September 2008 and May and October 2005, outside of the blooming period (Barale et al., 2008).

For the Chesapeake Bay, both the 5 MERIS RLw and 81 match-up field RLw spectra exhibit values less than 0.04 at all wavelengths, with 27 % of match-up field RLw(555) higher than 0.02.

From the North Sea site, most of the match-up field (Fig. 18), MERIS RLw (Fig. 19), and in situ spectra (Fig. 20) show a peak around 550–570 nm, not exceeding 0.05, with 50 % of the RLw(555) above 0.02. Lower reflectances are measured at shorter wavelengths (< 450 nm) associated with the higher CDOM and particles absorption in this spectral range (Fig. 12).

The match-up field RLw(555) measurements from the GBR region lie in a range comparable to that observed from the North Sea, but with a significantly different distribution: 86 % of RLw(555) measurements exceed 0.02. With respect to the North Sea RLw spectra, the spectral shapes and magnitudes of RLw from the GBR region are also markedly different in the blue spectral range. This may be attributed to the notable difference between the spectral shapes and magnitudes of phytoplankton absorption coefficients measured in the North Sea site (Fig. 12) and in the GBR region (Fig. 13), where the concentration of CHL (Chl a or TChl a) is generally 4 times lower than in the North Sea (Fig. 7).

Amongst the 47 match-up field reflectance measurements collected at the Gulf of Mexico site, 95 % of RLw measurements around 555 nm are higher than 0.02, 34 % range from 0.03 to 0.05, and one extreme value (> 0.15) is reported. The

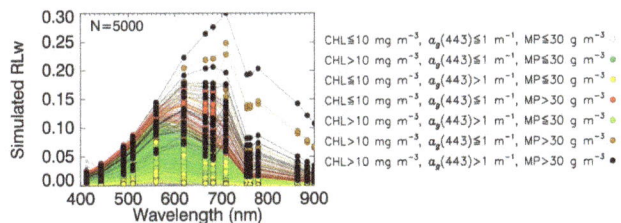

Figure 21. The simulated reflectance spectra in the CCRRv1. The colours represent the ranges of MP, CHL, and a_g(443) as reported in the key above.

peak of chlorophyll absorption is noticeable around 665 nm on the NOMAD reflectance spectra, which can be related to the generally high Chl a measurements (75 % of TChl a are higher than $7.6\,\mathrm{mg\,m^{-3}}$ and the median value is $17\,\mathrm{mg\,m^{-3}}$; see Fig. 7).

The 127 match-up field reflectance spectra collected from the Indonesian waters exhibit the highest variability in the amplitudes in the red and near-infrared spectra range, with some values exceeding 0.1 around 700 nm, likely due to high TSM concentrations as shown in Fig. 8 (TSM may exceed $100\,\mathrm{g\,m^{-3}}$, with a median value about $25\,\mathrm{g\,m^{-3}}$). At 555 nm, 96 % of the RLw data are above 0.02. Most of the spectra show a minimum around the chlorophyll absorption peak, which can be related to Chl a distribution in these waters with a median value about $7\,\mathrm{mg\,m^{-3}}$ (Fig. 7).

The 47 Oregon–Washington match-up field spectra exhibit a high variability in reflectance at 412 nm, with values ranging from 3×10^{-3} to 5×10^{-2}, with varying spectral shapes depicting the high spatial and time variability in phytoplankton concentrations noted earlier for that site (Sect. 3.2).

The simulated reflectance spectra in the CCRRv1 data set are presented in Fig. 21, and related to the ranges of the simulated CHL, MP, and a_g(443) via their colours (as indicated in the key).

The comparison between the CCRRv1 and the IOCCG Algorithm Working Group simulated data (IOCCG, 2006) indicates that the ranges of the total absorption coefficient at 440 nm, a(440), and the remote-sensing reflectance, Rrs(440), in CCRR are globally within those of IOCCG (Fig. 22a), with a few points of higher total absorption coefficient (maximum in CCRR is a(440) = $23.6\,\mathrm{m^{-1}}$). While more variability in the reflectance of CCRR for the mid- and high ranges of absorption is noted, the ranges of the reflectance band ratios 410:440 and 490:555 of CCRR are within those of the IOCCG data (Fig. 22b). The large variability in the CCRR reflectance is mainly due to the extended ranges of MP and CHL towards higher concentrations, yielding extended ranges of particle backscattering.

The distribution of reflectance products from the three CCRR data sets is examined through the following:

– RLw band ratio 490:555 vs. RLw band ratio 412:443 from reflectance measurements in the match-up field

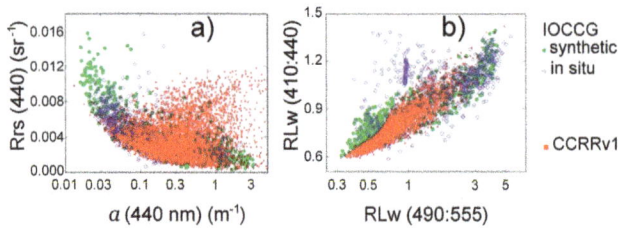

Figure 22. Comparison between IOCCG (green, reproduced from Fig. 2.3 in IOCCG, 2006), and CCRRv1 (red) simulated data sets. (a) Variations of the remote-sensing reflectance Rrs (440) with a(440), (b) variations in Rrs band ratio 410:440 with respect to Rrs band ratio 490:555. The blue diamonds represent the NOMAD (Werdell and Bailey, 2005) subset of in situ data extracted from the SeaBASS data set and used in the algorithm testing in IOCCG (2006).

data set (Fig. 23a) and vs. RLw band ratio 709:665 (Fig. 23b). Note that since most of the match-up field measurements contain RLw at 555 nm, that band is chosen instead of MERIS band 560 nm, where only few data are available.

– RLw band ratio 490:560 vs. RLw band ratio 412:443 (Fig. 23c) and vs. RLw band ratio 709:665 (Fig. 23d) from the MERIS RLw products of the match-up data set.

– RLw band ratio 490:560 vs. RLw band ratio 412:443 and vs. RLw band ratio 709:665 from the in situ reflectance data set (Fig. 23e, f).

From the match-up field data set, fewer measurements are available at band 709 nm (only 312 points). There is a general consistency in the distribution of RLw band ratios 709:665 (respectively 412:443) vs. 490:560 from the three data sets except for the in situ reflectances measured in the Benguela waters, which exhibits a high ratio of RLw(412) / RLw(443) > 1 in the lower range of reflectance band ratio 490:560, likely due to the hypertrophic nature of these waters.

Apart from the extreme ranges of reflectance ratios collected from the Benguela site, the large scatter of points observed in the in situ reflectance data set from the Gulf of Mexico, the Mediterranean Sea, and the North Sea (Fig. 23e and f) is drastically reduced in the match-up data set (that is, MERIS RLw) as shown in Fig. 23c and d. The distribution of the ratio RLw(709) / RLw(665) derived from the simulated data set better reproduces the ranges covered by the in situ data set (> 70 %) and by the match-up data set (> 95 %) than the distribution of the simulated RLw(412) / RLw(443). This is mainly attributable to the fact that the reduced variability in phytoplankton and CDOM inherent optical properties modelled in the simulated data set does not represent the large natural variability in these IOPs, which greatly affect RLw

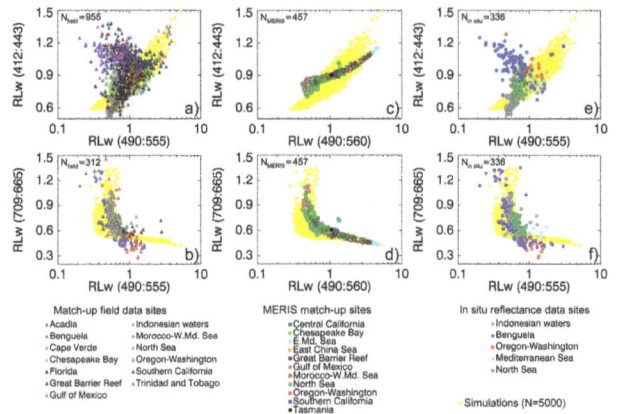

Figure 23. RLw band ratio 490:555 vs. RLw band ratio 412:443 (**a**) and vs. RLw band ratio 709:665 (**b**) within the match-up field data set, RLw band ratio 490:560 vs. RLw band ratio 412:443 (**c**) and vs. RLw band ratio 709:665 (**d**) within the MERIS RLw products of the match-up data set, and in the in situ reflectance data set (**e, f**). The yellow circles represent the simulated data set.

particularly at shorter wavelengths with the effect lessening at longer wavelengths.

3.7 Bio-optical relationships

For the comparison of MERIS and the match-up field data, only concurrent data (i.e. within ±1 h of MERIS overpass) are considered. Moreover, only match-up field data measured at depths less than 2 m are taken into account since in situ data collected at larger depths are not correlated with the surface remote-sensing signal (e.g. in the case of stratified waters).

The number of match-up field TSM and TChl a data measured above 2 m depth, concurrent with reflectance measurements, is 48 and 322 respectively. The number of concurrent in situ IOPs and RLw measurements is low (three from the North Sea and Morocco-W. Md. Sea sites, four from Acadia and the Indonesian waters, six from the Gulf of Mexico, and seven from the Cape Verde site), except from the southern California and Florida sites (313 and 66 data respectively). Furthermore, no IOP parameter is available from all these sites (see Table 3b). In the following, the analyses are focused on the distributions of CHL, TSM, and reflectance data and their relationships within the match-up, in situ, and simulated data sets.

3.7.1 CHL vs. RLw

An overview of the optical conditions and CHL ranges covered by the in situ measurements is given by Fig. 24.

Figure 24 presents a scatter plot of CHL vs. RLw band ratio 709:665 which shows that the highest CHL concentrations are exhibited during phytoplankton blooms in the

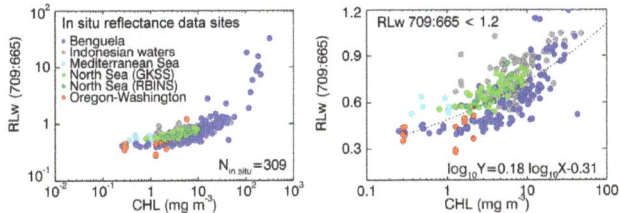

Figure 24. From the CCRR in situ reflectance data set: CHL (mg m^{-3}) vs. RLw band ratio 709:665 with a close-up on the lower range of RLw 709:665 < 1.2 presented in the left figure.

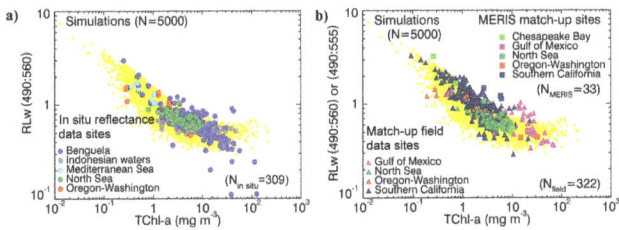

Figure 25. Reflectance band ratio 490:560 vs. CHL concentrations (mg m^{-3}) from **(a)** the in situ reflectance data set and from **(b)** MERIS RLw products and match-up field RLw measurements (using band 555 nm instead of 560 nm) with the associated match-up field TChl a concentrations, both superimposed on the simulated reflectance band ratio 490:560 vs. the simulated CHL (yellow circles). Match-up field TChl a data are restricted to measurements collected within 1 h after/before the time of MERIS overpass, and to the maximum measurement depth of 2 m.

Benguela waters, where the RLw band ratio 709:665 is the most sensitive to CHL variations. For CHL < 10 mg m^{-3}, the measurements from the Oregon–Washington and Benguela sites contain globally lower values of RLw band ratio than in the other sites (left graph in Fig. 24).

The relationship between CHL and reflectance band ratio 490:560 is quite consistent throughout the CCRR data sets (Fig. 25), except for the in situ measurements from the Benguela site (blue filled circles in Fig. 25a). This is due to the very high CHL > 100 mg m^{-3} present in the Benguela site, as previously noted in Fig. 24c, as well as the match-up field data set in Fig. 7b associated with low RLw ratios outlying the rest of data, and spanning from 0.2 down to 0.02.

3.7.2 TSM vs. RLw

Figure 26 shows the distribution of RLw at 620 nm as a function of TSM concentrations, plotted using different colours for each of the CoastColour sites: Indonesian waters, Mediterranean Sea, and the North Sea (light- and dark-green colours are used for the North Sea region to distinguish the data provider). Linear regression is applied and the associated equations are reported in the same figure. The slopes of the regression lines range from 0.48 to 0.81. The associated goodness of fit coefficients are 39, 59, 89, and 67 % for

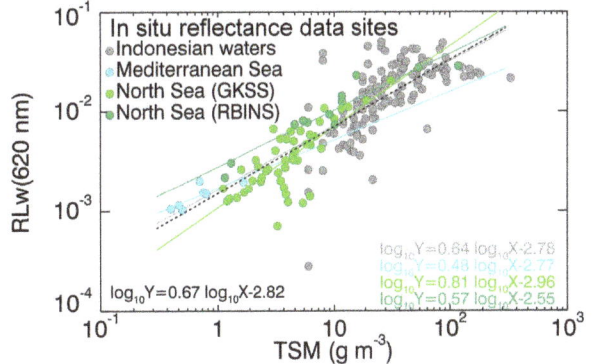

Figure 26. TSM (g m^{-3}) vs. RLw (620 nm) from the CCRR in situ reflectance data set and their associated regression lines.

the Indonesian, North Sea (GKSS), North Sea (RBINS), and Mediterranean Sea measurements respectively. In the GKSS data, most of the scatter occurs at low TSM ranges < 7 g m^{-3}, whereas the measurements from the Indonesian site, taken mainly in highly turbid waters (average and median values being 41.6 and 26 g m^{-3} respectively), show a global scatter. This scatter can be due to a high variability in the specific inherent optical properties of particles, caused by varying particles size, and/or composition within the sites. The scatter may also be impacted by mismatches of RLw and/or TSM measurements in water with high spatiotemporal variability.

The regression line fitting all the data is given by $\log_{10}(\text{RLw}(620)) = 0.67\log_{10}(\text{TSM}) - 2.82$ (shown by a dashed black line in Fig. 26) with $R^2 = 69\%$.

The distribution of the reflectance as a function of TSM concentrations in the three CCRR data sets is presented by the scatter plots of TSM vs. the reflectance at 665 nm (Fig. 27a, b) and of TSM vs. the reflectance band ratio 665:490 (Fig. 27c, d). The simulated TSM vs. RLw(665) scatter plots follow the trend of the match-ups and in situ data sets, with a scatter indicating either a variable particulate mass-specific backscattering coefficient different from that assumed in the simulations (Table 11) or measurement errors.

When using the reflectance band ratio, there is a significantly larger scatter of the simulated data due to the effects of CDOM and phytoplankton absorption affecting the reflectance at 490 nm, whereas this scatter is less noticeable in the in situ or match-up data of the five regions (Fig. 27c, d), which indicates less variability in CDOM absorption coefficient and phytoplankton concentrations in these measurements than in the modelled data set.

The general shift between the in situ reflectance data set and match-up data (Fig. 27a, b) is not noticeable in the reflectance ratios (Fig. 27c, d), suggesting similar 665:490 absorption coefficient ratios in the two data sets, at least in the North Sea regions; at lower TSM range ($< 10 \text{ g m}^{-3}$), the reduced scatter (in the reflectance band ratios vs. TSM, com-

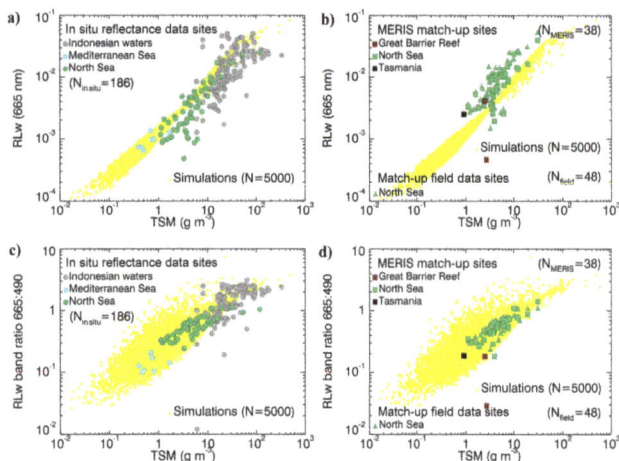

Figure 27. The relationships TSM vs. RLw(665) and TSM vs. RLw band ratio 665 : 490 in the in situ reflectance data set (non-yellow circles) plotted respectively in panels (**a**) and (**c**), in the MERIS and match-up field data sets (squares and triangles) respectively in panels (**b**) and (**d**), and in the simulated data set (yellow circles) in panels (**a–d**).

pared to RLw(665) vs. TSM) could also be due to a removal of spectrally white errors.

The distribution of the ranges of RLw(665) and RLw(665) / RLw(490) in terms of TSM and within the sites – Indonesian waters, the Mediterranean Sea, and the North Sea – is consistent with the distribution of CHL vs. TSM (Fig. 10) in these sites, especially at low TSM $< 10 \, \mathrm{g \, m^{-3}}$, where CHL is highly correlated with RLw(665).

4   Data repository

The match-up, in situ reflectance and simulated data sets are accessible from the PANGAEA website at http://doi.pangaea.de/10.1594/PANGAEA.841950. A description of files format and access follows.

The match-up field data for a site "SiteX" are stored in the compressed file at http://hs.pangaea.de/model/ccrr/Matchup_Dataset.zip, under directory /Match-up_Dataset/FieldData, in CSV files named following the classification of the parameters given in Table 2:

1. SiteX_metadata.csv, including the metadata and flags;

2. SiteX_biogeochem.csv, including the concentrations of the biogeochemical measurements;

3. SiteX_iops.csv, with the inherent optical properties;

4. SiteX_aops.csv, which includes the apparent optical properties.

For the North Sea region, two files are provided, having the names "North_Sea" and "North_Sea_Emeco", related to the

origin of the data: the North_Sea_Emeco data were downloaded from the EMECO website and North_Sea data from other data providers.

MERIS match-up products derived at a 5×5 pixel box around the locations of the match-up field measurements are provided as CSV files, with the headers as listed in Table 7, and stored in the compressed file http://hs.pangaea.de/model/ccrr/Matchup_Dataset.zip under directory /Match-up_Dataset/MERIS_5x5_L2R. The MERIS match-up products averaged at each location (from the 5×5 pixel box; see Sect. 2.1.2 for details) are stored in CSV files, with the headers listed in Table 8, and made available at directory /Match-up_Dataset/MERIS_average_L2R.

The in situ reflectance data are given in one CSV file, listing for each data provider (in the first column) the sample number, date, start and end of measurement time, latitude, longitude, the site identification number, the name of the location, the RLw spectra (nine columns for the nine MERIS selected bands), and CHL and TSM concentrations. These data are stored under the compressed directory InSituReflectance_Dataset.zip, accessible from the web address http://hs.pangaea.de/model/ccrr/InSituReflectance_Dataset.zip.

The simulated data are written in ASCII file format, and saved under the directory http://hs.pangaea.de/model/ccrr/Simulated_Dataset.zip. The concentrations of CHL, MP, and $a_g(443)$ are given in a separate file (named "Conc.txt"), where the simulation numbers going from 1 to 5000 are listed in the first column. Each entry (e.g. each simulation number or line) is associated with a given combination of CHL, MP, and $a_g(443)$.

The IOPs modelled for each entry, being the total absorption, a_{tot}, scattering, b_{tot}, and backscattering, b_{btot} coefficients excluding the pure water contributions; the absorption by phytoplankton pigments, a_{phy}; and the ratio of the total backscattering coefficient to the sum of the total absorption and backscattering coefficients $b_b/(a + b_b)$ – are provided in ASCII files called SPC_Atot.dat, SPC_Btot.dat, SPC_BBtot.data, SPC_Aphy.data and SPC_BBoABB.dat respectively, where SPC = "hyper" refers to hyperspectral input (from 350 to 900 nm, with a 5 nm step) and SPC = "maqua", "meris", or "swifs" to multispectral input at the band-centred wavelengths of the three sensors: MODIS-Aqua, MERIS, and SeaWiFS respectively.

The simulations generated hyperspectral and multispectral outputs specified in the prefix of the output filename. The three sun zenith angles ($x = 0, 40$, and $60°$), assumed successively for the set of the 5000 simulations are given in the output AOPs file names as suffixes "_sza$x$.dat". Separate files are provided for RLw and Kd, stored as SPC_RLw_sza$x$.dat and SPC_Kd_sza$x$.dat respectively. The column entry in the spectral data files gives the wavelength (in nm), and the line entry gives the simulation number.

The simulated data also include the photosynthetically available radiation PAR$_{\mathrm{Ed}}$ and PAR$_{\mathrm{Eo}}$, defined as the integration over 400 to 700 nm of the spectral downwelling irradiance Ed and of the scalar irradiance Eo respectively. PAR_{Ed} and PAR_{Eo} are profiled from 0 m above the water surface down to 80 m depth, at 27 depths listed along the columns (the line entry is related to the simulation number). The euphotic depths, Zeu_{Ed} and Zeu_{Eo}, defined as the depths where PAR_{Ed} and PAR_{Eo} have 1 % of their respective values at the water surface, are provided in the files called Zeu_from_PAR_Ed_sza$x$.dat and Zeu_from_PAR_Eo_sza$x$.dat respectively.

The concentrations, IOP and AOP spectral data, and PAR and Zeu data files include headers to facilitate reading the data. The IOPs and concentration files are stored under the subfolder "Input IOPs Concentrations" and the simulated RLw, Kd, PAR_{Ed}, PAR_{Eo}, Zeu_{Ed}, and Zeu_{Eo} under "Output AOPs".

5 Conclusion

The CCRR match-up, in situ, and simulated data sets form a large database covering a wide range of water types, from oligotrophic to hypertrophic, and from clear to very turbid waters with a high diversity of IOPs.

The data sets contain 336 in situ reflectance spectra (covering the spectral range 412 to 709 nm) from five CoastColour sites, 957 match-up field reflectance spectra from 13 sites and 457 MERIS RLw spectra from 11 sites which show global consistency over the match-up sites, despite the absence of harmonized protocols used for RLw measurements by the different laboratories. In total, 80 % of the MERIS RLw measurements are provided from the North Sea, matching various in situ water quality parameters collected at this site. This is balanced by the distribution of RLw measurements, throughout the CoastColour sites, given in the match-up field and in situ reflectance data sets where fewer than 5 % of RLw spectra are available from the North Sea, while 23 % of the match-up field IOP data (excluding turbidity) are provided for that site.

The high-quality reflectance data sets along with the biogeochemical (CHL, TSM) and inherent optical properties provided at 17 CoastColour sites, covering the period 2005 to 2010, are fully documented and made available publicly for use in ocean colour algorithm testing.

The simulated data set includes 5000 reflectance spectra and the associated concentrations and inherent optical properties of chlorophyll a, mineral particles, and CDOM. The simulated reflectance data have been compared to the in situ and match-up reflectance data, showing a global consistency and giving clues for the discrepancies noticed (e.g. variable inherent optical properties, measurement uncertainties).

The strengths and weaknesses of each individual data set are recognized; for example, an in situ measurement represents "sea truth" better than simulated data but is subject to measurement uncertainty and represents only a small volume. Testing of an algorithm on all three data sets using re-

spectively MERIS, in situ, and simulated RLw input has significantly added value: evaluating the robustness of an ocean colour algorithm against remote-sensing measurements uncertainties; identifying its domain of validity (e.g. the detection or saturation limits); and testing its performance on various regions, days and daytimes, sea and sky conditions, etc. Such exercises may point out the disadvantage/advantage of using an algorithm for a regional or global application.

Oceanographic databases have been built during the last few years and made available to the scientific community (e.g. open ocean phytoplankton data by O'Brien et al., 2013, and Buitenhuis et al., 2013), facilitating the sharing of data and stimulates collaboration between the research institutes. In this paper, the first public optical–biogeochemical database was established representing the core of an open resource dedicated to case 2 remote-sensing data validation and algorithm testing. With joint efforts from the research centres and laboratories, this database may be updated with extended in situ data for the existing sites and for new regions in coastal and inland waters, with extra information (e.g. data quality flags), and with artificial data sets covering extra ranges of optical properties (e.g. extremely absorbing waters, extremely turbid waters) and/or underlying new bio-optical models.

Acknowledgements. This work is part of the CoastColour project, funded by the European Space Agency. The CSIRO measurements were funded by the CSIRO Wealth from Oceans Flagship and the Australian Integrated Marine Observing System (IMOS).

We warmly thank the in situ data providers:

- Curtiss O. Davis, Ricardo M. Letelier, Angelique E. White, and Marnie Jo Zirbel for collecting and processing the CEOAS data set.

- Stewart Bernard, Hayley Evers-King, Mark Mattews, and Lisl Robertson for processing the CSIR data set over the Benguela region, with the support of the Department of Agriculture, Forestry and Fisheries, DAFF.

- Roland Doerffer, Wolfgang Schönfeld, and Friedhelm Schroeder for providing the GKSS data set.

- Rodney Forster for providing the Cefas Chl a data set, included in the EMECO data set.

- Aristomenis P. Karageorgis, Kalliopi A. Pagou, Dimitris Tsoliakos, and Christina Zeri for collecting and processing the HCMR data set. The data from HCMR was acquired in the framework of SESAME – EC FP6 Integrated Project: Southern European Seas: Assessing and Modelling Ecosystem changes; HERMES – EC FP6 Integrated Project: Hotspot Ecosystem Research on the Margins of European Seas; and SARONIKOS: Monitoring of the inner Saronikos Gulf ecosystems, under the influence of the Psittalia sewage treatment plant. Ministry of Environment of the Hellenic Government.

- Young Je Park for providing the KORDI data set.

- Syarif Budhiman for processing the ITC data set.

- Guy Westbrook for providing the MII data set.

- Bryan A. Franz for processing the NOMAD data set.

- Francis Gohin, Catherine Belin, and Alain Lefèbvre for the Ifremer (REPHY phytoplankton network) data set.

- Griet Neukermans for collecting and processing the RBINS data set.

- Xabier Guinda, Beatriz Echavarri, Isabel Santamaría, and Pablo Ruíz for the UNICAN data collection and processing.

- Elisabete Mota (formerly in the Center of Oceanography, FCUL) and Katharina Poser (formerly in Brockman Consult) are thanked for their participation in the collection and organization of the in situ data sets.

David Foley actively contributed to this paper through the NOAA data set, before he passed away in December 2013. May this work be a way to remember him for his rigorousness and generosity.

Special thanks to Frank Müller-Karger and two anonymous reviewers for their very appreciated suggestions and comments, and to the Topical Editor François Schmitt for his assistance throughout the reviewing process.

Edited by: F. Schmitt

References

Albert, A. and Mobley, C. D.: An analytical model for subsurface irradiance and remote sensing reflectance in deep and shallow Case 2 waters, Opt. Express, 11, 2873–2890, 2003.

Antoine, D., Siegel, D. A., Kostadinov, T. S., Maritorena, S., Nelson, N. B., Gentili, B., Vellucci, V., and Guillecheau, N.: Variability of optical particle backscattering in contrasting bio-optical oceanic regimes, Limnol. Oceanogr., 56, 3, 955–973, doi:10.4319/lo.2011.56.3.0955, 2011.

Arar, E. J. and Collins, G. B.: Method 445.0 – In vitro determination of chlorophyll a and pheophytin a in marine and freshwater algae by fluorescence, National Exposure Research Laboratory, Office of Research and Development, U.S. Environmental Protection Agency, Cincinnati, Ohio, USA, 1992.

Babin, M., Morel, A., Fournier-Sicre, V., Fell, F., and Stramski, D.: Light scattering properties of marine particles in coastal and open ocean waters as related to the particle mass concentration, Limnol. Oceanogr., 28, 843–859, 2003a.

Babin, M., Stramski, D., Ferrari, G. M., Claustre, H., Bricaud, A., Obolensky, G., and Hoepffner, N.: Variations in the light absorption coefficients of phytoplankton, nonalgal particles and dissolved organic matter in coastal waters around Europe, J. Geophys. Res., 108, 3211, doi:10.1029/2001JC000882, 2003b.

Barale, V., Jaquet, J.-M., and Ndiayé, M.: Algal blooming patterns and anomalies in the Mediterranean Sea as derived from the Sea-WiFS data set (1998–2003), Remote Sens. Environ., 112, 3300–3313, 2008.

Barlow, R. G., Cummings, D. G., and Gibb, S. W.: Improved resolution of mono and divinyl chlorophylls a and and zeaxanthin and lutein in phytoplankton extracts using reverse phase C-8 HPLC, Mar. Ecol.-Prog. Ser., 161, 303–307, 1997.

Blondeau-Patissier, D., Brando, V. E., Oubelkheir, K., Dekker, A. G., Clementson, L. A., and Daniel, P.: Bio-optical variability of the absorption and scattering properties of the Queensland inshore and reef waters, Australia, J. Geophys. Res.-Oceans, 114, C5, doi:10.1029/2008JC005039, 2009.

Bouvet, M. and Ramoino, F.: Equalization of MERIS L1b products from the 2nd reprocessing, European Space Agency, Technical Report TEC-EEP/2009.521/MB, Issue 1, Revision 3, 08/12/2009, 77 p., published by ESTEC, Noordwijk, the Netherlands, 2010.

Brewin, R. J. W., Sathyendranath, S., Mueller, D., Brockmann, C., Deschamps, P.-Y., Devred, E., Doerffer, R., Fomferra, N., Franz, B., Grant, M., Groom, S., Horseman, A., Hu, C., Krasemann, H., Lee, Z., Maritorena, S., Mélin, F., Peters, M., Platt, T., Regner, P., Smyth, T., Steinmetz, F., Swinton, J., Werdell, J., and White, G. N.: The Ocean Colour Climate Change Initiative: III. A round-robin comparison on in-water bio-optical algorithms, Rem. Sens. Environ., 162, 271–294, 2015.

Bricaud, A., Morel, A., Babin, M., Allali, K., and Claustre, H.: Variations of light absorption by suspended particles with chlorophyll a concentration in oceanic (Case 1) waters: Analysis and implications for bio-optical models, J. Geophys. Res., 103, 31033–31044, 1998.

Brodie, J., De'ath, G., Devlin, M., Furnas, M. J., and Wright, M.: Spatial and temporal patterns of near surface chlorophyll a in the Great Barrier Reef lagoon, Mar. Freshwater Res., 58, 342–353, 2007.

Buitenhuis, E. T., Vogt, M., Moriarty, R., Bednaršek, N., Doney, S. C., Leblanc, K., Le Quéré, C., Luo, Y.-W., O'Brien, C., O'Brien, T., Peloquin, J., Schiebel, R., and Swan, C.: MAREDAT: towards a world atlas of MARine Ecosystem DATa, Earth Syst. Sci. Data, 5, 227–239, doi:10.5194/essd-5-227-2013, 2013.

Chavez, F. P., Pennington, J. T., Castro, C. G., Ryan, J. P., Michisaki, R. P., Schlining, B., Walz, P., Buck, K. R., McFadyen, A., and Collins, C. A.: Biological and chemical consequences of the 1997–1998 El Niño in central California waters, Prog. Oceanogr., 54, 205–232, doi:10.1016/S0079-6611(02)00050-2, 2002.

Cherukuru, N., Brando, V. E., Schroeder, T., Clementson, L. A., and Dekker, A. G.: Influence of river discharge and ocean currents on coastal optical properties, Cont. Shelf Res., 84, 188–203, 2014.

Clesceri, L. S., Greenberg, A. E., and Eaton, A. D.: Standard methods for the examination of water and wastewater, 20th edition, American Public Health Association, 1325 pp., Washington D.C., 1998.

Doerffer, R.: Protocols for the validation of MERIS water products, ESA/GKSS Doc. No. PO-TN-MEL-GS-0043, 46 p., available as Appendix at: http://envisat.esa.int/workshops/mavt_2003, MERIS protocols, Issue 1.3.5, 2002.

Doerffer, R.: Alternative Atmospheric Correction Procedure for Case 2 Water Remote Sensing using MERIS, 2011.

Doerffer, R. and Schönfeld, W.: Validation transect between Cuxhaven and Helgoland: GKSS Technical Note, 18 February 2009, Helmholtz-Centrum Geesthacht, Geesthacht, 2009.

Dolan, J. R., Claustre, H., Carlotti, F., Plounevez, S., and Moutin, T.: Microzooplankton diversity: relationships of tintinnid ciliates with resources, competitors and predators from the Atlantic Coast of Morocco to the Eastern Mediterranean, Deep-Sea Res. Pt. I, 49, 1217–1232, 2002.

Freudenthal, T., Meggers, H., Henderiks, J., Kuhlmann, H., Moreno, A., and Wefer, G.: Upwelling intensity and filament activity off Morocco during the last 250,000 years, Deep-Sea Res. Pt. II, 49, 3655–3674, 2002.

Gohin, F.: Annual cycles of chlorophyll a, non-algal suspended particulate matter, and turbidity observed from space and in-situ in coastal waters, Ocean Sci., 7, 705–732, doi:10.5194/os-7-705-2011, 2011.

Groom, S. B., Herut, B., Brenner, S., Zodiatis, G., Psarra, S., Kress, N., Krom, M. D., Law, C. S., and Dracopoulos, P.: Satellite-derived spatial and temporal biological variability in the Cyprus Eddy, Deep-Sea Res. Pt. II, 52, 2990–3010, 2005.

Groom, S. B., Martinez-Vicente, V., Fishwick, J. R., Tilstone, G., Moore, G., Smyth, T. J., and Harbour, D.: The Western English Channel observatory: Optical characteristics of station 4, J. Marine Syst., 77, 278–295, 2009.

Guillocheau, N.: β-Correction Experiment Report, ICESS, University of California, Santa Barbara, CA, USA, 2003.

Harrison, A. W. and Coombes, C. A.: An opaque cloud cover model of sky short wavelength radiance, Sol. Energy, 41, 387–392, 1988.

Holm-Hansen, O., Lorenzen, C. J., Holmes, R. W., and Strickland, J. D. H.: Fluorometric determination of chlorophyll, Journal du Conseil Permanent International pour l'Exploration de la Mer, 30, 33–15, 1965.

Hooker, S. B., Van Heukelem, L., Thomas, C. S., Claustre, H., Ras, J., Barlow, R. G., Sessions, H., L., S., Perl, J., Trees, C. C., Stuart, V., Head, E., Clementson, L. A., Fishwick, J. R., Llewellyn, C., and Aiken, J.: The Second SeaWiFS HPLC Analysis Round-Robin Experiment (SeaHARRE-2), NASA Aeronautics and Space Information, Goddard Space Flight Center, Greenbelt, Maryland, USA, 2005.

IOCCG: Remote Sensing of Inherent Optical Properties: Fundamentals, Tests of Algorithms, and Applications, edited by: Lee, Z.-P., Reports of the International Ocean-Colour Coordinating Group, No. 5, IOCCG, Dartmouth, Canada, pp. 126, 2006.

IOC/UNESCO: Protocols for Joint Global Ocean Carbon Flux Study (JGOFS) Core Measurements, Paris, 1994.

Karageorgis, A. P., Georgopoulos, D., Kanellopoulos, T. D., Mikkelsen, O. A., Pagou, K., Kontoyiannis, H., Pavlidou, A., and Anagnostou, C.: Spatial and seasonal variability of particulate matter optical and size properties in the Eastern Mediterranean Sea, J. Marine Syst., 105–108, 123–134, 2012.

Kostadinov, T. S., Siegel, D. A., Maritorena, S., and Guillocheau, N.: Ocean color observations and modeling for an optically complex site: Santa Barbara Channel, California, USA, J. Geophys. Res., 112, C07011, doi:10.1029/2006JC003526, 2007.

Llewellyn, C. A., Fishwick, J. R., and Blackford, J. C.: Phytoplankton community assemblage in the English Channel: a comparison using chlorophyll a derived from HPLC-CHEMTAX and carbon derived from microscopy cell counts, J. Plankton Res., 27, 103–119, 2005.

Loisel, H. and Morel, A.: Light scattering and chlorophyll concentration in Case 1 waters: a re-examination., Limnol. Oceanogr. (Methods), 43, 847–857, 1998.

Loisel, H., Vantrepotte, V., Norkvist, K., Mériaux, X., Kheireddine, M., Ras, J., Pujo-Pay, M., Combet, Y., Leblanc, K., Dall'Olmo, G., Mauriac, R., Dessailly, D., and Moutin, T.: Characterization of the bio-optical anomaly and diurnal variability of particulate matter, as seen from scattering and backscattering coefficients, in ultra-oligotrophic eddies of the Mediterranean Sea, Biogeosciences, 8, 3295–3317, doi:10.5194/bg-8-3295-2011, 2011.

Maffione, R. A. and Dana, D. R.: Instruments and methods for measuring the backward-scattering coefficient of ocean waters, Appl. Optics, 36, 6057–6067, 1997.

Maritorena, S., Lee, Z.-P., Du, K.-P., Loisel, H., Doerffer, R., Roesler, C., Tanaka, A., Babin, M., and Kopelevich, O. V.: Syntheric and in situ data sets for algorithm testing. Remote Sensing of Inherent Optical Properties: Fundamentals, Tests of Algorithms, and Applications, IOCCG, Dartmouth, Canada, 73–79, 20065, 13–18, 2006.

Martinez-Vicente, V., Land, P. E., Tilstone, G. H., Widdicombe, C., and Fishwick, J. R.: Particulate scattering and backscattering related to water constituents and seasonal changes in the Western English Channel, J. Plankton Res., 32, 603–619, 2010.

Mobley, C. D.: Light and water: radiative transfer in natural waters, Academic Press, London, UK, 1994.

Mobley, C. D.: Estimation of the remote sensing reflectance from above-surface measurements, Appl. Optics, 38, 7442-7455, 1999.

Mobley, C. D. and Sundman, L. K.: HYDROLIGHT 5/ECOLIGHT 5 Technical Documentation, Sequoia Scientific, Inc., 2700 Richards Road, Suite 107, Bellevue, WA 98005, 2008.

Morel, A. and Prieur, L.: Analysis of variations in ocean color, Limnol. Oceanogr., 22, 709–722, 1977.

Morel, A., Antoine, D., and Gentili, B.: Bidirectional reflectance of oceanic waters: accounting for Raman emission and varying particle scattering phase function, Appl. Optics, 41, 6289–6306, 2002.

O'Brien, C. J., Peloquin, J. A., Vogt, M., Heinle, M., Gruber, N., Ajani, P., Andruleit, H., Arístegui, J., Beaufort, L., Estrada, M., Karentz, D., Kopczyńska, E., Lee, R., Poulton, A. J., Pritchard, T., and Widdicombe, C.: Global marine plankton functional type biomass distributions: coccolithophores, Earth Syst. Sci. Data, 5, 259–276, doi:10.5194/essd-5-259-2013, 2013.

Odermatt, D., Gitelson, A. A., Brando, V. E., and Schaepman, M.: Review of constituent retrieval in optically deep and complex waters from satellite imagery, Remote Sens. Environ., 118, 116–126, 2012.

O'Reilly, J. E., Maritorena, S., Mitchell, B. G., Siegel, D. A., Carder, K. L., Garver, S. A., Kahru, M., and McClain, C. R.: Ocean color chlorophyll algorithms for SeaWiFS, J. Geophys. Res., 103, 24937–24953, 1998.

Oubelkheir, K., Claustre, H., Sciandra, A., and Babin, M.: Bio-optical and biogeochemical properties of different trophic regimes in oceanic waters, Limnol. Oceanogr., 50, 1795–1809, 2005.

Oubelkheir, K., Clementson, L. A., Webster, I., Ford, P., Dekker, A. G., Radke, L., and Daniel, P.: Using inherent optical properties to investigate biogeochemical dynamics in a tropical macrotidal coastal system, J. Geophys. Res.-Oceans, 111, C7, doi:10.1029/2005JC003113, 2006.

Parsons, T. R., Maita, Y., and Lalli, C. M.: A manual of Chemical and Biological Methods for seawater Analysis, Pergamon Press, New York, USA, 173 pp., 1984.

Pegau, W. S., Zaneveld, J. R. V., Mitchell, B. G., Mueller, J. L., Kahru, M., Wieland, J., and Stramska, M.: Inherent Optical Properties: Instruments, Characterizations, Field Measurements and Data Analysis Protocols, volume IV, in: Ocean Optics Protocols for Satellite Ocean Color Sensor Validation, Revision 4, NASA Goddard Space Flight Center, Greenbelt, Maryland, USA, 76 pp., 2003.

Pitcher, G. C. and Probyn, T. A.: Anoxia in southern Benguela during the autumn of 2009 and its linkage to a bloom of the dinoflagellate Ceratium balechii, Harmful Algae, 11, 23–32, 2011.

Probyn, T. A.: Nitrogen uptake by size-fractionated phytoplankton populations in the southern Benguela upwelling system, Mar. Ecol.-Prog. Ser., 22, 249–258, 1985.

Ruddick, K., Brockmann, C., Doerffer, R., Lee, Z., Brotas, V., Fomfera, N., Groom, S. B., Krasemann, H., Martinez-Vicente, V., Sá, C., Santer, R., Sathyendranath, S., Stelzer, K., and Pinnock, S.: The CoastColour project regional algorithm round robin exercise, in: Remote Sensing of the Coastal Ocean, Land and Atmosphere Environment, SPIE, Incheon, Republic of Korea, 7858–7807, 2010.

Ruddick, K. G., De Cauwer, V., Park, Y., and Moore, G.: Seaborne measurements of near infrared water-leaving reflectance – the similarity spectrum for turbid waters, Limnol. Oceanogr., 51, 1167–1179, 2006.

Schaffelke, B., Carleton, J., Skuza, M., Zagorskis, I., and Furnas, M. J.: Water quality in the inshore Great Barrier Reef lagoon: Implications for long-term monitoring and management, Mar. Pollut. Bull., 65, 249–260, 2012.

Strickland, J. D. H. and Parsons, T. R.: A practical handbook of the sea water analysis, Fisheries Research Board of Canada, Ottawa, Bulletin 167, 311 p., 1972.

Tassan, S. and Ferrari, G. M.: An alternative approach to absorption measurements of aquatic particles retained on filters, Limnol. Oceanogr., 40, 1358–1368, 1995.

Tilstone, G., Moore, G., Sorensen, K., Doerffer, R., Rottgers, R., Ruddick, K. G., Pasterkamp, R., and Jorgensen, P. V.: Regional Validation of MERIS Chlorophyll products in North Sea coastal waters: REVAMP protocols, ENVISAT validation workshop European Space Agency, available as Appendix at: https://earth.esa.int/workshops/mavt_2003/, Frascatti, Italy, 20–24 October 2003.

Tilstone, G., Peters, S., van Der Woerd, H., Eleveld, M., Ruddick, K., Schönfeld, W., Krasemann, H., Martinez-Vicente, V., Blondeau-Patissier, D., Jorgensen, P. V., Rottgers, R., and Sorensen, K.: Variability in specific-absorption properties and their use in Ocean Colour Algorithms for MERIS in North Sea coastal waters, Remote Sens. Environ., 118, 320–338, 2012.

Toole, D. A., Siegel, D. A., Wenzies, D. W, Neumann, M. J., and Smith, R. C.: Remote-sensing reflectance determinations in the coastal ocean environment: impact of instrumental characteristics and environmental variability, Appl. Optics, 39, 3, 456–469, 2000.

Tzortziou, M., Subramaniam, A., Herman, J. R., Gallegos, C. L., Neale, P. J., and Harding, L. W. J.: Remote sensing reflectance and inherent optical properties in the mid Chesapeake Bay, Estuarine, Coastal and Shelf Science, 72, 16–32, 2007.

van der Linde, D. W.: Protocol for the determination of total suspended matter in oceans and coastal zones, Technical Note I.98.182, European Commission Joint Research Centre, Ispra, Italy, 1998.

Van Heukelem, L. and Thomas, C. S.: Computer-assisted high-performance liquid chromatography method development with applications to the isolation and analysis of phytoplankton pigments, J. Chromatogr. A, 910, 31–49, 2001.

Werdell, P. J.: An evaluation of inherent optical property data for inclusion in the NASA Bio-optical Marine Algorithm Data Set, NASA Ocean Biology Processing Group paper, NASA Goddard Space Flight Cent., Greenbelt, Md., available at: http://seabass.gsfc.nasa.gov/seabass/data/werdell_nomad_iop_qc.pdf, 2005.

Werdell, P. J. and Bailey, S. W.: An improved in situ bio-optical data set for ocean color algorithm development and satellite data product validation, Remote Sens. Environ., 98, 122–140, 2005.

Werdell, P. J., Bailey, S. W., Fargion, G. S., Pietras, K., Knobelspiesse, K. D., Feldman, G. C., and McClain, C.: Unique data repository facilitates ocean color satellite validation, EOS, Transactions, American Geophysical Union, 84, 377–392, 2003.

Werdell, P. J., Franz, B. A., Bailey, S. W., Feldman, G. C., Boss, E., Brando, V. E., Dowell, M. D., Hirata, T., Lavender, S. J., Lee, Z., Loisel, H., Maritorena, S., Mélin, F., Moore, T. S., Smyth, T. J., Antoine, D., Devred, E., Hembise Fanton d'Andon, O., and Mangin, A.: Generalized ocean color inversion model for retrieving marine inherent optical properties, Appl. Optics, 52, 2019–2037, 2013.

WET Labs ac-9 Protocol: Revision H, WET Labs, Inc., Philomath, OR, USA, 42 pp., 2003.

Wright, S. W., Jeffrey, S. W., Mantoura, R. F. C., Llewellyn, C. A., Bjornland, T., and Repeta, D. J.: Improved HPLC method for the analysis of chlorophylls and carotenoids from marine phytoplankton, Mar. Ecol.-Prog. Ser., 77, 183–196, 1991.

10

Distribution of known macrozooplankton abundance and biomass in the global ocean

R. Moriarty[1,2,*]**, E. T. Buitenhuis**[2]**, C. Le Quéré**[1,2,**]**, and M.-P. Gosselin**[1,***]

[1]British Antarctic Survey, High Cross, Madingley Road, Cambridge CB3 0ET, UK
[2]School of Environmental Sciences, University of East Anglia, Norwich Research Park, Norwich NR4 7TJ, UK
[*]Present address: School of Earth, Atmospheric and Environmental Sciences, University of Manchester, Williamson Building, Oxford Road, Manchester M13 9PL, UK
[**]Present address: Tyndall Centre for Climate Change Research, School of Environmental Sciences, University of East Anglia, Norwich Research Park, Norwich NR4 7TJ, UK
[***]Present address: Environment Agency, North West – Water Resources, Ghyll Mount, Gillan Way, Penrith 40 Business Park, Penrith CA11 9BP, UK

Correspondence to: R. Moriarty (r.moriarty@uea.ac.uk)

Abstract. Macrozooplankton are an important link between higher and lower trophic levels in the oceans. They serve as the primary food for fish, reptiles, birds and mammals in some regions, and play a role in the export of carbon from the surface to the intermediate and deep ocean. Little, however, is known of their global distribution and biomass. Here we compiled a dataset of macrozooplankton abundance and biomass observations for the global ocean from a collection of four datasets. We harmonise the data to common units, calculate additional carbon biomass where possible, and bin the dataset in a global 1×1 degree grid. This dataset is part of a wider effort to provide a global picture of carbon biomass data for key plankton functional types, in particular to support the development of marine ecosystem models. Over 387 700 abundance data and 1330 carbon biomass data have been collected from pre-existing datasets. A further 34 938 abundance data were converted to carbon biomass data using species-specific length frequencies or using species-specific abundance to carbon biomass data. Depth-integrated values are used to calculate known epipelagic macrozooplankton biomass concentrations and global biomass. Global macrozooplankton biomass, to a depth of 350 m, has a mean of $8.4 \mu g \, C \, L^{-1}$, median of $0.2 \mu g \, C \, L^{-1}$ and a standard deviation of $63.5 \mu g \, C \, L^{-1}$. The global annual average estimate of macrozooplankton biomass in the top 350 m, based on the median value, is 0.02 Pg C. There are, however, limitations on the dataset; abundance observations have good coverage except in the South Pacific mid-latitudes, but biomass observation coverage is only good at high latitudes. Biomass is restricted to data that is originally given in carbon or to data that can be converted from abundance to carbon. Carbon conversions from abundance are restricted by the lack of information on the size of the organism and/or the absence of taxonomic information. Distribution patterns of global macrozooplankton biomass and statistical information about biomass concentrations may be used to validate biogeochemical and plankton functional type models.

1 Introduction

Global ocean biogeochemical models representing lower-trophic ecosystems have been widely used to study the interactions and feedbacks between climate and marine bio-geochemistry. They have been applied to investigating the processes affecting atmospheric CO_2 concentration and climate (e.g. Lovenduski et al., 2008) and used to predict the possible effects of climate change on ecosystem structure, functioning and productivity. The most recent generation

of global biogeochemical models – called Dynamic Green
Ocean Models (DGOMs) – represent marine ecosystems using multiple plankton functional types (PFTs; Le Quéré et
al., 2005; Hood et al., 2006) and thus include basic trophic
structure and ecosystem diversity. Independent datasets are
required to validate the representation of each PFT in these
models, in particular their distribution in various ocean regions, depth profiles and concentrations. Observational data
on PFT-specific abundance and carbon biomass is sparse and
highly variable, and consequently model validation is challenging. The lack of coherent global PFT-specific datasets
is currently limiting the development of global models. The
data presented in this paper are part of a wider community effort known as MARine Ecosystem DATa (MAREDAT), and
cover data on a variety of major PFTs currently represented
in marine ecosystem models (Buitenhuis et al., 2013).

MAREDAT is a collection of global biomass datasets
gathered by marine ecosystem researchers. It contains data
on the global distribution of a variety of the major PFTs
currently represented in marine ecosystem models. These
include bacteria, picophytoplankton, nitrogen fixers, calcifiers, dimethyl sulphide (DMS)-producers, silicifiers, microzooplankton, foraminifera, mesozooplankton (mostly copepods; Moriarty and O'Brien, 2013), pteropods and macrozooplankton. MAREDAT is part of the MARine Ecosystem Model Inter-comparison Project (MAREMIP) and is responsible for this compilation of observation-based global
biomass datasets. The biomass data that populate MARE-
DAT are freely available for use in model evaluation and
development, and to the scientific community as a whole
(http://maremip.uea.ac.uk/.maredat.html).

Macrozooplankton are found throughout the global ocean
and epipelagic macrozooplankton in the sunlit portion. They
include the holoplanktic and meroplanktic members of
the thalicia, ctenophores, cnidaria, gastropoda, heteropoda,
pteropoda, chaetognatha, polychaetea, amphipods, stomatopods, mysids, decapods, and euphausiids, among others, each taxon containing many species. Macrozooplankton are commonly classified by size but classification varies
(see Sieburth et al., 1987; Schütt, 1892; Omori and Ikeda,
1984). Here we include all zooplankton whose adult size is
greater than 2 mm (Le Quéré et al., 2005). They are separated from nekton (i.e. fish) as they do not have the lo-
comotive capacity necessary to overcome ocean currents
(Omori and Ikeda, 1984). Copepods are an exception to
this definition. In order to try and avoid double counting
of copepods, the biomass of this group have been included
in a mesozooplankton/copepod-specific article (Moriarty and
O'Brien, 2013; see Sect. 3.2.1).

Macrozooplankton are involved in an intricate trophic
web, exerting a direct influence on all lower trophic levels
through variation in species-specific feeding behaviours and
prey preferences. They repackage autotrophic, heterotrophic
and detrital material in the surface ocean. Part of this repackaged material moves to higher trophic levels through the pre-

dation on macrozooplankton (Deibel, 1998), making them
an important link to higher trophic levels, especially fish.
In some areas of the ocean they are a crucial link between
lower and higher trophic levels, e.g. *Euphausia superba* in
the Southern Ocean. Repackaged material may also be exported from the surface to the intermediate and deep ocean
in the form of faecal pellets (Turner, 2002). In this manner macrozooplankton affect the cycling of carbon and nutrients in the ocean and have been included as a PFT in some
DGOMs (Le Quéré et al., 2005, 2013).

PFT modelling uses a coarse division of the ecosystem,
breaking the plankton into functional groups or plankton
functional types (PFTs). PFT models treat macrozooplankton
as a generalist group, linked by their common function in the
removal of carbon from the sunlit waters of the global ocean
to the deep ocean. Macrozooplankton demonstrate incredible
diversity when it comes to feeding preferences, feeding on
minute particles, bacteria, detritus, phytoplankton and other
zooplankton. We presented macrozooplankton as a unique
PFT compartment of the ecosystem, regardless of the position of individual species in the food web. This dataset is
designed for application in the study of global ocean ecosystems from a plankton function or PFT perspective.

This paper presents a global synthesis of macrozooplankton abundance and biomass data, including full details of the
biomass conversions used and the associated uncertainties.
Patterns in the spatial and temporal distribution of epipelagic
macrozooplankton abundance and biomass are examined.
There is no one data centre responsible for the collection and
synthesis of macrozooplankton abundance and biomass data.
The synthesis presented here is a step towards gathering all
available data to better understand the global distribution of
macrozooplankton and the mechanisms controlling them.

2 Data and methods

2.1 Origin of data

The data compiled in this study have come from four existing
databases and are summarised in Table 1a, b, c and d. Macrozooplankton data have been data mined from all databases
with the exception of the rawKRILLBASE database, which
already contained only macrozooplankton data.

The first two databases are from the Joint Global Ocean
Flux Study: the Hawaii Ocean Time-Series (HOTS; Landry
et al., 2001) and the Bermuda Atlantic Time-Series (BATS;
Steinberg et al., 2012; Table 1a). These are long-term time-
series stations representative of the oligotrophic subtropical
gyres in the North Pacific and North Atlantic, respectively.
Abundance and biomass data were collected by oblique net
tows on monthly cruises at Station ALOHA (A Long-term
Oligotrophic Habitat Assessment; 22°45′ N, 158°00′ W) for
HOTS and at the BATS station (31°50′ N, 64°10′ W). The
targeted maximum depth was between 150 and 200 m. At
both locations data for two macrozooplankton size fractions

Table 1a. Summary of data points for original macrozooplankton abundance and biomass as gathered from JEGOFS long-term time-series datasets.

Dataset	Principal Investigator	No. of data		Depth range (m)	Reference
		Abundance	Biomass		
HOTS	Landry	272	1330	0–271	Landry et al. (2001)
BATS	Steinberg	0	1730*	0–306	Steinberg et al. (2012)

* Biomass values are originally in wet weight and dry weight.

2000–5000 µm and >5000 µm were collected. The data presented in this study were collected between 1994 and 2010; data collection is still ongoing at both sites. The HOTS dataset includes 272 abundance (individual (ind.) m^{-2}) and 1330 biomass (mg carbon (C) m^{-2}) data. For the purposes of this study, abundance data were converted from ind. m^{-2} to ind. L^{-1} by averaging the sampling depth. Biomass data were converted from mg C m^{-2} to µg C L^{-1} in a similar manner. The BATS dataset includes no abundance and 1730 biomass (mg wet mass m^{-3} and mg dry mass m^{-3}) data; it does not provide carbon biomass. BATS abundance data were converted to carbon biomass using a dry mass to carbon conversion (see Sect. 2.2).

The rawKRILLBASE (Atkinson et al., 2004; Table 1b and c) database is a Southern Ocean dataset of krill and salp numerical densities and length frequencies. It includes 8192 abundance (ind. m^{-2}) measurements of postlarval *Euphausia superba* and 9719 abundance measurements of pooled individuals of aggregate and solitary forms of salps (mainly *Salpa thompsoni* but also *Ihlea racovitzai*). The data spans the periods 1926–1951 and 1976–2003. All data are from random hauls or those at pre-fixed locations (hauls specifically targeted on krill or salp aggregations were excluded). All net hauls were oblique or vertical and were taken mainly in the summer months. Full details of this database, including net sampling details, are provided in Atkinson et al. (2004, 2008, 2009; including the supplementary method appendices of the first two). The krill and salp density data taken from the rawKRILLBASE are used in their raw form, i.e. the data owners have standardised these densities to a common sampling method but have stipulated that the standardised data may not be used as part of our study. For the purposes of this study abundance data were converted from ind. m^{-2} to ind. L^{-1}, using the depth over which the samples were gathered. All the rawKRILLBASE krill and salp data were converted to carbon biomass (see Sect. 2.2).

The Coastal & Oceanic Plankton Ecology, Production, & Observation Database (COPEPOD; O'Brien, 2005; Table 1d) is a global database and includes a myriad of macrozooplankton taxa: salps, doliolids, pyrosomes, ctenophores, cnidaria, scyphozoa, hydrozoa, anthozoa, pelagic molluscs, pelagic polychaetes, chaetognatha, amphipods, mysids, stomatopods, decapods, euphausiids, and appendicularia. The COPEPOD database is maintained by the United States of America National Marine Fisheries Service (USA NMFS), and includes its own ecosystem survey and sampling programs, historical plankton data search and rescue work, institutional and project data, as well as individual submissions from researchers outside NMFS. In total there were 369 567 measurements of abundance within COPEPOD relating to macrozooplankton as defined above. Abundance measurements within COPEPOD were standardised to individuals per m^3 by NMFS. For the purposes of this study, abundance data were converted from ind. m^{-3} to ind. L^{-1}. It was possible to convert 15 297 abundance data to carbon biomass (see Sect. 2.2). COPEPOD is the only database considered here which included macrozooplankton abundance and biomass as a function of depth.

Where possible we have retained all taxonomic data associated with abundance and biomass values in the original raw datasets. While this is only directly important if we are converting abundance to biomass on a species-by-species level, we hope that this dataset will continue to be added to in terms of abundance, biomass and associated metadata and may be used for a wide variety of applications, a point of departure for research that requires more specific information than simple "macrozooplankton biomass" values.

2.2 Biomass conversion

The HOTS dataset contained biomass (mg C m^{-2}) data and it was only necessary to convert to common units and volume as described above. The BATS dataset contained both wet and dry mass biomass data. Dry mass data were converted using a dry mass to carbon conversion, $C = 0.36 \times$ dry weight (Madin et al., 2001, as documented in Steinberg et al., 2011).

For the rawKRILLBASE database the mean body mass of *Euphausia superba* was estimated as 140 mg dry mass ind.$^{-1}$ (A. Atkinson, personal communication, 2007) and was calculated from a length frequency distribution dataset containing 535 581 length measurements of *Euphausia superba* recovered from scientific hauls between October and April from 1926–1939 and 1976–2006 (Atkinson et al., 2009). A dry mass to carbon conversion of 0.45 (Pakhomov et al., 2002 and references therein) was used to convert dry mass to carbon per individual. The mean body mass, for both species of salp, was estimated at 120 mg dry mass ind.$^{-1}$ (Dubischar et al., 2006). A dry mass to carbon conversion of 0.2 (weighted

Table 1b. Summary of data points for *Euphausia superba* abundance in the Southern Ocean from the rawKRILLBASE dataset (Atkinson et al., 2004).

Principal Investigator	No. of abundance data	Depth range (m)	Country	Institute/Source/Reference
Loeb	1958	0–282	USA	Loeb et al. (2010)
				US Antarctic Marine Living Resources Field Season Reports
Atkinson	328	0–380	UK	BAS (unpublished data)
Atkinson	119	0–500	UK	Atkinson and Peck (1988)
Ward	59	0–200	UK	Ward et al. (2005)
Atkinson	55	0–400	UK	Ward et al. (2006)
				and BAS (unpublished data)
Chiba	25	0–268	Japan	Chiba et al. (1999)
				Japanese Antarctic Research Expedition
Chiba	35	0–150	Japan	Chiba et al. (2001)
				Japanese Antarctic Research Expedition
Pakhomov	1042	0–500	Russia	Pakhomov et al. (2002)
Hosie	66	0–200	Australia	Hosie et al. (2000)
				BROKE Survey, Australian Antarctic Division
Pakhomov	198	0–383	South Africa	Pakhomov et al. (2002)
Anadon	99	0–200	Spain	Anadon and Estrada (2002)
				FRUELA
Hosie	64	0–200	Australia	Hosie and Cochran (1994)
				Australian Antarctic Division
Hosie	83	0–200	Australia	Hosie et al. (1997)
				Australian Antarctic Division
Atkinson	1176	0–540	UK	Marr (1962); Atkinson et al. (2009);
				Foxton (1966)
				Discovery Expeditions
Atkinson	8	0–875	Japan	Casareto and Nemoto (1986)
				SIBEX BIOMASS Programme
Ross and Quetin	631	0–460	USA	Ross et al. (2008)
				US Palmer LTER Program
Atkinson	125	0–300	Poland	Jazdzewski et al. (1982)
Siegel	1692	0–1200	Germany	Siegel (2005)
Siegel	117	0–2700		Siegel et al. (2004)
				CCAMLR
Siegel	147	0–213	Germany	Hunt et al. (2011)

average of solitary and aggregate forms) (Dubischar et al., 2006) was used to convert dry mass to carbon per individual.

For the COPEPOD database, only species-specific abundance data that had a corresponding species-specific carbon conversion were converted to biomass. Body mass data were compiled for a number of the species found in the COPEPOD dataset (for individual species conversions see Table 2). We used the macrozooplankton biomass conversion dataset database to generate the biomass data from species-specific abundance data. Adult mean body mass of the species was used for the conversion of abundance (ind. L^{-1}) data to carbon. Species-specific conversions were preferred even though it was only possible to convert a small fraction ($\sim 4\%$) of the dataset to carbon biomass.

Data were gridded using the original entries for latitude, longitude and month from all datasets. The mean depth of the sampling depth range of each macrozooplankton concen-

tration was used as sample depth. Macrozooplankton concentrations in ind. L^{-1} and $\mu g C L^{-1}$ were binned on the 4-dimensional World Ocean Atlas grid. This is a monthly grid with horizontal resolution of $1° \times 1°$ and 33 vertical levels resolved to 5 m in surface waters, increasing to 500 m from 2000 m downwards. Only data that are gridded in the top 350 m of the ocean is used for calculation of global epipelagic macrozooplankton annual average biomass but all available data (all depths) on macrozooplankton have been included in both ungridded and gridded datasets.

2.3 Quality control

Chauvenet's criterion for data rejection was used for the removal of high outliers from both abundance and biomass data (Glover et al., 2011; Buitenhuis et al., 2013). A normal distribution of the data is assumed and data are rejected when the

Table 1c. Summary of data points for *Salpa thompsoni* and *Ihlea racovitzai* abundance in the Southern Ocean from the rawKRILLBASE dataset (Atkinson et al., 2004).

Principal Investigator	No. of data Abundance	Depth range (m)	Country	Institute/Source/Reference
Loeb	1130	0–210	USA	Loeb et al. (2010) US Antarctic Marine Living Resources Field Season Reports
Atkinson	169	0–2236	Australia	Australian ANARE Research Notes
Atkinson	97	0–2500	UK	Atkinson and Peck (1988) and BAS (unpublished data)
Atknison	409	0–2030	UK	BAS (unpublished data)
Atkinson	59	0–200	UK	Ward et al. (2005) and BAS (unpublished data)
Atkinson	55	0–400	UK	Ward et al. (2006) and BAS (unpublished data)
Chiba	24	0–268	Japan	Chiba et al. (1999) Japanese Antarctic Research Expedition
Chiba	44	0–150	Japan	Chiba et al. (2001) Japanese Antarctic Research Expedition
Pakhomov	66	0–1000	Russia	Pakhomov et al. (2002)
Hosie	66	0–200	Australia	Hosie et al. (2000) BROKE Survey, Australian Antarctic Division
Pakhomov	40	0–500	Germany	ANTARKTIS XIII5b
Pakhomov	9	0–394	South Africa	DEIMEC
Pakhomov	393	0–403	South Africa	Pakhomov et al. (2002)
Pakhomov	1423	0–1000	Russia	Pakhomov et al. (2002)
Anadon	99	0–200	Spain	Anadon and Estrada (2002) FRUELA
Hosie	147	0–200	Australia	Hosie and Cochran (1994) Australian Antarctic Division
Atkinson	2659	0–1300	UK	Foxton (1966) Discovery Expeditions
Atkinson	13	0–1050	Japan	Casareto and Nemoto (1986) SIBEX BIOMASS Programme
Nishikawa	50	0–99	Japan	Nishikawa et al. (1995)
Atkinson	32	0–500	Japan	Nishikawa and Tsuda (2001)
Ross and Quetin	633	0–460	USA	Ross et al. (2008) Palmer LTER Program
Atkinson	96	0–300	Poland	Jazdzewski et al. (1982)
	63	0–245	Poland	Witek et al. (1985)
Siegel	1731	10–200	Germany	Siegel (2005)
Siegel	119	0–2700		Siegel et al. (2004) CCAMLR
Siegel	93	0–200	Germany	Hunt et al. (2011)

probability of deviation from the mean is less than $1/(2n)$, where n is the number of data points. Chauvenet's criterion could not be applied directly because the macrozooplankton abundance and biomass data were not normal distributed but instead ranged from low or undetectable concentrations, to very high value bloom events. Chauvenet's criterion was applied to all non-zero log-transformed data, which had a near normal distribution. Data with zero values for abundance and biomass are included in the dataset as they represent an absence of macrozooplankton, but could not be included in the log-transformed data.

The mean $\bar{x}$ and the standard deviation σ of the log-transformed data are calculated and used to calculate the critical value x_c. One half of $1/(2n)$ is used because the Chauvenet's criterion is a two-tailed test; however, we only rejected data on one tail, the high side. All log-transformed data with values higher than $\bar{x} + x_c$ are rejected (Luo et al., 2012).

2.4 Database formats

The original datasets are available as Excel files with full details of data sources and conversions. We also provide a

Table 1d. Summary of data points for macrozooplankton abundance from the COPEPOD dataset O'Brien.

COPEPOD Dataset	No. of abundance data	Depth (m)		Project	Institute	NODC accession no.
		Range	Mean			
ALMIRANTE SALDANHA Collection	785	0–274	72	–	Max-Planck-Institut Fuer Meteorologie	0000942
AtlantNIRO plankton	268	0–100	50	–	Atlantic Research Institute of Fishing Economy and Oceanography	9600039
BCF – POFI	454	0–260	55	Pacific Oceanic Fisheries Investigations	Maritime Regional Administration of Hydrometeorology	0051848
Biological Atlas of the Arctic Seas 2000: Plankton of the Barents and Kara Seas	30346	0–2828	89	–	–	0000283
BIOMAN	1888	0–100	43	BIOMass of ANchovy	–	9700075
Brodskii (1950)	312	0–3884	536	–	–	0064569
CHIU LIEN Collection	236	0–150	40	–	National Taiwan University Institute of Fishery Biology, Taipei	0000095, 0000097
CINECA I	65	0–600	184	Cooperative Investigations of the Northern-part of the Eastern Central Atlantic	Cent Nat pour L'Exploit des Ocg Bur Nat des Donn Ocg	9000076
CINECA II	2954	0–100	50	Cooperative Investigations of the Northern-part of the Eastern Central Atlantic	Mediterranean Marine Sorting Center	0000088
CINECA IV	696	0	0	CORiolis-INDONesia	Office de la Recherche Scientifique et Technique Outre Mer	0000527
CSK	52121	0–1023	56	Cooperative Study of the Kuroshio and adjacent regions	Fisheries Research and Development Agency, Republic of Korea Institute of Marine Research, Jakarta Kagoshima University Faculty of Fisheries Nagasaki University Fisheries Institute Tokyo University of Fisheries Kominato Marine Biological Lab, Awa-kominato Shimonoseki University of Fisheries Tokyo University, Ocean Research Institute Hokkaido University Faculty of Fisheries, Hakodate Hong Kong Fisheries Research Station, Hong Kong Fisheries Biology Unit, University of Singapore Marine Fisheries Laboratory, Department of Fisheries Philippines Fisheries Commission, Department of Agriculture and Natural Resources Japan Meteorological Agency Fisheries Research Institute, Malaysia Fisheries Research Institute, Republic of Korea Marine Fisheries Laboratory, Thailand Pacific Scientific Research Institute of Fisheries & Oceanography, USSR National Oceanography Institute, Vietnam SouthEast Asian Fisheries DEvelopment Center, Singapore	9500141
Drift Station Alpha	123	0–2000	401	International Geophysical Year	Scripps Institution of Oceanography	0000810
EASTROPAC	7336	0–277	101	Eastern Tropical Pacific 1967–1968	Smithsonian Oceanographic Sorting Center	9500089, 9500090
EASTROPIC	823	0–427	103	–	Maritime Regional Administration of Hydrometeorology – Rosgidromet	9700300, 9700074
EcoMon-SOOP (Gulf of Maine)	2012	10	10	MARine Resources Monitoring; Assessment & Prediction 1977–1987	Bermuda Container Line Ltd.	0051894
EcoMon-SOOP (Mid-Atlantic Bight)	2773	10	10	–		
ECOSAR II	742	0–210	46	–	Federal University of Rio Grande	–
Finnish Baltic Sea Monitoring	2210	0–440	66	Helsinki Commission Baltic Monitoring	Finnish Environment Institute	–
GAVESHANI Collection	189	0–200	100	–	National Institute of Oceanography, Goa, India	0000941

Table 1d. Continued.

COPEPOD Dataset	No. of abundance data	Depth (m)		Project	Institute	NODC accession no.
		Range	Mean			
GILL Zooplankton Collection	10723	0–200	9	–	Woods Hole Oceanographic Institute & US Fish & Wildlife Service	9700101, 9700102, 9700103
Gulf of California 1983–1984	254	0–259	94	–	Hamburgische Schiffbauversuchsanstalt	0000911
HAKUHO MARU Collection	1003	0–2300	373	–	Japan Meteorological Agency	970003, 970005, 970007, 9700303
Historical data from the Japanese Oceanographic Data Center (JODC)	10167	0–300	98	–	–	9700311
IMECOCAL	13261	0–243	94	Investigaciones MExicanas de la COrriente del California	Hamburgische Schiffbauversuchsanstalt	0000911, 0000912, 0000913
IMR Norwegian Sea Survey	254	0–2680	121	–	Institute of Marine Research, Norway	0049894
INSTOP-6	2533	0–100	50	–	Mediterranean Marine Sorting Center	0000561
IIOE	91032	0–880	94	International Indian Ocean Expedition	–	9400059
IROP-4	797	0–100	50	–	Mediterranean Marine Sorting Center	0000561
JARE	19271	0–833	30	Japanese Antarctic Research Expedition	SAFHOS	0000039
JMA North Pacific Surveys	1795	0–999	73	–	–	0000051, 0000070, 0000398
Koyo Maru–Brazil	665	0–1300	68	–	Shimonoseki University of Fisheries	970003, 970005, 970007, 9700303
Koyo Maru Indian Ocean	75	0–199	41	–		
Marion Dufrense	44	0–300	71	–		0000940
Minoda 1967	175	0–200	74	Office of US Naval Research	–	0000978
NEWP	2587	0–515	114	NorthEast Water Polynya project	Rosenstiel School of Marine & Atmospheric Sciences, University of Miami	9700074, 9700300
NMFS-PFEL Zooplankton	88	0–140	29	–	Institute Experimental of Meteorology, Obninsk, Russia	9800046
NODC Zooplankton Database	47366			–	–	Multiple
North Pacific Survey	3482	0–400	87	Cooperative Survey of the North Pacific US-Japan-Canada	Fisheries Research Board of Canada – Pacific Oceanographic Group, Nanaimo, British Columbia, USA & Japan Hydrographic Association – Marine Information Research Center	9700074, 9700300
Pacific Salmon Investigations	158	0–500	115	Pacific Salmon Investigations	Fish & Wildlife Service, Seattle, Washington, USA	9700101
Pearl Harbour 1946	5727	0–10	5	–	–	0051848
Pelagic Ecosystems of the Indian Ocean	6417	0–262	51	–	Institute of Biology of the Southern Seas	0001310
Pelagic Ecosystems of the Mediterranean	29654	0–1710	63	–		
Pelagic Ecosystems of the Tropical Atlantic	38	0–2828	92	–		
Pioneer Cruise 66	6835	0–600	185	–	University of Hawaii, Honolulu	9800165
SAHFOS-CPR Atlantic Ocean	891	10	10	–	SAHFOS	0000301

Table 1d. Continued.

COPEPOD Dataset	No. of abundance data	Depth (m)		Project	Institute	NODC accession no.
		Range	Mean			
SSRF-312 North Pacific and Bering Sea Oceanography 1958	835	0–1710	113	–	Maritime Regional Administration of Hydrometeorology – Rosgidromet	9700074, 9700300
SSRF-377	504	0–212	45	–	–	
SSRF-619	600	0–1291	40	–	Japan Hydrographic Association – Marine Information Research Center	9600088
Sub-arctic Frontal Zone Zooplankton	223	0–150	73	–	Alaska Fisheries Science Center	9700074
TASC	529	0–710	207	Trans-Atlantic Study of Calanus	Western Washington State College	0000566
USCG Chelan	617	0–25	11	–	Scripps Institution of Oceanography,	9500110
Volcano-7 Zooplankton	1013	0–1500	246	–	University of Rhode Island – Graduate School of Oceanography, Narragansett, Rhode Island, USA	9500081
WEBSEC	569	0–533	22	Western Bering Sea Ecological Cruise	–	9700074, 9700300
Zulfiquar Cruises	417	0–212	44	–	Meteorological Department, Pakistan	9400163

gridded NetCDF data file on the World Ocean Atlas grid ($1° \times 1° \times 33$ depths $\times 12$ months), which contains both carbon biomass for evaluation of ocean biogeochemical models and a number of other variables, including abundance, standard deviations and non-zero biomass (see Buitenhuis et al., 2013 for details). It should be noted that all figures presented here have been created using data from the Excel files (pregridding) in order to showcase the original datasets. Very similar figures can be created using the gridded NetCDF data file.

We have included abundance data in the dataset for a variety of reasons: (1) most macrozooplankton data are recorded in abundance terms and abundance data are often used in the calculation of biomass; (2) abundance may be used as an indication of where the animals are, and in what quantities, i.e. it may be used in a qualitative sense; (3) additions to both the abundance dataset and the conversion dataset will make it possible to convert more abundance data to biomass in the future; and (4) we have carefully separated the data collection and data processing steps so that the data would be easily adaptable for purposes other than biogeochemical model validation.

3 Results and discussions

3.1 Results of quality control

Both abundance and biomass have distributions close to normal after the data are log-transformed. After applying Chauvenet's criterion to the log-transformed abundance data, only 49 data points from the abundance dataset and 32 data points from the biomass dataset are rejected as outliers, their values being higher than the critical values for both datasets. The reasoning behind the rejection of the higher values is thus: if two hypothetical databases were constructed that were as

similar as possible to each other, but in which one is stochastically skewed with respect to the other because of an unrepresentative number of extremely high values, then those values are rejected to remove the skew. So while we think the data are values that reflect reality, we are consciously working to remove any skew. Abundance and biomass outliers originate from bloom taxon/species and occur in Northern and Southern Hemisphere spring/summer. These are most likely real values associated with a bloom rather than a problem with the sampling or lack of metadata available.

Sampling protocols, handling, preservation and measurement techniques have not been considered when removing outliers. Within the HOTS, BATS and rawKRILLBASE datasets, and the various projects within COPEPOD data, these variables are assumed consistent but are most likely not uniform across datasets and projects. Issues related to sampling such as the inherent variability of field populations (Landry et al., 2001), net mesh size, type of net, net avoidance, seasonal/diel vertical migrations, sample handling (e.g. sample splitting), size fractionation, and sample analysis – i.e. all sources of random sampling error – were considered to have a greater effect than the sampling issues across projects/datasets.

3.2 Data description

3.2.1 Abundance data

A total of 387 750 abundance data points (280 631 non-zero) between all four datasets cover the Indian Ocean, Barents Sea, Southern Ocean, west Atlantic, north eastern and north western Atlantic, the Caribbean, eastern equatorial Pacific, western North Pacific and Hawaii. There are few abundance data in the tropical and temperate south Pacific, and in the tropical north Atlantic (Fig. 1).

Table 2. Mean body mass values used to convert COPEPOD species abundance (ind. L^{-1}) to biomass ($\mu mol \, CL^{-1}$). Notes: $n =$ number of data points per species, Min. = minimum body mass, Max. = maximum body mass, Stdev. = Standard deviation; all body mass units are $\mu mol \, C$. Data from Hirst (2003) and Moriarty (2009).

| Group | | | | Body mass | | |
Species	n	Min.	Max.	Mean	Stdev.	% Stdev.
Ctenophore						
Bolionopsis infundibulum	12	131	10764	3896	4062	104
Pleurobrachia pileus	31	13	571	166	181	109
Scyphozoa						
Aurelia aurita	394	0	97590	5728	11724	204
Cyaena capillata	25	114	135159	10283	28330	276
Hydrozoa						
Aglantha digitale	27	13	461	139	110	80
Eutonina indicans	24	30	1013	311	230	74
Philalidium gregarium	44	31	297	125	62	50
Sarsia princeps	8	39	230	101	73	73
Sarsia tubulosa	22	1.1	54	19	16	82
Stomotoca atra	26	75	392	164	75	46
Agalma elegans	7	0.5	4.6	1.6	1.4	89
Pelagic mollusc						
Clione limacina	31	12	1819	198	357	180
Diacrea trispinosa	1	293	293	293	–	–
Diphyes antarctica	3	378	779	520	225	43
Limacina helicina	5	13	25	17	4.9	28
Pelagic polychaete						
Tomopteris septentrionalis	1	174	174	174	–	–
Chaetognath						
Parasagitta elegans	530	0	187	54	44	82
Parasagitta enflata	3	18	38	31	11	36
Amphipod						
Cyphocaris challangeri	1	176	176	176	–	–
Hyperia galba	5	139	401	258	106	41
Parathemisto japonica	55	0.5	492	96	128	135
Themisto libellula	1	80	80	80	–	–
Phronima sedentaria	4	41	111	72	35	49
Themisto pacifica	1	47	47	47	–	–
Mysid						
Acanthomysis pseudomacropsis	2	89	124	106	25	23
Decapod						
Lucifer typus	1	2.3	2	2.30	–	–
Euphausiid						
Euphausia krohnii	2	8.2	9	8.60	1.0	11
Euphausia pacifica	234	0.2	1674	195	186	96
Meganyctiphanes norvegica	2	1477	1477	1476	–	–
Thysanoessa inermis	31	0.8	1495	373	377	101
Thysanoessa longipes	1	321	321	321	–	–
Thysanoessa raschi	5	51	309	166	108	65
Thysanoessa spinifera	6	26	1330	619	548	89
Thaliacia						
Salpa fusiformis	6	14.3	98	53.50	29.0	54
Salpa maxima	10	1.3	23	9.00	8.0	88
Thalia democratica	28	0.2	173	11.40	34.0	298
Appendicularia						
Fritillaria borealis sargassi	8	0	0.02	0.01	0.01	100
Fritillaria haplostomai	8	0	0.02	0.01	0.01	100
Oikopleura dioica	151	0	5.5	0.9	1.42	158
Oikopleura longicauda	18	0	1	0.20	0.0	–

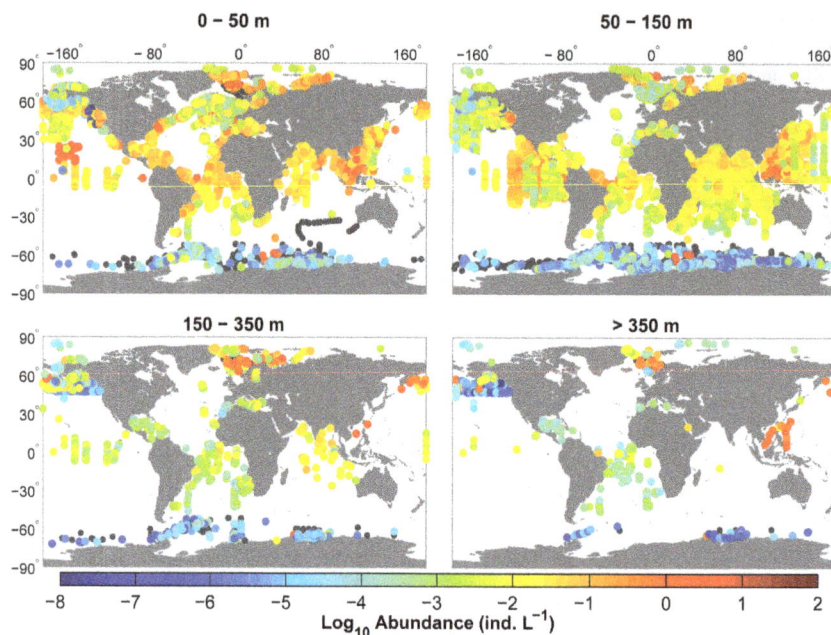

Figure 1. Global distribution of macrozooplankton abundance (ind. L^{-1}) at different depths. Grey points represent zero values.

The highest incidences of macrozooplankton abundance observations (Fig. 2a) are found at $\sim 60°$ in both the Northern and Southern Hemispheres. The number of observations falls off to the higher latitudes, with no observations of abundance south of 75° S. While the distribution of observations at high latitudes is the same in both hemispheres, there is an asymmetry in the lower latitudes (50° S–50° N) where the Southern Hemisphere has fewer observations, peaking at $\sim 30°$ S.

The temporal distribution of macrozooplankton abundance presented here covers 84 yr from 1926 to 2010 (Fig. 2b). Sampling did not occur between 1939 and 1950, with the exception of 1946; most observations where collected towards the late 1960s. There are far fewer observations between 2005 and 2010 (Fig. 2b), either because not all data have been archived within a data repository or in a database or there have been a decline in this type of sampling activity.

The mean sampling depth of macrozooplankton abundance presented here is 85 m (±90 m standard deviation) and ranges from the surface to ~ 2500 m (Fig. 2d). Most data are concentrated in the top 500 to 1000 m. Macrozooplankton may be found throughout the water column; species of macrozooplankton that live in the epipelagic, or sunlit surface waters of the ocean, are usually found in the top 350 m. The mean depth of sampling is well suited for investigating the concentrations of epipelagic macrozooplankton, the main focus of this study. Macrozooplankton abundance data have been collected in all months of the year in both the Northern and Southern Hemispheres (Fig. 2e and f).

The majority of the data in the Southern Ocean belongs to the three species *Euphausia superba*, *Salpa thompsoni* and

Ihlea racovitzai (data from the rawKRILLBASE dataset), whereas the majority of the data for the remainder of the global ocean is representative of the whole macrozooplankton community. This accounts partially for the difference in abundances in the northern and equatorial latitudes and those of the Southern Ocean. It is difficult to quantify the proportion of total global macrozooplankton abundance that is made up of the three species mentioned above. These species have distributions in the Southern Hemisphere but are predominant in the Antarctic waters of the Southern Ocean. They are not cosmopolitan in the global ocean. This is discussed further in Sect. 3.3.2.

There is potential ambiguity about whether large copepods are included in meso- or macrozooplankton sampling. Although we have used a cut-off of 2 mm adult body size for other taxonomic groups, previous work on mesozooplankton has used cut-off sizes between 5 and 30 mm to delimit mesozooplankton (Buitenhuis et al., 2006; see supplementary table 3). To prevent double counting with the MAREDAT mesozooplankton database (Moriarty and O'Brien, 2013), we have excluded copepod species that were available in the COPEPOD database. However, in the HOT and BATS databases, we only had access to the total macrozooplankton biomass data, which did include copepods greater in size than 2 mm. Large copepods can avoid nets with a small mesh size, such as are used for sampling small copepods (typically 200 μm < mesh size < 330 μm; Harris et al., 2000; Moriarty and O'Brien, 2013), but this under sampling has not been comprehensively quantified. We were therefore unable

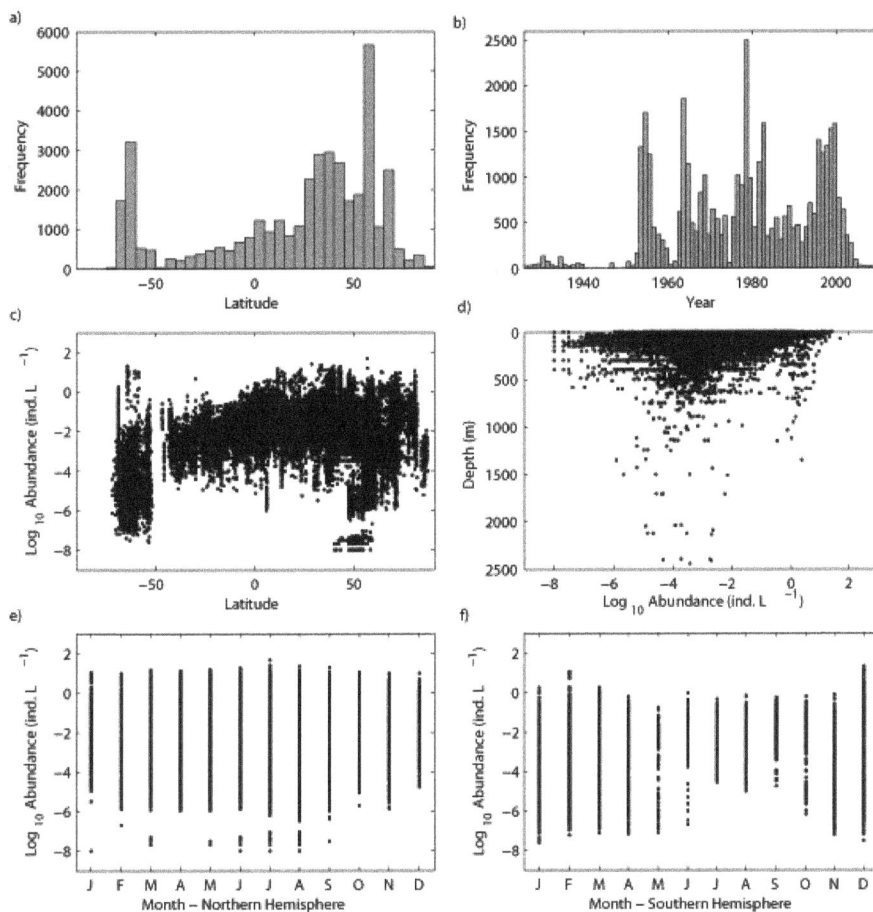

Figure 2. Description of macrozooplankton abundance: (**a**) latitudinal distribution of observations, (**b**) yearly distribution of observations, (**c**) latitudinal distribution of abundance, (**d**) depth distribution of abundance, (**e**) monthly abundance distribution in the Northern and (**f**) Southern Hemispheres.

to estimate whether there is double counting or a gap between the mesozooplankton and macrozooplankton datasets.

3.2.2 Biomass data

A total of 36 268 biomass data points (28 104 non-zero) (Fig. 3) between all four datasets leaves much of the global ocean uncharacterised, with the obvious exception of the Southern Ocean, Barents Sea, Hawaii and Bermuda. All rawKRILLBASE and BATS data were converted to biomass, along with the majority of HOTS data. Only ~ 15 000 COPE-POD abundance data were converted, which is why biomass has much less spatial coverage than abundance. In vast areas of the global ocean there is little or no information on biomass. This is a direct result of only converting abundance to biomass using species-specific conversions. This approach was necessary as bulk conversions of abundance to biomass as yet are not sophisticated enough to account for many of the variables, e.g. region, season, life history, food concentration and food quality, that are important to the amount of carbon in any particular individual or species.

Carbon values are a much more useful measurement than abundance data; however there are no published generic relations for the conversion of macrozooplankton abundance to biomass. This type of conversion has not been included in the analysis as the conversion factors are too general, and the large deviation of the bulk conversions from the species-specific conversions show the former would severely distort the results.

Efforts to assemble a comprehensive listing of conversions for macrozooplankton by groups such as the ICES Working Group on Zooplankton Ecology have been ongoing for years. The scientists involved, experts in the field, find this effort overwhelming, and the differences due to regions, seasons and life stage (length to body composition) make the equations hugely variable. A blanket global conversion, without a better conversion estimate, is not the best way to convert abundance to biomass. Without valid conversion equations from abundance (number of macrozooplankton per sample) to biomass (mass of biomass to sample), there is a need for length frequency data, mass data and carbon data. In the

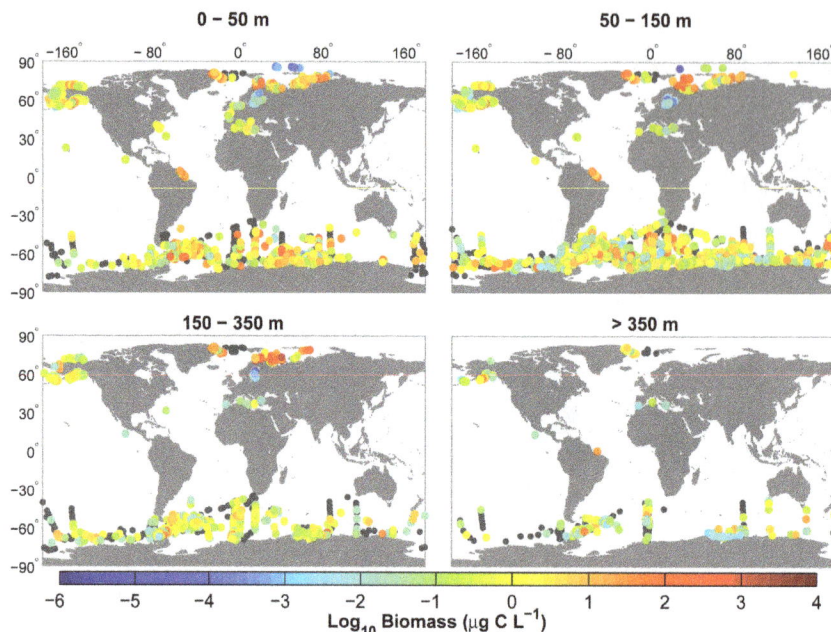

Figure 3. Global distribution of macrozooplankton biomass ($\mu g\,CL^{-1}$) at different depths. Grey points represent zero values.

majority of cases these data are not available for macrozooplankton species or the entire macrozooplankton size class or cohort of species.

Latitudinal distribution of biomass echoes that of the abundance data, with peaks in observations at ~ 60° in both hemispheres but with fewer in the north (Fig. 4a). There are greater gaps in the biomass data, with few data between 40° S and 20° N. The temporal distribution of macrozooplankton biomass also echoes largely what has been said above for the abundance data. There are, however, larger gaps in the distribution, with no data between 1939 and 1950, and few observations between the late 1950s and early 1970s (Fig. 4b). In the mid 1970s biomass observations increase and remain higher, only occasionally dropping down to pre-1975 values. Much work was done on the chemical composition of macrozooplankton species in the mid 1970s to early 1990s, which may be one explanation for the increase in biomass data associated with the end of the 20th century.

The mean sampling depth of macrozooplankton biomass is 88 m (±104 m standard deviation) and ranges from the surface to ~ 2500 m (Fig. 4c and d). Macrozooplankton biomass data have been collected in all months of the year in the Northern Hemisphere but there are no biomass data associated with the winter months (July, August and September) in the Southern Hemisphere (Fig. 4e and f).

3.3 Global estimates

Here, we use the gridded dataset to determine the depth integrated global values for macrozooplankton abundance and biomass in the top 350 m. Ninety-three percent of total abundance and ninety-nine percent of total biomass are found in the top 350 m of the global ocean. We have specifically chosen data gridded in the top 350 m of the ocean for use in these calculations of global epipelagic macrozooplankton annual average abundance and biomass as this is where macrozooplankton are usually found and because we have considerable coverage down to that depth (Figs. 2d and 4d).

3.3.1 Abundance

Global abundance to a depth of 350 m has a mean of $0.018\,ind.\,L^{-1}$, a median of $0.0006\,ind.\,L^{-1}$ and a standard deviation of $0.12\,ind.\,L^{-1}$ (Table 3a). The fact that the means are much higher than the median and that the standard deviation is high indicates that very high concentrations of abundance are occasionally observed (Luo et al., 2012). Abundance data show no clear latitudinal patterns. Mean and median latitudinal abundance values north of 15° S are of a similar range whereas the latitudinal abundances to the south have lower means and much lower medians. Differences in the number of samples (n) make it difficult to fully decipher if there are any broad latitudinal patterns in abundance concentrations.

3.3.2 Biomass

Global macrozooplankton biomass to a depth of 350 m has a mean of $8.4\,\mu g\,CL^{-1}$, a median of $0.2\,\mu g\,CL^{-1}$ and a standard deviation of $63.5\,\mu g\,CL^{-1}$ (Table 3b). Again, as in the abundance data, there are differences in the order of magnitude, within the mean, and within the median values between

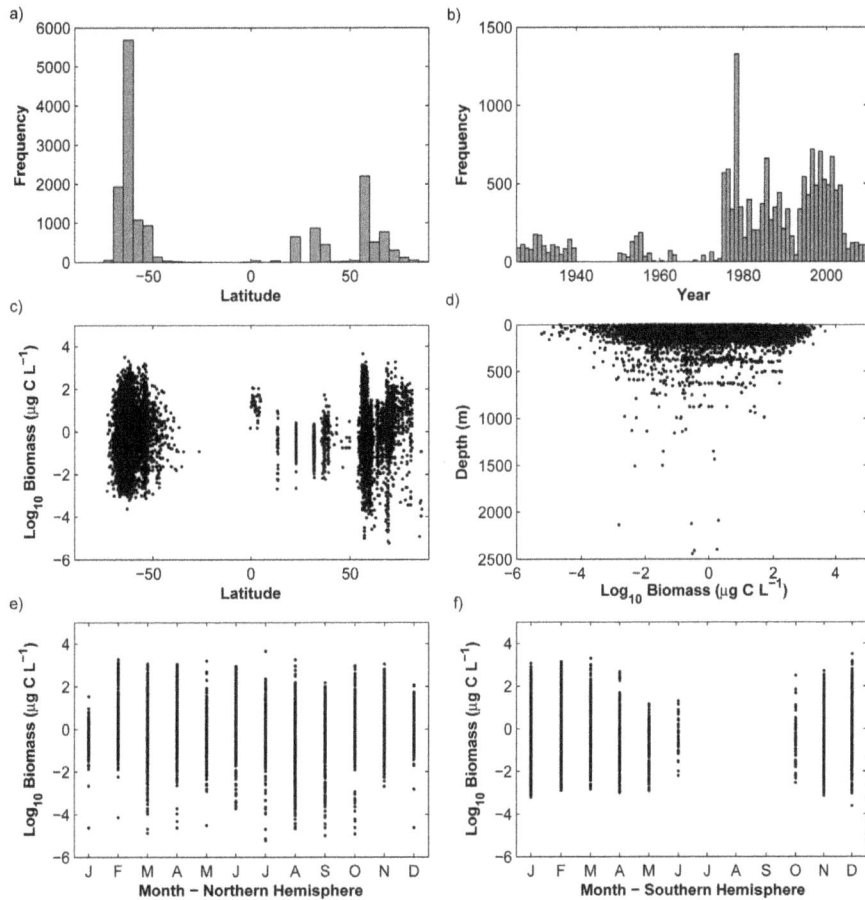

Figure 4. Description of macrozooplankton biomass: (**a**) latitudinal distribution of observations, (**b**) yearly distribution of observations, (**c**) latitudinal distribution of biomass, (**d**) depth distribution of biomass, (**e**) monthly biomass distribution in the Northern and (**f**) Southern Hemispheres. There are no data for biomass in the Southern Hemisphere winter months (July, August, September).

Table 3a. Global and latitudinal band values for the gridded macrozooplankton abundance data.

Latitude	n	Min.	Max.	Mean	Median	± std.
		Abundance (ind. L^{-1})				
Global	21 293	9.19×10^{-9}	5.11	0.018	0.0006	0.123
90–40° N	7216	1.00×10^{-8}	5.11	0.034	0.0017	0.179
40–15° N	3537	1.00×10^{-8}	2.67	0.023	0.0030	0.122
15° N–15° S	3547	2.39×10^{-7}	3.70	0.012	0.0015	0.095
15–40° S	1039	4.06×10^{-7}	0.07	0.001	0.0002	0.004
40–90° S	5954	9.19×10^{-9}	0.83	0.002	2.82×10^{-6}	0.023

latitudinal bands. Median biomass values to the south of 15° S are lower than their northern counterparts.

In Table 3b there is a difference of two orders of magnitude between the biomass values for 40 to 90° N and 40 to 90° S. This may be explained by differences in the type of data in the datasets associated with each of these regions. The rawKRILLBASE data, three species, are most of the data in the Southern Ocean, while the COPEPOD data, representa-

tive of the entire macrozooplankton community, are found throughout the global ocean. The rawKRILLBASE data are composed of three species that are the predominant macrozooplankton species in the Southern Ocean and can make up 90 % of the biomass (Witek et al., 1985). Depending on temporal and spatial scales, *Euphausia superba* and *Salpa thompsoni/Ihlea racovitzai* are estimated to account for between 30–90 % of the biomass in the Southern Ocean. Both

Table 3b. Global and latitudinal band values for the gridded macrozooplankton biomass data.

Latitude		Biomass (μgCL^{-1})				
	n	Min.	Max.	Mean	Median	$\pm$ std.
Global	8146	6.00×10^{-6}	3967	8.38	0.15	63.46
90–40° N	2147	6.00×10^{-6}	3967	16.62	0.42	114.39
40–15° N	270	0.0033	13.32	0.76	0.26	1.27
15° N–15° S	42	0.0026	116.8	18.06	10.23	25.96
15–40° S	44	1.60×10^{-2}	4.58	0.29	1.00×10^{-7}	0.88
40–90° S	5643	2.20×10^{-4}	582.3	5.60	0.08	28.14

species have bloom capabilities and patchy distributions and relatively high biomass in the Southern Ocean.

The median value, $0.2\,\mu$gCL^{-1}, for global epipelagic macrozooplankton biomass to a depth of 350 m has been used to estimate an annual average epipelagic macrozooplankton biomass of 0.02 PgC. There are two reasons why we have picked 350 m for this calculation: (1) biomass data are more evenly distributed at this depth, and (2) below this depth macrozooplankton have different metabolic rates. PFT models with a macrozooplankton component usually only consider macrozooplankton in the epipelagic surface ocean and this estimate has been tailored to this end. A number of caveats accompany this estimate of annual average epipelagic macrozooplankton biomass: (1) the data are not uniformly distributed spatially or temporally because some areas are not covered and because there is a slight bias in the biomass data against winter values in the Southern Hemisphere; and (2) data are not proportionally distributed between the various biomes of the global ocean. We have used the median value of the epipelagic data to calculate the annual average of epipelagic macrozooplankton biomass (the value for macrozooplankton from all depths is given in Table 2). We have chosen the median as an appropriate value to base the annual average biomass of epipelagic macrozooplankton on as it indicates a midpoint value rather than a value skewed by the occasional very high value, as indicated by the high values for the mean and standard deviation.

3.4 Effects of conversion factors

The limited availability of carbon conversion data is one of the major limitations of this dataset. A general conversion from dry mass to carbon mass was used for the BATS data. It was not thought to be appropriate to apply a general carbon conversion of this type to the abundance data (see above; see Mizdalski, 1988) because macrozooplankton span a range of diverse phyla, and there is huge variety within and between species; temperature, food quality, and life stage all affect the chemical composition and size of the organism. As a result the body mass of macrozooplankton spans at least 8 orders of magnitude. The uncertainty in biomass is greatest when there is no indication of body mass, body length or life stage.

There is difficulty in assessing the error on the global biomass values stated above. Only one dataset, HOTS, gives an indication of the combined sampling and conversion error (standard deviation) of 25 %. The standard deviation associated with the BATS and rawKRILLBASE dataset biomass conversion values are 17 % and 5 %, respectively. Conversion errors associated with the species-specific biomass conversions within COPEPOD range from 11 % (standard deviation as a percentage of the mean body mass) to 298 % (Table 2). The wide range in standard deviation associated with the COPEPOD species-specific conversions shows the wide range of body masses within macrozooplankton species.

4 Conclusions and recommendations

The global biomass of macrozooplankton is estimated and presented here alongside partial coverage for macrozooplankton abundance and biomass distribution. This work is presented as a first step towards a quantitative analysis of global distribution of macrozooplankton biomass. From the present dataset we estimate a biomass median of $0.2\,\mu$gCL^{-1} ($= 0.02$ PgC annual average epipelagic macrozooplankton biomass) and a standard deviation of $63.5\,\mu$gCL^{-1}. The global, latitudinal and depth estimates of biomass concentrations will be useful for understanding ocean biogeochemistry, and for evaluating global models that include macrozooplankton. Species level abundance data will be useful for understanding biodiversity, both globally and regionally, and will be of interest to researchers outside PFT and biogeochemical modelling. Although the dataset is not yet fully comprehensive in terms of taxonomic data or temporal and spatial distributions, it is a foundation upon which a comprehensive dataset can be based. The present database can act as a nucleus for a fully comprehensive dataset of macrozooplankton biodiversity in the global ocean, which will justify further details in their representation in models, e.g. inclusion of a separate representation for herbivorous and carnivorous macrozooplankton.

There is a requirement for the provision of guidelines for macrozooplankton abundance and biomass data and metadata collection. These guidelines may be consulted during planning stages of research cruises and supplementary data may be considered for collection. If detailed data on a taxonomic level, life stage, size, and chemical composition, was gathered this would augment the number of biomass data points from the COPEPOD dataset. Detailed carbon information on a species level along with environmental data would be incredibly useful. This would expand the supplementary conversion dataset to aid the accurate conversion of abundance data, length, wet, and dry mass data to carbon; a currency valued by the modelling community.

For the first time macrozooplankton data from national data centres have been collected in the original datasets, and synthesised in the gridded dataset, to create a macrozooplankton data product. The original datasets preserve all metadata that was received, including taxonomic information, although this information was not always detailed to species level. The gridded data includes amongst others abundance and biomass values. Both the original and gridded datasets will be of interest to researchers across biological oceanography and biogeochemical and PFT modelling. The taxonomic, abundance, conversion and biomass data may be extracted for a variety of uses. Although at present there are more biomass data at high latitudes, there are at least some data at low latitudes as well, so that the data can be used at both regional and global scales. Le Quéré et al. (2013) have shown the importance of macrozooplankton to the functioning of the lower trophic level in the ocean ecosystem and associated biogeochemistry, so we look forward to a wider interest in this group of organisms over the coming years.

Apart from COPEPOD, HOTS and BATS databases no national data centres are yet in a position to facilitate the provision of macrozooplankton data. Central data repositories are relatively new and time is required to gather, assess and supply accurate data and metadata. Communication between biogeochemical modellers, data managers and experimentalists is continually improving (Le Quéré and Pesant, 2009) and there is an ever increasing interest to combine expertise from the modelling and experimentalist communities to produce and share the data products necessary to parameterise and validate marine ecosystem models.

Acknowledgements. We thank Angus Atkinson and collaborators for their permission to use and reproduce the rawKRILLBASE database (Atkinson et al., 2004) and for discussions regarding length frequency distributions and body size of *Euphausia superba*, *Salpa thompsoni* and *Ihlea racovitzai*. We thank Todd O'Brien for compiling and collating macrozooplankton abundances in Coastal & Oceanic Plankton Ecology, Production & Observation Database (COPEPOD) (O'Brien, 2005). We thank Michael Landry and collaborators for permission to use and reproduce the macrozooplankton component of the Hawaii Ocean Time-Series (HOTS) dataset (Landry et al., 2001). We thank Deborah Steinberg and collaborators for permission to use and reproduce the macrozooplankton component of the Bermuda Atlantic Time-Series (BATS) dataset (Steinberg et al., 2012). Our most profound thanks go to all the people who participated in the collection, identification and analysis of the original zooplankton samples; without them and their efforts this work would not have been possible.

We thank Clare Enright for technical assistance, Stéphane Pesant for his support, and the members of the Dynamic Green Ocean Project for their input. We thank the European Union for funding to RM (FAASIS project MEST-CT-2004-514159) and M-PG (EurOcean project) and UK NERC for funding to ETB (MARQUEST project NE/C516079/1).

Edited by: D. Carlson

References

Anadon, R. and Estrada, M.: The FRUELA cruises. A carbon flux study in productive areas of the Antarctic Peninsula (December 1995–February 1996), Deep-Sea Res. Part II, 49, 567–583, doi:10.1016/s0967-0645(01)00112-6, 2002.

Atkinson, A. and Peck, J. M.: A summer-winter comparison of zooplankton in the oceanic area around South Georgia, Polar Biol., 8, 463–473, doi:10.1007/bf00264723, 1988.

Atkinson, A., Siegel, V., Pakhomov, E., and Rothery, P.: Long-term decline in krill stock and increase in salps within the Southern Ocean, Nature, 432, 100–103, doi:10.1038/nature02996, 2004.

Atkinson, A., Siegel, V., Pakhomov, E. A., Rothery, P., Loeb, V., Ross, R. M., Quetin, L. B., Schmidt, K., Fretwell, P., Murphy, E. J., Tarling, G. A., and Fleming, A. H.: Oceanic circumpolar habitats of Antarctic krill, Mar. Ecol.-Prog. Ser., 362, 1–23, 2008.

Atkinson, A., Siegel, V., Pakhomov, E. A., Jessopp, M. J., and Loeb, V.: A re-appraisal of the total biomass and annual production of Antarctic krill, Deep-Sea Res. Pt. I, 56, 727–740, doi:10.1016/j.dsr.2008.12.007, 2009.

Buitenhuis, E. T., Le Quéré, C., Aumont, O., Beaugrand, G., Bunker, A., Hirst, A., Ikeda, T., O'Brien, T., Piontkovski, S., and Straile, D.: Biogeochemical fluxes through mesozooplankton, Global Biogeochem. Cy., 20, GB2003, doi:10.1029/2005GB002511, 2006.

Buitenhuis, E. T., Vogt, M., Moriarty, R., Bednaršek, N., Doney, S. C., Leblanc, K., Le Quéré, C., Luo, Y.-W., O'Brien, C., O'Brien, T., Peloquin, J., Schiebel, R., and Swan, C.: MAREDAT: towards a world atlas of MARine Ecosystem DATa, Earth Syst. Sci. Data, 5, 227–239, doi:10.5194/essd-5-227-2013, 2013.

Casareto, B. E. and Nemoto, T.: Salps of the Southern Ocean (Australian sector) during the 1983–84 summer, with special reference to the species Salpa thompsoni, Foxton 1961, Memoirs of National Institute of Polar Research, 221–239 1986.

Chiba, S., Ishimaru, T., Hosie, G. W., and Wright, S. W.: Population structure change of Salpa thompsoni from austral mid-summer to autumn, Polar Biol., 22, 341–349, doi:10.1007/s003000050427, 1999.

Chiba, S., Ishimaru, T., Hosie, G. W., and Fukuchi, M.: Spatio-temporal variability of zooplankton community structure off east Antarctica (90 to 160 degrees E), Mar. Ecol.-Prog. Ser., 216, 95–108, doi:10.3354/meps216095, 2001.

Deibel, D.: Feeding and metabolism in Appendicularia, in: The Biology of Pelagic Tunicates, edited by: Bone, Q., Oxford University Press, New York, 139–150, 1998.

Dubischar, C. D., Pakhomov, E. A., and Bathmann, U. V.: The tunicate *Salpa thompsoni* ecology in the Southern Ocean II. Proximate and elemental composition, Mar. Biol., 149, 625–632, doi:10.1007/s00227-005-0226-8, 2006.

Foxton, P.: The distribution and life-history of *Salpa thompsoni* Foxton with observations on a related species, *Salpa gerlachei* Foxton, Discovery Report, 1–116, 1966.

Glover, D. M., Jenkins, W. J., and Doney, S. C.: Modeling Methods for Marine Science, Cambridge University Press, Cambridge, 588 pp., 2011.

Harris, R., Weibe, P., Lenz, J., Skjoldal, H.-R., and Huntley, M.: ICES Zooplankton Methodology Manual, Academic Press, London, 2000.

Hirst, A. G., Roff, J. C., and Lampitt, R. S.: A synthesis of growth rates in marine epipelagic invertebrate zooplankton, Adv. Mar. Biol., 44, 1–142, 2003.

Hood, R. R., Laws, E. A., Armstrong, R. A., Bates, N. R., Brown, C. W., Carlson, C. A., Chai, F., Doney, S. C., Falkowski, P. G., Feely, R. A., Friedrichs, M. A. M., Landry, M. R., Moore, J. K., Nelson, D. M., Richardson, T. L., Salihoglu, B., Schartau, M., Toole, D. A., and Wiggert, J. D.: Pelagic functional group modeling: Progress, challenges and prospects, Deep-Sea Res. Pt. II, 53, 459–512, 2006.

Hosie, G. W. and Cochran, T. G.: Mesoscale distribution patterns of macrozooplankton communities in Prydz Bay, Antarctica January to February 1991, Mar. Ecol.-Prog. Ser., 106, 21–39, doi:10.3354/meps106021, 1994.

Hosie, G. W., Cochran, T. G., Pauly, T., Beaumont, K. L., Wright, S. W., and Kitchener, J. A.: Zooplankton community structure of Prydz Bay, Antarctica, January-February 1993 (18th Symposium on Polar Biology), Proceedings of the NIPR Symposium on Polar Biology, 10, 90–133, 1997.

Hosie, G. W., Schultz, M. B., Kitchener, J. A., Cochran, T. G., and Richards, K.: Macrozooplankton community structure off East Antarctica (80–150 degrees E) during the Austral summer of 1995/1996, Deep-Sea Res. Pt. II, 47, 2437–2463, doi:10.1016/s0967-0645(00)00031-x, 2000.

Hunt, B. P. V., Pakhomov, E. A., Siegel, V., Strass, V., Cisewski, B., and Bathmann, U.: The seasonal cycle of the Lazarev Sea macrozooplankton community and a potential shift to top-down trophic control in winter, Deep-Sea Res. Pt. II, 58, 1662–1676, doi:10.1016/j.dsr2.2010.11.016, 2011.

Jazdzewski, K., Kittel, W., and Lotocki, K.: Zooplankton studies in the southern Drake Passage and in the Bransfield Strait during the austral summer (BIOMAS-FIBEX, February–March 1981), Pol. Polar Res., 3, 203–242, 1982.

Landry, M. R., Al-Mutairi, H., Selph, K. E., Christensen, S., and Nunnery, S.: Seasonal patterns of mesozooplankton abundance and biomass at Station ALOHA, Deep-Sea Res. Pt. II, 48, 2037–2061, doi:10.1016/s0967-0645(00)00172-7, 2001.

Le Quéré, C. and Pesant, S.: Plankton Functional Types in a new generation of biogeochemical models, EOS: Transactions American Geophysical Union, 2009.

Le Quéré, C., Harrison, S. P., Prentice, I. C., Buitenhuis, E. T., Aumont, O., Bopp, L., Claustre, H., Da Cunha, L. C., Geider, R., Giraud, X., Klaas, C., Kohfeld, K. E., Legendre, L., Manizza, M., Platt, T., Rivkin, R. B., Sathyendranath, S., Uitz, J., Watson, A. J., and Wolf-Gladrow, D.: Ecosystem dynamics based on plankton functional types for global ocean biogeochemistry models, Glob. Change Biol., 11, 2016–2040, 2005.

Le Quéré, C., Buitenhuis, E. T., Moriarty, R., Vogt, M., Sailley, S.,

Chollet, S., Stephens, N., Enright, C., Franklin, D., Larsen, S., Legendre, L., Platt, T., Rivkin, R. B., and Sathyendranath, S.: Role of plankton functional diversity for marine ecosystem services, in preparation, 2013.

Loeb, V., Hofmann, E. E., Klinck, J. M., and Holm-Hansen, O.: Hydrographic control of the marine ecosystem in the South Shetland-Elephant Island and Bransfield Strait region, Deep-Sea Res. Pt. II, 57, 519–542, doi:10.1016/j.dsr2.2009.10.004, 2010.

Lovenduski, N. S., Gruber, N., and Doney, S. C.: Toward a mechanistic understanding of the decadal trends in the Southern Ocean carbon sink, Global Biogeochem. Cy., 22, Gb3016, doi:10.1029/2007gb003139, 2008.

Luo, Y.-W., Doney, S. C., Anderson, L. A., Benavides, M., Berman-Frank, I., Bode, A., Bonnet, S., Boström, K. H., Böttjer, D., Capone, D. G., Carpenter, E. J., Chen, Y. L., Church, M. J., Dore, J. E., Falcón, L. I., Fernández, A., Foster, R. A., Furuya, K., Gómez, F., Gundersen, K., Hynes, A. M., Karl, D. M., Kitajima, S., Langlois, R. J., LaRoche, J., Letelier, R. M., Marañón, E., McGillicuddy Jr., D. J., Moisander, P. H., Moore, C. M., Mouriño-Carballido, B., Mulholland, M. R., Needoba, J. A., Orcutt, K. M., Poulton, A. J., Rahav, E., Raimbault, P., Rees, A. P., Riemann, L., Shiozaki, T., Subramaniam, A., Tyrrell, T., Turk-Kubo, K. A., Varela, M., Villareal, T. A., Webb, E. A., White, A. E., Wu, J., and Zehr, J. P.: Database of diazotrophs in global ocean: abundance, biomass and nitrogen fixation rates, Earth Syst. Sci. Data, 4, 47–73, doi:10.5194/essd-4-47-2012, 2012.

Madin, L. P., Horgan, E. F., and Steinberg, D. K.: Zooplankton at the Bermuda Atlantic Time-series Study (BATS) station: diel, seasonal and interannual variation in biomass, 1994-1998, Deep-Sea Res. Pt. II, 48, 2063–2082, doi:10.1016/s0967-0645(00)00171-5, 2001.

Marr, J. W. S.: The natural history and geography of the Antarctic krill (Euphausia superba Dana), Discovery Report, 32, 33–464, 1962.

Mizdalski, E.: Weight and length data of zooplankton in the Weddell Sea Antarctica in Austral Spring 1988 ANT V-3, Berichte zur Polarforschung, 1–72, 1988.

Moriarty, R.: Respiration rates in epipelagic macrozooplankton: a dataset, PANGAEA, 2009.

Moriarty, R. and O'Brien, T. D.: Distribution of mesozooplankton biomass in the global ocean, Earth Syst. Sci. Data, 5, 45–55, doi:10.5194/essd-5-45-2013, 2013.

Nishikawa, J. and Tsuda, A.: Diel vertical migration of the tunicate *Salpa thompsoni* in the Southern Ocean during summer, Polar Biol., 24, 299–302, doi:10.1007/s003000100227, 2001.

Nishikawa, J., Naganobu, M., Ichii, T., Ishii, H., Terazaki, M., and Kawaguchi, K.: Distribution of salps near the South Shetland Islands During Austral Summer, 1990–1991 with special reference to krill distribution, Polar Biol., 15, 31–39, 1995.

O'Brien, T. D.: COPEPOD: A global plankton database, US Dep. Commerce, NOAA Technical Memorandum, NMFS-F/SPO-73, 136, 2005.

Omori, M. and Ikeda, T.: Methods in marine zooplankton ecology., Wiley-Interscience Publications, John Wiley & Sons, Japan, 331 pp., 1984.

Pakhomov, E. A., Froneman, P. W., and Perissinotto, R.: Salp/krill interactions in the Southern Ocean: spatial segregation and implications for the carbon flux, Deep-Sea Res. Pt. II, 49, 1881–1907, doi:10.1016/s0967-0645(02)00017-6, 2002.

Ross, R. M., Quetin, L. B., Martinson, D. G., Iannuzzi, R. A., Stammerjohn, S. E., and Smith, R. C.: Palmer LTER: Patterns of distribution of five dominant zooplankton species in the epipelagic zone west of the Antarctic Peninsula, 1993–2004, Deep-Sea Res. Pt. II, 55, 2086–2105, 2008.

Schütt, F.: Analytische Planktonstudien, Lipsius and Tischer, Kiel, 117 pp., 1892.

Sieburth, J. M., Smetacek, V., and Lenz, J.: Pelagic ecosystem structure: heterotrophic compartments of the plankton and their relationship to plankton size fraction, Limnol. Oceanogr., 23, 1256–1263, 1987.

Siegel, V.: Distribution and population dynamics of Euphausia superba: summary of recent findings, Polar Biol., 29, 1–22, doi:10.1007/s00300-005-0058-5, 2005.

Siegel, V., Kawaguchi, S., Ward, P., Litvinov, F., Sushin, V., Loeb, V., and Watkins, J.: Krill demography and large-scale distribution in the southwest Atlantic during January/February 2000, Deep-Sea Res. Pt. II, 51, 1253–1273, 2004.

Steinberg, D. K., Lomas, M. W., and Cope, J. S.: Long-term increase in mesozooplankton biomass in the Sargasso Sea: Linkage to climate and implications for food web dynamics and biogeochemical cycling, Global Biogeochem. Cy., 26, GB1004, doi:10.1029/2010GB004026, 2012.

Turner, J. T.: Zooplankton fecal pellets, marine snow and sinking phytoplankton blooms, Aquat. Microb. Ecol., 27, 57–102, 2002.

Ward, P., Shreeve, R., Whitehouse, M., Korb, B., Atkinson, A., Meredith, M., Pond, D., Watkins, J., Goss, C., and Cunningham, N.: Phyto- and zooplankton community structure and production around South Georgia (Southern Ocean) during Summer 2001/02, Deep-Sea Res. Pt. I, 52, 421–441, doi:10.1016/j.dsr.2004.10.003, 2005.

Ward, P., Shreeve, R., Atkinson, A., Korb, B., Whitehouse, M., Thorpe, S., Pond, D., and Cunningham, N.: Plankton community structure and variability in the Scotia Sea: austral summer 2003, Mar. Ecol.-Prog. Ser., 309, 75–91, doi:10.3354/meps309075, 2006.

Witek, Z., Kittel, W., Czykieta, H., Zmijewska, M. I., and Presler, E.: Macrozooplankton in the southern Drake Passage and in the Bransfield Strait Antarctica during BIOMASS-SIBEX Dec. 1983–Jan. 1984, Pol. Polar Res., 6, 95–116, 1985.

11

Picoheterotroph (*Bacteria* and *Archaea*) biomass distribution in the global ocean

E. T. Buitenhuis[1], W. K. W. Li[2], M. W. Lomas[3], D. M. Karl[4], M. R. Landry[5], and S. Jacquet[6]

[1]Tyndall Centre for Climate Change Research and School of Environmental Sciences, University of East Anglia, Norwich NR4 7TJ, UK
[2]Fisheries and Oceans Canada, Bedford Institute of Oceanography, Dartmouth, Nova Scotia, Canada
[3]Bermuda Institute of Ocean Sciences, St. George's GE01, Bermuda
[4]Department of Oceanography, University of Hawaii, Honolulu, HI 96822, USA
[5]Scripps Institution of Oceanography, University of California San Diego, La Jolla, California, USA
[6]INRA, UMR CARRTEL, 75 Avenue de Corzent, 74200 Thonon-les-Bains, France

Correspondence to: E. T. Buitenhuis

Abstract. We compiled a database of 39 766 data points consisting of flow cytometric and microscopical measurements of picoheterotroph abundance, including both *Bacteria* and *Archaea*. After gridding with 1° spacing, the database covers 1.3 % of the ocean surface. There are data covering all ocean basins and depths except the Southern Hemisphere below 350 m or from April until June. The average picoheterotroph biomass is $3.9 \pm 3.6\,\mu g\,C\,l^{-1}$ with a 20-fold decrease between the surface and the deep sea. We estimate a total ocean inventory of about 1.3×10^{29} picoheterotroph cells. Surprisingly, the abundance in the coastal regions is the same as at the same depths in the open ocean. Using an average of published open ocean measurements for the conversion from abundance to carbon biomass of $9.1\,fg\,cell^{-1}$, we calculate a picoheterotroph carbon inventory of about 1.2 Pg C. The main source of uncertainty in this inventory is the conversion factor from abundance to biomass. Picoheterotroph biomass is ~ 2 times higher in the tropics than in the polar oceans.

1 Introduction

Picoheterotrophs are the main degraders of detritus in the ocean (Azam and Malfatti, 2007). The term picoheterotrophs was introduced by Le Quéré et al. (2005) to include heterotrophic *Bacteria* and *Archaea*, and exclude cyanobacteria. Most picoheterotrophs (> 95 %, Cho and Azam, 1988; Turley and Stutt, 2000) live on dissolved organic matter (DOM) as suspended/detached organisms, though in the deep sea the contribution from other energy sources such as reduced nitrogen could be significant (Herndl et al., 2005). Attached picoheterotrophs living in and on particulate detritus, although less abundant, have a higher specific activity (up to 12 % of picoheterotroph production, Turley and Stutt, 2000). Picoheterotrophs that spend part of their time attached to parti-

cles both attach and detach from particles on a timescale of hours (Kiørboe et al., 2002). They also produce ectoenzymes that solubilize POC to DOC that can be subsequently used by detached picoheterotrophs (Thor et al., 2003; Azam and Malfatti, 2007). Thus, the relative importance of attached picoheterotrophs may be higher still than their contribution to picoheterotroph production suggests.

Picoheterotrophs have a higher biomass than the metabolic theory of ecology would predict based on their small size (Brown et al., 2004). This may be due in part to the fact that they respire organic matter that is formed as losses at all trophic levels, i.e. that their trophic status is unrelated to their size. Furthermore, not all picoheterotrophs show the same activity, ranging from ghost cells with cell membranes but no internal structures, dead cells containing nucleic acids but

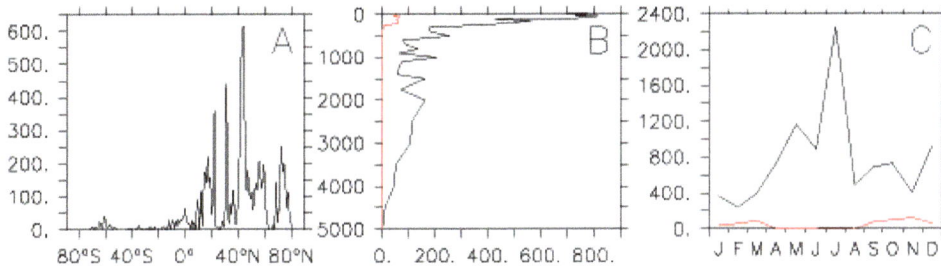

Figure 1. Number of grid points with data, as a function of (**A**) latitude, (**B**) depth, and (**C**) time. Red: Southern Hemisphere, black: total.

with compromised cell membranes, low nucleic acid cells with a lower specific activity and high nucleic acid cells (Gasol et al., 1999; Longnecker et al., 2006; Ortega-Retuerta et al., 2008; Morán et al., 2011). These dead or less active picoheterotrophs would contribute to a higher picoheterotroph biomass than the metabolic theory would predict.

Here, we present a database of picoheterotroph abundance and biomass in the global ocean. This is a contribution towards a world ocean atlas of plankton functional types (MAREDAT, this special issue), which we hope will help resolve some of the important issues on ecosystem functioning and its representation in models.

2 Data

Table 1 summarises the data that were compiled for this synthesis. Most of the data were obtained by flow cytometry. Cells were stained with nucleic acid stains, and therefore include (presumably recently) dead cells with compromised cell membranes, but not ghost cells. The data at BATS were stained with DAPI and counted microscopically, and could therefore include ghost cells. We treat *Bacteria* and *Archaea* as one group. Neither the DAPI stain used in microscopy nor the nucleic acid stains used in flow cytometry discriminate the two domains. *Archaea* make up about 5 % of picoheterotrophs in the surface, and typically about 50 % of the population that can be distinguished by domain-specific rRNA probes below 2000 m (Robinson et al., 2010 and references therein). In some cases, cyanobacteria will also have been included, especially *Prochlorococcus* near the surface, which have low red fluorescence and are therefore difficult to distinguish from picoheterotrophs. The data are available from PANGAEA (doi:10.1594/PANGAEA.779142) and the MAREDAT webpage (http://maremip.uea.ac.uk/.maredat.html).

2.1 Conversion factors

Table 2 gives abundance to carbon conversion factors from the literature. Picoheterotrophs have been shown to increase in size during incubation (Lee and Fuhrman, 1987). We therefore excluded measurements from cultures or incubated in situ samples. We also excluded conversion factor measurements from coastal waters. These have been shown to be higher than open ocean samples (Fukuda et al., 1998, Table 2), but not enough data are available to define the controlling factors for this increase or how it graduates to the open ocean value with distance from the coast. We are also unaware of measurements showing how the carbon content of picoheterotrophs varies with growth conditions. We therefore use a single conversion factor for the whole database. We calculated the conversion factor at BATS from the geometric mean cell volume and the relationship between cell volume and carbon content ($n = 164$) from Gundersen et al. (2002). We calculated the conversion factor as the average of the three studies in Table 2. The conservative conversion factor for incubated *Archaea* of 8.4 fg C cell^{-1} in Herndl et al. (2005) is similar to our conversion factor of 9.1 fg cell^{-1} for picoheterotrophs in the upper ocean, where the population is dominated by *Bacteria*.

2.2 Quality control

As a statistical filter for outliers, we applied the Chauvenet criterion (Glover et al., 2011; Buitenhuis et al., 2012) to the total carbon data. The data were not normally distributed, so we log transformed them, excluding 51 zero values. No high outliers were found by this criterion. The highest picoheterotroph biomass in the database is 74 µg C l^{-1}, measured near the coast of Oman.

3 Results

The database contains 39 766 data points. After gridding, we obtained 9284 points on the World Ocean Atlas grid ($1° \times 1° \times 33$ vertical layers $\times 12$ months), i.e. we obtain a coverage of vertically integrated and annually averaged biomass for 1.3 % of the ocean surface. Only 6 % of the data are from the Southern Hemisphere (58 % of the ocean surface; Fig. 1a); 24 % are from the tropics (43 % of the ocean surface), while 15 % are from the polar oceans (5 % of the ocean surface). Observations from the coast (bottom depth < 225 m) make up 12 % of the data (4.9 % of the ocean area, 0.13 % of the ocean volume). Observations in the upper 112.5 m make up 57 % of the data (Fig. 1b), while observations below 950 m make up 13 % of the data. There are no

Table 1. Data sources.

Cruise	Date	Area	Reference/Investigator
Li89003	Apr 1989	North Atlantic	Li et al. (2004)
HOT	1990–2008	Tropical Pacific	Campbell et al. (1997); Karl (unpublished data)
BATS	1990–2010	North Atlantic	DuRand et al. (2001); Lomas et al. (2010)
Li91001	Apr 1991	North Atlantic	Li et al. (2004)
EQPACTT007	Feb–Mar 1992	Equatorial Pacific	Landry et al. (1996)
EQPACTT008	Mar–Apr 1992	Equatorial Pacific	Binder et al. (1996)
EQPACTT011	Aug–Sep 1992	Equatorial Pacific	Landry et al. (1996)
Li92037	Sep 1992	North Atlantic	Li et al. (2004)
Li93002	May 1993	North Atlantic	Li et al. (2004)
NOAA93	Jul–Aug 1993	North Atlantic	Buck et al. (1996)
OLIPAC	Nov 1994	Equatorial Pacific	Neveux et al. (1999)
ArabianTTN043	Jan 1995	Arabian Sea	Campbell et al. (1998)
ArabianTTN045	Mar–Apr 1995	Arabian Sea	Campbell et al. (1998)
Delaware95	Apr 1995	North Atlantic	Li (unpublished data)
MINOS	Jun 1995	Mediterranean Sea	Vaulot, Marie, Partensky (unpublished data)
Chile95	Jun 1995	South Pacific	Li (unpublished data)
Lopez96	Jun 1995	Sargasso Sea	Li (unpublished data)
Li95016	Jul 1995	North Atlantic	Li and Harrison (2001)
ArabianTTN049	Jul–Aug 1995	Arabian Sea	Olson (unpublished data)
ArabianTTN050	Aug–Sep 1995	Arabian Sea	Campbell et al. (1998)
NOAA95	Sep–Oct 1995	Indian Ocean	Buck (unpublished data)
ArabianTTN054	Dec 1995	Arabian Sea	Campbell et al. (1998)
AZOMP	1995–2009	Labrador Sea	Li et al. (2004); Li (2009)
AZMP	1997–2009	North Atlantic	Li et al. (2004); Li (2009)
Kiwi6	Oct–Nov 1997	Antarctica	Landry (unpublished data)
Kiwi7	Dec 1997	Antarctica	Landry (unpublished data)
Almo-1	Dec 1997	Mediterranean Sea	Jacquet, Marie (unpublished data)
Almo-2	Jan 1998	Mediterranean Sea	Jacquet et al. (2010)
Kiwi8	Jan–Feb 1998	Antarctica	Landry (unpublished data)
Kiwi9	Feb–Mar 1998	Antarctica	Landry (unpublished data)
PROSOPE99	Sep 1999	Mediterranean Sea	Marie et al. (2006)
GLOBEC LTOP	Mar 2001–Sep 2003	North Pacific	Sherr et al. (2006)
JOIS	2002–2009	Arctic	Li et al. (2009)
C3O	2007–2008	Arctic	Li et al. (2009)

observations below 350 m in the Southern Hemisphere. Although there are some zero values in the raw database, presumably because of a detection limit in small samples, there are no zero values in the gridded dataset. There is some sampling bias towards the growing season, with 72 % of the data sampled during the spring and summer months (Fig. 1c).

The average abundance is $4.3 \times 10^8 \pm 3.9 \times 10^8$ picoheterotrophs l^{-1} with a median of 3.1×10^8 picoheterotrophs l^{-1}. The average biomass is $3.9 \pm 3.6\,\mu g\,C\,l^{-1}$ (Fig. 2) with a median of $2.8\,\mu g\,C\,l^{-1}$. The biomass decreases with depth, from $7.3 \pm 4.3\,\mu g\,C\,l^{-1}$ at the surface to $0.36 \pm 0.19\,\mu g\,C\,l^{-1}$ at 2750–4750 m depth (Fig. 3). The average biomass in the top 225 m is slightly higher in the northern temperate region (23–67° N, $5.5 \pm 3.7\,\mu g\,C\,l^{-1}$; Figs. 2, 3, 4) and tropics ($5.5 \pm 3.6\,\mu g\,C\,l^{-1}$) than in Antarctica ($3.2 \pm 1.9\,\mu g\,C\,l^{-1}$), the Arctic ($2.4 \pm 2.1\,\mu g\,C\,l^{-1}$) and southern temperate region ($3.1 \pm 1.9\,\mu g\,C\,l^{-1}$). The differences between most of these regions are significant (one-way

ANOVA with violated homogeneity of variances, Games Howell post-hoc test, $p < 0.001$), except for Antarctica, for which there are only 23 measurements in the upper 225 m, and which was only significantly different from the tropics ($p = 0.014$). There is no significant difference between abundance in coastal waters and in the upper 225 m of the open ocean (Fig. 3, t-test, $p = 0.86$).

If we calculate a total ocean picoheterotroph biomass based on the average profile with depth (Fig. 3) and multiply by the volume of ocean water at each depth, we calculate an inventory of 1.1 Pg C, of which 0.28 Pg C is found in the upper 225 m, 0.51 Pg C below 950 m, and only 0.0079 Pg C in the coastal ocean. If we calculate the inventory separately in the top 225 m for the 5 regions mentioned above, the inventory is higher at 0.35 Pg C due to the larger ocean volume at low latitudes. Since we do not have enough data to calculate regional differences in the deep sea, this would increase the total ocean picoheterotroph inventory to 1.2 Pg C.

Table 2. Conversion factors.

fg C cell^{-1}	reference
7.7 (5.5, 9.8)	oceanic, Antarctica Carlson et al. (1999)
12.4 ± 6.3 ($n = 6$)	oceanic, Pacific Fukuda et al. (1998)
30.2 ± 12.3 ($n = 5$)	coastal, Japan Fukuda et al. (1998)
7.1	oceanic, Atlantic, BATS Gundersen et al. (2002)
9.1	average (oceanic only)

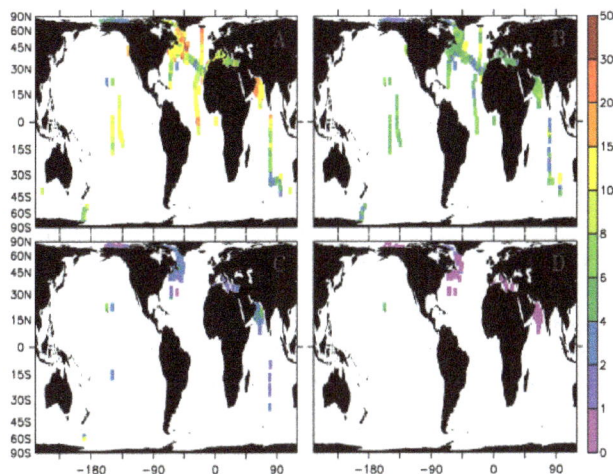

Figure 3. Picoheterotroph biomass averaged over all available longitudes, latitudes and months, as a function of depth, (black line) global average, (blue line) tropical oceans, (green line) temperate regions, (red line) polar oceans, (purple line) coastal ocean abundance × open ocean conversion factor.

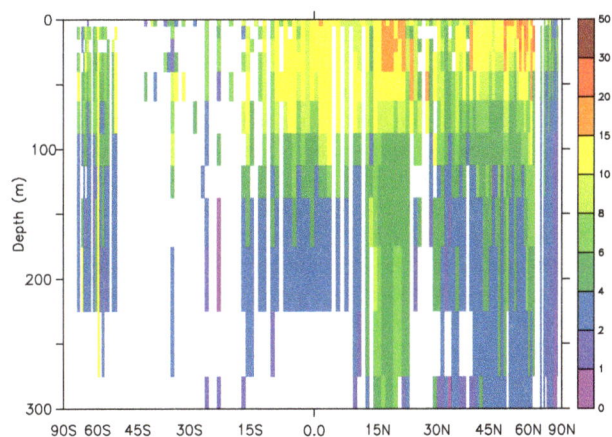

Figure 2. Picoheterotroph biomass (µg C l^{-1}) averaged over all available months. (**A**) 0–40 m, (**B**) 40–225 m, (**C**) 225–950 m, (**D**) ≥ 950 m.

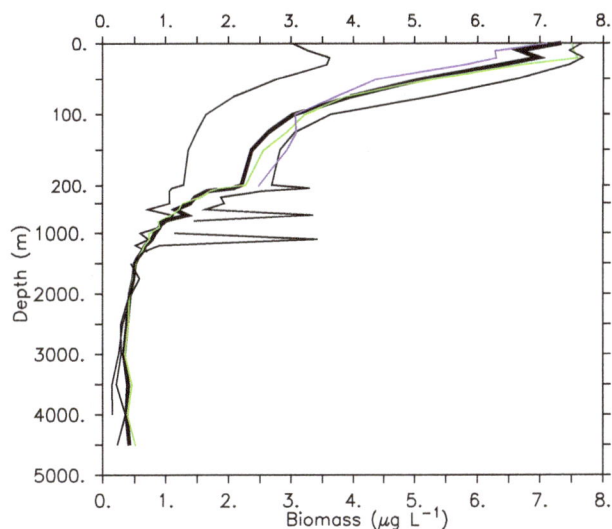

Figure 4. Picoheterotroph biomass (µg C l^{-1}) averaged over all available longitudes and months in the top 300 m.

4 Discussion

We could find only few measurements of carbon content of picoheterotrophs that were measured directly after collection, i.e. without incubation, from open ocean waters (Table 2). The range in these measurements is considerable, from 5.5 to 23.5 fg C cell^{-1}. Thus, there is a corresponding uncertainty in our conversion from cell abundance to carbon biomass.

In addition, a higher conversion factor has been found in coastal waters (Fukuda et al., 1998). However, it has not been established how far this higher conversion factor extends between the coastal bay waters and the open ocean. If we assume the higher conversion factor is valid up to a water depth to the bottom of 225 m (i.e. the continental shelf), then, based on the average profile of picoheterotroph biomass (Fig. 3), increasing the conversion factor from 9.1 to 30.2 fg cell^{-1} would only add 0.02 Pg C to the global inventory. Thus, at present the main sources of uncertainty in picoheterotroph biomass appear to be the open ocean conversion factor and lack of spatial coverage, and not the increase in the conver-

sion factor near the coast. All of the open ocean conversion factors in Table 1 were measured on samples from the upper 250 m, so whether the conversion factor changes with depth is yet to be resolved.

Whitman et al. (1998) estimated the global ocean picoheterotroph inventory at 2.0 Pg C. This higher estimate is entirely due to their use of a higher conversion factor of 20 fg C cell^{-1}. In fact, the present database gives a 20 % higher inventory of global picoheterotroph abundance of 1.2×10^{29} cells based on an averaged depth profile, or 30 % higher, 1.3×10^{29} cells, based on regional inventories in the upper 225 m, but a considerably lower biomass inventory of 1.1–1.2 Pg C. Despite the uncertainties that we discuss above, we judge that the direct measurements of cellular carbon contents for open ocean picoheterotrophs that we have used here are the most precise conversion factors. For applications where biomass rather than abundance of picoheterotrophs is relevant (most notably in biogeochemical models), the database that is presented here has the largest coverage and the best estimates that are available at present.

Acknowledgements. We thank Liam Aspin for help with the statistics, and NERC for funding (NE/G006725/1) to ETB. We thank the three reviewers for their helpful comments.

Edited by: S. Pesant

References

Azam, F. and Malfatti, F.: Microbial structuring of marine ecosystems, Nat. Rev. Microb., 5, 782–791, 2007.

Binder, B. J., Chisholm, S. W., Olson, R. J., Frankel, S. L., and Worden, A. Z.: Dynamics of picophytoplankton, ultraphytoplankton and bacteria in the central equatorial Pacific, Deep-Sea Res. Pt. II, 43, 907–931, 1996.

Brown, J. H., Gillooly, J. F., Allen, A. P., Savage, V. M., and West, G. B.: Toward a metabolic theory of ecology, Ecology, 85, 1771–1789, 2004.

Buck, K. R., Chavez, F. P., and Campbell, L.: Basin-wide distributions of living carbon components and the inverted trophic pyramid of the central gyre of the North Atlantic Ocean, summer 1993, Aquat. Microb. Ecol., 10, 283–298, 1996.

Buitenhuis, E. T., Vogt, M., Bednarsek, N., Doney, S. C., Leblanc, K., Le Quéré, C., Luo, Y.-W., Moriarty, R., O'Brien, C., O'Brien, T., Peloquin, J., and Schiebel, R.: MAREDAT: Towards a World Ocean Atlas of MARine Ecosystem DATa, Earth Syst. Sci. Data Discuss., in preparation, 2012.

Campbell, L., Liu, H. B., Nolla, H. A., and Vaulot, D.: Annual variability of phytoplankton and bacteria in the subtropical North Pacific Ocean at Station ALOHA during the 1991–1994 ENSO event, Deep-Sea Res. Pt. I, 44, 167–192, 1997.

Campbell, L., Landry, M. R., Constantinou, J., Nolla, H. A., Brown, S. L., Liu, H., and Caron, D. A.: Response of microbial community structure to environmental forcing in the Arabian Sea, Deep-Sea Res. Pt. II, 45, 2301–2325, 1998.

Carlson, C. A., Bates, N. R., Ducklow, H. W., and Hansell, D. A.: Estimation of bacterial respiration and growth efficiency in the Ross Sea, Antarctica, Aquat. Microb. Ecol., 19, 229–244, 1999.

Cho, B. C. and Azam, F.: Major Role of Bacteria in Biogeochemical Fluxes in the Oceans Interior, Nature, 332, 441–443, 1988.

DuRand, M. D., Olson, R. J., and Chisholm, S. W.: Phytoplankton population dynamics at the Bermuda Atlantic Time-series station in the Sargasso Sea, Deep-Sea Res. Pt. II, 48, 1983–2003, 2001.

Fukuda, R., Ogawa, H., Nagata, T., and Koike, I.: Direct determination of carbon and nitrogen contents of natural bacterial assemblages in marine environments, Appl. Environ. Microbiol., 64, 3352–3358, 1998.

Gasol, J. M., Zweifel, U. L., Peters, F., Fuhrman, J. A., and Hagstrom, A.: Significance of size and nucleic acid content heterogeneity as measured by flow cytometry in natural planktonic bacteria, Appl. Environ. Microbiol., 65, 4475–4483, 1999.

Glover, D. M., Jenkins, W. J., and Doney, S. C.: Modeling Methods for Marine Science, Cambridge University Press, Cambridge, UK, 2011.

Gundersen, K., Heldal, M., Norland, S., Purdie, D. A., and Knap, A. H.: Elemental C, N, and P cell content of individual bacteria collected at the Bermuda Atlantic Time-Series Study (BATS) site, Limnol. Oceanogr., 47, 1525–1530, 2002.

Herndl, G. J., Reinthaler, T., Teira, E., van Aken, H., Veth, C., Pernthaler, A., and Pernthaler, J.: Contribution of Archaea to total prokaryotic production in the deep Atlantic Ocean, Appl. Environ. Microbiol., 71, 2303–2309, 2005.

Jacquet, S., Prieur, L., Nival, P., and Vaulot, D.: Structure and variability of the microbial community associated to the Alboran Sea frontal system (Western Mediterranean) in winter, J. Oceanogr., Research and Data, 3, 47–75, 2010.

Kiørboe, T., Grossart, H. P., Ploug, H., and Tang, K.: Mechanisms and rates of bacterial colonization of sinking aggregates, Appl. Environ. Microbiol., 68, 3996–4006, 2002.

Landry, M. R., Kirshtein, J., and Constantinou, J.: Abundances and distributions of picoplankton populations in the central equatorial Pacific from 12° N to 12° S, 140° W, Deep-Sea Res. Pt. II, 43, 871–890, 1996.

Le Quéré, C., Harrison, S. P., Prentice, I. C., Buitenhuis, E. T., Aumont, O., Bopp, L., Claustre, H., Da Cunha, L. C., Geider, R., Giraud, X., Klaas, C., Kohfeld, K. E., Legendre, L., Manizza, M., Platt, T., Rivkin, R. B., Sathyendranath, S., Uitz, J., Watson, A. J., and Wolf-Gladrow, D.: Ecosystem dynamics based on plankton functional types for global ocean biogeochemistry models, Glob. Change Biol., 11, 2016–2040, 2005.

Lee, S. and Fuhrman, J. A.: Relationships between Biovolume and Biomass of Naturally Derived Marine Bacterioplankton, Appl. Environ. Microbiol., 53, 1298–1303, 1987.

Li, W. K. W.: From cytometry to macroecology: a quarter century quest in microbial oceanography, Aquat. Microb. Ecol., 57, 239–251, 2009.

Li, W. K. W. and Harrison, W. G.: Chlorophyll, bacteria and picophytoplankton in ecological provinces of the North Atlantic, Deep-Sea Res. Pt. II, 48, 2271–2293, 2001.

Li, W. K. W., Head, E. J. H., and Harrison, W. G.: Macroecological limits of heterotrophic bacterial abundance in the ocean, Deep-Sea Res. Pt. I, 51, 1529–1540, 2004.

Li, W. K. W., McLaughlin, F. A., Lovejoy, C., and Carmack, E. C.: Smallest Algae Thrive As the Arctic Ocean Freshens, Science, 326, 539–539, 2009.

Lomas, M. W., Steinberg, D. K., Dickey, T., Carlson, C. A., Nelson, N. B., Condon, R. H., and Bates, N. R.: Increased ocean carbon export in the Sargasso Sea linked to climate variability is countered by its enhanced mesopelagic attenuation, Biogeosciences, 7, 57–70, doi:10.5194/bg-7-57-2010, 2010.

Longnecker, K., Sherr, B. F., and Sherr, E. B.: Variation in cell-specific rates of leucine and thymidine incorporation by marine bacteria with high and with low nucleic acid content off the Oregon coast, Aquat. Microb. Ecol., 43, 113–125, 2006.

Marie, D., Zhu, F., Balague, V., Ras, J., and Vaulot, D.: Eukaryotic picoplankton communities of the Mediterranean Sea in summer assessed by molecular approaches (DGGE, TTGE, QPCR), FEMS Microbiol. Ecol., 55, 403–415, 2006.

Moran, X. A. G., Ducklow, H. W., and Erickson, M.: Single-cell physiological structure and growth rates of heterotrophic bacteria in a temperate estuary (Waquoit Bay, Massachusetts), Limnol. Oceanogr., 56, 37–48, 2011.

Neveux, J., Lantoine, F., Vaulot, D., Marie, D., and Blanchot, J.: Phycoerythrins in the southern tropical and equatorial Pacific Ocean: Evidence for new cyanobacterial types, J. Geophys. Res.-Oceans, 104, 3311–3321, 1999.

Ortega-Retuerta, E., Reche, I., Pulido-Villena, E., Agusti, S., and Duarte, C. M.: Exploring the relationship between active bacterioplankton and phytoplankton in the Southern Ocean, Aquat. Microb. Ecol., 52, 99–106, 2008.

Robinson, C., Steinberg, D. K., Anderson, T. R., Aristegui, J., Carlson, C. A., Frost, J. R., Ghiglione, J.-F., Hernandez-Leon, S., Jackson, G. A., Koppelmann, R., Queguiner, B., Ragueneau, O., Rassoulzadegan, F., Robison, B. H., Tamburini, C., Tanaka, T., Wishner, K. F., and Zhang, J.: Mesopelagic zone ecology and biogeochemistry – a synthesis, Deep-Sea Res. Pt. II, 57, 1504–1518, 2010.

Sherr, E. B., Sherr, B. F., and Longnecker, K.: Distribution of bacterial abundance and cell-specific nucleic acid content in the Northeast Pacific Ocean, Deep-Sea Res. Pt. I, 53, 713–725, 2006.

Thor, P., Dam, H. G., and Rogers, D. R.: Fate of organic carbon released from decomposing copepod fecal pellets in relation to bacterial production and ectoenzymatic activity, Aquat. Microb. Ecol., 33, 279–288, 2003.

Turley, C. M. and Stutt, E. D.: Depth-related cell-specific bacterial leucine incorporation rates on particles and its biogeochemical significance in the Northwest Mediterranean, Limnol. Oceanogr., 45, 419–425, 2000.

Whitman, W. B., Coleman, D. C., and Wiebe, W. J.: Prokaryotes: The unseen majority, P. Natl. Acad. Sci. USA, 95, 6578–6583, 1998.

12

In situ measurement of the biogeochemical properties of Southern Ocean mesoscale eddies in the Southwest Indian Ocean, April 2014

S. de Villiers, K. Siswana, and K. Vena

Oceans and Coastal Research, Department of Environmental Affairs, Cape Town, South Africa

Correspondence to: S. de Villiers (steph.devilliers@gmail.com)

Abstract. Several open-ocean mesoscale features – a "young" warm-core (anti-cyclonic) eddy at 52° S, an "older" warm-core eddy at 57.5° S and an adjacent cold-core (cyclonic) eddy at 56° S – were surveyed during a R/V *S.A. Agulhas II* cruise in April 2014. The main aim of the survey was to obtain hydrographical and biogeochemical profile data for contrasting open-ocean eddies in the Southern Ocean, which will be suitable for comparative study and modelling of their heat, salt and nutrient characteristics, and the changes that occur in these properties as warm-core eddies migrate from the polar front southwards. The major result is that the older warm-core eddy at 57.5° S is, at its core, 2.7 °C colder than a younger eddy at 52° S, while its dissolved silicate levels are almost 500 % higher and accompanied by chlorophyll *a* levels that are more than 200 % higher than that in the younger eddy. A total of 18 CTD stations were occupied in a sector south of the Southwest Indian Ridge, along three transects crossing several mesoscale features identified from satellite altimetry data prior to the cruise. The CTD data, as well as chlorophyll *a* and dissolved nutrient data (for NO_3^-, NO_2^-, PO_4^{3-} and SiO_2), have been processed, quality controlled and made available via the PANGAEA Data Archiving and Publication database at doi:10.1594/PANGAEA.848875.

1 Introduction

The circulation and thermohaline structure of the Southern Ocean is of critical importance to global exchanges of heat, freshwater and biogeochemical constituents such as nutrients and CO_2. A detailed understanding of the role of mesoscale eddy transport in these processes is still lacking. It has only fairly recently been established that mesoscale eddies contain most of the kinetic energy of ocean circulation (Fu et al., 2010; Ferrari and Wunsch, 2009) and that the global zonal eddy volume transport is comparable in magnitude to that of the large-scale wind- and thermohaline-driven circulation (Zhang et al., 2014). It is estimated that in the open ocean most of the vertical transport of biogeochemical properties, such as nutrients, takes place at the sub-mesoscale and is associated with eddies (Klein and Lapeyre, 2009; McGillicuddy Jr. et al., 2007; Lévy et al., 2001). At the global scale, areas of enhanced eddy kinetic energy usually also ex-

hibit elevated levels of marine primary productivity (Chelton et al., 2011; Falkowski et al., 1991; Siegel et al., 2011). However, our understanding of the global significance of coincident large-scale patterns of enhanced open-ocean productivity and mesoscale activity, as well as the importance of eddy-induced nutrient transport, is still in its infancy.

Progress in this field, including the incorporation of biogeochemical cycles into eddy-resolving general circulation models, is severely limited by scarce in situ data, collected with the specific aim of improving our understanding of the physical and biogeochemical processes associated with mesoscale features such as eddies (Joyce et al., 1981; Mahadevan and Archer, 2010; Ansorge et al., 2010; Lehahn et al., 2011; Stramma et al., 2013; Chen et al., 2015). This scarcity of in situ data is particularly pronounced in the Southern Ocean. Despite the significance of the Southern Ocean to ocean–atmosphere CO_2 exchange and global climate, and the important role of ocean eddies to these pro-

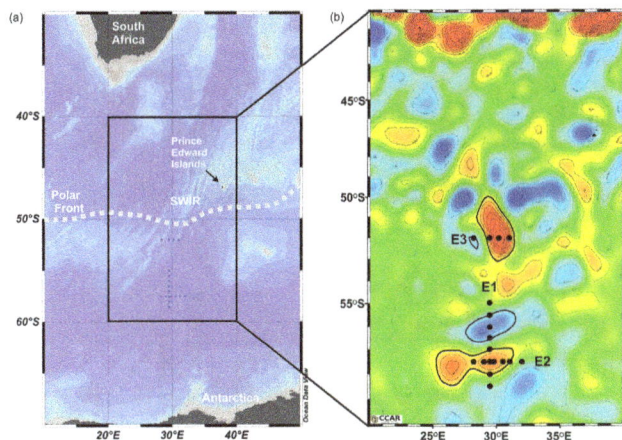

Figure 1. (a) Seafloor bathymetry in the southwestern Indian Ocean sector of the Southern Ocean to indicate the position of the Southwest Indian Ridge (SWIR); shown in grey is the average position of the polar front (adapted from Dong et al., 2006). **(b)** The CTD station locations along transect lines E1 to E3 are shown as black dots superimposed upon a satellite altimetry map; red indicates positive SSHAs and blue indicates negative SSHAs. SSHA contour intervals are 5 cm, and the −10 cm and +10 cm contour lines that the transect lines cross are highlighted with solid black lines.

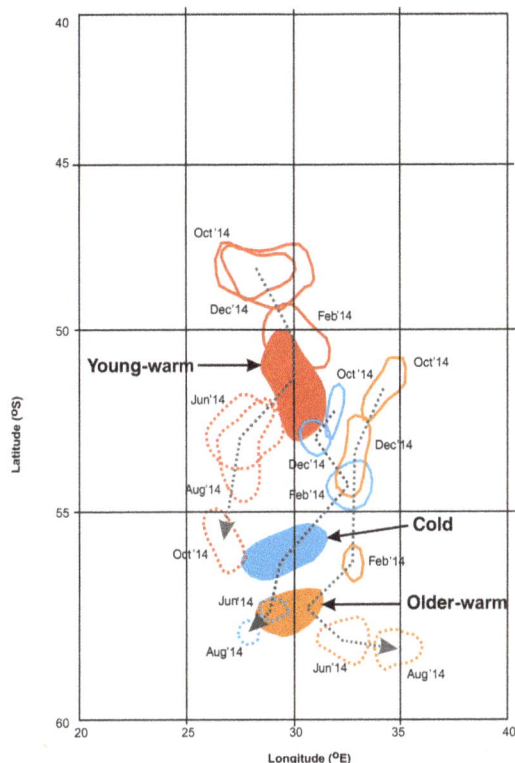

Figure 2. Eddy migration tracks, as inferred from SSHA archive data. Solid red, blue and orange areas respectively reflect the area inside the 10 cm SSHA contour for the young, warm-core, cold-core and the older, warm-core eddies at the time of sampling. Solid red, blue and orange lines, similarly, indicate the position of the eddies (and the extent of the eddy within the 10 cm SSHA contour) prior to sampling, at 2-month intervals, as indicated by the labels. Broken red, blue and orange lines indicate the position of the eddies subsequent to sampling, again at 2-month intervals. Black arrows are used to highlight the southerly migration of the eddies.

cesses (Frenger et al., 2013; Sheen et al., 2014; Morrow et al., 1994), it remains a remote, hostile and under-sampled ocean environment.

The objective of this paper is to present an overview of in situ data that had been collected in the southern hemispheric autumn, across a number of distinct mesoscale features in the southwestern Indian sector of the Southern Ocean (Fig. 1), and to make this data set available to the scientific community. The interaction of the Antarctic Circumpolar Current with the shallow topographic features of the Southwest Indian Ocean Ridge plays an important role in the generation of open-ocean eddies just south of the polar front (Gouretski and Danilov, 1994; Pollard and Read, 2001; Durgadoo et al., 2011; Ansorge et al., 2015). The subsequent movement of these eddies in a southerly direction, into the Southern Ocean proper, represents an ideal natural laboratory for the in situ observation and study of the eddy transport of heat, salt and chemicals across strong frontal zones in the Southern Ocean. To date, detailed studies of the chemical characteristics of such eddies and the evolution of these properties over time and distance have not been carried out. This represents an important knowledge gap, particularly with regards to understanding Southern Ocean nutrient transport processes and carbon cycling.

2 Sampling survey design

The survey cruise was conducted from 2 April to 6 May 2014 (EXPOCODE 91AH20140402), the austral autumn, as part of the Department of Environmental Affairs' 2014 Marion Relief Voyage 011 on the M/V *S.A. Agulhas II* to its base in the subantarctic Prince Edward Islands (Fig. 1a). The M/V *S.A. Agulhas II* is a relatively new (commissioned in 2012) polar research and supply vessel and is equipped with a moon pool that can be used as a CTD launch area, even in the event of severe weather conditions.

Several months prior to the ship survey, evaluation of satellite altimetry sea surface height anomaly (SSHA) data was initiated to identify and track the position of eddies suitable for study (Fig. 2). Composite SSHA satellite altimetry data, representing the sampling period, were obtained from the online data viewer of the Colorado Center for Astrodynamics Research (CCAR) (http://eddy.colorado.edu/ccar/data_viewer/index) (Fig. 1b). The global Historical Gridded SSH data viewer was used, which is typically a composite of ±10 days of Topex/Poseidon, Jason-a and Jason-2/OSTM data.

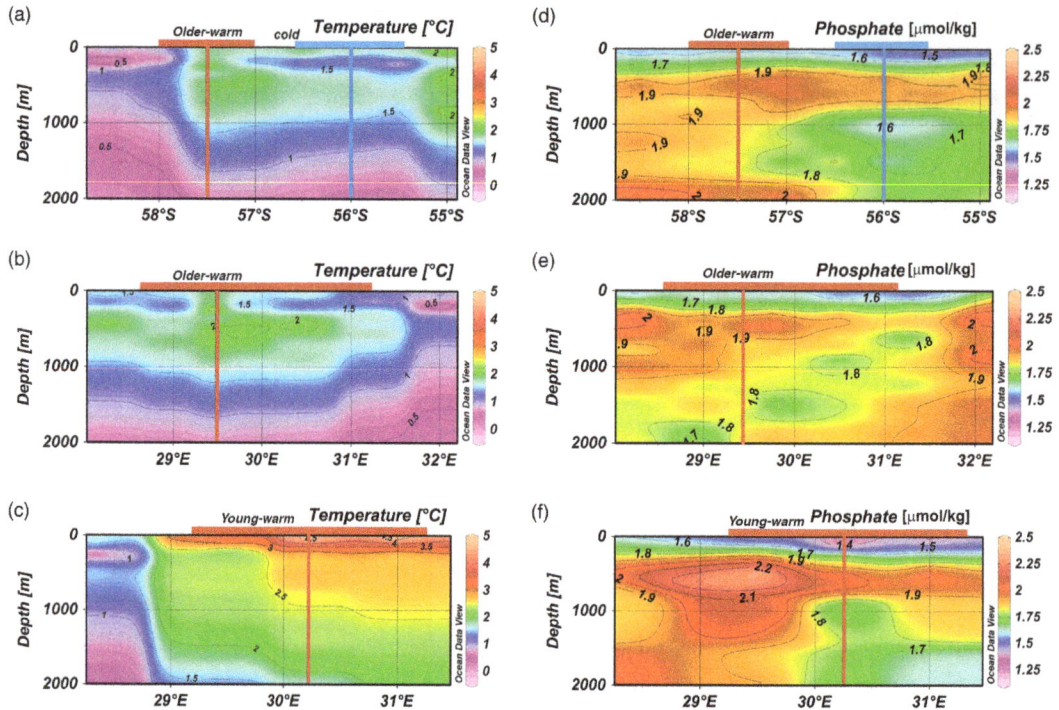

Figure 3. Temperature and dissolved phosphate profile data to a depth of 3000 m, for Transect E1 (**a** and **d**), Transect E2 (**b** and **e**) and Transect E3 (**c** and **f**). Vertical red lines (at 57.5° S in (**a**) and (**d**), 29.5° E in (**b**) and (**e**) and 30.25° E in **c** and **f**) indicate the approximate position of the centre of the older and young warm-core eddies respectively and vertical blue lines (at 56° S in **a** and **d**) similarly indicate the position of the cold-core eddy. The horizontal red and blue bars above these vertical red lines, in turn, indicate the estimated horizontal extent of the eddies, within the 10 cm SSHA contour intervals.

Mesoscale features with positive SSHA values, identified from satellite altimetry, were assumed to represent anti-cyclonic (counterclockwise rotation) eddies (Fig. 1b). Similarly, features with negative SSHAs were assumed to be cyclonic (clockwise rotation) eddies. In the Southern Hemisphere, the centre or core of anti-cyclonic eddies is warm and sea surface height is elevated, whereas the core of cyclonic eddies is cold and characterised by negative SSHAs (Chelton, 2013). Downwelling in the core of warm-core eddies and upwelling in the core of cold-core eddies have been inferred from isopycnal displacements (Zhang et al., 2014).

Observation of the evolution of the SSHA characteristics of mesoscale features over several months suggested that intense positive SSHA values can be assumed to represent younger, more recently formed anti-cyclonic eddies (Fig. 2). Ship-based ADCP data (M. van den Berg, unpublished cruise report contribution) confirmed the direction of flow around these (Fig. 1b) mesoscale features.

On the basis of satellite SSHA images (Figs. 1b, 2), the following three main mesoscale features were identified for detailed study prior to the start of CTD transects, within the constraints of the ship time available (based on approximate eddy core positions and SSHA at time of survey, Table 1):

– a "young, warm-core" anti-cyclonic eddy at 52° S, 30.2° E (core SSHA > 40 cm);

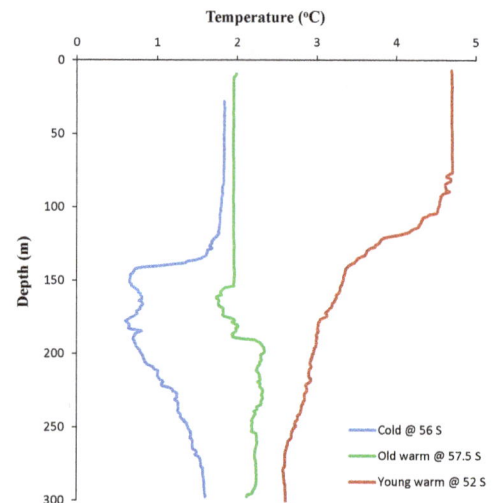

Figure 4. Upper 300 m water column temperature profiles from CTD stations located at the approximate cores of the three main mesoscale features present along the transect lines: in blue the cold-core eddy (CTD station E1-3 in Table 1), in green the older warm-core eddy (E1-6) and in red the young warm-core eddy (E3-3).

Table 1. CTD station locations and identification of CTD station positions relative to eddy cores, as inferred from altimetry data. E1, E2 and E3 refer to the three transect lines shown in Fig. 1. Based on interpretation of relative SSH anomalies, E1-6 is the core of the older warm-core eddy, E1-3 that of a cold-core eddy and E3-3 that of the young warm-core eddy.

Ship station	ID	Position relative to eddy core	Sampling date	Latitude (° S)	Longitude (° E)	Depth (m)
AM00264	E1-1	1° N of cold core	15 April 2014	55.0	29.5	4506
AM00265	E1-2	0.5° N of cold core	15 April 2014	55.5	29.5	5296
AM00266	E1-3	Cold core	15 April 2014	56.0	29.5	5581
AM00267	E1-4	0.5° S of cold core	15 April 2014	56.5	29.5	5288
AM00268	E1-5	0.5° N of warm core	16 April 2014	57.0	29.5	5275
AM00269	E1-6	Warm core	16 April 2014	57.5	29.5	5565
AM00270	E1-7	0.5° S of warm core	16 April 2014	58.0	29.5	5261
AM00271	E1-8	1° S of warm core	16 April 2014	58.5	29.5	5534
AM00272	E2-1	2.5° E of warm core	17 April 2014	57.5	32.0	5550
AM00273	E2-2	1.75° E of warm core	18 April 2014	57.5	31.3	5262
AM00274	E2-3	1° E of warm core	18 April 2014	57.5	30.5	5270
AM00275	E2-4	0.25° E of warm core	18 April 2014	57.5	29.8	5269
AM00276	E2-5	0.5° W of warm core	19 April 2014	57.5	29.0	5268
AM00277	E2-6	1.25° W of warm core	19 April 2014	57.5	28.3	5135
AM00281	E3-1	1.9° W of warm core	21 April 2014	52.0	28.3	5481
AM00282	E3-2	0.73° W of warm core	22 April 2014	52.0	29.5	5379
AM00283	E3-3	Warm core	22 April 2014	52.0	30.2	3817
AM00284	E3-4	0.78° E of warm core	22 April 2014	52.0	31.0	5154

- an "older, warm-core" anti-cyclonic eddy at 57.5° S, 29.5° E (core SSHA > 20 cm);

- a "cold-core" cyclonic eddy feature at 56° S, 29.5° E (core SSHA < −20 cm).

The propagation path of each of the three these eddies were tracked using archived SSHA data (Fig. 2). The results show that about 6 months before the survey cruise, the "older" warm eddy occupied a similar latitude (about 52° S) than the "young" eddy sampled during the cruise. Also, the "cold-core" eddy had a similar southerly migration route and life history than the "older warm-core" eddy. It is also interesting to note that 6 months after the survey, the "young" warm eddy ended up at approximately the same latitude as that of the "older" warm eddy at the time of sampling (i.e. about 57.5° S).

Based on the identification of these three mesoscale features, the following three transects (E1 to E3 in Fig. 1b) were chosen for detailed CTD profiling (station locations in Table 1) and water column sampling for chemical analysis.

- Transect E1 (15 to 16 April 2014): north-to-south transect from 55° S to 58.5° S, along 29.5° E. Eight CTD stations were occupied at 0.5° latitude intervals; the main features along this transect were the "mature" warm-core eddy and a cold-core cyclonic eddy just north of it.

- Transect E2 (17 to 19 April 2014): east-to-west transect from 32° E to 28.25° E, along 57.5° S. Six CTD stations

were occupied along this transect at 0.75° longitude intervals; the main feature along this transect is the mature warm-core eddy.

- Transect E3 (21 to 22 April 2014): west-to-east transect along 52° S, from 28.3° E to 31° E. Four CTD stations were occupied along this transect at approximately 0.75° longitude intervals; the main feature along this transect was the relatively young, warm-core eddy.

Along all three transects the spacing of the stations (0.5° latitude and 0.75° longitude intervals) was constrained by the available ship time and weather conditions; although higher resolution spacing would have been more ideal, the spacing is sufficient to resolve the general structure of the eddies.

3 Seawater sampling and analysis

Two SBE 9plus CTD systems and a moon pool CTD with a 24 20 L Niskin bottle rosette, or alternatively a 12 10 L Niskin bottle rosette for over-the-side deployment, were used for water column profiling and discrete water sampling at standard depths (20, 30, 40, 50, 75, 100, 150, 200, 300, 400, 500, 600, 700, 800, 900, 1000, 1250, 1500, 2000, 2500, 3000, 4000, 5000 m). The CTD deployments along the most southerly transects (E1 and E2) were conducted with the moon pool CTD. The moon pool CTD provides more and higher volume samples but does not sample the upper 20 m of the water column. Unless sampling the upper 20 m is essential, the moon pool CTD alone is sufficient, preferable and

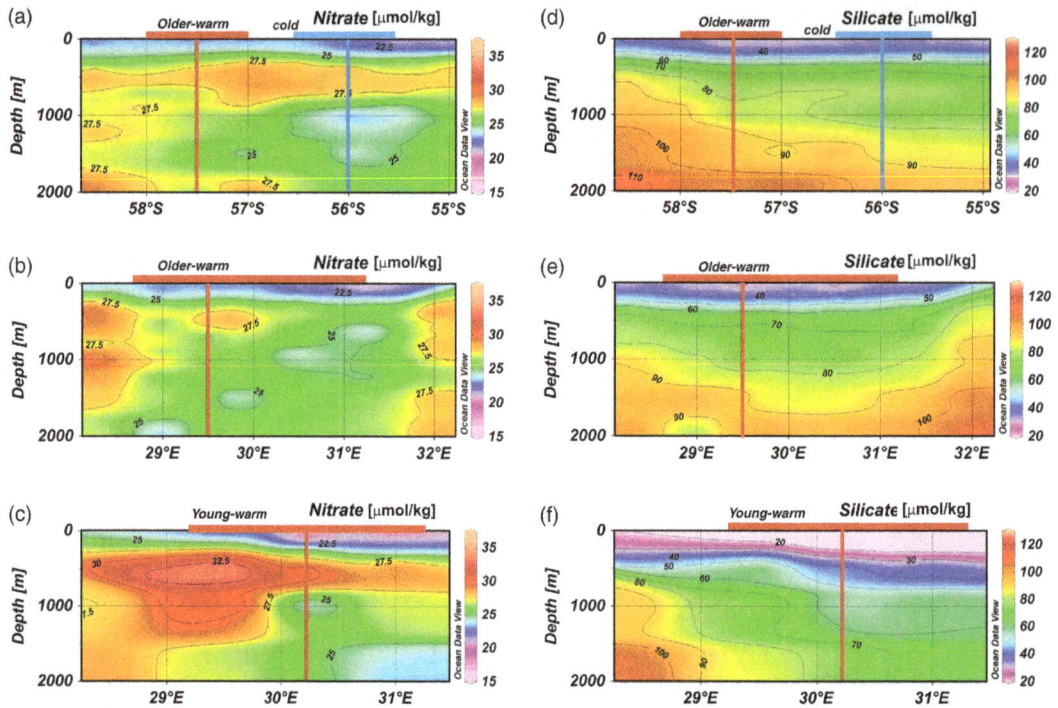

Figure 5. Dissolved nitrate and silicate profiles, plotted to a depth of 3000 m, for Transect E1 (**a** and **d**), Transect E2 (**b** and **e**) and Transect E3 (**c** and **f**). Vertical red and blue lines represent the eddy cores as described for Fig. 3. The data available online can be used to reproduce these Ocean Dataviewer plots to shallower or deeper depths.

Table 2. Average upper ocean (surface to 100 m depth) physical and chemical water column characteristics, obtained from CTD profiles, in and around warm and cold eddies. Samples for nutrient and chlorophyll *a* analysis were not collected at CTD stations E2-1 and E3-1.

ID	Position relative to eddy core	T (°C)	S (PSU)	SiO_2 (μmol kg^{-1})	PO_4^{3-} (μmol kg^{-1})	NO_3^- (μmol kg^{-1})	Chl a (μg L^{-1})
E1-1	1° N of cold core	2.31	33.97	29.9	1.50	20.9	0.14
E1-2	0.5° N of cold core	1.75	33.98	29.6	1.37	20.4	0.33
E1-3	Cold core	1.83	33.98	30.8	1.47	21.6	0.35
E1-4	0.5° S of cold core	1.99	33.95	29.3	1.58	23.6	0.36
E1-5	0.5° N of warm core	1.77	33.96	27.6	1.53	22.7	0.58
E1-6	Warm core	1.96	33.98	25.7	1.56	22.7	0.57
E1-7	0.5° S of warm core	0.88	33.87	34.8	1.56	23.3	0.31
E1-8	1° S of warm core	1.04	33.93	35.2	1.51	22.3	0.23
E2-1	2.5° E of warm core	1.09	33.92				
E2-2	1.75° E of warm core	1.27	33.99	31.2	1.46	20.1	0.44
E2-3	1° E of warm core	1.71	33.96	26.2	1.51	21.5	0.49
E2-4	0.25° E of warm core	1.62	33.97	29.0	1.56	22.5	0.33
E2-5	0.5° W of warm core	1.60	33.96	29.7	1.60	22.7	0.44
E2-6	1.25° W of warm core	1.60	33.98	29.1	1.49	22.4	0.42
E3-1	1.9° W of warm core	1.69	34.04				
E3-2	0.73° W of warm core	3.45	33.80	13.7	1.39	21.9	0.24
E3-3	Warm core	4.70	33.77	4.4	1.34	20.2	0.18
E3-4	0.78° E of warm core	4.76	33.77	3.5	1.34	19.5	0.18

often the only option in extremely rough seas. If time and conditions allow, both CTDs can be deployed, in sequence, to maximise sampling resolution. The moon pool CTD was lost after completion of transects E1 and E2 as a result of damage sustained in very rough seas, and transect E3 was carried out with the smaller CTD only. The CTD temperature sensors were calibrated before and after the cruise to ensure accuracy and achievement of the manufacturer's stated measurement precision of $\pm 0.0001\,°C$. Salinity was computed from CTD data in practical salinity units (PSU) and the conductivity sensor calibrated using a combination of international standards and discrete samples, for every CTD cast, with the salinometer's stated measurement precision ± 0.0001 PSU. The CTD oxygen sensor was calibrated with oxygen measurements obtained from discrete samples at selected depths for each CTD cast, applying the Winkler titration method using an electronic stand (Hansen, 1999). The precision of the oxygen titration was $\pm 0.45\,\mu mol\,kg^{-1}$. Turbidity, given in nephelometric turbidity units, was measured with a sensor connected to the CTD system, using the original calibration provided by the manufacturer.

Seawater samples for chemical analyses were collected from the Niskin bottles at the standard depths detailed earlier. Samples for dissolved oxygen analysis were collected from the Niskin bottles via silicone tubing, taking care not to introduce or trap air bubbles. For dissolved inorganic nutrient analysis, acid-washed 15 mL polypropylene tubes were thoroughly rinsed with sample water before filling and frozen at $-80\,°C$ prior to ship-based analysis using an Astoria Auto-Analyser Series 300, expanded to four channels. Dissolved nitrate (NO_3^-), nitrite (NO_2^-) and silicate (SiO_2) were determined following the methods of Armstrong et al. (1967) and phosphate (PO_4^{3-}) according to the methods of Bernhardt and Wilhelms (1967), with typical in-run precision of $\pm 0.1\,\mu mol\,L^{-1}$ for NO_3^- and NO_2^-, $\pm 0.02\,\mu mol\,L^{-1}$ for PO_4^{3-} and $\pm 0.24\,\mu mol\,L^{-1}$ for SiO_2.

Samples for chlorophyll a analysis were taken at four depths (near-surface, above the fluorescence maximum (F-max), at F-max and below the F-max), with the depth of F-max established from the downcast, continuous CTD fluorescence profile. Seawater samples (200 mL) for chlorophyll a analysis were collected in pre-rinsed plastic bottles. Samples were immediately filtered, under vacuum onto 25 mm Whatman™ GF/F glass fibre filter papers (Parsons et al., 1984). The filter papers were frozen in aluminium foil pouches for storage until analysis. Chlorophyll a was measured fluorometrically on a Turner Designs 10-AU fluorometer after extraction in 90 % acetone (Welschmeyer and Waterhouse, 1995), with a precision of approximately 5 %. The fluorometer was calibrated with chlorophyll a standard (Sigma Chemical Co., USA) in 90 % acetone solution with a GBC Cintra 404 spectrophotometer and an extraction coefficient of $87.67\,Lg^{-1}\,cm^{-1}$.

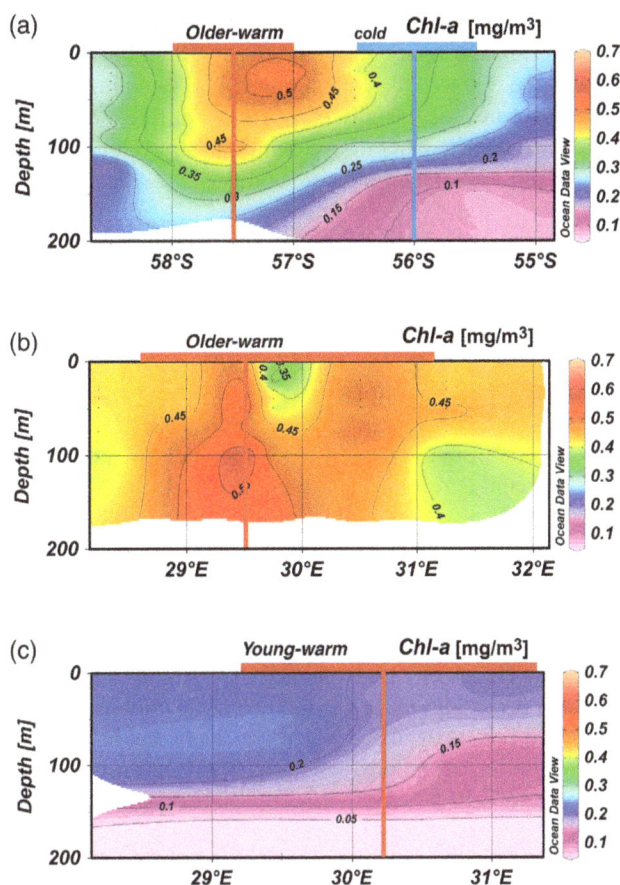

Figure 6. Chlorophyll a profiles, plotted to a depth of 200 m, for Transect E1 (**a** and **d**), Transect E2 (**b** and **e**) and Transect E3 (**c** and **f**). Vertical red and blue lines represent the eddy cores as described for Fig. 3. The data available online can be used to reproduce these Ocean Dataviewer plots to shallower or deeper depths.

4 Data overview and discussion

In situ CTD profiling data (Figs. 3a, b, c and 4) confirmed the presence and position of the anti-cyclonic and cyclonic eddies, targeted for this study based on SSHA observational data (Fig. 1). The average values of physical and chemical parameters for the upper 100 m of the water column is summarised for all CTD stations in Table 2 and profile data for temperature, salinity, nutrients and chlorophyll a are shown in Figs. 3, 5 and 6. These data demonstrate a number of significant differences between the eddies, most notably

– the old, warm (or anti-cyclonic) eddy at 57.5° S (Table 1) has a maximum core temperature of 2.2 to 2.35 °C, which is observed in the 200 to 300 m depth range (Fig. 4). In the core of the cold-core eddy located 1.5° north of the core of the older warm-core feature, the temperature reaches a value of almost 1.85 °C in the upper 100 m of the water column. The young, warm-core eddy much further north at 52° S, in contrast, is characterised by core temperatures of around 4.7 °C in the well-mixed upper 100 m layer and around 3 to 2.6 °C in the 200 to 300 m depth range. The sampling survey, therefore, successfully captured the different temperature characteristics of cyclonic and anti-cyclonic eddies at different stages in their maturity.

- In the upper 100 m, the young warm-core eddy is 2.74 °C warmer, less saline by 0.21 PSU, with silicate levels ~ 80 % lower and chlorophyll a levels more than a factor of 2 lower than those in its older warm-core equivalent further south eddy at the time of sampling (i.e. about 57.5° S).

- In contrast, the upper ocean water characteristics of the cold-core and older warm-core eddies, at 56 and 57.5° S respectively, are much more similar to each other than is the case for the younger versus older warm-core eddies (Table 2). The cold eddy is, on average, only 0.13 °C colder than the older warm-core eddy, with approximately the same upper ocean salinity values. The older warm-core eddy has slightly higher upper ocean phosphate and nitrate concentrations than the cold-core eddy, 6 and 5 % respectively, but has significantly higher chlorophyll-a (+63 %) and lower silicate levels (−17 %).

The data show that the most notable differences between the young and older warm-core eddies are their surface ocean dissolved silicate and chlorophyll a characteristics. One possible explanation is that, as warm-core eddies migrate in a southerly direction (Fig. 2), lateral and/or vertical mixing with adjacent silicate-rich water masses results in increasing surface silicate levels. These higher silicate levels, together with potential additional controls such as light and iron availability, then results in dramatically increased levels of productivity. Bongo net tows conducted during the survey (unpublished results), confirmed that the increased levels of chlorophyll a, south of the polar front, are associated with diatom productivity. Alternatively, if it is more appropriate to think of eddies as local (but persistent) perturbations in the density structure, then higher silicate values in the older, warm-core eddy are simply consistent with the characteristics of water masses south of the polar front compared to further north.

In summary, the in situ nutrient and chlorophyll a profile data presented here provide valuable input data for the modelling of the complex biogeochemical processes associated with mesoscale activity in the Southern Ocean.

Author contributions. S. de Villiers determined the sampling strategy, collected samples and oversaw and collated the measurements; K. Siswana and K. Vena assisted in sample collection, chemical analysis and data presentation; S. de Villiers prepared the manuscript.

Acknowledgements. All ship-based participants, crew and research staff, in some way or another, contributed to the collection of this data set, and their contributions are gratefully acknowledged. The Chief Scientist for the cruise was H. Verheye, and the cruise and associated research activities were financed by the Department of Environmental Affairs.

Edited by: R. Key

References

Ansorge, I. J., Pakhomov, E. A., Kaehler, S., Lutjeharms, J. R. E., and Durgadoo, J. V.: Physical and biological coupling in eddies in the lee of the South-West Indian Ridge, Polar Biol., 33, 747–759, 2010.

Ansorge, I. J., Jackson, J. M., Reid, K., Durgadoo, J. V., Swart, S., and Eberenz, S.: Evidence of a southward eddy corridor in the South-West Indian ocean, Deep-Sea Res. II, 119, 69–76, 2015.

Armstrong, F. A. J., Stearns, C. A., and Strickland, J. D. H.: The measurement of upwelling and subsequent biological processes by means of the Technicon Autoanalyzer and associated equipment, Deep-Sea Res., 14, 381–389, 1967.

Bernhardt, H. and Wilhelms, A.: The continuous determination of low level iron, soluble phosphate and total phosphate with the AutoAnalyzer, Technicon Symposia I, 385–389, 1967.

Chelton, D.: Mesoscale eddy effects, Nat. Geosci., 6, 594–594, 2013.

Chelton, D. B., Gaube, P., Schlax, M. G., Early, J. J., and Samelson, R. M.: The influence of nonlinear mesoscale eddies on near-surface oceanic chlorophyll, Science, 334, 328–332, 2011.

Chen, Y. L., Chen, H.-Y., Jan, S., Lin, Y.-H., Kuo, T.-H., and Hung, J.-J.: Biologically active warm-core anticyclonic eddies in the marginal seas of the western Pacific Ocean, Deep Sea Res. Pt. I, 106, 68–84, 2015.

Dong, S., Sprintall, J., and Gille, S. T.: Location of the Antarctic Polar Front from AMSR-E satellite sea surface temperature measurements, J. Phys. Ocean., 36, 2075–2089, 2006.

Durgadoo, J. V., Ansorge, I. J., de Cuevas, B. A., Lutjeharms, J. R. E., and Coward, A. C.: Decay of eddies at the South-West Indian Ridge, S. Afr. J. Sci., 107, 673, doi:10.4102/sajs.v107i11/12.673, 2011.

Falkowski, P., Ziemann, D., Kolber, Z., and Bienfang, P.: Role of eddy pumping in enhancing primary production in the ocean, Nature, 352, 55–58, 1991.

Ferrari, R. and Wunch, C.: Ocean circulation kinetic energy: reservoirs, sources, and sinks, Ann. Rev. Fluid Mech., 41, 253–282, 2009.

Frenger, I., Gruber, N., Knutti, R., and Münnich, M.: Imprint of Southern Ocean eddies on winds, clouds and rainfall, Nat. Geosci., 6, 608–612, 2013.

Fu, L.-L., Chelton, D. B., Le, P.-Y., and Morrow, R.: Eddy dynamics from satellite altimetry, Oceanography, 23, 14–25, 2010.

Gouretski, V. V. and Danilov, A. I.: Characteristic of warm rings in the African sector of the Antarctic Circumpolar current, Deep-Sea Res., 41, 1131–1157, 1994.

Hansen, H. P.: Determination of oxygen, in: Methods of Seawater analysis, edited by: Grasshoff, K. K. and Ehrhardt, M., Wiley-VCH, Weinheim, 75–89, 1999.

Joyce, T. M., Patterson, S. L., and Millard Jr., R. C.: Anatomy of a cyclonic ring in the Drake Passage, Deep-Sea Res., 28, 1265–1287, 1981.

Klein, P. and Lapeyre, G.: The oceanic vertical pump induced by mesoscale and submesoscale turbulence, Ann. Rev. Mar. Sci., 1, 351–375, 2009.

Lehahn, Y., d'Ovidio, F., Lévy, M., Amitai, Y., and Heifetz, E.: Long range transport of a quasi isolated chlorophyll patch by an Agulhas ring, Geophys. Res. Lett., 38, L16610, doi:10.1029/2011GL048588, 2011.

Lévy, M., Klein, P., and Treguier, A.-M.: Impact of sub-mesoscale physics on production and subduction of phytoplankton in an oligotrophic regime, J. Mar. Res., 59, 535–565, 2001.

Mahadevan, A. and Archer, D.: Modeling the impact of fronts and mesoscale circulation on the nutrient supply and biogeochemistry of the upper ocean, J. Geophys. Res., 105, 1209–1225, 2000.

McGillicuddy Jr., D. J., Anderson, L. A., Bates, N. R., Bibby, T., Buesseler, K. O., Carlson, C. A., Davis, C. S., Ewart, C., Falkowski, P. G., Goldtwait, S. A., Hansell, D. A., Jenkins, W. J., Johnson, R., Kosnyrev, V. K., Ledwell, J. R., Li, Q. P., Siegel, D. A., and Steinberg, D. K.: Eddy/wind interactions stimulate extraordinary mid-ocean plankton blooms, Science, 316, 1021–1026, 2007.

Morrow, R. A., Coleman, R., Church, J. A., and Chelton, D. B.: Surface eddy momentum flux and velocity variance in the Southern Ocean from GEOSAT altimetry, J. Phys. Ocean., 24, 2050–2071, 1994.

Parsons, T. R., Maita, Y., and Lalli, C. M.: A Manual of Chemical and Biological Methods for Seawater Analysis, Pergamon Press, Oxford, 173 pp., 1984.

Pollard, R. T. and Read, J. F.: Circulation pathways and transports of the Southern Ocean in the vicinity of the Southwest Indian Ridge, J. Geophys. Res., 106, 2881–2898, doi:10.1029/2000JC900090, 2001.

Sheen, K. L., Garabato, A. C. N., Brearley, J. A., Meredith, M. P., Polzin, K. L., Smeed, D. A., Forryan, A., King, B. A., Sallee, J.-B., St. Laurent, L., Thurnherr, A. M., Toole, J. M., Waterman, S. N., and Watson, A. J.: Eddy-induced variability in Southern Ocean abyssal mixing on climatic timescales, Nat. Geosci., 7, 577–582, 2014.

Siegel, D., Peterson, P., McGillicuddy, D., Maritorena, S., and Nelson, N.: Bio-optical footprints created by mesoscale eddies in the Sargasso Sea, Geophys. Res. Lett., 38, L13608, doi:10.1029/2011GL047660, 2011.

Stramma, L., Bange, H. W., Czeschel, R., Lorenzo, A., and Frank, M.: On the role of mesoscale eddies for the biological productivity and biogeochemistry in the eastern tropical Pacific Ocean off Peru, Biogeosciences, 10, 7293–7306, doi:10.5194/bg-10-7293-2013, 2013.

Welschmeyer, N. A. and Waterhouse, T. Y.: Taxon-specific analysis of microzooplankton grazing rates and phytoplankton growth rates, Limnol. Oceanogr., 40, 827–834, 1995.

Zhang, Z., Wang, W., and Qui, B.: Oceanic mass transport by mesoscale eddies, Science, 345, 322–324, 2014.

13

Global database of surface ocean particulate organic carbon export fluxes diagnosed from the ^{234}Th technique

F. A. C. Le Moigne[1], S. A. Henson[1], R. J. Sanders[1], and E. Madsen[*]

[1]Ocean Biogeochemistry and Ecosystems, National Oceanography Centre, Southampton, UK
[*]formerly at: School of Applied Sciences, Cranfield University, Cranfield, UK

Correspondence to: F. A. C. Le Moigne (f.lemoigne@noc.ac.uk)

Abstract. The oceanic biological carbon pump is an important factor in the global carbon cycle. Organic carbon is exported from the surface ocean mainly in the form of settling particles derived from plankton production in the upper layers of the ocean. The large variability in current estimates of the global strength of the biological carbon pump emphasises that our knowledge of a major planetary carbon flux remains poorly constrained. We present a database of 723 estimates of organic carbon export from the surface ocean derived from the ^{234}Th technique. The dataset is archived on the data repository PANGEA® (www.pangea.de) under doi:10.1594/PANGAEA.809717. Data were collected from tables in papers published between 1985 and early 2013. We also present sampling dates, publication dates and sampling areas. Most of the open ocean provinces are represented by multiple measurements. However, the western Pacific, the Atlantic Arctic, South Pacific and the southern Indian Ocean are not well represented. There is a variety of integration depths ranging from surface to 300 m. Globally the fluxes ranged from 0 to 1500 mg C m^{-2} d^{-1}.

1 Introduction

The concept of the biological carbon pump, dating from the late 1970s (Eppley and Peterson, 1979), quantifies the importance of oceanic primary production in the global carbon cycle. The biological carbon pump can be divided into three stages: the production of organic matter (and biominerals) in surface waters, the sinking of these particles into the deep ocean, and the subsequent decomposition of the settling (or settled) particles in the water column or the seabed. In this way the coupling of production and export processes allows the ocean to store CO_2 away from the atmosphere and contributes to the buffering of the global climate system. Without the oceanic biological carbon pump, atmospheric CO_2 concentrations would be almost twice their current levels (Sarmiento and Toggweiler, 1984). Recent studies have highlighted the challenge of quantifying the magnitude of the biological carbon pump with estimates ranging from 5 to 20 GtC yr^{-1} (Henson et al., 2011).

There are several ways by which downward export fluxes can be estimated. We can divide the techniques into two groups: (1) indirect estimates based on nutrient uptake (Sanders et al., 2005; Henson et al., 2006; Pondaven et al., 2000), oxygen utilization (Jenkins, 1982), radioisotopes (Buesseler et al., 1998; Cochran and Masque, 2003; Rutgers Van Der Loeff et al., 1997b; Le Moigne et al., 2012, 2013) or by synthesising numerous biological rate processes (Boyd and Newton, 1999), and (2) direct measurements from sediments traps (Lampitt et al., 2008).

Here we focus on the ^{234}Th technique, which has the advantage that its fundamental operation allows a downward flux rate to be determined from a single water column profile of thorium coupled to an estimate of the POC/^{234}Th ratio in sinking matter (POC is particulate organic carbon; Buesseler et al., 1992). This is highly advantageous in that it removes the complications associated with sediment trap deployments and provides an integrated estimate of export (over a timescale of weeks) rather than a snapshot of export rates (Lampitt et al., 2008).

Although several comprehensive worldwide datasets of POC flux from sediment traps have been published (e.g.

Honjo et al., 2008), to date only one thorium derived export dataset has been published (Henson et al., 2011). As part of the SeasFX project (Seasonal Variability in the Efficiency of Upper Carbon Export, http://www.seasfx.info, funded by the UK National Environment Research Council), we compiled a global database of ^{234}Th-derived POC export from the surface ocean (0–300 m). It comprises 723 data points from 1985 to 2013 covering most oceanic provinces. The dataset is archived on the data repository PANGEA® (www.pangea.de) under doi:10.1594/PANGAEA.809717.

2 Data

2.1 The crux of the ^{234}Th technique

The radioactive short-lived thorium-234 (^{234}Th, $t_{1/2} = 24.1$ d) has been used as a tracer of several transport processes and particle cycling in aquatic systems by different techniques (Van der Loeff et al., 2006). The most widespread application of the ^{234}Th approach is to estimate how much POC is exported into the deep ocean (Waples et al., 2006). ^{234}Th is the daughter isotope of naturally occurring Uranium-238 (^{238}U, $t_{1/2} = 4.47 \times 10^9$ yr) that is conservative in seawater and proportional to salinity in well-oxygenated environments (Ku et al., 1977; Chen et al., 1986). Unlike ^{238}U, ^{234}Th is insoluble in seawater and is particle reactive in the water column (i.e. ^{234}Th adheres to particles as they form). As particles with ^{234}Th sink through the water column, a radioactive disequilibrium is formed between ^{238}U and ^{234}Th that can be used to quantify the rate of particle export from the surface ocean.

Export rates of ^{234}Th from the surface ocean can be calculated using a one-box model (Coale and Bruland, 1987; Buesseler et al., 1992, 1998; Cochran et al., 2000; Cochran and Masque, 2003; Savoye et al., 2006; Benitez-Nelson et al., 2001a; Verdeny et al., 2008). Assuming steady state (SS) conditions, $\frac{\partial A_2}{\partial t} = 0$ where the total ^{234}Th activity does not change with time, and no supply of ^{234}Th from physical processes (e.g. advection), the ^{234}Th flux (dpm m^{-2} d^{-1}), P, is calculated through the water column as

$$P = \gamma \sum_{z=0}^{z=h} (A_2 - A_1) \cdot dz. \tag{1}$$

A_1 is the total parent activity concentration (dpm m^{-3}) for ^{238}U; A_2 is the total ^{234}Th activity concentration (dpm m^{-3}); λ is the decay constant of the daughter (d^{-1}); h is the sample depth; and P is the loss of the daughter due to sinking particles (dpm m^{-2} d^{-1}). In ^{234}Th studies generally advection effects are neglected, as shown in Morris et al. (2007), with the exception of upwelling regions or areas of strong advection (Murray et al., 1992; Buesseler et al., 1998). Using the SS model from a single profile of ^{234}Th activity needs to be justified as we assume that the initial activity does not change with time (Savoye et al., 2006). If several profiles of ^{234}Th activities are measured at the same site over a certain period of

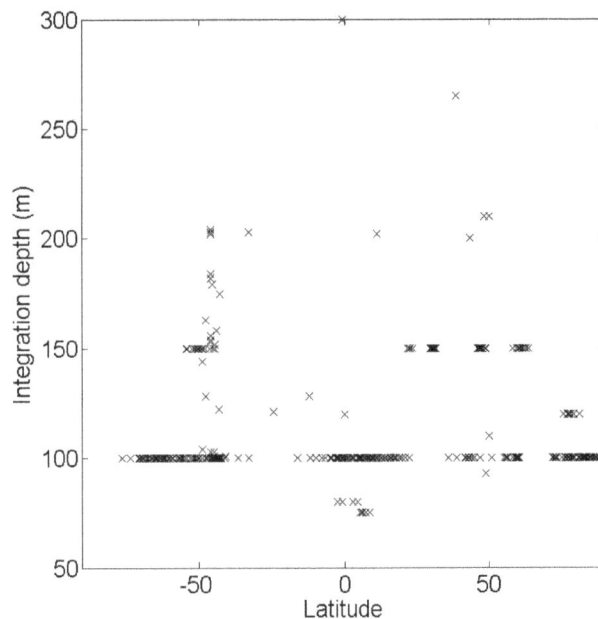

Figure 1. Integration depth of the ^{234}Th fluxes versus latitude.

time (weeks or months), a non-steady-state (NSS) model has to be applied. The NSS model may also be used during temporally variable periods with high particle flux events, such as the onset of a bloom (Buesseler et al., 1998). The NSS model factors in the term $\frac{\partial A_2}{\partial t}$ that is set to zero in the SS model (Eq. 2 below).

$$P = \lambda \sum_{z=0}^{z=h} [(A_2 - A_1)] - \frac{\partial A_2}{\partial t} dz \tag{2}$$

We report ^{234}Th fluxes from both SS and NSS models in our database. Reported ^{234}Th fluxes were integrated from depths ranging from the surface down to 300 m (Fig. 1 and Table 1). The vast majority of fluxes are integrated to between 100 and 150 m. A few studies report ^{234}Th integrated over greater depths, but not more than 300 m depth (Table 1). In the final stage of the thorium methodology, the estimated ^{234}Th flux is converted to POC export by applying the ratio of POC to particulate ^{234}Th activity.

2.2 Determination of POC : ^{234}Th ratio of sinking particles

The accuracy of the Th method relies critically on estimating the POC/^{234}Th ratio of material sinking from the upper ocean (Buesseler et al., 2006). This estimate is most frequently achieved by assuming that sinking carbon is contained within large particles, often greater than 50 μm in size (or 53 μm, depending on the mesh supplier), whereas organic carbon within small particles is suspended in the water column, and is therefore assumed to be insufficiently large and/or dense to sink (Bishop et al., 1977; Fowler and Knauer, 1986). Size

Table 1. Sampling year, area, number of samples (*N*), model used (see text, Sect. 2.1), C:Th size fraction ("Part." refers to the entire particulate fraction) and reference of studies used in the database. "Traps" is indicated when C:Th ratio were measured in sediment traps and "Equ." refers to equilibrium depth.

Date	Area	N	Model	Integration depth (m)	C:Th ratio size fraction (µm)	Reference/investigator
1987	Equatorial Pacific	4	SS	80	Traps	Murray et al. (1989)
1992	Equatorial Pacific	65		100	> 53	Buesseler et al. (1995)
1992	Equatorial Pacific	24	SS	100	> 53	Murray et al. (1996)
1992	Southern Ocean	1	SS	100	Part.	Shimmield and Ritchie (1995)
1992	Equatorial Pacific	2	SS	120	> 53	Bacon et al. (1996)
1992	Southern Ocean	10	NSS	100	Part.	Rutgers van der Loeff et al. (1997a)
1992	Equatorial Pacific	16	SS	100	> 53	Buesseler (1998)
1993–1994	Middle Atlantic Bight	7	SS	200	Traps	Santschi et al. (1999)
1995	Arabian Sea	56	NSS	100	> 53	Buesseler et al. (1998)
1996	Equatorial Atlantic	12	NSS	100	> 53	Charette and Moran (1999)
1996	Subartic Pacific	3	SS	110–210	> 53	Charette et al. (1999)
1996	Southern Ocean	6	NSS	100	Part.	Friedrich and van der Loeff (2002)
1997	Gulf of Maine	7	SS	150	> 53	Charette et al. (2001)
1997	Southern Ocean	25	NSS/SS	100	> 70	Cochran et al. (2000)
1997	China Sea	1	SS	100	Part.	Cai et al. (2001)
1997–1998	Southern Ocean	41	NSS/SS	100	> 70	Buesseler et al. (2001)
1997–1998	Southern Ocean	28	NSS	100	> 70	Buesseler et al. (2003)
1998–1999	Arctic	15	SS	100	> 70	Amiel et al. (2002)
1999	North Pacific	4	SS	100	Part.	Chen et al. (2003)
1999	Southern Ocean	8	SS	100	> 60	Coppola et al. (2005)
1999	Labrador Sea	3	SS	100	> 53	Moran et al. (2003)
2003	North Pacific	22	SS	100	Part.	Kawakami et al. (2007)
2003–2005	Arctic	8	SS	60–120	> 53	Lalande et al. (2008)
2003	Antarctic	6	NSS/SS	100	> 70	Rodriguez y Baena et al. (2008)
2004	Arctic	8	SS	100	> 53	Lalande et al. (2007)
2004	Atlantic gyres	10	SS	Equ. depth	> 53	Thomalla et al. (2008)
2004	China Sea	36	SS	100	> 1	Cai et al. (2008)
2004	Mediteranean Sea	4	SS	200	> 70	Stewart et al. (2007)
2004–2005	Southern Ocean	20	SS	Equ. depth	> 53	Morris et al. (2007)
2004–2005	Atlantic	64	SS	150	> 53	Buesseler et al. (2008a)
2004–2005	Pacific	45	SS	150	> 53	Buesseler et al. (2009)
2005	Southern Ocean	5	SS	100	< 210	Savoye et al. (2008)
2007	Arctic	36	SS	100	Part.	Cai et al. (2010)
2007	North Atlantic	10	SS	Euphotic zone depth	> 53	Sanders et al. (2010)
2007	Southern Ocean	14	NSS/SS	60–120	> 54	Jacquet et al. (2011)
2008	Southern Ocean	27	SS	100	> 50	Rutgers van der Loeff et al. (2011)
2008	South-west Pacific	25	SS	100	Part.	Zhou et al. (2012)
2008	Southern Ocean	11	SS	100	> 53	Planchon et al. (2013)
2009	PAP site	10	SS	150	> 53	Le Moigne et al. (2013)
2010	North Atlantic	20	SS	150	> 53	Le Moigne et al. (2012)

fractions for the POC/^{234}Th ratios used in the database are given in Table 1.

There is a considerable body of literature on how and why POC/^{234}Th ratios vary with particle size and depth (see review in Buesseler et al., 2006); however, there is little consensus on the most appropriate ratio to use. Numerous processes can impact POC/^{234}Th ratios in the ocean including particle surface-area-to-volume ratios (Santschi et al., 2006), solution chemistry issues (Guo et al., 2002; Hung et al., 2004), the chemical composition of particles and their affinity for ^{234}Th (Szlosek et al., 2009), POC assimilation by food webs (Buesseler and Boyd, 2009), particle aggregation (Burd et al., 2000) and fragmentation (Maiti et al., 2010) and Th decay (Cai et al., 2006).

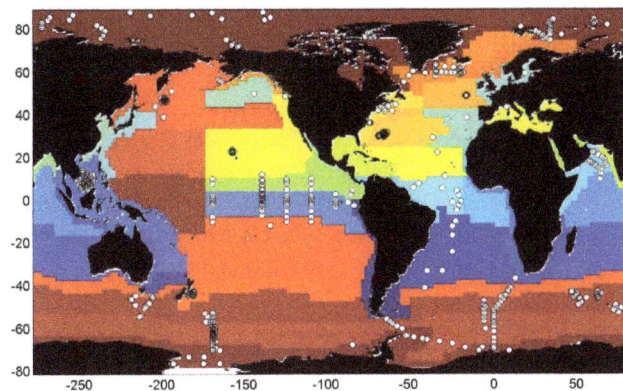

Figure 2. Map showing the distribution of sampling stations. Longhurst oceanic (Longhurst, 2006) provinces are represented in different colours.

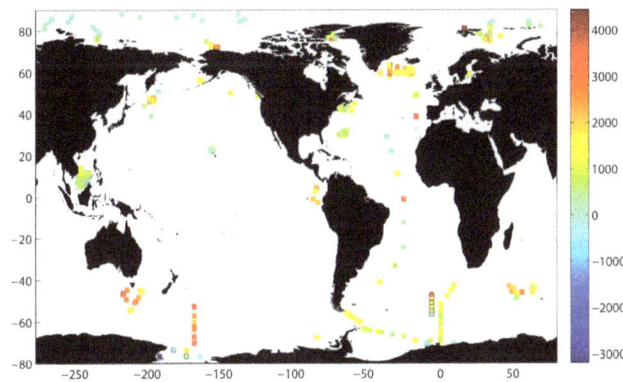

Figure 3. Global distribution of ^{234}Th export fluxes (in dpm m^{-2} d^{-1}). SS model (see text) derived fluxes are squares and NSS model derived fluxes are circles.

3 Results and discussion

3.1 Data sources

The dataset is archived on the data repository PANGEA® (www.pangea.de) under doi:10.1594/PANGAEA.809717. Latitude, longitude, date, POC flux, primary production (when available), integration depth and references are given as metadata. All fluxes were converted to mgC m^{-2} d^{-1} if not already reported in these units. Th-derived POC export has been reported at 723 stations globally (Fig. 2). Some stations were part of transect cruises whereas others were part of small-scale surveys or reoccupation at different seasons and years. Sampling date, sampling area and reference investigator are given in Table 1 in addition to the literature reference. The ^{234}Th fluxes derived from both SS and NSS model are presented in Fig. 3. Because of the uncertainties associated with POC/^{234}Th ratios, examining the Th fluxes prior to conversion to POC fluxes provides a robust picture of the variability in particle flux on the global scale. The lowest and highest ^{234}Th flux are both measured in the Arctic Ocean.

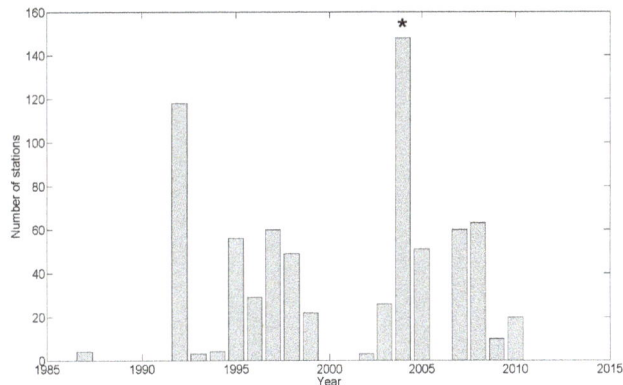

Figure 4. Histogram of datapoints presented in Table 1 and published since 1987. The star indicates the year 2004 when the VERTIGO study was undertaken.

Also, it is worth mentioning that on small scales the ^{234}Th flux can be quite variable, e.g. in the Iceland and Irminger basins (Fig. 3). Generally, the patchiness of export, which can affect the robustness of point observations, is greater in region of high eddy kinetic energy (Resplandy et al., 2012).

Our database covers measurements published between 1985 and 2013. We do not include unpublished data here and therefore assume that the originating authors and editors have undertaken steps necessary to control data quality. Fig. 4 shows the number of thorium-derived export data per year published from 1985 to 2013. In years 1992, 1998 and 2002, the number of ^{234}Th measurements increased. This is likely due to significant improvements in the ^{234}Th methodology such as the introduction of the small volume technique (Benitez-Nelson et al., 2001b), and it also highlights dedicated carbon export programmes such as the VERTIGO voyages in the Pacific Ocean (Buesseler et al., 2008b, 2009). It is important to mention that our database only references papers where Th-derived export data are presented in tables, rather than only graphically.

More POC fluxes are reported in the Northern Hemisphere ($\sim 60\%$ of the database) than in the Southern Hemisphere (Fig. 5). In the Northern Hemisphere, each month of the year has been sampled (Fig. 5). Springtime (May) and summertime have been most frequently sampled. In the Southern Hemisphere, although stations are more evenly distributed in time, no ^{234}Th-derived POC export numbers are reported for winter months (July and August, Fig. 5).

3.2 Global POC export and ocean provinces

^{234}Th-derived POC export estimates are reported in 32 out of 56 Longhurst provinces (Longhurst, 2006) that are based on the prevailing role of physical forcing as a regulator of phytoplankton distribution, with measurements in most of the large open ocean biomes (Fig. 2). Figure 6 shows the mean ^{234}Th-derived POC export (mg m^{-2} d^{-1}) in each

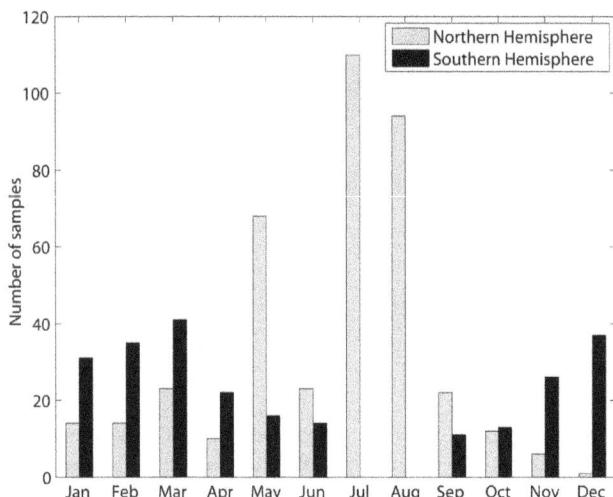

Figure 5. Number of samples collected in each month, separated into Northern Hemisphere and Southern Hemisphere.

Longhurst province that has been sampled at least once. Only four provinces are represented with one measurement (NASE, CCAL, CHIL and NECS; see Table 2 for details and provinces names).

Our dataset exhibits similar global patterns of POC export as those estimated with other methods (e.g. Laws et al., 2000; Schlitzer, 2004), with highest daily POC export rate occurring in the high-latitude North Atlantic, the Arctic and the Southern Ocean. NASE and WTRA provinces located in the subtropical and equatorial Atlantic (Fig. 6 and Table 2) are exceptions to this trend with POC export of 450 (but note that $n = 1$) and $250 \pm 200\,\mathrm{mg\,m^{-2}\,d^{-1}}$ reported in Thomalla et al. (2008) and Charette and Moran (1999), respectively. Some regions are relatively well sampled, such as the Arctic Ocean (Longhurst's BPLR), which is represented by 72 stations which display high spatial variability. For instance, POC flux associated with Arctic shelf regions is large while the POC flux in the central Arctic is very low (Cai et al., 2010). This implies that the magnitude of export is not necessarily a simple function of temperature in high-latitude regions.

Some regions show unexpectedly high POC flux, such as the NASE, where Thomalla et al. (2008) suggest that the occurrence of a short-lived bloom triggered by nutrient injection into the surface from a local upwelling event resulted in very high POC flux (however, note that $n = 1$ in this region). Alternatively, Charette and Moran (1999) propose that scavenging of ^{234}Th by inorganic particles may have overestimated the POC flux in the WTRA region, as also observed by Le Moigne et al. (2013); Brew et al. (2009).

A comparison of Th-derived export with direct measurements of surface export (from free drifting sediment traps to avoid any problem due to overcollection of horizontally advected material) would be useful at this stage. However,

surface POC fluxes from direct measurements are scarce. The few studies that have examined the discrepancy between ^{234}Th-derived estimates and direct measurements of POC export (e.g. Le Moigne et al., 2013; Stewart et al., 2007) suggest that ^{234}Th-derived estimates in most cases overestimate the direct POC flux. This may be due to a mismatch in timescales over which different methods estimate export. ^{234}Th deficits persist after an export event, whereas free-drifting sediment traps capture only the instantaneous export flux.

3.3 Towards better understanding of the ocean's biological carbon pump

A portion of this database has already been used to extrapolate the local measurements to a global scale by correlation with satellite sea surface temperature fields (Henson et al., 2011). The resulting estimates of global integrated carbon export were significantly lower than those derived from new production measurements, at just $\sim 5\,\mathrm{GtC\,yr^{-1}}$ compared to $12\,\mathrm{Gt\,C\,yr^{-1}}$ (Laws et al., 2000). However, the parameterisation of the export ratio presented in Henson et al. (2011) has relatively large uncertainty at cold sea surface temperature (SST) (see their Fig. 2). As the type of phytoplankton present in the upper ocean may also influence the export ratio (because large, dense phytoplankton cells sink rapidly and export more efficiently than smaller plankton), the variability in export ratio at low temperatures could be due to large seasonal shifts in phytoplankton community structure at high latitudes.

In high-latitude regions, simultaneous measurements of upper ocean particulate organic carbon flux and phytoplankton community structure could help to assess how seasonal variability of the phytoplankton bloom alters the export ratio. The knowledge gained from this approach could then be applied to our global dataset, combining satellite-derived data on SST, bloom stage and phytoplankton community structure. Ultimately, a revised parameterisation of the export ratio, including relevant seasonal information, could be used to calculate a new global estimate of the magnitude of the biological carbon pump.

3.4 Significant gaps in the global dataset

Globally the Th-derived POC fluxes ranged from 0 to $1500\,\mathrm{mg}$ of $\mathrm{C\,m^{-2}\,d^{-1}}$ (Fig. 7). In this database, some areas such as the equatorial Pacific, Arabian Sea, South China Sea and the high-latitude North Atlantic are fairly well represented (Fig. 7). However, there are significant gaps that could potentially bias estimates of the global carbon export. Most notably, ^{234}Th-derived POC fluxes are not reported for the Benguela system (BENG), the Mauritanian upwelling (CNRY; ETRA), the entire western Pacific (consisting of numerous Longhurst provinces), and the southern Indian Ocean (ISSG).

Table 2. Mean POC flux ($mg\,m^{-2}\,d^{-1}$) per Longhurst province (Longhurst, 2006).

Province number	Province name	Mean POC flux ($mgC\,m^{-2}\,d^{-1}$)	Standard deviation in POC flux	Number of stations
2	CHIL – Chile-Peru Current Coastal	176.00	0	1
5	SATL – South Atlantic Gyre	37.44	54.38	7
14	PEQD – Pacific Equatorial Divergence	46.20	40.43	71
15	MONS – Indian Monsoon Gyre	62.16	54.26	10
19	ARAB – NW Arabian Upwelling	85.18	82.45	48
20	WTRA – Western Tropical Atlantic	251.39	178.19	10
22	NECS – NE Atlantic Shelves	53.52	0	1
23	NASE – North Atlantic Subtropical Gyre (East)	488.40	0	1
24	PSAE – Pacific Subarctic Gyre (East)	49.00	32.39	6
26	INDE – East India Coastal	31.60	15.66	3
28	PNEC – North Pacific Equatorial Countercurrent	48.65	50.29	24
30	INDW – West India Coastal	33.60	52.43	18
32	NPTG – North Pacific Tropical Gyre	26.27	23.57	21
33	NATR – North Atlantic Tropical Gyre	91.20	128.98	2
34	MEDI – Mediterranean Sea, Black Sea	115.80	78.70	4
35	CCAL – California Upwellinf Coastal	14.76	0	1
36	NWCS – NW Atlantic Shelves	127.08	97.75	14
37	NASW – North Atlantic Subtropical Gyre (West)	25.90	19.41	65
39	NADR – North Atlantic Drift	87.56	51.22	11
41	ARCT – Atlantic Arctic	242.34	122.65	16
42	SARC – Atlantic Subarctic	214.67	127.06	14
44	SSTC – South Subtropical Convergence	114.34	136.16	7
45	SPSG – South Pacific Subtropical Gyre	28.79	13.73	10
47	BERS – North Pacific Epicontinental	146.00	24.04	3
50	ANTA – Antarctic	156.95	100.90	67
51	SANT – Subantarctic	126.54	116.71	107
53	APLR – Austral Polar	195.07	233.32	41
54	BPLR – Boreal Polar	171.05	298.48	72

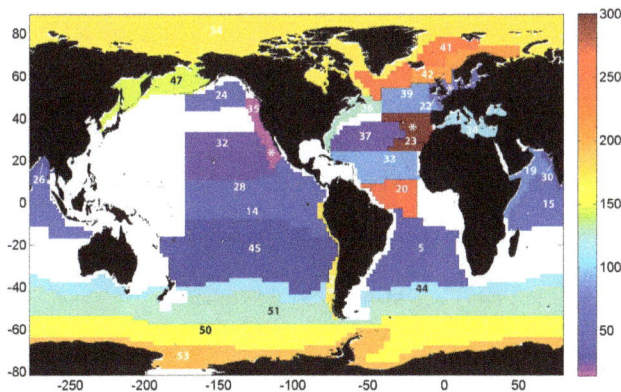

Figure 6. Mean POC export ($mg\,m^{-2}\,d^{-1}$) in Longhurst provinces (provinces with only one measurement are marked with a star). Areas in white represent areas where no data have been collected. Numbers on map indicate Longhurst province (cf. Table 2).

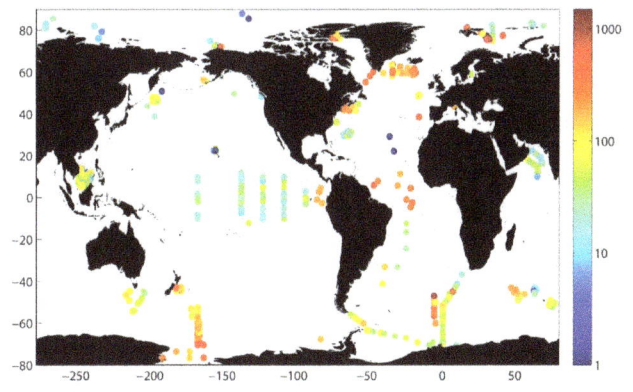

Figure 7. Global distribution of POC export fluxes derived from the ^{234}Th technique (in $mg\,m^{-2}\,d^{-1}$).

Some of these areas are deemed to be high production and export regions due to the occurrence of upwelling. For example, deep ($\sim 2000\,m$) sediment trap measurements of POC export for the Mauritanian upwelling suggest that POC flux can peak at 5 to $25\,mg\,m^{-2}\,d^{-1}$ (Fischer et al., 2009), and is therefore presumably higher in the upper water column. Also, in the Benguela system POC export has been estimated to be $550\,mg\,m^{-2}\,d^{-1}$ on the basis of nutrient uptake (Waldron et al., 1992). Provinces such as KURO and PSAW in the north-west Pacific may also export a significant amount of POC ($\sim 120\,\mathrm{mg\,m^{-2}\,d^{-1}}$ averaged over one year based on a modelling study (Schlitzer, 2004). Although these regions represent a small percentage of the global surface area of the ocean, the lack of data in these high export areas could potentially result in estimates of global POC export that are biased low.

We suggest that future studies should investigate ^{234}Th-derived POC export flux in regions that are currently unsampled or undersampled. However, in upwelling regions where advective current velocities are high, the influence of advection and diffusion on the ^{234}Th model should be carefully assessed and accounted for in the calculation of POC flux, as done, for example, in Morris et al. (2007), Buesseler et al. (1998), and Charette et al. (1999).

4 Conclusions

Here we provide a global database of 723 published estimates of POC export derived from the ^{234}Th technique spanning 1985–2013. The observed pattern of POC fluxes reflects the expected dynamics of primary production and export. Some notable gaps in the dataset are the Benguela system, the Mauritanian upwelling, the western Pacific, and the southern Indian Ocean. This database could be used to provide revised and more robust estimates of the ocean's biological carbon pump.

Acknowledgements. This work was supported by NERC grants NE/G013055/1 and NE/J004383/1 to SAH. Ken Buesseler (WHOI) is acknowledged for sharing data. We thank K. Cochran, M. Rutgers van der Loeff and the editor, Robert Key, for providing useful comments on this paper.

Edited by: R. Key

References

Amiel, D., Cochran, J. K., and Hirschberg, D. J.: Th-234/U-238 disequilibrium as an indicator of the seasonal export flux of particulate organic carbon in the North Water, Deep-Sea Res. Pt. II, 49, 5191–5209, 2002.

Bacon, M. P., Cochran, J. K., Hirschberg, D., Hammar, T. R., and Fleer, A. P.: Export flux of carbon at the equator during the EqPac time-series cruises estimated from ^{234}Th measurements, Deep-Sea Res. Pt. II, 43, 1133–1153, 1996.

Benitez-Nelson, C., Buesseler, K. O., Karl, D. M., and Andrews, J.: A time-series study of particulate matter export in the North Pacific Subtropical Gyre based on Th-234 : U-238 disequilibrium, Deep-Sea Res. Pt. I, 48, 2595–2611, 2001a.

Benitez-Nelson, C. R., Buesseler, K. O., van der Loeff, M. R., Andrews, J., Ball, L., Crossin, G., and Charette, M. A.: Testing a new small-volume technique for determining Th-234 in seawater, J. Radioanal. Nucl. Ch., 248, 795–799, 2001b.

Bishop, J. K. B., Edmond, J. M., Ketten, D. R., Bacon, M. P., and Silker, W. B.: The chemistry, biology, and vertical flux of par-

ticulate matter from the upper 400 m of the equatorial Atlantic Ocean, Deep-Sea Res., 24, 511–548, 1977.

Boyd, P. W. and Newton, P. P.: Does planktonic community structure determine downward particulate organic carbon flux in different oceanic provinces?, Deep-Sea Res. Pt. I, 46, 63–91, 1999.

Brew, H. S., Moran, S. B., Lomas, M. W., and Burd, A. B.: Plankton community composition, organic carbon and thorium-234 particle size distributions, and particle export in the Sargasso Sea, J. Mar. Res., 67, 845–868, 2009.

Buesseler, K., Ball, L., Andrews, J., Benitez-Nelson, C., Belastock, R., Chai, F., and Chao, Y.: Upper ocean export of particulate organic carbon in the Arabian Sea derived from thorium-234, Deep-Sea Res. Pt. II, 45, 2461–2487, 1998.

Buesseler, K. O.: The decoupling of production and particulate export in the surface ocean, Global Biogeochem. Cy., 12, 297–310, 1998.

Buesseler, K. O. and Boyd, P. W.: Shedding light on processes that control particle export and flux attenuation in the twilight zone of the open ocean, Limnol. Oceanogr., 54, 1210–1232, 2009.

Buesseler, K. O., Bacon, M. P., Cochran, J. K., and Livingston, H. D.: Carbon and nitrogen export during the JGOFS North Atlantic Bloom Experiment estimated from ^{234}Th:^{238}U disequilibria, Deep-Sea Res. Pt. I, 39, 1115–1137, 1992.

Buesseler, K. O., Andrews, J. A., Hartman, M. C., Belastock, R., and Chai, F.: Regional estimates of the export flux of particulate organic carbon derived from thorium-234 during the JGOFS EqPac program, Deep-Sea Res. Pt. II, 42, 777–804, 1995.

Buesseler, K. O., Benitez-Nelson, C. R., Rutgers van der Loeff, M., Andrews, J., Ball, L., Crossin, G., and Charette, M. A.: An intercomparison of small- and large-volume techniques for thorium-234 in seawater, Mar. Chem., 74, 15–28, 2001.

Buesseler, K. O., Barber, R. T., Dickson, M. L., Hiscock, M. R., Moore, J. K., and Sambrotto, R.: The effect of marginal ice-edge dynamics on production and export in the Southern Ocean along 170 degrees W, Deep-Sea Rese. Pt. II, 50, 579–603, doi:10.1016/s0967-0645(02)00585-4, 2003.

Buesseler, K. O., Benitez-Nelson, C. R., Moran, S. B., Burd, A., Charette, M., Cochran, J. K., Coppola, L., Fisher, N. S., Fowler, S. W., Gardner, W., Guo, L. D., Gustafsson, O., Lamborg, C., Masque, P., Miquel, J. C., Passow, U., Santschi, P. H., Savoye, N., Stewart, G., and Trull, T.: An assessment of particulate organic carbon to thorium-234 ratios in the ocean and their impact on the application of Th-234 as a POC flux proxy, Mar. Chem., 100, 213–233, doi:10.1016/j.marchem.2005.10.013, 2006.

Buesseler, K. O., Lamborg, C., Cai, P., Escoube, R., Johnson, R., Pike, S., Masque, P., McGillicuddy, D., and Verdeny, E.: Particle fluxes associated with mesoscale eddies in the Sargasso Sea, Deep-Sea Res. Pt. II, 55, 1426–1444, doi:10.1016/j.dsr2.2008.02.007, 2008a.

Buesseler, K. O., Trull, T. W., Steinber, D. K., Silver, M. W., Siegel, D. A., Saitoh, S. I., Lamborg, C. H., Lam, P. J., Karl, D. M., Jiao, N. Z., Honda, M. C., Elskens, M., Dehairs, F., Brown, S. L., Boyd, P. W., Bishop, J. K. B., and Bidigare, R. R.: VERTIGO (VERtical Transport in the Global Ocean): A study of particle sources and flux attenuation in the North Pacific, Deep-Sea Res. Pt. II, 55, 1522–1539, doi:10.1016/j.dsr2.2008.04.024, 2008b.

Buesseler, K. O., Pike, S., Maiti, K., Lamborg, C. H., Siegel, D. A., and Trull, T. W.: Thorium-234 as a tracer of spatial, temporal and vertical variability in particle flux in the North Pacific, Deep-Sea

Res. Pt. I, 56, 1143–1167, doi:10.1016/j.dsr.2009.04.001, 2009.

Burd, A. B., Moran, S. B., and Jackson, G. A.: A coupled adsorption-aggregation model of the POC/Th-234 ratio of marine particles, Deep-Sea Res. Pt. I, 47, 103–120, 2000.

Cai, P., Dai, M., Chen, W., Tang, T., and Zhou, K.: On the importance of the decay of ^{234}Th in determining size-fractionated C/^{234}Th ratio on marine particles, Geophys. Res. Lett., 33, L23602, doi:10.1029/2006GL027792, 2006.

Cai, P., Chen, W., Dai, M., Wan, Z., Wang, D., Li, Q., Tang, T., and Lv, D.: A high-resolution study of particle export in the southern South China Sea based on ^{234}Th:^{238}U disequilibrium, J. Geophys. Res., 113, CB4019, doi:10.1029/2007JC004268, 2008.

Cai, P., Rutgers van der Loeff, M., Stimac, I., Nothig, E. M., Lepore, K., and Moran, S. B.: Low export flux of particulate organic carbon in the central Artic Ocean as revealed by ^{234}Th:^{238}U disequilibrium, J. Geophys. Res., 115, C10037, doi:10.1029/2009JC005595, 2010.

Cai, P. H., Huang, Y. P., Chen, M., Liu, G. S., and Qiu, Y. S.: Export of particulate organic carbon estimated from Th-234-U-238 disequilibria and its temporal variation in the South China Sea, Chinese Sci. Bull., 46, 1722–1726, 2001.

Charette, M. A. and Moran, S. B.: Rates of particle scavenging and particulate organic carbon export estimated using ^{234}Th as a tracer in the subtropical and equatorial Atlantic Ocean, Deep-Sea Res. Pt. II, 46, 885–906, 1999.

Charette, M. A., Moran, S. B., and Bishop, J. K. B.: Th-234 as a tracer of particulate organic carbon export in the subarctic northeast Pacific Ocean, Deep-Sea Res. Pt. II, 46, 2833–2861, 1999.

Charette, M. A., Moran, S. B., Pike, S. M., and Smith, J. N.: Investigating the carbon cycle in the Gulf of Maine using the natural tracer thorium 234, J. Geophys. Res.-Oceans, 106, 11553–11579, 2001.

Chen, J. H., Edwards, R. L., and Wasserburg, G. J.: ^{238}U, ^{234}U and ^{232}Th in seawater, Earth Planet. Sc. Lett., 80, 241–251, 1986.

Chen, M., Huang, Y. P., Cai, P. G., and Guo, L. D.: Particulate organic carbon export fluxes in the Canada Basin and Bering Sea as derived from Th-234/U-238 disequilibria, Arctic, 56, 32–44, 2003.

Coale, K. H. and Bruland, K. W.: Oceanic stratified euphotic zone as elucidated by ^{234}Th:^{238}U disequilibria, Limnol. Oceanogr., 32, 189–200, 1987.

Cochran, J. K. and Masque, P.: Short-lived U/Th series radionuclides in the ocean: Tracers for scavenging rates, export fluxes and particle dynamics, Uranium-Series Geochemistry, 52, 461–492, 2003.

Cochran, J. K., Buesseler, K. O., Bacon, M. P., Wang, H. W., Hirschberg, D. J., Ball, L., Andrews, J., Crossin, G., and Fleer, A.: Short-lived thorium isotopes (Th-234, Th-228) as indicators of POC export and particle cycling in the Ross Sea, Southern Ocean, Deep-Sea Res. Pt. II, 47, 3451–3490, 2000.

Coppola, L., Roy-Barman, M., Mulsow, S., Povinec, P., and Jeandel, C.: Low particulate organic carbon export in the frontal zone of the Southern Ocean (Indian sector) revealed by ^{234}Th, Deep-Sea Res. Pt. I, 52, 51–68, 2005.

Eppley, R. W. and Peterson, B. J.: Particulate organic matter flux and planktonic new production in the deep ocean, Nature, 282, 677–680, 1979.

Fischer, G., Karakas, G., Blaas, M., Ratmeyer, V., Nowald, N., Schlitzer, R., Helmke, P., Davenport, R., Donner, B., Neuer, S.,

and Wefer, G.: Mineral ballast and particle settling rates in the coastal upwelling system off NW Africa and the South Atlantic, Int. J. Earth Sci., 98, 281–298, doi:10.1007/s00531-007-0234-7, 2009.

Fowler, S. W. and Knauer, G. A.: Role of large particles in the transport of elements and organic compounds through the oceanic water column, Prog. Oceanogr., 16, 147–194, 1986.

Friedrich, J. and van der Loeff, M. M. R.: A two-tracer (Po-210-Th-234) approach to distinguish organic carbon and biogenic silica export flux in the Antarctic Circumpolar Current, Deep-Sea Res. Pt. I, 49, 101–120, 2002.

Guo, L., Hung, C.-C., Santschi, P. H., and Walsh, I. D.: ^{234}Th scavenging and its relationship to acid polysaccharide abundance in the Gulf of Mexico, Mar. Chem., 78, 103–119, 2002.

Henson, S., Sanders, R., Madsen, E., Morris, P., Le Moigne, F., and Quartly, G.: A reduced estimate of the strength of the ocean's bioloical carbon pump, Geophys. Res. Lett., 38, L04606, doi:10.1029/2011GL046735, 2011.

Henson, S. A., Sanders, R., Holeton, C., and Allen, J. T.: Timing of nutrient depletion, diatom dominance and a lower-boundary estimate of export production for Irminger Basin, North Atlantic, Mar. Ecol.-Prog. Ser., 313, 73–84, 2006.

Honjo, S., Manganini, S. J., Krishfield, R. A., and Francois, R.: Particulate organic carbon fluxes to the ocean interior and factors controlling the biological pump: A synthesis of global sediment trap programs since 1983, Prog. Oceanogr., 76, 217–285, doi:10.1016/j.pocean.2007.11.003, 2008.

Hung, C. C., Guo, L. D., Roberts, K. A., and Santschi, P. H.: Upper ocean carbon flux determined by size-fractionated 234Th data and sediment traps in the Gulf of Mexico, Geochem. J., 38, 601–611, 2004.

Jacquet, S. H. M., Lam, P. J., Trull, T., and Dehairs, F.: Carbon export production in the subantarctic zone and polar front zone South of Tasmania, Deep-Sea Res. Pt. II, 58, 2277–2292, 2011.

Jenkins, W. J.: Oxygen Utilization Rates in North-Atlantic Sub-Tropical Gyre and Primary Production in Oligotrophic Systems, Nature, 300, 246–248, 1982.

Kawakami, H., Honda, M. C., Wakita, M., and Watanabe, S.: Time-series observation of POC fluxes estimated from ^{234}Th in the northwestern North Pacific, Deep-Sea Res. Pt. I, 54, 1070–1090, 2007.

Ku, T. L., Knauss, K. G., and Mathieu, G. G.: Uranium in open ocean: concentration and isotopic composition, Deep-Sea Res., 24, 1005–1017, 1977.

Lalande, C., Lepore, K., Cooper, L. W., Grebmeier, J. M., and Moran, S. B.: Export fluxes of particulate organic carbon in the Chukchi Sea: A comparative study using ^{234}Th/ ^{238}U disequilibria and drifting sediment traps, Mar. Chem., 103, 185–196, 2007.

Lalande, C., Moran, S. B., Wassmann, P., Grebmeier, J. M., and Cooper, L. W.: ^{234}Th-derived particulate organic carbon fluxes in the northern Barents Sea with comparison to drifting sediment trap fluxes, J. Mar. Syst., 73, 103–113, 2008.

Lampitt, R. S., Boorman, B., Brown, L., Lucas, M., Salter, I., Sanders, R., Saw, K., Seeyave, S., Thomalla, S. J., and Turnewitsch, R.: Particle export from the euphotic zone: Estimates using a novel drifting sediment trap, Th-234 and new production, Deep-Sea Res. Pt. I, 55, 1484–1502, doi:10.1016/j.dsr.2008.07.002, 2008.

Laws, E. A., Falkowski, P. G., Smith, W. O., Ducklow, H., and McCarthy, J. J.: Temperature effects on export production in the open ocean, Global Biogeochem. Cy., 14, 1231–1246, 2000.

Le Moigne, F. A. C., Sanders, R. J., Villa-Alfageme, M., Martin, A. P., Pabortsava, K., Planquette, H., Morris, P. J., and Thomalla, S. J.: On the proportion of ballast versus non-ballast associated carbon export in the surface ocean, Geophys. Res. Lett., 39, L15610, doi:10.1029/2012GL052980, 2012.

Le Moigne, F. A. C., Villa-Alfageme, M., Sanders, R. J., Marsay, C. M., Henson, S., and Garcia-Tenorio, R.: Export of organic carbon and biominerals derived from ^{234}Th and ^{210}Po at the Porcupine Abyssal Plain, Deep-Sea Res. Pt. I, 72, 88–101, doi:10.1016/j.dsr.2012.10.010, 2013.

Longhurst, A. R.: Ecological Geography of the Sea (Second Edition), Elsevier Inc., ISBN: 978-0-12-455521-1, 2006.

Maiti, K., Benitez-Nelson, C. R., and Buesseler, K.: Insights into particle formation and remineralization using the short-lived radionuclide, Thorium-234, Geophys. Res. Lett., 37, L15608, doi:10.1029/2010GL044063, 2010.

Moran, S. B., Weinstein, S. E., Edmonds, H. N., Smith, J. N., Kelly, R. P., Pilson, M. E. Q., and Harrison, W. G.: Does ^{234}Th/^{238}U disequilibrium provide an acurate record of the export flux of particulate organic carbon from the upper ocean?, Limnol. Oceanogr., 48, 1018–1029, 2003.

Morris, P. J., Sanders, R., Turnewitsch, R., and Thomalla, S.: Th-234-derived particulate organic carbon export from an island-induced phytoplankton bloom in the Southern Ocean, Deep-Sea Res. Pt. II, 54, 2208–2232, doi:10.1016/j.dsr2.2007.06.002, 2007.

Murray, J. W., Downs, J. N., Strom, S., Wei, C. L., and Jannasch, H. W.: Nutrient assimilation, export production and ^{234}Th scavenging in the eastern equatorial Pacific, Deep-Sea Res., 36, 1471–1489, 1989.

Murray, J. W., Leinen, M. W., Feely, R. A., Toggweiler, J. R., and Wanninkhof, R.: EqPac: A process study in the Central Equatorial Pacific, Oceanography, 5, 134–142, 1992.

Murray, J. W., Young, J., Dunne, J. P., Chapin, T., and Paul, B.: Export flux of particulate organic carbon frm the Central Equatorial Pacific determined using a combined drifting trap-^{234}Th approach, Deep-Sea Res. Pt. II, 43, 1095–1132, 1996.

Planchon, F., Cavagna, A.-J., Cardinal, D., André, L., and Dehairs, F.: Late summer particulate organic carbon export and twilight zone remineralisation in the Atlantic sector of the Southern Ocean, Biogeosciences, 10, 803–820, doi:10.5194/bg-10-803-2013, 2013.

Pondaven, P., Ragueneau, O., Treguer, P., Hauvespre, A., Dezileau, L., and Reyss, J. L.: Resolving the 'opal paradox' in the Southern Ocean, Nature, 405, 168–172, 2000.

Resplandy, L., Martin, A. P., Le Moigne, F., Martin, P., Aquilina, A., Memery, L., Levy, M., and Sanders, R.: How does dynamical spatial variabilty impact ^{234}Th-derived estimates of organic export?, Deep-Sea Res. Pt. I, 68, 24–45, 2012.

Rodriguez y Baena, A. M., Boudjenoun, R., Fowler, S. W., Miquel, J. C., Masque, P., Sanchez-Cabeza, J. A., and Warnau, M.: ^{234}Th-based carbon export during the ice-edge bloom: Sea-ice algae as a likely bias in data interpretation, Earth Planet. Sc. Lett., 269, 596–604, 2008.

Rutgers van der Loeff, M., Friedrich, J., and Bathmann, U.: Carbon export during the spring bloom at the Antarctic Polar Front, determined with the natural tracer ^{234}Th, Deep-Sea Res. Pt. II, 44, 457–478, 1997a.

Rutgers Van Der Loeff, M. M., Bathmann, U. V., and Buesseler, K. O.: Export production measured with the natural tracer Th-234 and sediment traps, Berichte zur Polarforschung, 0, 113–115, 1997b.

Rutgers van der Loeff, M., Cai, P., Stimac, I., Bracher, A., Middag, R., Klunder, M., and van Heuven, S.: ^{234}Th in surface waters: distribution of particle export flux across the Antarctic Circumpolar Current and in the Weddell Sea during the GEOTRACES expedition ZERO and DRAKE, Deep-Sea Res. Pt. II, 58, 2749–2766, 2011.

Sanders, R., Brown, L., Henson, S., and Lucas, M.: New production in the Irminger Basin during 2002, J. Marine Syst., 55, 291–310, doi:10.1016/j.jmarsys.2004.09.002, 2005.

Sanders, R., Morris, P. J., Poulton, A. J., Stinchcombe, M. C., Charalampopoulou, A., Lucas, M. I., and Thomalla, S. J.: Does a ballast effect occur in the surface ocean?, Geophys. Res. Lett., 37, L08602, doi:10.1029/2010gl042574, 2010.

Santschi, P. H., Guo, L., Walsh, I. D., Quigley, M. S., and Baskaran, M.: Boundary exchange and scavenging of radionuclides in continental margin waters of the middle Atlantic Bight. Implications for carbon fluxes, Cont. Shelf Res., 19, 609–636, 1999.

Santschi, P. H., Murray, J. W., Baskaran, M., Benitez-Nelson, C. R., Guo, L. D., Hung, C. C., Lamborg, C., Moran, S. B., Passow, U., and Roy-Barman, M.: Thorium speciation in seawater, Mar. Chem., 100, 250–268, 2006.

Sarmiento, J. L. and Toggweiler, J. R.: A new model for the role of the oceans in determining atmospheric pCO$_2$, Nature, 308, 621–624, 1984.

Savoye, N., Benitez-Nelson, C., Burd, A. B., Cochran, J. K., Charette, M., Buesseler, K. O., Jackson, G. A., Roy-Barman, M., Schmidt, S., and Elskens, M.: Th-234 sorption and export models in the water column: A review, Mar. Chem., 100, 234–249, doi:10.1016/j.marchem.2005.10.014, 2006.

Savoye, N., Trull, T. W., Jacquet, S. H. M., Navez, J., and Dehairs, F.: ^{234}Th-based export fluxes during a natural iron fertilization experiemnt in the Southern Ocean (KEOPS), Deep-Sea Res. Pt. II, 55, 841–855, 2008.

Schlitzer, R.: Export production in the equatorial and North Pacific derived from dissolved oxygen, nutrient and carbon data, J. Oceanogr., 60, 53–62, 2004.

Shimmield, G. B. and Ritchie, G. R.: The impact of marginal ice zone processes on the distribution of ^{210}Pb, ^{210}Po and ^{234}Th and implications for new production in the Bellingshausen Sea, Antarctica, Deep-Sea Res. Pt. II, 42, 1313–1335, 1995.

Stewart, G., Cochran, J. K., Miquel, J. C., Masque, P., Szlosek, J., Baena, A., Fowler, S. W., Gasser, B., and Hirschberg, D. J.: Comparing POC export from Th-234/U-238 and Po-210/Pb-210 disequilibria with estimates from sediment traps in the northwest Mediterranean, Deep-Sea Res. Pt. I, 54, 1549–1570, doi:10.1016/j.dsr.2007.06.005, 2007.

Szlosek, J., Cochran, J. K., Miquel, J. C., Masque, P., Armstrong, R. A., Fowler, S. W., Gasser, B., and Hirschberg, D. J.: Particulate organic carbon-Th relationships in particles separated by settling velocity in the northwest Mediterranean Sea, Deep-Sea Res. Pt. II, 56, 1519–1532, 2009.

Thomalla, S. J., Poulton, A. J., Sanders, R., Turnewitsch, R., Holligan, P. M., and Lucas, M. I.: Variable export fluxes and efficiencies for calcite, opal, and organic carbon in the Atlantic Ocean: A ballast effect in action?, Global Biogeochem. Cy., 22, Gb1010, doi:10.1029/2007gb002982, 2008.

Van der Loeff, M. R., Sarin, M. M., Baskaran, M., Benitez-Nelson, C., Buesseler, K. O., Charette, M., Dai, M., Gustafsson, O., Masque, P., Morris, P. J., Orlandini, K., Baena, A. R. Y., Savoye, N., Schmidt, S., Turnewitsch, R., Voge, I., and Waples, J. T.: A review of present techniques and methodological advances in analyzing Th-234 in aquatic systems, Mar. Chem., 100, 190–212, doi:10.1016/j.marchem.2005.10.012, 2006.

Verdeny, E., Masque, P., Maiti, K., Garcia-Orellana, J., Brauch, J. M., Mahaffey, C., and Benitez-Nelson, C. R.: Particle export within cyclonic Hawaiian lee eddies derived from Pb-210-Po-210 disequilibrium, Deep-Sea Res. Pt. II, 55, 1461–1472, doi:10.1016/j.dsr2.2008.02.009, 2008.

Waldron, H. N., Probyn, T. A., Lutjeharms, J. R. E., and Shillington, F. A.: Carbon export associated with the Benguela upwelling system, S. Afr. J. Marine Sci., 12, 369–374, 1992.

Waples, J. T., Benitez-Nelson, C., Savoye, N., Van der Loeff, M. R., Baskaran, M., and Gustafsson, O.: An introduction to the application and future use of Th-234 in aquatic systems, Mar. Chem., 100, 166–189, doi:10.1016/j.marchem.2005.10.011, 2006.

Zhou, K., Nodder, S. D., Dai, M., and Hall, J. A.: Insignificant enhancement of export flux in the highly productive subtropical front, east of New Zealand: a high resolution study of particle export fluxes based on [234]Th: [238]U disequilibria, Biogeosciences, 9, 973–992, doi:10.5194/bg-9-973-2012, 2012.

14

The global distribution of pteropods and their contribution to carbonate and carbon biomass in the modern ocean

N. Bednaršek[1], J. Možina[2], M. Vogt[3], C. O'Brien[3], and G. A. Tarling[4]

[1]NOAA Pacific Marine Environmental Laboratory, 7600 Sand Point Way NE, Seattle, WA 98115, USA
[2]University of Nova Gorica, Laboratory for Environmental Research, Vipavska 13, Rožna Dolina, 5000 Nova Gorica, Slovenia
[3]Institute for Biogeochemistry and Pollutant Dynamics, ETH Zürich, Universitaetstrasse 16, 8092 Zürich, Switzerland
[4]British Antarctic Survey, Natural Environment Research Council, High Cross, Madingley Road, Cambridge CB3 0ET, UK

Correspondence to: N. Bednaršek (nina.bednarsek@noaa.gov)

Abstract. Pteropods are a group of holoplanktonic gastropods for which global biomass distribution patterns remain poorly described. The aim of this study was to collect and synthesise existing pteropod (Gymnosomata, Thecosomata and Pseudothecosomata) abundance and biomass data, in order to evaluate the global distribution of pteropod carbon biomass, with a particular emphasis on temporal and spatial patterns. We collected 25 939 data points from several online databases and 41 scientific articles. These data points corresponded to observations from 15 134 stations, where 93 % of observations were of shelled pteropods (Thecosomata) and 7 % of non-shelled pteropods (Gymnosomata). The biomass data has been gridded onto a $360 \times 180°$ grid, with a vertical resolution of 33 depth levels. Both the raw data file and the gridded data in NetCDF format can be downloaded from PANGAEA, doi:10.1594/PANGAEA.777387. Data were collected between 1950–2010, with sampling depths ranging from 0–2000 m. Pteropod biomass data was either extracted directly or derived through converting abundance to biomass with pteropod-specific length to carbon biomass conversion algorithms. In the Northern Hemisphere (NH), the data were distributed quite evenly throughout the year, whereas sampling in the Southern Hemisphere (SH) was biased towards winter and summer values. 86 % of all biomass values were located in the NH, most (37 %) within the latitudinal band of 30–60° N. The range of global biomass values spanned over four orders of magnitude, with mean and median (non-zero) biomass values of $4.6\,\mathrm{mg\,C\,m^{-3}}$ (SD = 62.5) and $0.015\,\mathrm{mg\,C\,m^{-3}}$, respectively. The highest mean biomass was located in the SH within the 70–80° S latitudinal band ($39.71\,\mathrm{mg\,C\,m^{-3}}$, SD = 93.00), while the highest median biomass was in the NH, between 40–50° S ($0.06\,\mathrm{mg\,C\,m^{-3}}$, SD = 79.94). Shelled pteropods constituted a mean global carbonate biomass of $23.17\,\mathrm{mg\,CaCO_3\,m^{-3}}$ (based on non-zero records). Total biomass values were lowest in the equatorial regions and equally high at both poles. Pteropods were found at least to depths of 1000 m, with the highest biomass values located in the surface layer (0–10 m) and gradually decreasing with depth, with values in excess of $100\,\mathrm{mg\,C\,m^{-3}}$ only found above 200 m depth.

Tropical species tended to concentrate at greater depths than temperate or high-latitude species. Global biomass levels in the NH were relatively invariant over the seasonal cycle, but more seasonally variable in the SH. The collected database provides a valuable tool for modellers for the study of marine ecosystem processes and global biogeochemical cycles. By extrapolating regional biomass to a global scale, we established global pteropod biomass to add up to 500 Tg C.

1 Introduction

The phylum Mollusca comprises at least 100 000 species, of which only 4000 species inhabit the upper ocean, principally those in the class Gastropoda. Approximately 140 species are holoplanktonic, meaning that they do not inhabit the seabed during any stage of their life cycle. The holoplanktonic lifestyle is facilitated by adaptations such as the development of swimming appendages and the reduction or loss of the calcareous shell. The pteropods are holoplanktonic gastropods that are widespread and abundant in the global ocean (Lalli and Gilmer, 1989). They consist of two orders: the Thecosomata (shelled pteropods) and the Gymnosomata (naked pteropods). The two orders are taxonomically separated not only by their morphology and behaviour, but also by their trophic position within the marine food web, with the former consisting mainly of herbivores and detritivores (Hopkins, 1987; Harbison and Gilmer, 1992) and the latter of carnivores (Lalli, 1970). A further systematic detail divides order Thecosomata into two suborders, the Euthecosomes and Pseudothecosomes. The two suborders have similar lifestyles, but they are set apart by their anatomical characteristics, most notably a gelatinous internal pseudoconch in Pseudothecosomes that replaces the external shell present in Euthecosomes (Lalli and Gilmer, 1989).

Pteropods have high ingestion rates that are in the upper range for mesozooplankton (Perissinotto, 1992; Pakhomov and Perissinotto, 1997). Although pteropods constitute, on average, only 6.5 % of the total abundance density of grazers in areas such as the Southern Ocean, they contribute on average 25 % to total phytoplankton grazing and consume up to 19 % of daily primary production (Hunt et al., 2008). Pteropods themselves are also an important prey item for many predators, such as larger zooplankton as well as herring, salmon and birds (Hunt et al., 2008; Karnovsky et al., 2008).

Pteropods are also involved in numerous pathways of organic carbon export. They contribute to the downward flux of carbon through the production of negatively buoyant faecal pellets. A number of pteropods also produce pseudo-faeces, i.e. accumulations of rejected particles expelled in mucous strings (Gilmer, 1990). Pteropods feed using mucous webs that trap fine particles and small faecal pellets, which form fast sinking colloids when abandoned (Jackson et al., 1993; Gilmer and Harbison, 1991). Pteropods actively transport carbon downwards during the descent phase of nycthemeral migrations, mostly from the shallow euphotic zone into the deeper twilight zone, where they respire and defecate.

In terms of inorganic carbon, pteropods are one of only a few taxa that make their shells out of aragonite as opposed to the calcite form of calcium carbonate. The biogeochemical importance of aragonite production by pteropods has been shown in a number of studies (Berner and Honjo, 1981; Acker and Byrne, 1989). Their aragonite shell not only contributes to the transfer of inorganic material into the deep ocean (Tréguer et al., 2003) but also increases the weight of pteropods as settling particles and hence their sinking speed (Lochte and Pfannkuche, 2003). Ontogenetic (or seasonal) migration, often followed by mass mortality, transports both organic and inorganic carbon to depth (Tréguer et al., 2003). On a global scale, aragonite production by pteropods might constitute at least 12 % of the total carbonate flux worldwide (Berner and Honjo, 1981).

Although the ecological and biogeochemical importance of pteropods has been well recognised, essential details on their global biomass distribution remain poorly resolved. Such information is required for modellers to be able to incorporate this group as a plankton functional type within ecosystem models and to allow the quantification of their contribution to carbon export in biogeochemical models.

The Marine Ecosystem Model Inter-comparison Project (MAREMIP) has been launched as an initiative to compare current plankton functional type models, and to collect data necessary for their validation. In 2009, MAREMIP launched the MARine Ecosystem DATa project, with the aim to construct a database based on field measurements for the biomass of ten major plankton functional types (PTFs) currently represented in marine ecosystem models (Le Quéré et al., 2005). The resulting biomass databases include diatoms (silicifiers), *Phaeocystis* (DMS producers), coccolithophores (calcifying phytoplankton), diazotrophs (nitrogen fixers), picophytoplankton, bacterioplankton, mesozooplankton, macrozooplankton and pteropods and foraminifera (calcifying zooplankton). All MAREDAT data sets of global biomass distribution are publicly available and will serve marine ecosystem modellers for model evaluation, development and future model inter-comparison studies. This study will present and evaluate the seasonal and temporal distribution of pteropod carbon biomass, with a particular emphasis on the seasonal and vertical biomass patterns. Finally, global estimates of pteropod biomass and productivity will be presented.

2 Data

2.1 Origin of data

The sources of the data were several online databases (PANGEA, ZooDB, NMFS127 COPEPOD) and 41 scientific articles. The full data set is comprised of 25 939 data points (Table 1). Each data point includes the following information: Year, Month, Day, Longitude, Latitude, Sampling Depth (m), Mesh size (µm) Abundance (ind. m^{-3}) and Biomass (mg C m^{-3}) and the data source. All data points presenting abundance measurements were later converted to biomass values. Zero biomass values were included as biologically valid data points in the data set. Some data sets included multiple samples at several stations, which would bias the global biomass estimates if not suitably treated. Thus, when repeat sampling of the same station location

Table 1. The list of data contributors in alphabetical order, with the two major online databases listed at the end of the list.

Entry No.	Principal Investigator	Database	Year (data collection)	Region
1	Andersen et al. (1997)	PANGEA	1991–1992	NE tropical Atlantic
2	Bednaršek et al. (2012)	–	1996–2010	Southern Ocean (Scotia Sea)
3	Bernard and Froneman (2005)	–	2004	Southern Ocean (west-Indian sector of the Polar Frontal Zone)
4	Bernard and Froneman (2009)		2002/2004/2005	Indian sector PFZ
5	Blachowiak-Samolyk et al. (2008)	–	2003	Arctic (N Svalbard waters)
6	Boysen-Ennen et al. (1991)	–	1983	Antarctica (Weddell Sea)
7	Broughton and Lough (2006)	–	1997	North Atlantic (Georges Bank)
8	Clarke and Roff (1990)	–	1986	Caribbean Sea (Lime Cay)
9	Daase and Eiane (2007)	–	2002–2004	Arctic (N Svalbard waters)
10	Dvoretsky and Dvoretsky (2009)	–	2006	E Barents Sea (Novaya Zemlya)
11	Elliot et al. (2009)	–	2006–2007	Antarctica (McMurdo Sound)
12	Flores et al. (2011)	–	2004–2008	Southern Ocean (Lazarev Sea)
13	Foster (1987)	–	1985	Antarctica (McMurdo Sound)
14	Froneman et al. (2000)	–	1998	Southern Ocean (Prince Edward Archipelago)
15	Hunt and Hosie (2006)	–	2001–2002	Southern Ocean (south of Australia)
16	Koppelmann et al. (2004)	PANGEA	1999	Eastern Mediterranean Sea
17	Marrari et al. (2011)	–	2001/2002	W Antarctic (Marguerite Bay)
18	Mazzocchi et al. (1997)	PANGEA	1991–2002	Eastern Mediterranean Sea
19	Mileikovsky (1970)	–	1966	North Atlantic, Subarctic and North Pacific Ocean
20	Moraitou-Apostolopoulou et al. (2008)	PANGEA	1994	Eastern Mediterranean Sea
21	Mousseau et al. (1998)	–	1991–1992	NW Atlantic (Scotian Shelf)
22	Nishikawa et al. (2007)	–	2000–2002	Pacific Ocean (Sulu Sea, Celebes Sea, South China Sea)
23	Pakhomov and Perissinotto (1997)	–	1993	Southern Ocean (Subtropical Convergence)
24	Pane et al. (2004)	–	1995	Antarctica (Ross Sea)
25	Fernandez de Puelles et al. (2007)	–	1994–2003	Western Mediterranean
26	Ramfos et al. (2008)	PANGEA	2000	Eastern Mediterranean
27	Rogachev et al. (2008)	–	2004	W Pacific Ocean (Academy Bay, Sea of Okhotsk)
28	Schalk (1990)	–	1984–1999	Indo-Pacific waters (E Banda Sea, W Arafura Sea)
29	Schiebel et al. (2002)		1997/1999	S of Azores Islands
30	Schnack-Schiel and Cornils (2009)	PANGEA	2005	Pacific Ocean (Java Sea)
31	Siokou-Frangou et al. (2008)	PANGEA	1987–1997	Eastern Mediterranean
32	Solis and von Westernhagen (1978)	–	1972	Philippines (Hilutangan Channel)
33	Swadling et al. (2011)	–	2004–2008	E Antarctica (Dumont d'Urville Sea)
34	Volkov (2008)	–	1984–2006	Okhotsk Sea, Bering Sea, NWP
35	Ward et al. (2007)	–	2004–2005	Southern Ocean (S&W of Georgia)
36	Wells Jr. (1973)	–	1972	N Atlantic Ocean (Barbados)
37	Werner (2005)	–	2003	Arctic (W Barents Sea)
38	Wormuth (1985)	–	1975–1977	N Atlantic Ocean (NW Sargasso Sea)
39	Zervoudaki et al. (2008)	PANGEA	1997–2000	Eastern Mediterranean
40	NMFS-COPEPOD (2011) NOAA (National Oceanic and Atmospheric Administration)	COPEPOD – The global plankton database	1953–2001	Global data set
41	ZooDB (2011), Ohman	ZooDB – Zooplankton database	1951–1999	Pacific Ocean (Southern and Central California)

was conducted in a single day (for instance through sampling both night and day or with different mesh-sized nets), a mean biomass at that station was calculated and used in subsequent processing. As the sampling methodology can in-troduce major errors in the biomass estimates, a systematic characterisation of the sampling gear, was also included to allow sources of error to be identified. In addition, all details on pteropod species composition and life stages were

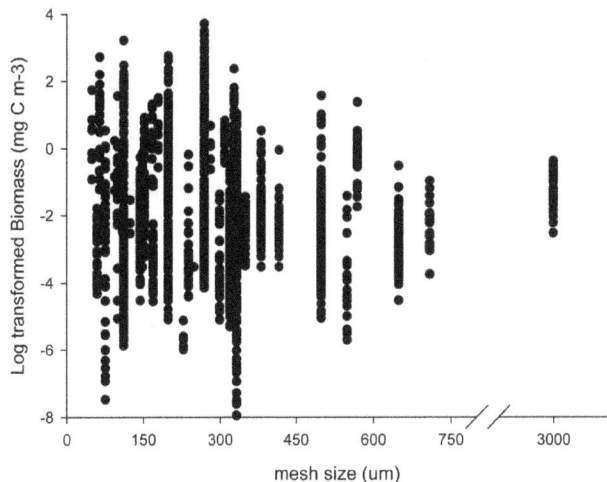

Figure 1. The relationship between net mesh size and pteropod biomass.

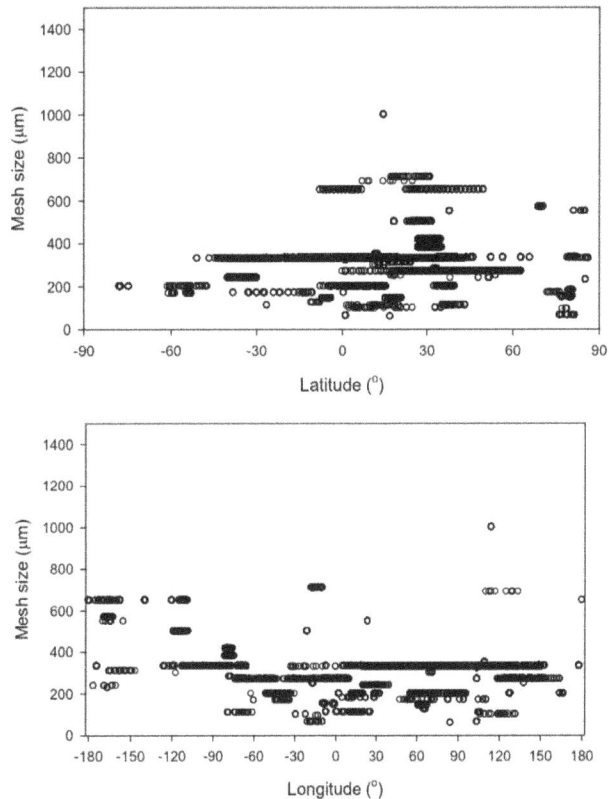

Figure 2. Net-mesh size versus longitude (above) and latitude (below). Data was excluded if multiple mesh sizes were reported.

documented within the database. Where there were a number of species identified per station, we also provided summary statistics of total pteropod biomass per station ($n = $ 14 136). The database included both Gymnosomata and Thecosomata, encompassing all genera included in the taxonomic tree, which was taken from Marine Species Identification Portal (http://species-identification.org) presented in Fig. 3. Further subspecies levels (or formae) were not resolved within the database. No observations of the suborder Pseudothecosomata were reported in the source data sets.

2.2 Quality control

The identification and rejection of statistical outliers in the summarised biomass data set was performed using Chauvenet's criterion (Glover et al., 2011; Buitenhuis et al., 2012). Based on this statistical analysis, none of the stations were excluded as outliers (two sided z-score = 4.1257).

2.3 Methodology for biomass conversion

Of the data sets obtained, the majority only reported values for abundance (ind. m^{-3}), with very few providing biomass values (mg m^{-3}). Furthermore, abundance data was collected with varying mesh sizes and net-sampling strategies, which might introduce uncertainties. Therefore, we have reported the mesh sizes and net sampling strategy in the database whenever this information was available (PANGEA Table). In certain cases, multiple mesh-sized samplers were used, of which we have included all descriptions available. No data were excluded on the basis of mesh size and we examine the influence of mesh size in the Results section (Figs. 1 and 2).

Where direct biomass values were not available, we calculated biomass as a product of abundance and dry weight (DW, mg). To estimate DW, the length (L, mm) of organisms

was first converted to wet weight (WW, mg) using various conversions (see below), with subsequent conversion to dry weight (Table 2).

For many pteropod species, specific length-to-wet weight conversions were not available so more general length-weight conversions for pteropods were applied based on those used by the GLOBal ocean ECosystems dynamics (GLOBEC) data management program. In GLOBEC, wet weights (WW) of different pteropod families were calculated based on their specific body geometry and length (Little and Copley, 2003). The GLOBEC conversions covered the barrel shaped *Clione* family of naked pteropods, the cone shaped family of *Styliola*, the low-spire (globular) family of *Limacina* spp., and the pyramidally shaped family of *Clio* spp. Accordingly, we assorted groups or species into respective geometric shapes and then applied the GLOBEC L to WW conversions. Although species-specific conversions are lacking for many of the groups (Table 2), we believe that this approach provides a reasonable first order approximation of individual biomass for the purpose of the present analysis. More specific details of these conversions are given below:

Equation (1) was used to convert all non-shelled (naked) taxa, including barrel-and oval- shaped families of *Spongiobranchia* spp., *Pneumodermopsis* and *Paedoclione* and class Gymnosomata (Little and Copley, 2003). Equation (2) was

Table 2. Length to weight equations for different pteropod groups based on the geometric shapes.

SPECIES	Group	Equation source	Conversion	Equation name	Equation (size-weight relationship)	Equation (Davis and Wiebe, 1985)	
Limacina helicina	Round/cylindrical/ globular	Bednaršek et al. (2012)	Diameter→DW		$DW = 0.137 \times D^{1.5005}$		
Limacina spp.	Round/cylindrical/ globular	GLOBEC	Diameter→DW		$WW = 0.000194 \times L^{2.5473}$	WW→DW	WW×0.28
Clione spp.	Barell/oval-shaped (naked)	GLOBEC	Length→WW	Pteropod (naked: Clione)	$WW = 10^{(2.533 \times \log(L) - 3.89095) \times 10^3}$	WW→DW	WW×0.28
Hyalocylis spp.	Cone/needle/ tube/bottle-shaped	GLOBEC	Length→WW	Pteropod (cone-shaped: Styliola)	$WW = PI \times L^{3 \times 3/25}$	WW→DW	WW×0.28
Styliola spp.	Cone/needle/ tube/bottle-shaped	GLOBEC	Length→WW	Pteropod (cone-shaped: Styliola)	$WW = PI \times L^{3 \times 3/25}$	WW→DW	WW×0.28
Spongiobranchaea spp.	Barell/oval-shaped (naked)	GLOBEC	Length→WW	Pteropod (naked: Clione)	$WW = 10^{(2.533 \times \log(L) - 3.89095) \times 10^3}$	WW→DW	WW×0.28
Pneumodermopsis and *Paedoclione*	Barell/oval-shaped (naked)	GLOBEC	Length→WW	Pteropod (naked: Clione)	$WW = 10^{(2.533 \times \log(L) - 3.89095) \times 10^3}$	WW→DW	WW×0.28
Cavolinia spp.	Triangular/ pyramidal	GLOBEC	Length→DW	Pteropod (Clio)	$WW = 0.2152 \times L^{2.293}$	WW→DW	WW×0.28
Clio spp.	Triangular/ pyramidal	GLOBEC	Length→WW	Pteropod (Clio)	$WW = 0.2152 \times L^{2.293}$	WW→DW	WW×0.28
Creseis spp.	Cone/needle/ tube/bottle-shaped	GLOBEC	Length→WW	Pteropod (cone-shaped: Styliola)	$WW = PI \times L^{3 \times 3/25}$	WW→DW	WW×0.28
Cuvierina spp.	Cone/needle/ tube/bottle-shaped	GLOBEC	Length→WW	Pteropod (cone-shaped: Styliola)	$WW = PI \times L^{3 \times 3/25}$	WW→DW	WW×0.28
Diacria spp.	Triangular/ pyramidal	GLOBEC	Length→WW	Pteropod (Clio)	$WW = 0.2152 \times L^{2.293}$	WW→DW	WW×0.28
Thecosomata	Shelled	Davis and Wiebe (1985)	Length→WW		$WW = 0.2152 \times L^{2.293}$	WW→DW	WW×0.28
Gymnosomata	Naked	Davis and Wiebe (1985)	Length→WW		$WW = 10^{(2.533 \times \log(L) - 3.89095) \times 10^3}$	WW→DW	WW×0.28
Pteropoda	Shelled	Davis and Wiebe (1985)	Length→WW		$WW = 0.2152 \times L^{2.293}$	WW→DW	WW×0.28

applied to *Clione* spp., being a genus species conversion equation originally derived by Böer et al. (2005):

$$WW = 10^{(2.533 \times \log(L) - 3.89095) \times 10^3}, \tag{1}$$

$$DW = 1.6146^{e0.0088 \times L}. \tag{2}$$

Three different shapes were distinguished within the shelled taxa, each with their own L to WW conversions:

$$WW = WW = 0.2152 \times L^{2.293} \text{ triangular/pyramidal shaped}$$
(Davis and Wiebe, 1985) $\tag{3}$

$$WW = 0.000194 \times L^{2.5473} \text{ round/cylindrical/globular shaped}$$
(Little and Copley, 2003) $\tag{4}$

$$WW = PI \times L^{3 \times 3/25} \text{ cone/needle/bottle-shaped}$$
(Little and Copley, 2003). $\tag{5}$

Limacinidae were one of the most abundant taxa within our database, for which there are several published L to DW conversions in the literature:

$$DW = 0.257 L^{2.141} \quad \text{(Gannefors et al., 2005)} \tag{6}$$

$$\log DW = 0.685 L^{-2.222} \quad \text{(Fabry, 1989)} \tag{7}$$

$$DW = 0.1365 L^{1.501} \quad \text{(Bednaršek et al., 2012).} \tag{8}$$

Gannefors et al. (2005), Fabry (1989) and Bednaršek et al. (2012) fitted the respective functions to differing size ranges of *Limacinidae*, so we compared their performance across a uniform size range to consider their suitability for more broad scale application (Appendix B, Fig. B1). The functional form of Fabry (1989), although optimal for animals in a size range between 1 and 4 mm, became exponentially large at shell diameters above this range so was considered unsuitable for the present analysis. The Gannefors et

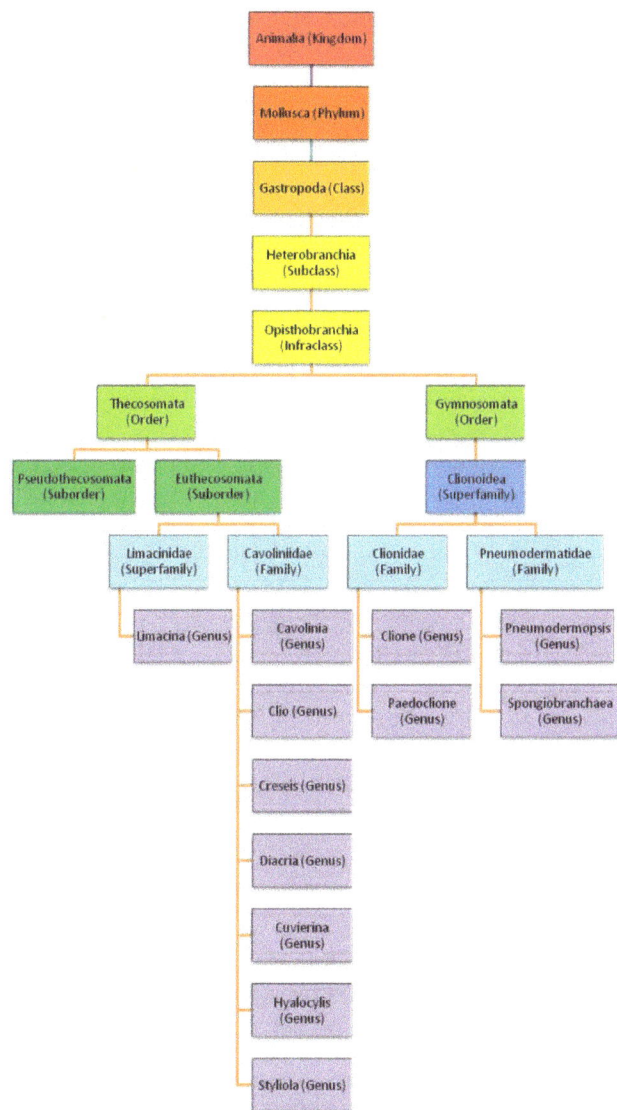

Figure 3. Taxonomy of pteropods.

al. (2005) and Bednaršek et al. (2012) functional forms performed similarly well and realistically (Appendix B) across the shell diameter size ranges encountered in the present study (0.01 to 50 mm). We chose the Bednaršek et al. (2012) function given that its estimate of dry weight between 1 and 4 mm shell diameter fell midway between the estimates of the Fabry (1989) and Gannefors et al. (2005) algorithms, combined with the fact that its behaviour remained realistic at larger size categories.

In cases, where the data-source referred to orders or classes rather than species, Eq. (3) was applied since the taxa were principally non-*Limacinidae* shelled species.

In the case of juveniles, the above length to weight conversions were used according to their respective taxa or body shape, but the length of the veligers and larvae set at 10 % of the adult average size, which is based on our own comparisons of average juvenile and adult sizes.

2.3.1 Calculation of length for the individual pteropod species

For some data records, only the species and abundance was recorded without any indication of individual size or weight. Individual shell diameter was therefore inferred in order to calculate biomass. Our first step was to determine size of adult specimens of each species using information from the Marine Species Identification Portal (http: //species-identification.org/), of which results are presented in Appendix C (Table C1), along with the body shape, length and mean size.

Where the abundance data was given for a higher taxonomic level than species (e.g. class, suborder, order), the average length across all species within that respective taxa was determined (Table C1). Because this procedure only took account of adult sizes, we were aware that this would result in an overestimation of biomass. This was compensated for in two ways: firstly, by taking into account data points where a juvenile status was indicated (283 in total, representing 2 % of entire database) in which case length was assumed to be 10 % of adult size (see above). Secondly, where the data was not species-specific (but family- or higher order-specific), the average length across all species within the taxon was calculated, so preventing extreme bias from very large or very small species.

Unfortunately, the lack of data points where both biomass and abundance values were reported made it impossible to do a quantitative comparison of the performance of our L to W conversions.

2.3.2 Calculation of dry weight and carbon biomass from wet weight

Wet weight was converted to dry weight using Davis and Wiebe (1985):

$$DW = WW \times 0.28. \tag{9}$$

Biomass was subsequently transformed to carbon using a conversion factor of 0.25, following Larson (1986).

2.3.3 Global contribution of shelled pteropods to carbonate biomass

Once conversions from abundance to carbon biomass had been completed, we considered the global biomass distribution of both shelled and non-shelled pteropod taxa. Separating out the shelled pteropod taxa allows the global carbonate distribution resulting from pteropods to be assessed, so permitting the evaluation of their contribution to the global carbonate budget. Bednaršek et al. (2012) have calculated inorganic carbon as a percentage of total organic subtracted from total carbon, deducing the PIC/POC ratio of 0.27 vs. 0.73.

Table 3. Mean, median, maximum and minimum and standard deviation (SD) of pteropod biomass ($mg\,C\,m^{-3}$) determined (i) for all global data, (ii) all non-zero data points, (iii) all non-zero Northern Hemisphere (NH) data-points and (iv) all non-zero Southern Hemisphere (SH) data-points.

summed biomass data	mean	median	max	min	SD
all global data	4.09	0.008	5.05e+003	0.00	59.06
non-zero global data	4.58	0.0145	5.05e+003	1.00e-006	62.46
for the NH non-zero data	4.04	0.0145	5.05e+003	1.00e-006	64.84
for the SH non-zero data	8.15	0.001	608.35	2.00e-006	45.36

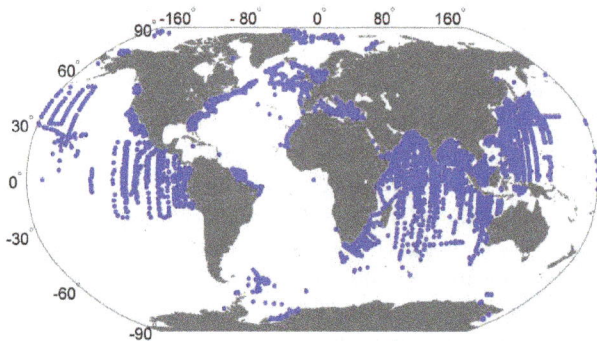

Figure 4. Global distribution of quality-controlled pteropod data.

Assuming that all inorganic carbon is in the form of calcium carbonate, the amount of calcium carbonate can be estimated as follows:

$$CaCO_3\ (\%) = [TC\ (\%) - TOC\ (\%)] \times 8.33 \qquad (10)$$

where the constant 8.33 represents the molecular mass ratio of carbon to calcium carbonate.

3 Results

3.1 Global data distribution of biomass data

Altogether, we collected 25 939 data entries across all oceanic regions, which corresponded to 15 134 samples of total pteropod biomass (Fig. 4). Out of these, 14 136 data points (93 %) represented shelled pteropods (Thecosomes), and the remaining 7 % represented non-shelled pteropods (Gymnosomes). Within the whole data set, 1608 data points (11 % of all values) were reported as zero values for all pteropod groups.

Although pteropod observations were available for all ocean basins, there was a clear bias of the data towards observations in the Northern Hemisphere (NH) (77 % of non-zero entries), with the remaining 23 % in the Southern Hemisphere (SH, Table 3 and 6, Fig. 4). With respect to latitude, the most entries (37 %) were collected within the latitudinal band of 10–60° N (Table 4).

The maximum net sampling depth was 2000 m but 83 % of all nets were sampled to a maximum depth of 200 m (Ta-

ble 5). Across all observations, 62 % of all biomass occurred within the top 200 m, with the remaining biomass (38 %) being relatively evenly distributed down to 2000 m. The deepest occurrence of pteropods in our database was 2000 m, located at 81° N,163° E. The highest biomass for shelled pteropods ($2980\,mg\,C\,m^{-3}$) was recorded at the surface in the NH temperate region, at 42° N,70° W. The highest biomass for the non-shelled pteropods ($5045\,mg\,C\,m^{-3}$) was recorded in the same region (42° N,66° W). There were very few direct measurements of pteropod biomass (see Sect. 2.3), but of those, the highest recorded values were in the Sea of Okhotsk (54° N,138° E), where biomass reached $538\,mg\,C\,m^{-3}$ (Rogachev et al., 2008).

3.2 The effect of nets and mesh sizes on global pteropod biomass

Mesh size will influence the size range of organisms captured by nets. In assembling this database, we decided to include all net-catch data, irrespective of mesh size. This will undoubtedly create error, particularly in the undersampling of smaller individuals by larger meshed nets through the lack of retention and of larger, more motile individuals by finer meshed nets through avoidance. For the purpose of the present analysis, with a focus mainly on comparative patterns, it is important that these errors do not generate bias, since this could distort any discerned geographic trends. We considered this in two ways. In Fig. 1, we compared the biomass to net mesh size across 19 671 samples. The figure shows a peak in biomass towards the mid-size meshes ($\sim 300\,\mu m$). This demonstrates that the majority of biomass lay within organisms with an equivalent spherical diameter of $300\,\mu m$ or greater, and that the undersampling of smaller organisms by some studies is unlikely to have a considerable impact on biomass estimates. Equally, the figure is indicating that the average biomass is similar, regardless of the mesh size used for sampling.

In Fig. 2, net mesh-size was compared to latitude. Although this illustrates the considerable variety of meshes used within the present database, it also shows there was no apparent bias towards certain mesh size being used at some latitudes more than others. Therefore, although the use of different meshes between studies is undoubtedly a source of

Table 4. Latitudinal distribution of abundance data in ten degree latitudinal bands (90° to 90°). Mean, maximum (max), median and standard deviation (SD) of biomass (mg C m^{-3}) per latitudinal band, calculated from non-zero data points.

Latitude	Entries	Mean (mg C m^{-3})	SD	Max (mg C m^{-3})	Min (mg C m^{-3})	Median (mg C m^{-3})
90 to 80° S	0	–	–	–	–	–
80 to 70° S	72	27.20	98.44	557.41	0.001	0.19
70 to 60° S	59	0.09	0.42	2.63	2.00e-006	0
60 to 50° S	90	13.93	35.55	168.47	0.01	0.48
50 to 40° S	90	0.25	2.27	21.53	8.00e-006	1.32e-004
40 to 30° S	127	0.02	0.07	0.64	2.83e-006	8.80-005
30 to 20° S	167	0.01	0.05	0.45	5.33e-006	2.18e-004
20 to 10° S	310	0.02	0.08	0.86	3.25e-006	6.14e-004
10° N to 0°	1007	11.93	53.98	608.35	3.50e-006	0
0° to 10° N	1078	0.06	0.26	4.30	4.67e-006	0.01
10° to 20° N	2044	1.47	8.91	226.66	1.00e-006	0.01
20° to 30° N	1725	0.06	0.49	9.85	8.00e-006	0.003
30° to 40° N	2958	4.51	21.65	362.89	1.00e-006	0.01
40° to 50° N	744	34.76	248.13	5.05e+003	2.90e-005	0.09
50° to 60° N	1960	1.26	17.26	538	0.003	0.40
60° to 70° N	896	0.31	0.46	11.82	0.003	026
70° to 80° N	77	17.31	61.97	517.05	1.75e-004	0.69
80° to 90° N	177	4.60	10.63	34.33	1.00e-006	0.01

Table 5. Depth distribution of non-zero biomass values. Mean, maximum (max), median and standard deviation (SD) of biomass (mg C m^{-3}) per depth interval, calculated from non-zero data points.

depth range (m)	entries	Mean (mg C m^{-3})	Max (mg C m^{-3})	Min (mg C m^{-3})	Median (mg C m^{-3})	SD
0–10	1806	20.65	5.45e+003	0	0.02	157.81
10–25	612	14.44	557.41	0	0.04	57.53
25–50	1296	3.25	434.37	0	0.002	18.26
50–200	7508	0.65	308.47	0	0.02	5.74
200–500	2028	0.19	9.85	0	0.002	1.04
500–2000	276	0.02	3.20	0	0.004	0.18

error, it is not a major source of bias in our analyses of geographic trends in pteropod biomass distribution.

3.3 Temporal distribution of data

Our database spans the period 1950–2010, and temporally, the data was fairly evenly distributed across all decades, with at least one sampling peak per decade. Several sampling peaks were recorded in the late 1950s, then in the 1960s–1970s, followed by high numbers of data from the early 1990s and 2000s. We recorded fewer samples in the 1980s (Fig. 6). To check for seasonal biases, the data was divided into four seasons for each hemisphere (Table 7). While in the NH, the data was distributed evenly across the four seasons (24 % in 335 spring, 23 % in summer, 24 % in autumn and 30 % in winter), sampling in the SH was biased towards winter and summer (30 % and 25 %, respectively), with much

lower coverage during the other seasons (19 and 16 % in spring and fall, respectively).

3.4 Global biomass characteristics for all pteropod groups and for shelled-pteropods only

For all pteropod groups combined, the range of global biomass concentrations was wide, spanning over four orders of magnitude (Fig. 8a), with a mean and median biomass of 4.1 mg C m^{-3} (SD = 59.1) and 0.0083 mg C m^{-3} for all data points, and 4.6 mg C m^{-3} (SD = 62.5) and 0.0145 mg C m^{-3} for non-zero biomass values, respectively. In the NH, the mean biomass was 4.0 mg C m^{-3} (SD = 64.8) and the median biomass, 0.02 mg C m^{-3}. In the SH, the mean biomass was 8.15 mg C m^{-3} (SD = 45.4) and the median biomass 0.001 mg C m^{-3} (Table 3). Although the median biomass in the SH was one order of magnitude smaller than in the NH, the mean biomass in the SH was twice that of the NH.

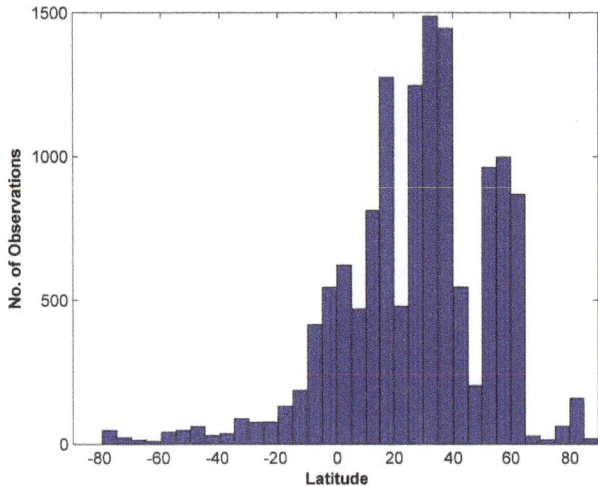

Figure 5. Number of pteropod observations as a function of latitude for the period 1950–2010.

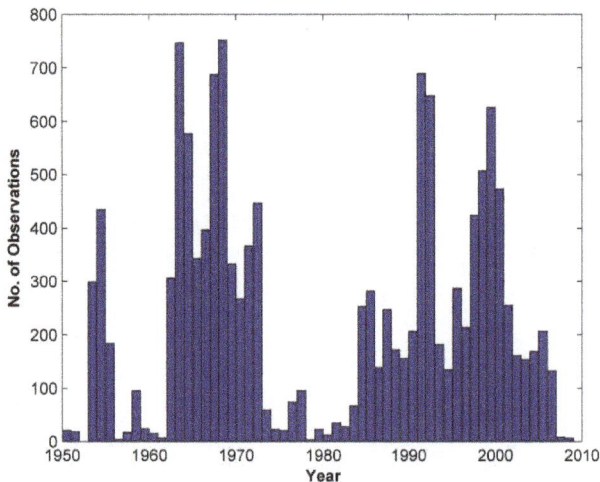

Figure 6. Number of observations per year, for the years 1950–2010.

For shelled pteropod groups only, the mean and median biomass for non-zero values was $3.81\,\mathrm{mg\,C\,m^{-3}}$ ($SD = 40.24$), and $0.0078\,\mathrm{mg\,C\,m^{-3}}$, respectively, and the maximum biomass was $2979.7\,\mathrm{mg\,C\,m^{-3}}$. Considering the mean biomass of shelled and non-shelled pteropods, shelled pteropods constitute 83 % to the total pteropod biomass, the remainder being made up of non-shelled pteropod taxa. When considered in terms of median biomass, 54 % was made up of shelled-pteropods and 46 % made up of non-shelled pteropods, indicating that the dominance of shelled-pteropods is in part due to the fact that they sometimes occur at very high concentrations.

Through assuming, firstly, an inorganic to organic carbon ratio of $0.27:0.73$ (Bednaršek et al., 2012) and secondly an inorganic carbon to calcium carbonate molecular

Table 6. Percentage distribution of non-zero data entries with respect to month for the Northern (NH) and Southern (SH) Hemispheres.

months	entries	NH season	SH season	% NH non-zero data	% SH non-zero data
January	1185	winter	summer	8.4	11.7
February	1457	winter	summer	9.4	20.7
March	998	spring	autumn	7.4	6.1
April	1298	spring	autumn	9.5	9.0
May	876	spring	autumn	6.9	3.7
June	802	summer	winter	6.4	4.1
July	1352	summer	winter	10.4	7.1
August	1790	summer	winter	13.1	13.8
September	1143	autumn	spring	8.4	9.0
October	1049	autumn	spring	8.4	3.7
November	859	autumn	spring	6.8	3.7
December	806	winter	summer	5.4	10.2

mass ratio of 8.33 (Eq. 10) gave a mean global carbonate biomass of $23.17\,\mathrm{mg\,CaCO_3\,m^{-3}}$, and a maximum biomass was $1.81\,\mathrm{g\,CaCO_3\,m^{-3}}$. These estimates were derived from non-zero biomass records only.

3.4.1 Latitudinal biomass distribution

Pteropods were found at all latitudes at which samples were taken (Figs. 5, 8a). The highest maximum, mean and median biomass values were located in the NH between 40 and 50° N (mean biomass of $5.42\,\mathrm{mg\,C\,m^{-3}}$ ($SD = 79.94$), median biomass of $0.06\,\mathrm{mg\,C\,m^{-3}}$). The highest mean and median biomass values in the SH were located between 70 and 80° S ($39.71\,\mathrm{mg\,C\,m^{-3}}$ ($SD = 93.00$) and $0.009\,\mathrm{mg\,C\,m^{-3}}$, respectively; Table 3). However, relatively high biomasses were not restricted to a particular latitude or ocean basin but were widespread, including high-latitudinal, temporal and equatorial regions in the both hemispheres. The only exception was the latitudinal band between 20 and 40° in the NH and SH, where biomass was considerably lower (Fig. 8). There was a difference in latitudinal trends between hemispheres (Fig. 9a, b), with highest biomass values in the NH being at mid-latitudes decreasing towards the equator and the poles, while, in the SH, highest biomass values were seen at the poles and steadily decreasing through the mid-latitudes towards the equator. Biomass values at both poles were within the same order of magnitude.

3.4.2 Depth distribution

Pteropods were observed at all depths down to 2000 m, although the funnel-shaped biomass pattern from the surface towards the depth indicates a sharp decrease in biomass below 200 m (Fig. 8b). The highest values were recorded at the surface (0–10 m), with a mean biomass of $20.65\,\mathrm{mg\,C\,m^{-3}}$ ($SD = 157.81$) and median biomass of $0.02\,\mathrm{mg\,C\,m^{-3}}$. Mean

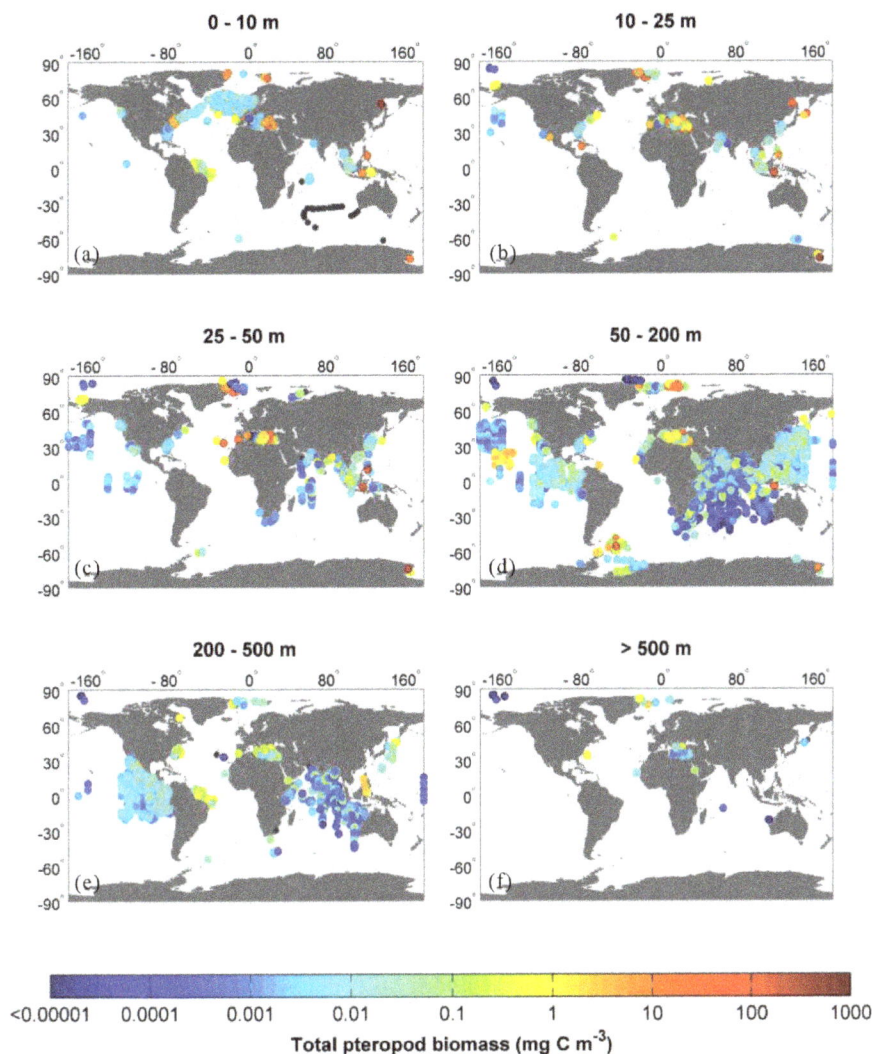

Figure 7. Pteropod carbon biomass (mg C m^{-3}) for six depth intervals: (**a**) surface (0–10 m), (**b**) 10–25 m, (**c**) 25–50 m, (**d**) 50–200 m, (**e**) 200–500 m, (**f**) ≥ 500 m.

and median biomass gradually decrease with the depth by one order of magnitude from 10 to 200 m, and by two orders of magnitude between the 10–200 m and 200–2000 m depth bands (Table 5, Fig. 8b).

The pattern of pteropod distribution demonstrates that higher abundances are closely related to continental shelves and areas of high productivity or nutrient loads (Fig. 7). This can be particularly exemplified in the eastern North Pacific central water, which is a rather small area affected by the inflow from the more productive transitional and equatorial adjacent areas (Longhurst, 2007), with a three to four magnitude higher biomass, in comparison to the surrounding areas.

In all ocean basins, biomass levels above 100 mg C m^{-3} only occurred in the 0–200 m depth layers. However, in tropical regions, some of the highest biomass levels were found in the 200–500 m depth strata, where concentrations typically reached between 1 and 10 mg C m^{-3} (Fig. 7). This sug-

gests that tropical species concentrate at deeper depths than temperate and high-latitude species. Such geographic patterns in the depth distribution of pteropods have previously been noted by Solis and von Westernhagen (1978), Wormuth (1981) and Almogi-Labin et al. (1998).

3.4.3 Seasonal distribution of pteropod biomass

Seasonal variations in biomass values were much more extreme in the SH compared to the NH, although it is to be noted that sample coverage was comparatively greater in the NH (Table 7, Fig. 9). In both hemispheres, mean biomass peaked in the spring. However, the peak was an order of magnitude higher in the SH compared to the NH (Table 7). The ratio between spring and winter biomass was approximately 2 : 1 in the NH, but around 1300 : 1 in the SH. The difference in ratios is mainly explained by the virtual disappearance of

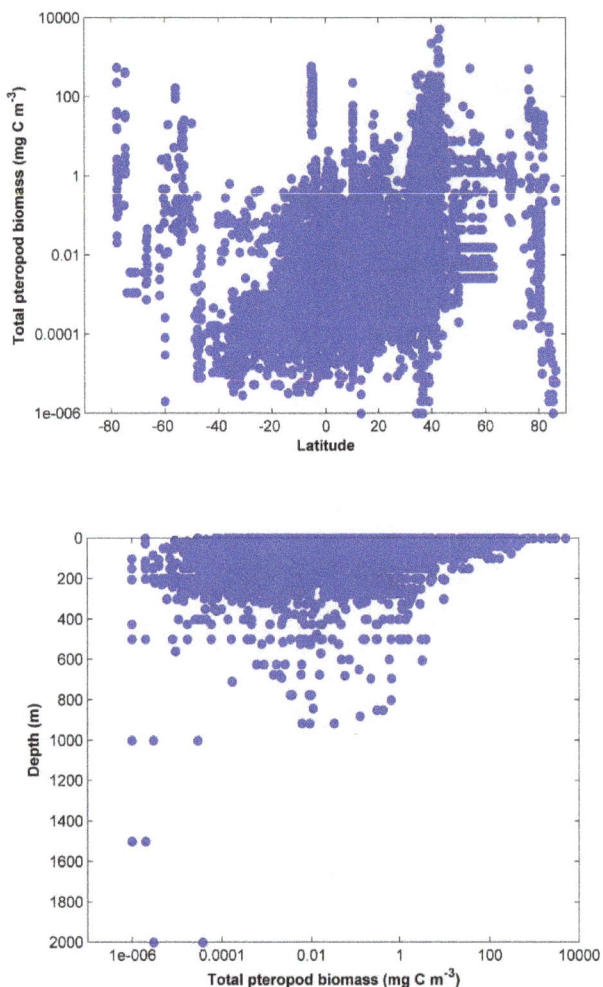

Figure 8. (Above) the distribution of pteropod biomass $(mg\,C\,m^{-3})$ as a function of latitude; (below) the relationship between pteropod biomass and net-capture depth.

pteropods in the SH during winter. Biomass levels were relatively similar between the NH and SH during summer and autumn. The seasonal peaks and troughs in mean biomass in both hemispheres correspond to a life-history pattern of spring spawning, probably in response to seasonal pulses of productivity, as described by Hunt et al. (2008) and Bednaršek et al. (2012).

Despite the seasonal peaks and troughs in biomass, a residual biomass level was always present (Fig. 9). This indicates that there must be a degree of overlap in generations (Bednaršek et al., 2012). In the higher latitudes, where there is likely just a single recruitment event per year, meaning that these pteropods must have a life-cycle that extends into a second year. In the Southern Ocean, Bednaršek et al. (2012) proposed that some *Limacina helicina ant.* lived for more than 2 yr and, although small in number, these individuals may be vital for future recruitment. Strong seasonality increases the vulnerability of early life-stages of pteropods that rely on

pulses of production to thrive (Bernard and Froneman, 2009; Seibel and Dierssen, 2003). An overlap of generations gives populations greater stability in temporally variable environments.

3.4.4 Global estimates of the pteropod biomass stock and productivity

Given representative data coverage at the both hemispheres, global mean pteropod biomass of $0.0046\,g\,C\,m^{-3}$ (SD = 62.5) was calculated for any point of time. To extrapolate from regional to global pteropod biomass, pteropod depth distribution and absolute area of the global ocean are required. With regards to depth distribution, Fig. 8a is indicative of pteropod biomass to be uniformly distributed within the upper 300 m, and two orders of magnitude less abundant below 300 m. The 300-m depth level was hence taken as a conservative estimate of their overall occurrence. Considering the absolute surface area of the global deep ocean (Milliman and Droxler, 1996; total area equals $362.03 \times 10^6\,km^2$ cf. Dietrich et al., 1975), two values were taken to determine global pteropod biomass: the global ocean surface excluding shelf seas $(322 \times 10^6\,km^2)$ was taken as a minimum area inhibited by pteropods, while the total ocean surface area was determined as a maximum $(362.03 \times 10^6\,km^2)$. Considering minimum and maximum area inhabited by pteropods, global pteropod biomass ranges from 444 to 505 Tg of C at any point in time. This range of estimates, based on the observational results is similar to pteropod productivity estimate of $0.87\,Pg\,C\,yr^{-1}$ obtained through modelling work by Gangstø et al. (2008). Lebrato et al. (2010) estimated global carbon productivity budget to range between 0.96 and $2.56\,Pg\,C\,yr^{-1}$. This indicates that pteropods contribute 20–42 % towards global carbonate budget.

The average turnover time is known to be different for various species, shorter (several months) for tropical species and longer (more than one year) for the high-latitudinal species (Lalli and Gilmer, 1989). Here, as reported in several papers (Van der Spoel, 1973; Wells Jr., 1976; Hunt et al., 2008; etc.), the average pteropods turnover time was assumed to be one year, with high latitudinal species to be exceptions (e.g. Bednaršek et al., 2012) and recorded the life cycle of *Limacina helicina antarctica* to span over 3 yr. At a global scale, and an average annual distribution, the entire pteropod production would hence amount to $444–505\,Tg\,C\,yr^{-1}$, which is about five times the estimated planktic foraminifers biomass production (Schiebel and Movellan, 2012: $25–100\,Tg\,C\,yr^{-1}$), more than double of the estimated diazotroph biomass (Luo et al., 2012: $40–200\,Tg\,C$), and around one fifth of the total diatom production (Leblanc et al., 2012: $500–3000\,Tg\,C$). Comparing global pteropod to coccolithophorid carbon productivity (Balch et al., 2007), coccolithophorid production are approximately 1.5 to 3 times higher than our estimated pteropod production.

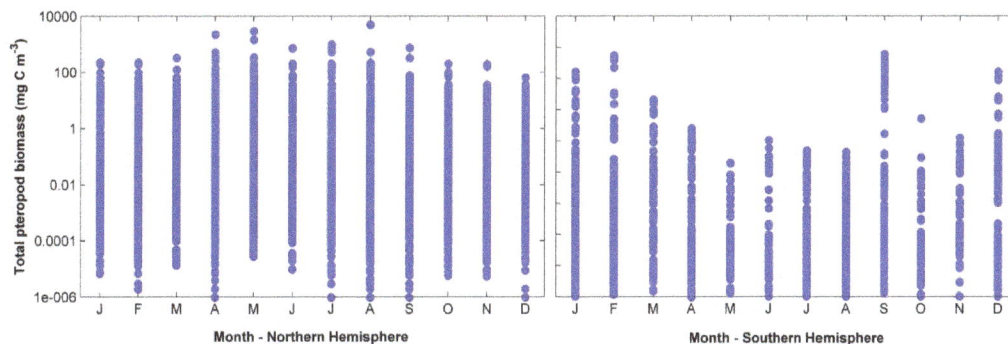

Figure 9. Distribution of pteropod biomass values (mg C m^{-3}) with respect to month, in the Northern Hemisphere (left) and Southern Hemisphere (right).

Table 7. Biomass (mg C m^{-3}) with respect to season for the Northern (NH) and Southern (SH) Hemispheres, showing the calculated mean, standard deviation (SD), median, minimum (min) and maximum (max). Biomass statistics are based on non-zero data entries only.

	NH mean	NH SD	NH median	NH min	NH max	SH mean	SH SD	SH median	SH min	SH max
winter	2.77	15.63	0.02	1e-006	557.41	0.03	0.09	4.54e-004	2.00e-006	1.06
spring	5.42	79.94	0.06	1e-006	3.0e+003	39.71	93.00	0.009	7.50e-006	608.35
summer	4.32	92.69	0.02	1e-006	5.05e+003	3.73	32.83	0.002	3.00e-006	557.41
autumn	2.44	18.39	0.03	1e-006	765.24	0.51	2.47	7.28e-004	3.30e-006	21.05

4 Discussion and conclusions

The aim of this study was to collect and synthesise available existing abundance and biomass data to generate the first global pteropod biomass database. Most studies reported abundance rather than biomass data, making it necessary to estimate carbon biomass using length to weight conversions and introducing levels of uncertainty as a result. Further uncertainties in the biomass estimates in this study will result from sampling errors such as net-escapement and net-avoidance, the variation in size classes between different pteropod species and generations. Further considerations around these uncertainties are discussed below.

With regards to the sampling error, the use of different nets for different pteropod size classes generates uncertainty, as the capture and filtering efficiencies differ between nets. Furthermore, sampling issues such as net-avoidance behaviour, extrusion of animals through mesh and clogging of the net (Harris et al., 2000) will influence abundance measurements. In addition, there is generally an insufficient use of smaller meshed nets to estimate population size. Wells Jr. (1973) proposed that there was a clear underestimation of the fraction of the pteropod population smaller than 100 μm. As they constitute by far the most numerous part of the natural population (Fabry, 1989), there is a clear under-representation of this cohort in the scientific literature and thus of their importance within the microzooplankton community (Dadon and Masello, 1999). When sampling with small vertical nets, which preferentially catch small or sluggish taxa, additional

sampling errors arise from the fact that the nets can be avoided by larger plankton. On the other hand, nets with larger mesh size can miss the mesozooplankton size fractions including pteropods (Boysen-Ennen et al., 1991). We tried to address potential biases through systematic examination of mesh sizes, net types and sampling strategies (wherever available in the literature) relative to biomass estimates. Our analyses indicated, firstly, that most biomass lay within the mid-size ranges, meaning that the undersampling of smaller organisms by some studies is unlikely to have a large impact on biomass estimates. Secondly, there was no geographic bias in the use of different nets and meshes, indicating that sampling error is unlikely to bias analyses of geographic trends in biomass. Overall, we conclude that the documented variation in mesh size between studies included within the database was not a source of a large-scale bias within global biomass patterns. Therefore, although users of the database must be vigilant with regards to this potential source of error, we believe that the inclusion of all data, irrespective of the mesh size and sampling strategy used, maximises the potential insights that can be gained from this database.

There were a number of sources of uncertainty in deriving biomass values from the majority of studies within the database that only provided abundance data. To convert from abundance to biomass requires knowledge of the length distribution of specimens but neither this data, nor the respective life-stages of specimens were commonly reported. Where such information was not given, we assumed that all specimens were adults and used literature based estimated of body length. This approach probably resulted in an overestimation of biomass, given that at least part of the sampled population may have been smaller juvenile stages. Furthermore, where sizes were reported, there was often a lack of further statistical descriptors such as minimum or maximum length, so preventing levels of variance in biomass to be estimated. For some species, there was no available length to weight conversions and so more generic algorithms were applied based on the shape and morphological features (shelled/non-shelled) of the organisms, following the approach of GLOBEC (Little and Copley, 2003). This approach no doubt introduced further errors although there is little alternative to the use of such generic functions until a more systematic documentation of the length and weight characteristics of a wider range of pteropod species is undertaken.

The seasonal spread of sampling was much more even in the NH compared to the SH. Whereas we were able to document how patterns of biomass shifted geographically between seasons in the NH, our ability to achieve this was far more constrained in the SH. In particular, sampling in winter and spring was particularly sparse in the SH. It is important that future sampling efforts in that hemisphere concentrate on these less sampled times of year.

This study has enabled estimates of global pteropod biomass across a number of spatial and temporal scales. Furthermore, it has revealed some global patterns of pteropod biomass, only possible due to the wealth of data available in our data sets. Also, calculating the biomass of shelled pteropods only, we have estimated the contribution of this group to the global carbonate inventory. This database has the potential be a valuable tool for future modelling work, both of ecosystem processes and for the study of global biogeochemical cycles, since pteropods are a major contributor to organic and inorganic carbon fluxes. It can also make a timely contribution to the assessment of the effects of ocean acidification, particularly in terms of the vulnerability of calcifying species, since it provides a benchmark against which model projections and future sampling efforts can be compared.

Appendix A

A1 Available dataset at PANGEA

A full data set containing all abundance/biomass data points can be downloaded from the data archive PANGAEA, The data file contains longitude, latitude, sampling depth (m), date (Year, Month, Day in ISO format), taxon/species/body size, abundance (ind. m^{-3}), biomass (C mg m^{-3}), mesh size (μm), sampling strategy and full data reference list (doi/journal/database) doi:10.1594/PANGAEA.777387.

A2 Gridded NetCDF biomass product

The biomass data has been gridded onto a $360 \times 180°$ grid, with a vertical resolution of 33 WOA depth levels. Data has been converted to NetCDF format for easy use in model evaluation exercises. The NetCDF file can be downloaded from PANGAEA (doi:10.1594/PANGAEA.777387). It contains data on longitude, latitude, sampling depth (m), month, abundance (ind. m^{-3}) and biomass (mg C m^{-3}).

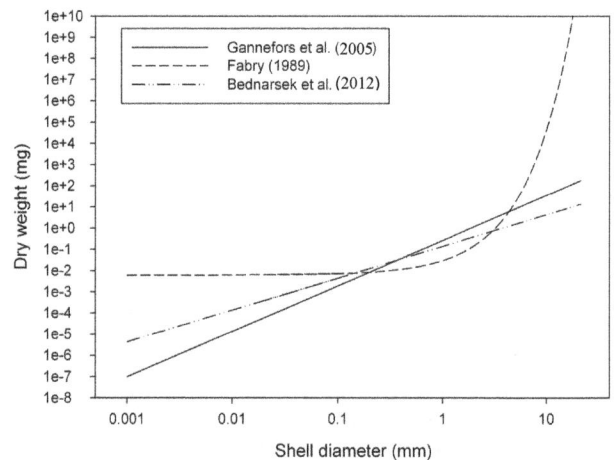

Figure B1. Shell diameter to dry weight relationships for *Limacina helicina* derived by three different studies.

Table C1. Body dimensions and shapes of a range of shelled and non-shelled pteropod species (source: Marine identification portal (http: //species-identification.org/), except for *Clione limacina** – Böer et al., 2005).

Order	Suborder	Taxon	Subspecies/ Formae	Mean shell length (mm)	Mean shell width (mm)	Body length (mm)	Shell/body shape	Additional information	Group
Thecosomata	Euthecosomata	*Limacina helicina*	*helicina helicina*	6	8		round	left coiled shell, moderately highly spired, aperture higher than wide, height/diameter ration=0.75	Round/cylindrical/ globular
Thecosomata	Euthecosomata	*Limacina helicina*	*helicina pacifica*	5	2				Round/cylindrical/ globular
Thecosomata	Euthecosomata	*Limacina retroversa*	*retroversa*	2.5	2.6		round	small, left coiled shell, no umbilical keel, spire moderately highly coiled	Round/cylindrical/ globular
Thecosomata	Euthecosomata	*Limacina bulimoides*		2	1.4		round	highly coiled spire	Round/cylindrical/ globular
Thecosomata	Euthecosomata	*Limacina inflata*			1.3		round	coiled nearly in one level; average shell diameter =0.86, aperture length=0.68 mm, diameter of operculum=0.31 mm, aperture breadth=0.5 mm	Round/cylindrical/ globular
Thecosomata	Euthecosomata	*Limacina helicina*	*antarctica*		5		round	left coiled, spire variable	Round/cylindrical/ globular
Thecosomata	Euthecosomata	*Limacina helicina*	*antarctica antarctica rangii*	2	3.5				Round/cylindrical/ globular
Thecosomata	Euthecosomata	*Limacina trochiformis*		1	0.8		round	left coiled, apical angle 75–96°	Round/cylindrical/ globular
Thecosomata	Euthecosomata	*Limacina helicina* spp. *average*			4.22				Round/cylindrical/ globular
Thecosomata	Euthecosomata	*Limacina trochiformis*		1	0.8		round	left coiled, apical angle 75–96°	Round/cylindrical/ globular
Thecosomata	Euthecosomata	*Limacina lesueuri*		0.8	1		round	flatly left coiled, spire depressed; max diameter of operculum = 0.6 mm and length/width ratio=2/3	Round/cylindrical/ globular
Thecosomata	Euthecosomata	*Limacina* spp.			2.98			the length calculated as the average of all species	Round/cylindrical/ globular
Gymnosomata		*Clione limacina*	*limacina antarctica*		25	Up to 40	barrel	body pointed posteriorly	Barrel/oval-shaped (naked)
Gymnosomata		*Clione limacina*	*limacina meridionalis*		21	20	barrel	Cone elongated	Barrel/oval-shaped (naked)
Gymnosomata		*Clione limacina**			12				Barrel/oval-shaped (naked)
Gymnosomata		*Clione limacina larvae*			0.3				Barrel/oval-shaped (naked)
Gymnosomata		*Clione* spp.			14.57			the length calculated as the average of all species	Barrel/oval-shaped (naked)
Thecosomata	Euthecosomata	*Hyalocylis striata*		8		up to 8	cylindrical	uncoiled, cross-section round, shell curved faintly dorsally; rear angle of adult shell 24°	Cone-shaped (needle/tube/bottle)

Table C1. Continued.

Order	Suborder	Taxon	Subspecies/ Formae	Mean shell length (mm)	Mean shell width (mm)	Body length (mm)	Shell/body shape	Additional information	Group
Thecosomata	Euthecosomata	*Styliola subula*		13		13	needle-like	shell is (conical), uncoiled, the cross-section is round, long, tubular, not curved; rear angle of shell is 11°	Cone-shaped (needle/tube/bottle)
Gymnosomata		*Spongiobranchaea australis*		20		max 22	oval	long body	Barrel/oval-shaped (naked)
Gymnosomata		*Spongiobrachaea australis juv.*		10					Barrel/oval-shaped (naked)
Gymnosomata		*Spongiobranchaea* spp.		15					Barrel/oval-shaped (naked)
Gymnosomata		*Pneumodermopsis*	teschi			up to 9.1	barrel		Barrel/oval-shaped (naked)
Gymnosomata		*Pneumodermopsis*	pulex			up to 8	barrel		Barrel/oval-shaped (naked)
Gymnosomata		*Pneumodermopsis*	macrochira			up to 2	barrel		Barrel/oval-shaped (naked)
Gymnosomata		*Pneumodermopsis*	ciliata			up to 15	barrel	slender body	Barrel/oval-shaped (naked)
Gymnosomata		*Pneumodermopsis*	spoeli			up to 3 (2.6)	barrel	body rounded then contracted	Barrel/oval-shaped (naked)
Gymnosomata		*Pneumodermopsis*	simplex			up to 5 (4.5)	barrel		Barrel/oval-shaped (naked)
Gymnosomata		*Pneumodermopsis*	paucidens			up to 5	barrel		Barrel/oval-shaped (naked)
Gymnosomata		*Pneumodermopsis*	canephora			up to 12	barrel		Barrel/oval-shaped (naked)
Gymnosomata		*Pneumodermopsis*	polycotyla			up to 5	barrel		Barrel/oval-shaped (naked)
Gymnosomata		*Pneumodermopsis* spp.				6.5		the length calculated as the average of all species	Barrel/oval-shaped (naked)
Gymnosomata		*Paedocline*	doliiformis	1.5				elongate oval to cylindrical shape	Barrel/oval-shaped (naked
Thecosomata	Euthecosomata	*Cavolinia globulosa*		6	4.5		globular		Triangular/pyramidal
Thecosomata	Euthecosomata	*Cavolinia inflexa*	inflexa	7	5	6	triangular		Triangular/pyramidal
Thecosomata	Euthecosomata	*Cavolinia inflexa*	imitans	8			triangular		Triangular/pyramidal
Thecosomata	Euthecosomata	*Cavolinia inflexa*	labiata	8	5.5		triangular		Triangular/pyramidal
Thecosomata	Euthecosomata	*Cavolinia longirostris*	f. longirostris	6.2	6.8–4.9	7	triangular	accepted name *Dicavolinia longirostris*	Triangular/pyramidal
Thecosomata	Euthecosomata	*Cavolinia longirostris*	f. angulosa	3.9	3.7–2.3	5	triangular	accepted name *Dicavolinia longirostris*	Triangular/pyramidal
Thecosomata	Euthecosomata	*Cavolinia longirostris*	f. strangulata	4	4.1–2.7	5	triangular	accepted name *Dicavolinia longirostris*	Triangular/pyramidal
Thecosomata	Euthecosomata	*Cavolinia uncinata*	uncinata uncinata	6.5	4.0–6.6	8	triangular	uncoiled shell	Triangular/pyramidal
Thecosomata	Euthecosomata	*Cavolinia uncinata*	uncinata f. pulsatapusilla	6.1	9.5		triangular		Triangular/pyramidal
Thecosomata	Euthecosomata	*Cavolinia* spp.		6.2				the length calculated as the average of all species	Triangular/pyramidal
Thecosomata	Euthecosomata	*Clio convexa*		8	4.5	up to 8	pyramidal	shell uncoiled	Triangular/pyramidal

Table C1. Continued.

Order	Suborder	Taxon	Subspecies/ Formae	Mean shell length (mm)	Mean shell width (mm)	Body length (mm)	Shell/body shape	Additional information	Group
Thecosomata	Euthecosomata	*Clio cuspidata*		20	30	up to 20	pyramidal	shell uncoiled	Triangular/pyramidal
Thecosomata	Euthecosomata	*Clio piatkowskii*		13.5	16	14	broad pyramidal		Triangular/pyramidal
Thecosomata	Euthecosomata	*Clio pyramidata*		20	10		pyramidal		Triangular/pyramidal
Thecosomata	Euthecosomata	*Clio pyramidata*	*martensi*	17					Triangular/pyramidal
Thecosomata	Euthecosomata	*Clio pyramidata*	*antarctica*	17					Triangular/pyramidal
Thecosomata	Euthecosomata	*Clio pyramidata*	*lanceolata*	20					Triangular/pyramidal
Thecosomata	Euthecosomata	*Clio pyramidata* spp.		18.5					Triangular/pyramidal
Thecosomata	Euthecosomata	*Clio* spp.		16.5				the length calculated as the average of all species	Triangular/pyramidal
Thecosomata	Euthecosomata	*Creseis acicula*	*acicula*	33	1.5		tube	shell is not curved, cross-section circular, extremely long and narrow, aperture rounded, rear angle of shell 13–14°	Cone-shaped (+needle/tube/bottle)
Thecosomata	Euthecosomata	*Creseis acicula*	*clava*	6					Cone-shaped (+needle/tube/bottle)
Thecosomata	Euthecosomata	*Creseis acicula* spp.		19.5					Cone-shaped (+needle/tube/bottle)
Thecosomata	Euthecosomata	*Creseis virgula*	*virgula*	6	max 2	6	tube	shell is curved (distinctly curved dorsally), uncoiled, long and narrow	Cone-shaped (+needle/tube/bottle)
Thecosomata	Euthecosomata	*Creseis virgula*	*conica*	7	aperture-diameter =1 mm	up to 7	tube	shell curved and slender, cross-section is round	Cone-shaped (+needle/tube/bottle)
Thecosomata	Euthecosomata	*Creseis virgula*	*constricta*	3.5	0.4	4	tube	uncoiled shell, cross-section round, short and narrow, slightly curved	Cone-shaped (+needle/tube/bottle)
Thecosomata	Euthecosomata	*Creseis virgula* spp.		5.5	0.2		tube		Cone-shaped (+needle/tube/bottle)
Thecosomata	Euthecosomata	*Creseis* spp.		11.5				the length calculated as the average of all species	Cone-shaped (+needle/tube/bottle)
Thecosomata	Euthecosomata	*Cuvierina columnella*	*columnella*	10	3	up to 10	bottle-shaped	the greatest shell width is found at less than 173 of the shell length from posterior	Cone-shaped (+needle/tube/bottle)
Thecosomata	Euthecosomata	*Diacria costata*		2.3	1.7–2.2	3	globular	shell uncoiled	Triangular/pyramidal
Thecosomata	Euthecosomata	*Diacria danae*		1.7	1.1–1.7	2	globular	shell uncoiled	Triangular/pyramidal
Thecosomata	Euthecosomata	*Diacria quadridentata*		3	1.8–2.5	2	globular	shell uncoiled	Triangular/pyramidal

Table C1. Continued.

Order	Suborder	Taxon	Subspecies/Formae	Mean shell length (mm)	Mean shell width (mm)	Body length (mm)	Shell/body shape	Additional information	Group
Thecosomata	Euthecosomata	*Diacria rampali*		9.5	9	9	cone-shaped	bilateral symmetrical, uncoiled shell, slender, long caudal spine; spine mark width=0.95 mm, aperture height=0.95.	Triangular/pyramidal
Thecosomata	Euthecosomata	*Diacria trispinosa*	*trispinosa*	8	10	1	cone-shaped	bilateral symmetrical, uncoiled shell, long caudal spine; the ration upperlip-spine tip/spine tip-membrane=1.3, spine mark width=1.5 mm, aperture height=0.9 mm.	Triangular/pyramidal
Thecosomata	Euthecosomata	*Diacria major*		10.7	11			uncoiled bilateral symmetrical, long caudal spine; ratio upperlip-spine tip/spine-tip membrane=1.65 mm, spine mark width=1.2 mm, aperture height=1 mm;	Triangular/pyramidal
Thecosomata	Euthecosomata	*Diacria* spp.		5.9				the length calculated as the average of all species	Triangular/pyramidal
THECOSOMATA COMBINED				8.1				shelled	
GYMNOSOMATA COMBINED						12.0		naked	
PTEROPODA COMBINED				8.9				shelled	

Acknowledgements. We thank K. Blachowiak-Samolyk, E. Boysen-Ennen, E. A. Broughton, C. Clarke, M. Daase, V. G. Dvoretsky, T. D. Elliot, H. Flores, B. A. Foster, P. W. Froneman, B. P. V. Hunt, L. Mousseau, J. Nishikawa, M. D. Ohman, T. O'Brien, E. A. Pakhomov, L. Pane, M. Fernandez de Puelles, K. A. Rogachev, P. H. Schalk, N. Solis, K. M. Swadling, A. F. Volkov, P. Ward, I. Werner, J. H. Wormuth for a permission to use and republish their data on pteropods in the MAREDAT project and this paper.

The lead author is grateful to R. A. Feely (NOAA) and R. Schiebel (Université d'Angers-BIAF) for their invaluable advice and ideas. Thanks also go to Marko Vuckovic from the University of Nova Gorica for the help with the Matlab software changes. The work at ETH for NB was partly funded through an Erasmus scholarship. GT was supported by the Ecosystems core research programme at the British Antarcttic Survey. The research leading to these results has received funding from the European Community's Seventh Framework Programme (FP7 2007–2013) under grant agreement no. 238366.

Edited by: W. Smith

References

Almogi-Labin, A., Hemleben, C., and Meischner, D.: Carbonate preservation and climatic changes in the central Red Sea during the last 380 kyr as recorded by pteropods, Marine Micropaleontology, 33, 87–107, doi:10.1016/S0377-8398(97)00034-0, 1998.

Acker, J. G. and Byrne, R. H.: The influence of surface state and saturation state on the dissolution kinetics of biogenic aragonite in seawater, Am. J. Sci., 289, 1098–1116, 1989.

Andersen, V., Sardou, J., and Gasser, B.: Macroplankton and micronekton in the northeast tropical Atlantic: abundance, community composition and vertical distribution in relation to different trophic environments, Deep-Sea Res. Pt. I, 44, 193–222, doi:10.1016/S0967-0637(96)00109-4, 1997.

Bednaršek, N., Tarling, G., Fielding, S., and Bakker, D.: Population dynamics and biogeochemical significance of *Limacina helicina antartica* in the Scotia Sea (Southern Ocean), Deep-Sea Res. Pt. II, 59–60, 105–116, doi:10.1016/j.dsr2.2011.08.003, 2012.

Bernard, K. S. and Froneman, P. W.: Trophodynamics of selected mesozooplankton in the west-Indian sector of the Polar Frontal Zone, Southern Ocean, Polar Biol., 28, 594–606, doi:10.1007/s00300-005-0728-3, 2005.

Bernard, K. S. and Froneman, P. W.: The sub-Antarctic euthecosome pteropod, *Limacina retroversa*: Distribution patterns and trophic role, Deep-Sea Res. Pt. I, 56, 582–598, doi:10.1016/j.dsr.2008.11.007, 2009.

Berner, R. A. and Honjo, S.: Pelagic sedimentation of aragonite: its geochemical significance, Science, 211, 940–942, 1981.

Blachowiak-Samolyk, K., Søreide, J. E., Kwasniewski, S., Sundfjord, A., Hop, H., Falk-Petersen, S., and Hegseth, E. N.: Hydrodynamic control of mesozooplankton abundance and biomass in northern Svalbard waters (79–81° N), Deep-Sea Res. Pt. II, 55, 2210–2224, 2008.

Böer, M., Gannefors, C., Kattner, G., Graeve, M., Hop, H., and Falk-Petersen, S.: The Arctic pteropod *Clione limacina*: seasonal lipid dynamics and life-strategy, Mar. Biol., 147, 707–717, 2005.

Boysen-Ennen, E., Hagen, W., Hubold, G., and Piatkowski, U.: Zooplankton biomass in the ice-covered Weddell Sea, Antarctica, Mar. Biol., 111, 227–235, 1991.

Broughton, E. A. and Lough, R. G.: A direct comparison of MOCNESS and Video Plankton Recorder zooplankton abundance estimates: Possible application for augmenting net sampling with video systems, Deep-Sea Res. Pt. II, 53, 2789–2807, doi:10.1016/j.dsr2.2006.08.013, 2006.

Buitenhuis, E. T., Vogt, M., Moriarty, R., Bednaršek, N., Doney, S. C., Leblanc, K., Le Quéré, C., Luo, Y.-W., O'Brien, C., O'Brien, T., Peloquin, J., Schiebel, R., and Swan, C.: MAREDAT: towards a World Ocean Atlas of MARine Ecosystem DATa, Earth Syst. Sci. Data Discuss., 5, 1077–1106, doi:10.5194/essdd-5-1077-2012, 2012.

Clarke, C. and Roff, J. C.: Abundance and Biomass of Herbivorous Zooplankton off Kingston, Jamaica, with Estimates of their Annual Production, Estuar. Coast. Shelf S., 31, 423–437, 1990.

Daase, M. and Eiane, K.: Mesozooplankton distribution in northern Svalbard waters in relation to hydrography, Polar Biol., 30, 969–981, doi:10.1007/s00300-007-0255-5, 2007.

Dadon, J. R. and Masello, J. F.: Mechanisms generating and maintaining the admixture of zooplanktonic molluscs (Euthecosomata: Opistobranchiata: Gastropoda) in the Subtropical Front of the South Atlantic, Mar. Biol., 135, 171–179, 1999.

Davis, C. S. and Wiebe, P. H.: Macrozooplankton Biomass in a Warm-Core Gulf Stream Ring: Time Series Changes in Size Structure, Taxonomic Composition, and Vertical Distribution, J. Geophys. Res., 90, 8871–8884, 1985.

Dietrich, G., Kalle, K., Krauss, W., and Siedler, G.: Allgemeine Meereskunde, 3, Auflage, Gebruder Borntrager, Berlin, Stuttgart, 593 pp., 1975.

Dvoretsky, V. G. and Dvoretsky, A. G.: Summer mesozooplankton distribution near Novaya Zemlya (eastern Barents Sea), Polar Biol., 32, 719–731, doi:10.1007/s00300-008-0576-z, 2009.

Elliot, T. D., Tang, K. W., and Shields, A. R.: Mesozooplankton beneath the summer sea ice in McMurdo Sound, Antarctica: abundance, species composition and DMSP content, Polar Biol., 32, 113–122, doi:10.1007/s00300-008-0511-3, 2009.

Fabry, V. J.: Aragonite production by pteropod molluscs in the subarctic Pacific, Deep-Sea Res., 36, 1735–1751, doi:10.1016/0198-0149(89)90069-1, 1989.

Fernandez de Puelles, M. L., Alemany, F., and Jansa, J.: Zooplankton time-series in the Balearic Sea (Western Mediterranean): Variability during the decade 1994–2003, Prog. Oceanogr., 74, 329–354, doi:10.1016/j.pocean.2007.04.009, 2007.

Flores, H., van Franeker, J.-A., Cisewski, B., Leach, H., van de Putte, A. P., Meesters, E. H. W. G., Bathmann, U., and Wolff, W. J.: Macrofauna under sea ice and in the open surface layer of the Lazarev Sea, Southern Ocean, Deep-Sea Res. Pt. II, 58, 1948–1961, doi:10.1016/j.dsr2.2011.01.010, 2011.

Foster, B. A.: Composition and Abundance of Zooplankton Under the Spring Sea Ice of McMurdo Sound, Antarctica, Polar Biol., 8, 41–48, 1987.

Froneman, P. W., Pakhomov, E. A., and Treasure, A.: Trophic importance of the hyperiid amphipod, *Themisto gaudichaudi*, in the Prince Edward Archipelago (Southern Ocean), Polar Biol., 23, 429–436, 2000.

Gannefors, C., Böer, M., Kattner, G., Graeve, M., Eiane, K., Gulliksen, B., Hop, H., and Falk-Petersen, S.: The Arctic sea butterfly *Limacina helicina*: lipids and life strategy, Mar. Biol., 147, 169–177, doi:10.1007/s00227-004-1544-y, 2005.

Gangstø, R., Gehlen, M., Schneider, B., Bopp, L., Aumont, O., and Joos, F.: Modeling the marine aragonite cycle: changes under rising carbon dioxide and its role in shallow water $CaCO_3$ dissolution, Biogeosciences, 5, 1057–1072, doi:10.5194/bg-5-1057-2008, 2008.

Gilmer, R. W.: In situ observations of feeding in thecosomatous pteropod molluscs, Am. Malacol. Bull., 8, 53–59, 1990.

Gilmer, R. W. and Harbison, G. R.: Diet of *Limacina helicina* (Gastropoda: Thecosomata) in Arctic waters in midsummer, Mar. Ecol.-Prog. Ser., 77, 125–134, 1991.

Glover, D. M., Jenkinds, W. J., and Doney, S. C.: Modelling Methods for Marine Science, Cambridge University Press, Cambridge, UK, ISBN 978-0-521-86783-2, 2011.

Harbison, G. R. and Gilmer, R. W.: Swimming, buoyancy and feeding in shelled pteropods: a comparison of field and laboratory observations, J. Mollus. Stud., 58, 337–339, doi:10.1093/mollus/58.3.337, 1992.

Harris, R. P., Wiebe, P. H., Lenz, J., Skjoldal, H.-R., and Huntley, M.: Zooplankton methodology manual, Elsevier Academic Press, London, UK, 684 pp., 2000.

Hopkins, T. L.: Midwater food web in McMurdo Sound, Ross Sea, Antarctica, Mar. Biol., 96, 93–106, doi:10.1007/BF00394842, 1987.

Hunt, B. P. V. and Hosie, G. H.: The seasonal succession of zooplankton in the Southern Ocean south of Australia, part I: The seasonal ice zone, Deep-Sea Res. Pt. I, 53, 1182–1202, doi:10.1016/j.dsr.2006.05.001, 2006.

Hunt, B. P. V., Pakhomov, E. A., Hosie, G. W., Siegel, V., Ward, P., and Bernard, K.: Pteropods in Southern Ocean ecosystems, Prog. Oceanogr., 78, 193–221, doi:10.1016/j.pocean.2008.06.001, 2008.

Jackson, G. A., Najjar, R. G., and Toggweiler, J. R.: Flux feeding as a mechanism for zooplankton grazing and its implications for vertical particulate flux, Limnol. Oceanogr., 38, 1328–1332, 1993.

Karnovsky, N. J., Hobson, K. A., Iverson, S., and Hunt Jr., G. L.: Seasonal changes in the diets of seabirds in the North Water Polynya: a multiple-indicator approach, Mar. Ecol.-Prog. Ser., 357, 291–299, 2008.

Koppelmann, R., Weikert, H., Halsband-Lenk, C., and Jennerjahn, T. C.: Mesozooplankton community respiration and its relation to particle flux in the oligotrophic eastern Mediterranean, Global Biogeochem. Cy., 18, GB1039, doi:10.1029/2003GB002121, 2004.

Lalli, C. M.: Structure and function of the buccal apparatus of *Clione limacina* (Phipps) with a review of feeding in gymnosomatous pteropods, J. Exp. Mar. Biol. Ass. UK, 4, 101–118, 1970.

Lalli, C. M. and Gilmer, R. W.: Pelagic snails: the biology of holoplanktonic gastropod molluscs, Stanford, Stanford University Press, California, 1989.

Larson, R. J.: Water content, organic content, and carbon and nitrogen composition of medusae from the northeast Pacific, J. Exp. Mar. Biol. Ecol., 99, 107–120, doi:10.1016/0022-0981(86)90231-5, 1986.

Leblanc, K., Aristegui, J., Armand, L., Assmy, P., Beker, B., Bode, A., Breton, E., Cornet, V., Gibson, J., Gosselin, M.-P., Kopczynska, E., Marshall, H., Peloquin, J., Piontkovski, S., Poulton, A. J., Queguiner, B., Schiebel, R., Shipe, R., Stefels, J., van Leeuwe, M. A., Varela, M., Widdicombe, C., and Yallop, M.: A global diatom database – abundance, biovolume and biomass in the world ocean, Earth Syst. Sci. Data Discuss., 5, 147–185, doi:10.5194/essdd-5-147-2012, 2012.

Le Quéré, C., Harrison, S. P., Prentice, C., Buitenhuis, E. T., Aumonts, O., Bopp, L., Claustre, H., da Cunha, L. C., Geider, R., Giraud, X., Klaas, C., Kohfeld, K. E., Legendre, L., Manizza, M., Plattss, T., Rivkin, R., Sathyendranath, S., Uitz, J., Watson, A. J., Wolf-Gladrow, D.: Ecosystem dynamics based on plankton functional types for global ocean biochemistry models, Glob. Change Biol., 11, 2016–2040, 2005.

Little, W. S. and Copley, N. J.: WHOI Silhouette DIGITIZER, Version 1.0, Users Guide, WHOI Technical Report, Woods Hole Oceanographic Institution, 2003.

Lochte, K. and Pfannkuche, O.: Processes driven by the small sized organisms at the water-sediment interface, in: Ocean Margin Systems, edited by: Wefer, G., Billet, D., Hebbeln, D., Jørgensen, B. B., Schlüter, M., and Weering, T. C., Springer, Heidelberg, Germany, 2003.

Longhurst, A.: Ecological Geography of the Sea, Academic Press, USA, 2007.

Luo, Y.-W., Doney, S. C., Anderson, L. A., Benavides, M., Bode, A., Bonnet, S., Bostrom, K. H., Bottjer, D., Capone, D. G., Carpenter, E. J., Chen, Y. L., Church, M. J., Dore, J. E., Falcon, L. I., Fernandez, A., Foster, R. A., Furuya, K., Gomez, F., Gundersen, K., Hynes, A. M., Karl, D. M., Kitajima, S., Langlois, R. J., LaRoche, J., Letelier, R. M., Maranon, E., McGillicuddy Jr., D. J., Moisander, P. H., Moore, C. M., Mourino-Carballido, B., Mulholland, M. R., Needoba, J. A., Orcutt, K. M., Poulton, A. J., Raimbault, P., Rees, A. P., Riemann, L., Shiozaki, T., Subramaniam, A., Tyrrell, T., Turk-Kubo, K. A., Varela, M., Villareal, T. A., Webb, E. A., White, A. E., Wu, J., and Zehr, J. P.: Database of diazotrophs in global ocean: abundances, biomass and nitrogen fixation rates, Earth Syst. Sci. Data Discuss., 5, 47–106, doi:10.5194/essdd-5-47-2012, 2012.

Marrari, M., Daly, K. L., Timonin, A., and Semenova, T.: The zooplankton of Marguerite Bay, Western Antarctic Peninsula – Part I: Abundance, distribution, and population response to variability in environmental conditions, Deep-Sea Res. Pt. II, 58, 1599–1613, doi:10.1016/j.dsr2.2010.12.007, 2011.

Mazzocchi, M. G., Christou, E., Rastaman, N., and Siokou-Frangou, I.: Mesozooplankton distribution from Sicily to Cyprus (Eastern Mediterranean): I. General aspects, Oceanol. Acta, 20, 521–535, 1997.

Mileikovsky, S. A.: Breeding and larval distribution of the pteropod *Clione limacina* in the North Atlantic, Subarctic and North Pacific Oceans, Mar. Biol., 6, 317–334, 1970.

Milliman, J. D. and Droxler, A. W.: Neritic and pelagic carbonate sedimentation in the marine environment: Ignorance is not bliss, Geologische Rundschau, 85, 496–504, 1996.

Moraitou-Apostolopoulou, M., Zervoudaki, S., and Kapiris, K.: Mesozooplankton abundance in waters of the Aegean Sea, Hellenic Center of Marine Research, Institut of Oceanography, Hydrodynamics and Biogeochemical Fluxes in the Straits of the Cretan Arc (project), Greece, 2008.

Mousseau, L., Fortier, L., and Legendre, L.: Annual production of fish larvae and their prey in relation to size-fractioned primary production (Scotian Shelf, NW Atlantic), ICES J. Mar. Sci., 55, 44–57, 1998.

Nishikawa, J., Matsuura, H., Castillo, L. V., Wilfredo, L. C., and Nishida, S.: Biomass, vertical distribution and community structure of mesozooplankton in the Sulu Sea and its adjacent waters, Deep-Sea Res. Pt. II, 54, 114–130, doi:10.1016/j.dsr2.2006.09.005, 2007.

NMFS-COPEPOD: The Global Plankton Database, NOAA National Oceanic and Atmospheric Administration, US Department of Commerce (http://www.st.nfms.noaa.gov/plankton), 2011.

Pakhomov, E. A. and Perissinotto, R.: Mesozooplankton community structure and grazing impact in the region of the Subtropical Convergence south of Africa, J. Plankton Res., 19, 675–691, 1997.

Pane, L., Feletti, M., Francomacaro, B., and Mariottini, G. L.: Summer coastal zooplankton biomass and copepod community structure near Italian Terra Nova Base (Terra Nova Bay, Ross Sea, Antarctica), J. Plankton Res., 26, 1479–1488, doi:10.1093/plankt/fbh135, 2004.

Perissinotto, R.: Mesozooplankton size-selectivity and grazing impact on the phytoplankton community of the Prince Edward Archipelago (Southern Ocean), Mar. Ecol.-Prog. Ser., 79, 243–258, 1992.

Ramfos, A., Isari, S., and Rastaman, N.: Mesozooplankton abundance in water of the Ionian Sea (March 2000), Department of Biology, University of Patras, 2008.

Rogachev, K. A., Carmack, E. C., and Foreman, M. G. G.: Bowhead whales feed on plankton concentrated by estuarine and tidal currents in Academy Bay, Sea of Okhotsk, Cont. Shelf Res., 28, 1811–1826, doi:10.1016/j.csr.2008.04.014, 2008.

Schalk, P. H.: Spatial and seasonal variation in pteropods (Mollusca) of Indo-Malayan waters related to watermass distribution, Mar. Biol., 105, 59–71, 1990.

Schiebel, R. and Movellan, A.: First-order estimate of the planktic foraminifer biomass in the modern ocean, Earth Syst. Sci. Data, 4, 75–89, doi:10.5194/essd-4-75-2012, 2012.

Schiebel, R., Waniek, J., Zeltner, A., and Alves, M.: Impact of the Azores Front on the distribution of planktic foraminifers, shelled gastropods, and coccolithophorids, Deep-Sea Res. Pt. II, 49, 4035–4050, 2002.

Schnack-Schiel, S. B. and Cornils, A.: Zooplankton abundance measured on Apstein net samples during cruise CISKA2005,

Alfred Wegener Institute for Polar and Marine Research, Science for the Protection of Indonesian Coastal Environment (project), Bremerhaven, 2009.

Seibel, B. A. and Dierssen, H. M.: Cascading trophic impacts of reduced biomass in the Ross Sea, Antarctica: Just the tip of the iceberg?, Biol. Bull., 205, 93–97, 2003.

Siokou-Frangou, I., Christou, E., and Rastman, N.: Mesozooplankton abundance in waters of the Ionian Sea, Hellenic Center of Marine Research, Institut of Oceanography, Physical Oceanography of the Eastern Mediterranean (project), Greece, 2008.

Solis, N. B. and von Westernhagen, H.: Vertical distribution of Euthecosomatous Pteropods in the Upper 100 m of the Hilutangan Channel, Cebu, The Philippines, Mar. Biol., 48, 79–87, 1978.

Swadling, K. M., Penot, F., Vallet, C., Rouyer, A., Gasparini, S., Mousseau, L., Smith, M., and Goffart, A.: Interannual variability of zooplankton in the Dumont d'Urville sea (139° E–146° E), east Antarctica, 2004–2008, Polar Sci., 5, 118–133, doi:10.1016/j.polar.2011.03.001, 2011.

Tréguer, P., Legendre, L., Rivkin, R. T., Raueneau, O., and Dittert, N.: Water column biogeochemistry below the euphotic zone, Ocean biogeochemistry: The role of the ocean carbon cycle in global change, edited by: Fasham, M. J. R., Global Change – The IGBP Series (closed), 145–156, doi:10.1007/978-3-642-55844-3_7, 2003.

Van der Spoel, S.: Growth, reproduction and vertical migration in *Clio pyramidata* Linne, 1767 forma *lunceolata* (Lesueur, 1813), with notes on some other Cavoliniidae (Mollusca, Pteropoda), Beaufortia, 281, 117–134, 1973.

Volkov, A. F.: Mean Annual Characteristics of Zooplankton in the Sea of Okhotsk, Bering Sea and Northwestern Pacific (Annual and Seasonal Biomass Values and Predominance, ISSN 1063-0740, Russ. J. Mar. Biol., 34, 437–415, doi:10.1134/S106307400807002X, 2008.

Ward, P., Whitehouse, M., Shreeve, R., Thorpe, S., Atkinson, A., Korb, R., Pond, D., and Young, E.: Plankton community structure south and west of South Georgia (Southern Ocean): Links with production and physical forcing, Deep-Sea Res. Pt. I, 54, 1871–1889, doi:10.1016/j.dsr.2007.08.008, 2007.

Wells Jr., F. E.: Effects of Mesh Size on Estimation of Population Densities of Tropical Euthecosomateus Pteropods, Mar. Biol., 20, 347–350, 1973.

Wells Jr., F. E.: Seasonal patterns of abundance and reproduction of euthecosomatous pteropods off Barbados, West Indies, Veliger, 18, 241–248, 1976.

Werner, I.: Living conditions, abundance and biomass of under-ice fauna in the Storfjord area (Western Barents Sea, Arctic) in late winter (March 2003), Polar Biol., 28, 311–318, 2005.

Wormuth, J. H.: Vertical distributions and diel migrations of Euthecosomata in the northwest Sargasso Sea, Deep-Sea Res., 28, 1493–1515, doi:10.1016/0198-0149(81)90094-7, 1981.

Wormuth, J. H.: The role of cold-core Gulf Stream rings in the temporal and spatial patterns of euthecosomatous pteropods, Deep-Sea Res., 32, 773–788, 1985.

Zervoudaki, S., Christou, E., Siokou-Frangou, I., and Zoulias, T.: Mesozooplankton abundance in water of the Aegean Sea, Hellenic Center of Marine Research, Institut of Oceanography, Invest. of new marine biol. resources in deep waters of Ionian and Aegean Seas, Greece, 2008.

ZooDB – Zooplankton database, Plankton sample analysis supported by NSF grants to M. D. Ohman, Scripps Institution of Oceanography and by the SIO Pelagic Invertebrates Collection, http://oceaninformatics.ucsd.edu/zoodb/#, 2011.

15

Global open-ocean biomes: mean and temporal variability

A. R. Fay[1] and G. A. McKinley[2]

[1]Space Science and Engineering Center, University of Wisconsin–Madison, 1225 W. Dayton St., Madison, Wisconsin 53706, USA
[2]Department of Atmospheric and Oceanic Sciences, University of Wisconsin–Madison, 1225 W. Dayton St., Madison, Wisconsin 53706, USA

Correspondence to: A. R. Fay (arfay@wisc.edu)

Abstract. Large-scale studies of ocean biogeochemistry and carbon cycling have often partitioned the ocean into regions along lines of latitude and longitude despite the fact that spatially more complex boundaries would be closer to the true biogeography of the ocean. Herein, we define 17 open-ocean biomes classified from four observational data sets: sea surface temperature (SST), spring/summer chlorophyll a concentrations (Chl a), ice fraction, and maximum mixed layer depth (maxMLD) on a $1° \times 1°$ grid (available at doi:10.1594/PANGAEA.828650). By considering interannual variability for each input, we create dynamic ocean biome boundaries that shift annually between 1998 and 2010. Additionally we create a core biome map, which includes only the grid cells that do not change biome assignment across the 13 years of the time-varying biomes. These biomes can be used in future studies to distinguish large-scale ocean regions based on biogeochemical function.

1 Introduction

In recent decades, many studies have partitioned the pelagic environment into gyre- or subgyre-scale regions in order to investigate biogeochemical processes over large ocean regions (Longhurst, 1995; Sarmiento et al., 2004; Gurney et al., 2008; Reygondeau et al., 2013). Recent studies of the terrestrial carbon cycle have moved away from the division of the landmasses into latitudinally defined regions (Gurney et al., 2008). Despite the limitation of this latitudinal-defined approach in the oceans, recent observational and modeling studies have generally used such definitions (Gruber et al., 2009; Schuster et al., 2013). This is, at least in part, due to the lack of an alternative biome map available from the peer-reviewed literature. Efforts to address this need date back to the 1980s when Emery and Meincke (1986) released a paper outlining global water masses based on temperature and salinity.

Biogeography is a discipline that seeks to identify ecosystem distributions across space and time (Cox and Moore, 2010). High costs and the three dimensions of space challenge our ability to observe the ocean, and thus most oceanographic sampling is sparse and heterogeneous. This, in turn, means that our knowledge of the detailed biogeography of the global oceans is more elementary than that for the terrestrial biosphere, though satellite-based estimates of surface ocean chlorophyll have helped to remedy this since the late 1990s. Ocean biogeochemistry is organized, to first order, by the large-scale ocean circulation, with frontal zones acting as boundaries, especially for the surface ocean (Longhurst, 2007). In this study, we take advantage of satellite chlorophyll and physical variables associated with large-scale circulation to define 17 surface ocean biomes that capture patterns of large-scale biogeochemical function at the basin scale. This work builds upon previous biome definitions (McKinley et al., 2011; Fay and McKinley, 2013), with the addition of ice fraction criteria. Additionally, in this study biome boundaries shift annually due to variability in the physical state and surface chlorophyll.

In contrast to previous studies by Reygondeau et al. (2013) and Longhurst (1995, 2007) which consider subbasin-scale ocean provinces, the biomes presented here are for the open ocean and do not include coastal regions. These biomes are substantially larger than the provinces proposed initially by Longhurst (1995) in order to address the first-order differences in biogeochemical function at the scale of ocean gyres. Our goal is to partition the surface ocean into regions of common biogeochemical function at the largest possible scale. This goal is analogous to previous water-mass-boundaries work (Emery and Meincke, 1986; Iudicone et al., 2011), and is consistent with a variety of observational and modeling studies assessing air–sea CO_2 fluxes and primary productivity (Sarmiento et al. 2004; Gurney et al., 2008; Gruber et al., 2009; Takahashi et al., 2009; Schuster et al., 2013).

Coastal oceans are important and diverse regions that constitute the transition zones between terrestrial and open-ocean regions. These regions are impacted by terrestrial runoff, tidal mixing and close coupling with the benthos. They receive from the land, and produce in situ, large amounts of organic matter. Coastal regions are very biogeochemically active. Thus, as has been traditional in biogeochemical oceanography, the coastal oceans are not grouped with open-ocean regions in this study (Sarmiento and Gruber, 2006). With respect to quantifying coastal carbon cycling, several groups are currently active. A North American effort is led by the Coastal Carbon Synthesis (CCARS) (http://www.whoi.edu/website/ccars/) and the European effort is included in the Blue Carbon Initiative (http://thebluecarboninitiative.org/).

The ocean biomes presented here are of a similar scale to those used in RECCAP (the Regional Carbon Cycle Assessment Project), a global effort to establish the mean carbon balance and component fluxes of large regions of the globe (Wanninkhof et al., 2013). The regions used in RECCAP (available at http://transcom.lsce.ipsl.fr) are taken from those used in the TransCom atmospheric inversion intercomparison project (Gurney et al., 2008) and have become a standard for global ocean carbon research (Jacobson et al., 2007; Mikaloff-Fletcher et al., 2007; Gruber et al., 2009; Canadell et al., 2011; Lenton et al., 2013; Schuster et al., 2013). While there is substantial similarity between the biomes and the RECCAP regions, we will argue that the biomes presented here are preferable because they are defined by relevant environmental parameters instead of by lines of latitude. Going forward, these biomes could be used as a new basis for a wide range of analyses and intercomparison studies in ocean biogeochemistry and carbon cycling.

Herein, we present time-varying biomes for the global ocean, spanning from 1998 to 2010 (limited by chlorophyll a data availability). We also present a mean biome map as well as a core biome map that assigns a biome classification only for those ocean grid cells that retain the same biome classification for all 13 years of the time-varying biomes.

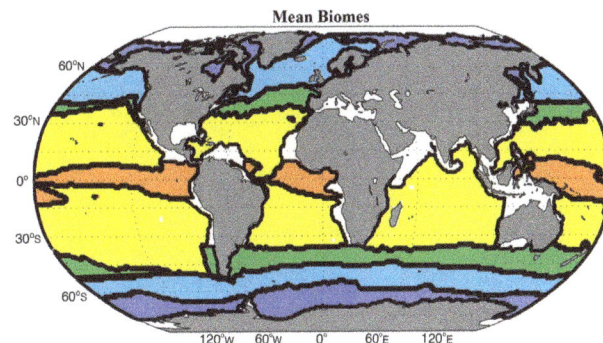

Figure 1. Mean biome map created from mean climatologies of maxMLD, SST, summer Chl a, and maximum ice fraction. Dark blue: ice biome (ICE); cyan: subpolar seasonally stratified biome (SPSS); green: subtropical seasonally stratified biome (STSS); yellow: subtropical permanently stratified biome (STPS); orange: equatorial biome (EQU). White indicates ocean areas that do not fit the criteria for any biome and are excluded from further analysis.

2 Methodology

We create physical, biologically defined regions or "biomes" delineated based on four climatological criteria: maximum mixed layer depth (maxMLD), spring/summer chlorophyll a concentration (Chl a), sea surface temperature (SST), and sea ice fractional coverage (Table 1). SST has substantial latitudinal variation, and thus helps to separate subtropical from subpolar regions. SST also indicates major upwelling zones in the equatorial regions. maxMLD and Chl a together indicate the amplitude of vertical mixing and the resulting seasonality in biogeochemical processing. Chl a also separates the relatively unproductive subtropical regions from other regions. Sea ice fractional coverage is used to distinguish ice-influenced biomes from subpolar biomes. The use of these observables as criteria is consistent with previous studies of ocean biogeography (Longhurst 1995, 2007; Sarmiento et al., 2004; Reygondeau et al., 2013; D'Ortenzio and d'Alcala, 2009).

The global open ocean is first divided into ocean basins (Atlantic, Pacific, Indian, and Southern Ocean) and then further classified into biomes, using the criteria in Table 1, starting from the poles: the ice biome (ICE), the subpolar seasonally stratified biome (SPSS), the subtropical seasonally stratified biome (STSS), the subtropical permanently stratified biome (STPS) and the equatorial biome (EQU). Biome assignments are made from pole to Equator for each hemisphere, beginning with the Northern Hemisphere (NH). This order is important as some biome's criteria are not mutually exclusive (specifically the EQU and South Pacific and Atlantic STPS biomes).

Provided here are maps for mean biomes, created using the mean climatology for years 1998–2010 (Fig. 1). Figure 2 shows the 13 years (1998–2010) of time-varying biomes, created using the annual data. Additionally, we create a core

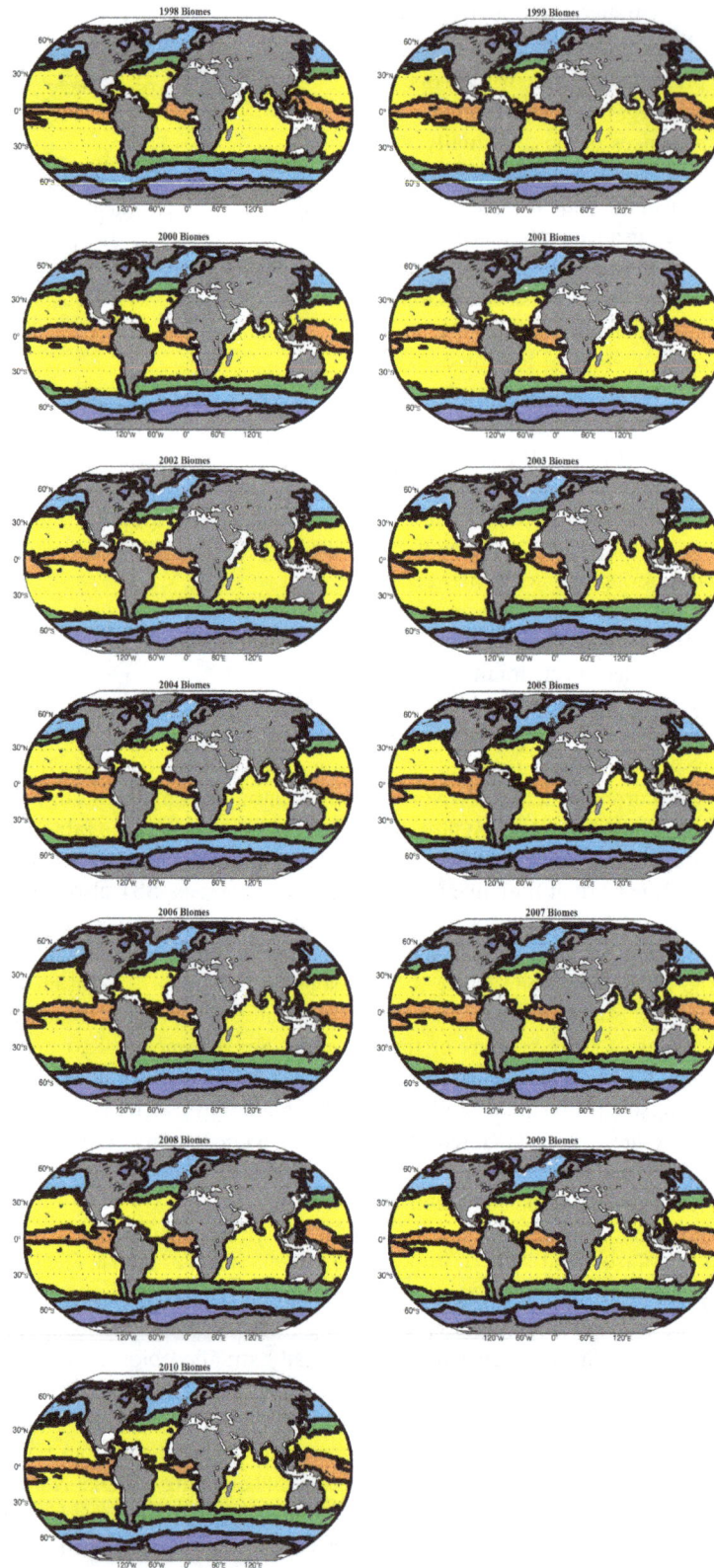

Figure 2. Time-varying biomes for years 1998–2010. Created from annual climatology of SST, summer mean Chl a, maximum ice fraction, and climatological maxMLD. Dark blue: ICE biome; cyan: SPSS biome; green: STSS biome; yellow: STPS biome; orange: EQU biome. White indicates ocean areas that do not fit the criteria for any biome and are excluded from further analysis.

Table 1. Characteristics for the environmental envelopes defined for each biome. All criteria must be met for the biome to be assigned except where otherwise noted (STSS biomes). Chlorophyll is based on summer chlorophyll values.

Biome	Sea ice fraction	SST (°C)	Chl a (mg m^{-3})	maxMLD (m)	Notes
NH ICE	$x \geq 0.5$				
N Pacific SPSS		$x < 14$	$x \geq 0.25$		
N Atlantic SPSS		$x < 14$	$x \geq 0.4$		
NH STSS		$11 \leq x < 29$	$0.16 \leq x < 0.4$	$x > 125$	Either Chl a or maxMLD; lat $\geq 25°$ N
NH STPS		$14 \leq x < 29$	$x < 0.16$	$x \leq 125$	
IND STPS		$x \geq 11$	$x < 0.25$		
West Pacific EQU		$x \geq 29$	$x < 0.25$		$15°$ S $\leq$ latitude $\leq 15°$ N
East Pacific EQU		$19 \leq x < 29$	$0.16 \leq x < 0.7$		$10°$ S $\leq$ latitude $\leq 10°$ N
Atlantic EQU		$19 \leq x < 29$	$0.16 \leq x < 0.7$		$10°$ S $\leq$ latitude $\leq 10°$ N
SH STPS		$x \geq 8$	$x < 0.25$	$x \leq 150$	
SH STSS		$x \geq 8$	$x \geq 0.16$	$x > 150$	either Chl a or maxMLD
SH SPSS		$x < 8$			
SH ICE	$x \geq 0.5$				

biome map (Fig. 3), which only includes biome assignments for grid cells that retain the same biome assignment for all 13 years of the time-varying biomes.

2.1 Biome descriptions

Using the four criteria discussed in detail in Sects. 2.2–2.4, the ocean is divided into biomes. Each biome is characterized by a range of values from the observational fields (Table 1) and, if multiple criteria define a biome, all must be met for the grid cell to be assigned to that biome. The one exception to this is for the STSS biomes, which require either the Chl a or the maxMLD criteria to be met. In all cases, the specific criteria used to define each biome have been developed to capture large-scale biogeochemical functioning while limiting the number of undefined regions between biome boundaries.

Cooler, polar waters that have at least a 50 % ice-cover fraction during some part of the year are grouped into the marginal sea ice (ICE) biome.

SPSS biomes have divergent surface flow driven by the positive wind stress curl and thus upwelling from below, allowing for higher summer Chl a concentrations due to continual nutrient resupply. The Pacific and Atlantic oceans have different Chl a constraints, with the Pacific being a high nutrient, low chlorophyll (HNLC) region (Sarmiento and Gruber, 2006; Table 1).

The STSS biome is an area of downwelling due to negative wind stress curl, but intermediate chlorophyll concentrations due to deep winter maxMLDs. The STPS biome experiences negative wind stress curl, leading to convergence and year-round stratification, such that maxMLDs are shallow and Chl a is low. In the North Atlantic, spring runoff from the Amazon and Orinoco rivers cause a plume of high Chl a levels in the eastern Caribbean area, excluding this region from STPS biome assignment.

EQU biomes are defined by SST and Chl a constraints, coupled with latitudinal bounds, which are required because the criteria have similarity to coastal points at the edges of the STSS biomes (moderate temperatures and intermediate Chl a). The Pacific equatorial biome is separated into east and west biomes, with the western EQU biome having warmer temperatures and lower Chl a than the eastern. As easterlies blow across the equatorial region, warm water pools in the western half of the basin, causing warm temperatures and stratified waters with low Chl a. In the east Pacific, cool, high-nutrient waters are upwelled from depth, leading to higher Chl a. Similarly, in the Atlantic Ocean, warmer SSTs and moderate Chl a levels define the EQU biome. In the Indian Ocean, the equatorial region is grouped in with the STPS biome due to seasonally varying physical ocean circulation patterns associated with the monsoon.

Ocean areas not defined by any of these biomes are largely coastal or influenced by coastal upwelling. Gulfs, bays, and seas such as the Mediterranean Sea and the Gulf of Mexico are not included in these biomes because these areas do not meet these biome criteria, designed for the open-ocean, due to their particular circulation patterns and/or strong influence from the land. A few open-ocean points also cannot be categorized into biomes by our criteria, and these points are omitted.

For calculation of the mean biomes, 1998–2010 mean Chl a, SST and sea ice fraction are used with climatological maxMLD.

2.2 Monthly ice fraction and sea surface temperature

The Hadley Centre Meteorological Office provides monthly mean gridded, global fractional sea ice coverage and SST from 1870 to present (HadISST – Hadley Centre Sea Ice and Sea Surface Temperature data set). The data set is described and its quality assessed by Rayner et al. (2003). This product is derived from gridded in situ observations with data-sparse regions filled using reduced space optimal interpolation (RSOI; Kaplan et al., 1997). For the period 1998–2010, the RSOI technique was applied to combined in situ and satellite data, with additional interpolation and smoothing procedures performed for the Southern Ocean. SST standard deviations for HadISST are shown for several regions in Sect. 6 of Rayner et al. (2003). Variance maximum occurs in the eastern equatorial Pacific and the coastal and Arctic regions of the global ocean. These patterns and magnitudes of variance are not unlike other SST products such as NOAA's OI (Optimum Interpolation) v2 product (Reynolds et al., 2002), while HadISST has reduced variance for other areas, such as the Indian Ocean.

HadISST sea ice coverage is reported as a fraction of each $1° \times 1°$ cell. A minimum threshold of 0.5 fractional coverage in any month of the year designates the ICE biome. This contrasts to Sarmiento et al. (2004), who define their ICE region as having any sea ice coverage in any part of the year. We use a more stringent ice criteria in order to ensure only waters influenced by ice year-round are included in the ICE biome. At the same time, we note that there is a relatively sharp transition (five grid cells or less) from no ice coverage ($< 10\%$) to full coverage ($> 90\%$) in HadISST, which means that a choice of a higher or lower percentage of coverage would have only limited impact on ICE biome extents.

For mean SST and sea ice fraction criteria, annual climatologies for years 1998–2010 are averaged. Specific criteria for each biome are presented in Table 1.

2.3 Monthly chlorophyll a

Chlorophyll a concentration (Chl a) is a proxy for the abundance of marine phytoplankton and offers a first-order quantification of rates of biogeochemical cycling in the surface ocean. Global ocean color data from the Sea-viewing Wide Field-of-view Sensor (SeaWiFS) satellite have been used to estimate Chl a concentrations using NASA's OC4 (ocean color) algorithm (O'Reilly et al., 1998) (available at: http://oceandata.sci.gsfc.nasa.gov/). Monthly binned climatology products at $9\,km \times 9\,km$ resolution are provided by NASA beginning in September 1997 and ending in December 2010. Smigen, a NASA-provided program (available at http://seadas.gsfc.nasa.gov/doc/smigen/smigen.html) was used to recalculate monthly Chl a to $1° \times 1°$ resolution in a manner consistent with other SeaWiFS products.

The prelaunch accuracy target for SeaWiFS Chl a was 35% for the range 0.05–0.50 mgm^{-3} (Hooker et al., 1992), which is a common accuracy standard for satellite Chl a products. O'Reilly et al. (1998) find the median percentage error for operational SeaWiFS compared to available open ocean in situ measurements to be 26%.

In order to avoid bias due to high cloud coverage in winter, all biome selection is based on mean spring/summer Chl a rather than an annual mean. The months April through

September are used in the Northern Hemisphere (NH), and December through March in the Southern Hemisphere (SH). Summer/spring is used because these months have the most consistent cloud-free coverage (over 75% of the Northern (Southern) Hemisphere has at least 5 (3) of the spring/summer months with coverage). The selected Chl a criteria thresholds presented in Table 1 reflect this choice by being higher values than what would be used if annual mean Chl a criteria had been selected.

We extend the Northern Hemisphere definition (April–September) to $10°$ S latitude in the construction of the biomes in order to have a consistent Chl a field for the equatorial biome assignment. Equatorial chlorophyll has a strong seasonal cycle. If the north-to-south switch in Chl a definition were made at the Equator, regions just south of the Equator would reflect a 6-month phase shift in Chl a values despite similar physics as regions just north of the Equator. This choice does not influence the Southern Hemisphere STPS biome because of the very small seasonal cycle in the STPS regions (the April–September mean is very similar to the December–March mean).

SeaWiFS data are not available for January–March 2008, which could impact Southern Hemisphere biomes for 2008 because December 2007 would be the only month with data. To avoid this potential bias, climatological summer Chl a is used for the Southern Hemisphere biome classification for 2008.

2.4 Climatological maximum mixed layer depth

Climatological maxMLD indicates the amplitude of seasonality in biogeochemical processing, particularly in the transition regions between the subpolar and subtropical gyres where deep mixing provides nutrients to the otherwise oligotrophic surface ocean. Criteria for maxMLD are noted in Table 1. maxMLD does not vary annually because data are insufficient for global coverage.

Criteria for definition of the mixed layer and methods for finding its depth are numerous. We use the Argo mixed layer depth climatology (Holte et al., 2010) calculated with the density algorithm (Holte and Talley, 2009). Other threshold methods (de Boyer Montegut et al., 2004; Sarmiento et al., 2004) were also considered. The selected algorithm builds on traditional threshold and gradient methods by drawing its estimate of the MLD from physical features in the profile and by considering a pool of various MLDs from which the algorithm selects the final MLD estimate (Holte and Talley, 2009). The Argo climatology is based on data from years 2002–2008 that overlap reasonably well with the satellite chlorophyll record used here for our time-varying biomes (1998–2010). The de Boyer Montegut et al. (2004) climatology is based on earlier data (1941–2002) that have less overlap.

In some regions, most notably in the Southern and Arctic oceans, MLDs for some grid cells are not defined in the

Argo climatology. These missing pixels are filled with a latitudinal mean from the same ocean basin. This filling occurs mostly in the higher latitudes and near continental shelves, where the maxMLD criterion is not applied for biome selection (Table 1). The maxMLD criterion is applied only to divide the STPS and STSS biomes, both of which have good coverage prior to filling.

2.5 Smoothing

After initial processing, each biome map is smoothed to limit the number of outlier grid cells within the biomes. An iterative smoothing process is used, cycling through the map numerous times and changing only the grid cells that are bordered on at least three sides by another biome classification.

The result of this automated, iterative smoothing is that each of the 13 years has between 350 and 467 grid cells change biome classification. For the 13 years of time-varying biomes, the maximum number of $1° \times 1°$ grid cells smoothed is 467 in the year 2002, which is 1.3 % of the total grid cells in the global ocean. No more than 6 % of the grid cells in any one biome change due to smoothing. Smoothing of the mean biomes changes 263 grid cells (0.73 % of total ocean grid cells and no more than 5 % of any one biome). These smoothed areas occur most frequently in the intergyre regions (STSS), a region of strong interannual variability.

After smoothing, the core biomes are created by selecting only the grid cells that have the same biome definition in all of the 13 time-varying biomes.

2.6 Data format and availability

The biome maps are provided in netCDF-4 (network common data form) format (with an accompanying readme file) and can be found at the PANGAEA web page doi:10.1594/PANGAEA.828650 (or doi:10.1594/PANGAEA.828650). This contains files for the mean and core biome map boundaries as well as maps for each year of the time-varying biomes. An animation of the 13 time-varying biome maps is available at http://oceancarbon.aos.wisc.edu/biomes-2014/ and as a Supplement with this manuscript.

3 Biomes

Seventeen global biomes are represented on each of our biome maps using the criteria outlined in Table 1.

3.1 Mean biomes

Mean biomes are created using the climatological SST, Chl a, and ice fraction criteria for years 1998–2010, and climatological maxMLD with the criteria in Table 1. These biomes are presented in Fig. 1 with their areas listed in Table 2. From the poles, there are the marginal sea ice (ICE)

Table 2. Size (in 10^6 km^2) of the mean biomes and core biomes.

Biome	Mean biome area (10^6 km^2)	Core biome area (10^6 km^2)
NP ICE	4.5852	3.9112
NP SPSS	12.838	8.7063
NP STSS	6.8257	4.3398
NP STPS	41.048	31.305
Pac EQU W	11.593	4.9040
Pac EQU E	14.890	7.7117
SP STPS	52.705	43.067
NA ICE	5.4750	4.6123
NA SPSS	10.062	7.8048
NA STSS	5.9744	4.6178
NA STPS	17.464	13.746
Atl EQU	7.4147	3.1972
SA STPS	18.055	16.056
IND STPS	35.936	32.383
SO STSS	29.692	23.670
SO SPSS	30.628	25.875
SO ICE	18.678	16.169

biome and the subpolar gyres, or SPSS. Next come the intergyre regions between the subtropics and subpolar gyres, labeled STSS biomes. The subtropical gyres are classified as STPS biomes and the equatorial regions are included in the EQU biomes. The Indian Ocean is entirely a STPS biome. Going southward from the equatorial Atlantic and Pacific, we find the Southern Hemisphere subtropical gyres (STPS), and then the Southern Ocean regions that we define as STSS, SPSS and ICE biomes. Respectively, these three Southern Ocean biomes are comparable to the Subantarctic Zone (SAZ), the Polar Frontal Zone (PFZ) and Antarctic Zone (AZ) (Lovenduski et al., 2007).

3.2 Time-varying biomes

In Fig. 2 biome maps spanning years 1998–2010 are presented. Chl a product availability is the limitation on the years of analysis. In each year, the criteria as listed in Table 1 are applied to annual data. Changes between these 13 maps are due to the combined impacts of changes in Chl a, SST, and ice fraction from year to year. Due to lack of an interannually varying data product, maxMLD remains a climatological variable.

3.3 Core biomes

Core biomes are defined as the grid cells that maintain the same biome assignment for all 13 years (Fig. 3). With respect to analysis of variability or trends in biogeochemical variables, core biomes would be a conservative region for analysis because there is a strict consistency with the biome definition for each individual year between 1998 and 2010. Core biome areas are included in Table 2.

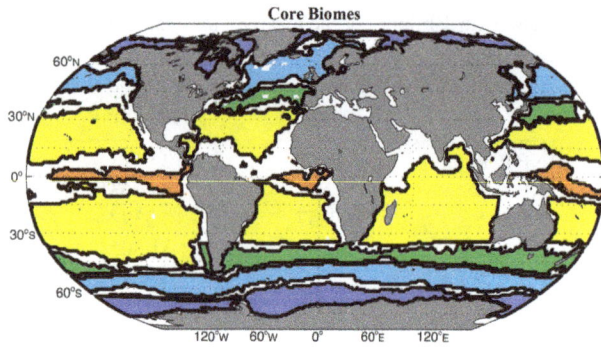

Figure 3. Core biome map. Core biomes are the portions of the 13 time-varying biomes that do not shift between one biome and another from year to year but remain as a single biome for all 13 years. Dark blue: ICE biome ; cyan: SPSS biome; green: STSS biome; yellow: STPS biome; orange: EQU biome . White indicates ocean areas that do not fit the criteria for any biome and also areas that shift biome assignments at least once between 1998 and 2010.

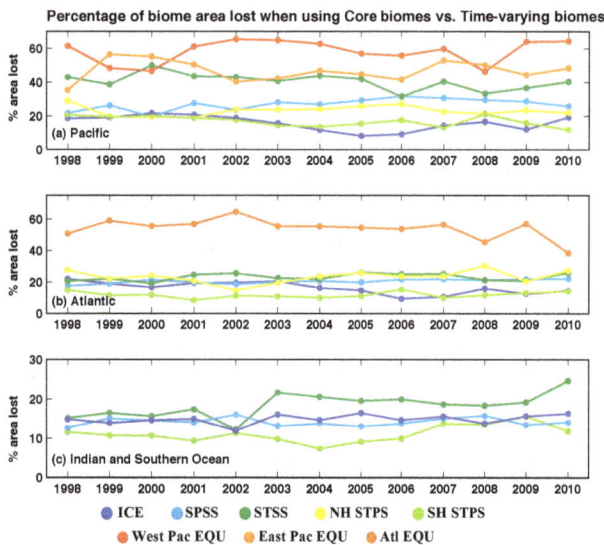

Figure 4. Percentage of biome area lost each year if using core biomes as compared to time-varying biomes for **(a)** Pacific Ocean, **(b)** Atlantic Ocean, and **(c)** Southern and Indian oceans. Dark blue: ICE; cyan: SPSS; green: STSS; yellow: Northern Hemisphere STPS; orange: EQU; light green: Southern Hemisphere STPS.

Undefined regions in the core biome map are caused by variability of the time-varying biomes. A comparison of the area of core biomes to the time-varying biomes indicates the degree of year-to-year variability in biome extent, with high percentages indicating more change (Fig. 4). The Pacific Basin shows the highest percentage of areas lost as compared to other basins (Fig. 4a). The greatest fraction of area omitted by the use of core biomes occurs in the equatorial, subtropical seasonally stratified, and subpolar biomes (>25 % area excluded).

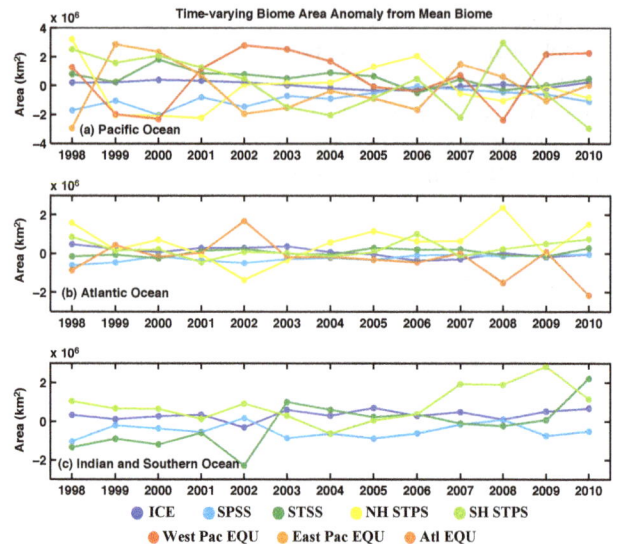

Figure 5. Biome area anomaly as the residual from the corresponding mean biome area for years 1998–2010 in **(a)** Pacific Ocean; **(b)** Atlantic Ocean; and **(c)** Southern and Indian oceans. Dark blue: ICE; cyan: SPSS; green: STSS; yellow: Northern Hemisphere STPS; orange: EQU; light green: Southern Hemisphere STPS.

A complimentary measure of interannual variability is the change in total biome area each year (Fig. 5). This is complimentary because latitudinal migration without change in total area allows for a small area anomaly (Fig. 5) and at the same time a significant percentage of area lost by use of the core biomes (Fig. 4).

Between the North Pacific SPSS and STSS biomes, the southern Gulf of Alaska shifts biomes year to year due to changing annual temperatures. Low productivity in some years also contributes to the northward recession of the North Pacific SPSS biome. This is reflected by the relatively high variability of the Pacific SPSS biome area (Fig. 5a) and high areas lost with the core biome for both biomes (Fig. 4a).

In the North Atlantic (NA) ICE biome, variation in ice fraction causes changing designation between ICE and SPSS, and therefore area lost in the core biomes (Fig. 4b). A significant inverse correlation ($r = -0.83$) between NA ICE and NA SPSS biome areas further demonstrates this shift over the 13 years (Fig. 5b). Biome area changes are also large in the equatorial biomes (Fig. 5a, b) resulting in a high percentage of area loss in the core biomes (Fig. 4a, b).

Decreasing Chl a in the northwestern Indian Ocean (IND) allows for the expansion of the IND STPS biome in the late 2000s (Fig. 5c). This could be related to weakening winds of the summer monsoon. Interannual change in Southern Ocean (SO) circulation impacts temperature and productivity causing biome areas to have substantial variability, particularly for SO STSS (Fig. 5c). This also results in a loss of area for the core biomes (Fig. 4c).

4 Discussion

As first discussed by Longhurst (1995), ocean biogeography does not organize itself along lines of latitude and longitude. For example, in the surface ocean pCO_2 climatology of Takahashi et al. (2009), it is clear in the North Atlantic and North Pacific that the subtropical–subpolar boundary follows the major ocean currents. Going forward, it will be advantageous to use biogeochemically relevant biomes in studies of large-scale ocean biogeochemistry and carbon cycling so as to avoid the limitations of square regions (Takahashi et al., 2006; Canadell et al., 2011). Schuster et al. (2013) note that the limited agreement in the seasonal cycle of North Atlantic air–sea CO_2 fluxes from a range of methodologies is partially driven by the use of boundaries defined by latitude.

A visual comparison of the mean and core biomes to the Longhurst provinces and to the RECCAP regions is presented in Fig. 6. Largely, the biomes created here follow the outlines of one or of several Longhurst provinces. Major differences occur mostly in the intergyre regions of the North Atlantic and Pacific, in the South Pacific–Antarctic transition, and in the equatorial Atlantic. While Longhurst defines 56 provinces in his maps, many of these are specifically designed to capture marginal seas such as the Gulf of Mexico and the Arabian Sea or coastal regions. Only 30 Longhurst provinces are defined for the open ocean, and are thus most comparable to our 17 open-ocean biomes.

A comparison of the mean Chl a and pCO_2 to the mean values at each $1° \times 1°$ grid cell contained within the biomes and those within the RECCAP regions confirms that the use of biome boundaries rather than latitude lines results in better large-scale spatial coherence of the fields. As an example, in the midlatitude North Atlantic Gulf Stream transition zone, the standard deviation of annual Chl a is an order of magnitude larger when using RECCAP region 3 (Gruber et al., 2009; location indicated in Fig. 7c) vs. biome NA STSS (0.40 ± 1.29 vs. 0.30 ± 0.12, respectively). This tendency persists in the subtropical North Atlantic (0.21 ± 0.76 in RECCAP region 4 vs. 0.11 ± 0.07 in biome NA STPS) and North Pacific (0.25 ± 0.41 in RECCAP region 13 vs. 0.09 ± 0.044 in biome NP STPS) as well as in the equatorial Pacific (0.10 ± 0.22 in RECCAP region 16 vs. 0.10 ± 0.046 in the Pac EQU W biome; 0.20 ± 0.42 in RECCAP region 17 vs. 0.22 ± 0.1 in the Pac EQU E biome). In the Southern Ocean, interannual mean Chl a variation is similar whether these biomes or the RECCAP regions are used, consistent with the biomes being more longitudinal. A similar comparison of the standard deviation of the long-term mean pCO_2 between grid cells contained within biomes or RECCAP regions shows that the standard deviations of the individual points around the biome averages are smaller than around the RECCAP region averages in all cases except the Southern Ocean.

The seasonal cycles of climatological chlorophyll and pCO_2 for two areas are shown in Fig. 7. The seasonal cycles

of pCO_2 for all grid cells within the NA STSS biome vs. grid cells within the RECCAP region 3 (Gruber et al., 2009) show a similar mean cycle, but there is more spread around the mean for the RECCAP region ($139\,\mu$atm) than for the biome ($128\,\mu$atm; Fig. 7a, b). A comparison of Chl a seasonal cycles shows the same tendency – more variation around the mean for the RECCAP boundary coupled with a higher spring peak (RECCAP is 0.31 mg m^{-3}, biome is 0.16 mg m^{-3}; Fig. 7e, f).

In the western equatorial Pacific, the pCO_2 seasonal cycle is smaller than at high latitudes, but still varies with the choice of regional definition (Fig. 7c, d). When using the biomes, the pCO_2 climatology peaks during the winter months, whereas in RECCAP region 16, pCO_2 peaks in late summer. The standard deviation around the mean is larger for the RECCAP ($162\,\mu$atm) than for the biome ($149\,\mu$atm). Considering Chl a, the standard deviation around the mean Chl a cycle for the RECCAP region is nearly double that of the variation around the biome mean cycle (RECCAP is 0.083 mg m^{-3}, biome is 0.047 mg m^{-3}; Fig. 7g, h).

These comparisons are representative of other biome to RECCAP region comparisons for the spatial coherence of the mean and seasonal cycles of pCO_2 and Chl a within a large ocean area. Together, they suggest that biogeography is better defined by biomes than by lines of latitude.

Previous studies have proposed that the subtropical ocean regions are expanding and that warming ocean temperatures are contributing to reduced productivity in these important ocean gyres (Polovina et al., 2008; Behrenfeld et al., 2006). We find no significant trend in the area of any subtropical biome (STPS) over the years 1998–2010. In both the North Pacific and North Atlantic, the SPSS biome is expanding over the 1998–2010 time frame (9.27 ± 7.48 10^4 km yr^{-1}; 3.78 ± 1.86 10^4 km yr^{-1}). The SO STSS biome is also expanding (1.98 ± 1.45 10^5 km yr^{-1}) as increasing Chl a values at the South Pacific–Southern Ocean boundary allow it to extend further equatorward (Figs. 2, 5c). This SO STSS expansion persists for all time series longer than 10 years with end year 2010 (e.g., 1998–2010, 1999–2010, 2000–2010), however the signal disappears for trends ending prior to 2010. Likewise, a negative trend in the area of the North Pacific ICE biome appears for all trends longer than 10 years if the final year is prior to 2010 (Fig. 5a). When the area time series is extended to include 2010, this signal of declining area of the Pacific ICE biome disappears.

The North Atlantic ICE biome has a significant declining area trend for 1998–2010 (-4.84 ± 2.90 10^4 km yr^{-1}) and a corresponding positive trend in the North Atlantic SPSS biome (3.78 ± 1.89 10^4 km yr^{-1}; Fig. 5b). The statistical significance of these trends persists for any 10-year or longer time series considered (i.e., 1998–2008, 1999–2009, 2000–2010, 1998–2009, etc.). As mentioned above, the 1998–2010 trends in NA ICE and SPSS biome areas are highly correlated ($r = -0.83$). All trends reported here include their 95 % confidence intervals calculated following Wilks (2006). As more data become available, significant secular trends in areas may

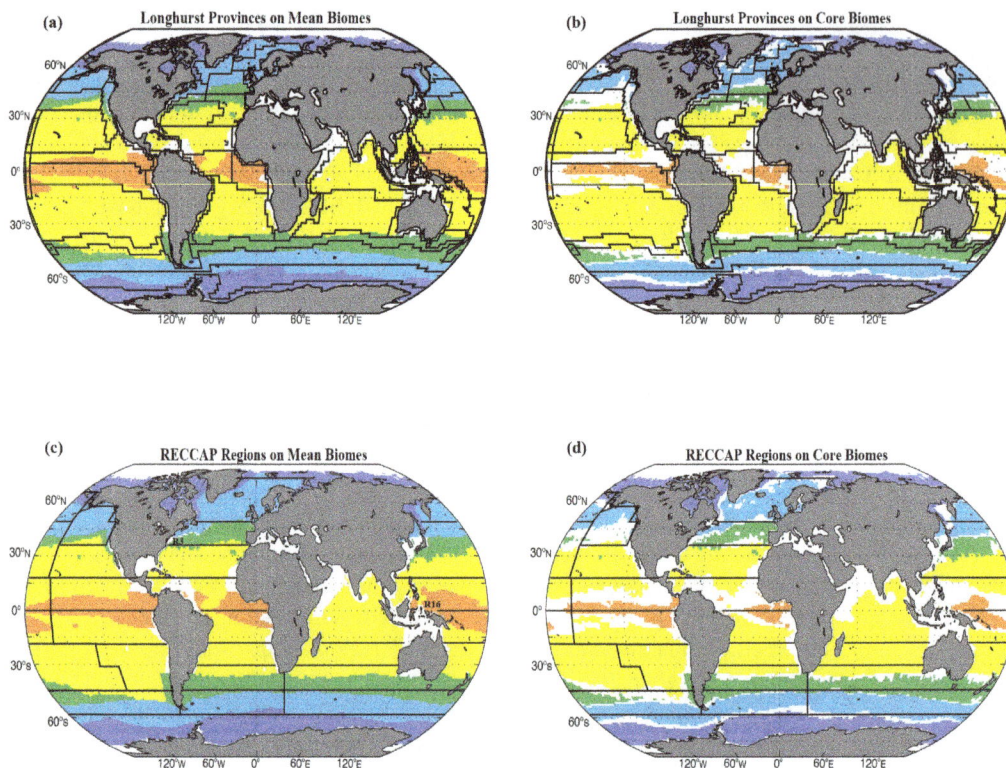

Figure 6. Longhurst's 56 provinces overlaid on top of (**a**) mean biomes and (**b**) core biomes; RECCAP ocean regions overlaid on top of (**c**) mean biomes and (**d**) core biomes. RECCAP regions 3 and 16 are labeled to facilitate interpretation of Fig. 7.

be revealed, but at present we find greater evidence for inter-annual variability.

The global biomes previously used to study surface ocean pCO_2 trends (McKinley et al., 2011; Fay and McKinley, 2013) varied slightly from those presented here. Time-varying biomes were not used in these analyses. For the products presented here, biome criteria have been updated with the most current and complete data sets and products available. This results in some noticeable differences between the mean biome map presented here and that used in previous work by Fay and McKinley (2013). Changes in the North Atlantic biomes are primarily due to an improved mixed layer depth climatology. Changes in the extent of the ICE biomes are due to the inclusion of ice fraction as a criterion.

The data used as biome criteria have limitations. The main constraint is the lack of coverage for year-to-year change in mixed layer depth estimates. Mixed layer depth estimates should improve as Argo coverage expands. Extension of the time-varying biomes will be possible with updated HadISST products and new satellite Chl a estimates that extend beyond the SeaWiFS period. We have tried to use the current version of the GlobColour product in this analysis, but found substantial secular trends in equatorial biome areas after 2010 due to strong trends in GlobColour Chl a. 2010 is the point where the SeaWiFS record ends, and these trends have not

been validated in the literature. Thus, we did not use Glob-Colour in this analysis.

Biomes created with an alternate SST and sea ice concentration product (OI v2; Reynolds et al., 2002) result in smaller ice biomes and correspondingly larger SPSS biomes than those presented here. Other biomes had no significant change for either the mean or time-varying versions ($< 1\%$ of grid cells change for any biome). The core biome map has no notable differences when using this alternative SST and ice product.

Other recent works have defined many smaller bioregions over single-ocean basins or seas including the Southern Ocean (Grant et al., 2006) the Mediterranean Sea (D'Ortenzio and d'Alcala, 2009) and the Indian Ocean (Leìvy et al., 2007). These studies also used observed satellite chlorophyll, sea surface temperature, and other variables to define bioregions. These efforts have been detailed and focused on understanding the relatively small part of the ocean that they have subdivided into dozens of bioregions. However, these locally defined bioregions cannot be easily applied to other basins, as they are unique to the specific ocean. Regional definitions based on water masses are a relatively new and promising method of studying surface-to-deep connections in the ocean and climate modeling community, and have shown utility for global-scale comparison with respect to heat uptake and biogeochemical change due to climate

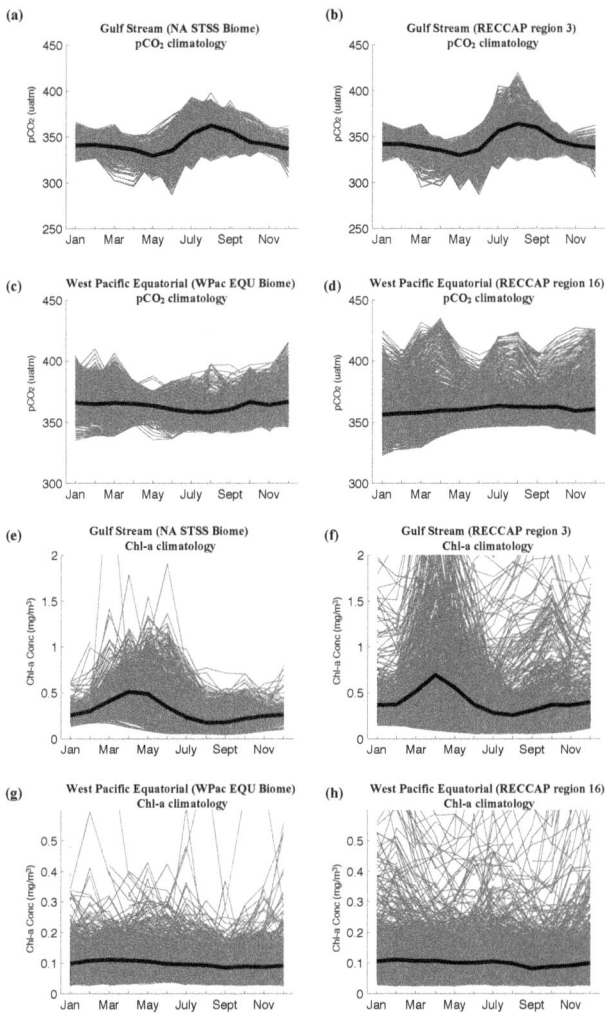

Figure 7. Comparison of biogeochemical variables in biomes vs. RECCAP regions 3 and 16. Seasonal pCO_2 climatology (**a–d**) and Chl a climatology (**e–h**) for two example regions: the North Atlantic Gulf Stream (**a–b; e–f**) and the west Pacific equatorial region (**c–d; g–h**).

tially, defining model-specific biomes may be useful for comparison of trends and variability within and between models. The concise biome criteria presented should facilitate application to modeling studies, as the input data sets are typically standard output fields for global-scale models.

5 Conclusions

We offer three versions of environmentally defined biomes to be used as an alternative to the current standard of latitudinally defined ocean regions for biogeochemical and carbon cycle studies. The 17 mean biomes offer nearly full coverage of the open ocean and are based on mean data for years 1998–2010. Also presented are time-varying ocean biomes for each year from 1998 to 2010 that should be of use to studies focused in this period. Finally, core biomes can be utilized in analyses that wish to be most conservative in their definition by avoiding any points where the biome to which that point is assigned is not the same in all years between 1998 and 2010.

Opportunities for use of these biomes in future studies range from the aggregation of sparse data to the large-scale analysis of trends and seasonal cycles of surface biogeochemical properties. Clear distinction between biogeochemically different regions should improve comparisons of data and numerical models. Atmospheric and ocean inversion studies could use the mean biomes for their regional discretization (Gurney et al., 2008; Gruber et al., 2009). If the presented biomes are as widely adopted as the latitudinally defined regions have been, this advancement in biogeography may enrich future carbon-cycle intercomparison studies such as RECCAP (Canadell et al., 2011).

Acknowledgements. A. R. Fay and G. A. McKinley are supported by NASA grants 07-NIP07-0036, NNX/11AF53G, and NNX/13AC53G. We thank the editor and reviewers for their constructive comments that improved the manuscript.

Edited by: R. Key

change globally (Bopp et al., 2013) and in the Southern Ocean (Séférian et al., 2012; Sallée et al., 2013). Iudicone et al. (2008, 2011) discuss partitioning the domains using specific neutral density ranges to classify the water masses following Sloyan and Rintoul (2001). The biomes presented in this paper are meant to capture biogeochemical functioning at the surface ocean at the largest possible scale. Future comparisons to these previous studies offering regional definitions or water-mass-based partitioning will be of value.

These time-varying biomes offer a new basis for aggregation and analysis of biogeochemical data for global openocean studies. Utilizing these biomes may provide new insights on the response of the ocean biogeochemistry to climate change. For model intercomparison studies, where model physics such as currents and gyres can vary substan-

References

Behrenfeld, M. J., O'Malley, R. T., Siegel, D. A, McClain, C. R., Sarmiento, J. L., Feldman, G. C., Milligan, A. J., Falkowski, P. G., Letelier, R. M., and Boss, E. S.: Climate-driven trends in contemporary ocean productivity, Nature, 444, 752–755, doi:10.1038/nature05317, 2006.

Bopp, L., Resplandy, L., Orr, J. C., Doney, S. C., Dunne, J. P., Gehlen, M., Halloran, P., Heinze, C., Ilyina, T., Séférian, R., Tjiputra, J., and Vichi, M.: Multiple stressors of ocean ecosystems in the 21st century: projections with CMIP5 models,

Biogeosciences, 10, 6225–6245, doi:10.5194/bg-10-6225-2013, 2013.

Canadell, J. G., Ciais, P., Gurney, K., Le Quéré, C., Piao, S., Raupach, M. R., and Sabine, C. L.: An International Effort to Quantify Regional Carbon Fluxes, Eos Trans. AGU, 92(10), 81, 2011.

Cox, C. B. and Moore, P. D.: Biogeography: An Ecological and Evolutionary Approach, John Wiley & Sons, Hoboken, NJ, 5–7, ISBN: 0470637943, 2010.

de Boyer Montégut, C., G. Madec, A. S. Fischer, Lazar, A., and Iudicone, D.: Mixed layer depth over the global ocean: An examination of profile data and a profile-based climatology, J. Geophys. Res., 109, C12003, doi:10.1029/2004JC002378, 2004.

D'Ortenzio, F. and Ribera d'Alcalà, M.: On the trophic regimes of the Mediterranean Sea: a satellite analysis, Biogeosciences, 6, 139–148, doi:10.5194/bg-6-139-2009, 2009.

Emery, W. J. and Meincke, J.: Global water masses-summary and review, Oceanol. Acta, 9, 383–391, 1986.

Fay, A. R. and McKinley, G. A.: Global trends in surface ocean pCO_2 from in situ data, Global Biogeochem. Cy., 27, 541–557, doi:10.1002/gbc.20051, 2013.

Grant S., Constable, A., Raymond, and B., Doust, S.: Bioregionalisation of the Southern Ocean, Report of Experts Workshop, WWF- Australia and ACE CRC, Hobart, September, 2006.

Gruber, N., Gloor, M., Mikaloff Fletcher, S. E., Doney, S. C., Dutkiewicz, S., Follows, M. J., Gerber, M., Jacobson, A. R., Joos, F., Lindsay, K., Menemenlis, D., Mouchet, A., Müller, S. A., Sarmiento, J. L., and Takahashi, T.: Oceanic sources, sinks, and transport of atmospheric CO_2, Global Biogeochem. Cy., 23, GB1005, doi:10.1029/2008GB003349, 2009.

Gurney, K. R., Baker, D., Rayner, P., and Denning, S.: Interannual variations in continental-scale net carbon exchange and sensitivity to observing networks estimated from atmospheric CO_2 inversions for the period 1980 to 2005, Global Biogeochem. Cy., 22, GB3025, doi:10.1029/2007GB003082, 2008.

Holte, J. and Talley, L.: A New Algorithm for Finding Mixed Layer Depths with Applications to Argo Data and Subantarctic Mode Water Formation, J. Atmos. Oceanic Tech., 26, 1920–1939, doi:10.1175/2009JTECHO543.1, 2009.

Holte, J., Gilson, J., Talley, L., and Roemmich D.: Argo Mixed Layers, Scripps Institution of Oceanography/UCSD, http://mixedlayer.ucsd.edu (last access: 14 March 2014), 2010.

Hooker, S. B., Esaias, W. E., Feldman, G. C., Gregg, W. W., and McClain, C. R.: An overview of SeaWiFS and ocean color, NASA Tech. Memo. 104566, Vol. 1, 24 pp., 1992.

Iudicone, D., Madec, G., and McDougall, T. J.: Water-mass transformations in a neutral density framework and the key role of light penetration, J. Phys. Oceanogr., 38, 1357–1376, 2008.

Iudicone, D., Rodgers, K. B., Stendardo, I., Aumont, O., Madec, G., Bopp, L., Mangoni, O., and Ribera d'Alcala', M.: Water masses as a unifying framework for understanding the Southern Ocean Carbon Cycle, Biogeosciences, 8, 1031–1052, doi:10.5194/bg-8-1031-2011, 2011.

Jacobson, A. R., Mikaloff-Fletcher, S. E., Gruber, N., Sarmiento, J. L., and Gloor, M.: A joint atmosphere-ocean inversion for surface fluxes of carbon dioxide: 2. Regional results, Global Biogeochem. Cy., 21, GB1019, doi:10.1029/2006GB002703, 2007.

Kaplan, A., Kushnir, Y., Cane, M. A., and Blumenthal, M. B.: Reduced space optimal analysis for historical data sets: 136 years

of Atlantic sea surface temperatures, J. Geophys. Res.-Oceans, 102, 27835–27860, 1997.

Lenton, A., Tilbrook, B., Law, R. M., Bakker, D., Doney, S. C., Gruber, N., Ishii, M., Hoppema, M., Lovenduski, N. S., Matear, R. J., McNeil, B. I., Metzl, N., Mikaloff Fletcher, S. E., Monteiro, P. M. S., Rödenbeck, C., Sweeney, C., and Takahashi, T.: Sea-air CO_2 fluxes in the Southern Ocean for the period 1990–2009, Biogeosciences, 10, 4037–4054, doi:10.5194/bg-10-4037-2013, 2013.

Leìvy, M., Shankar, D., Andreì, J.-M., Shenoi, S. S. C., Durand, F., and de Boyer Monteìgut, C.: Basin-wide seasonal evolution of the Indian Ocean's phytoplankton blooms, J. Geophys. Res., 112, C12014, doi:10.1029/2007JC004090, 2007.

Longhurst, A.: Seasonal cycles of pelagic production and consumption, Prog. Oceanogr., 36, 77–167, doi:10.1016/0079-6611(95)00015-1, 1995.

Longhurst, A. R.: Ecological geography of the sea, Academic Press, Elsevier Inc., Burlington, MA, Pg: 35–40, 89–113, ISBN: 978-0-12-455521-1, 2007.

Lovenduski, N. S., Gruber, N., Doney, S. C., and Lima, I. D.: Enhanced CO_2 outgassing in the Southern Ocean from a positive phase of the Southern Annular Mode, Global Biogeochem. Cy., 21, doi:10.1029/2006GB002900, 2007.

McKinley, G. A., Fay, A. R., Takahashi, T., and Metzl, N.: Convergence of atmospheric and North Atlantic carbon dioxide trends on multidecadal timescales, Nat. Geosci., 4, 606–610, doi:10.1038/ngeo1193, 2011.

Mikaloff-Fletcher, S. E., Gruber, N., Jacobson, A. R., Gloor, M., Doney, S. C., Dutkiewicz, S., Gerber, M., Follows, M., Joos, F., Lindsay, K., Menemenlis, D., Mouchet, A., Müller, S. A., and Sarmiento, J. L.: Inverse estimates of the oceanic sources and sinks of natural CO_2 and the implied oceanic carbon transport, Global Biogeochem. Cy., 21, GB1010, doi:10.1029/2006GB002751, 2007.

O'Reilly, J. E., Maritorena, S., Mitchell, B. G., Siegel, D. A., Carder, K. L., Garver, S. A., Kahru, M., and McClain, C.: Ocean color chlorophyll algorithms for SeaWiFS, J. Geophys. Res., 103, 24937, doi:10.1029/98JC02160, 1998.

Polovina, J. J., Howell, E. A., and Abecassis, M.: Ocean's least productive waters are expanding, Geophys. Res. Lett., 35, L03618, doi:10.1029/2007GL031745, 2008.

Rayner, N. A., Parker, D. E., Horton, E. B., Folland, C. K., Alexander, L. V., Rowell, D. P., Kent, E. C., and Kaplan, A.: Global analyses of sea surface temperature, sea ice, and night marine air temperature since the late nineteenth century, J. Geophys. Res., 108, 4407, doi:10.1029/2002JD002670, 2003.

Reygondeau, G., Longhurst, A., Martinez, E., Beaugrand, G., Antoine, D., and Maury, O.: Dynamic biogeochemical provinces in the global ocean, Global Biogeochem. Cy., 27, 1046–1058, doi:10.1002/gbc.20089, 2013.

Reynolds, R. W., Rayner, N. A., Smith, T. M., Stokes, D. C., and Wang, W.: An improved in situ and satellite SST analysis for climate, J. Climate, 15, 1609–1625, 2002.

Sallée, J. B., Shuckburgh, E., Bruneau, N., Meijers, A. J. S., Bracegirdle, T. J., Wang, Z., and Roy, T.: Assessment of Southern Ocean water mass circulation and characteristics in CMIP5 models: Historical bias and forcing response, J. Geophys. Res.-Oceans, 118, 1830–1844, doi:10.1002/jgrc.20135, 2013.

Sarmiento, J. L., Slater, R., Barber, R., Bopp, L., Doney, S. C., Hirst, A. C., Kleypas, J., Matear, R., Mikolajewicz, U., Monfray, P., Soldatov, V., Spall, S. A., and Stouer, R.: Response of ocean ecosystems to climate warming, Global Biogeochem. Cy., 18, GB3003, doi:10.1029/2003GB002134, 2004.

Sarmiento, J. L. and Gruber, N.: Ocean Biogeochemical Cycles, Princeton University Press, Princeton, NJ, 503 pp., 2006.

Schuster, U., McKinley, G. A., Bates, N., Chevallier, F., Doney, S. C., Fay, A. R., González-Dávila, M., Gruber, N., Jones, S., Krijnen, J., Landschützer, P., Lefèvre, N., Manizza, M., Mathis, J., Metzl, N., Olsen, A., Rios, A. F., Rödenbeck, C., Santana-Casiano, J. M., Takahashi, T., Wanninkhof, R., and Watson, A. J.: An assessment of the Atlantic and Arctic sea–air CO_2 fluxes, 1990–2009, Biogeosciences, 10, 607–627, doi:10.5194/bg-10-607-2013, 2013.

Séférian, R., Iudicone, D., Bopp, L., Roy, T. and Madec, G.: Water Mass Analysis of Effect of Climate Change on Air–Sea CO_2 Fluxes: The Southern Ocean, J. Climate, 25, 3894–3908, doi:10.1175/JCLI-D-11-00291.1, 2012.

Sloyan, B. M. and Rintoul, S. R.: The Southern Ocean limb of the global deep overturning circulation, J. Phys. Oceanogr., 31, 143–173, 2001.

Takahashi, T., Sutherland, S. C., Feely, R. A., and Wanninkhof, R.: Decadal change of the surface water pCO_2 in the North Pacific: A synthesis of 35 years of observations, J. Geophys. Res., 111, C07S05, doi:10.1029/2005JC003074, 2006.

Takahashi, T., Sutherland, S. C., Wanninkhof, R., Sweeney, C., Feely, R. A., Chipman, D. W., Hales, B., Friederich, G., Chavez, F., Sabine, C., Watson, A., Bakker, D. C. E., Schuster, U., Metzl, N., Yoshikawa-Inoue, H., Ishii, M., Midorikawa, T., Nojiri, Y., Körtzinger, A., Steinho, T., Hoppema, M., Olafsson, J., Arnarson, T. S., Tilbrook, B., Johannessen, T., Olsen, A., Bellerby, R., Wong, C. S., Delille, B., Bates, N. R., and de Baar, H. J. W.: Climatological mean and decadal change in surface ocean pCO_2, and net sea–air CO_2 flux over the global oceans, Deep Sea Res. Pt II, 56, 554–577, doi:10.1016/j.dsr2.2008.12.009, 2009.

Wanninkhof, R., Park, G.-H., Takahashi, T., Sweeney, C., Feely, R., Nojiri, Y., Gruber, N., Doney, S. C., McKinley, G. A., Lenton, A., Le Quéré, C., Heinze, C., Schwinger, J., Graven, H., and Khatiwala, S.: Global ocean carbon uptake: magnitude, variability and trends, Biogeosciences, 10, 1983–2000, doi:10.5194/bg-10-1983-2013, 2013.

Wilks, D. S.: Statistical Methods in the Atmospheric Sciences, 627 pp., Elsevier, Amsterdam, 2006.

16

Calibration procedures and first dataset of Southern Ocean chlorophyll *a* profiles collected by elephant seals equipped with a newly developed CTD-fluorescence tags

C. Guinet[1], X. Xing[2,3,4], E. Walker[5], P. Monestiez[5], S. Marchand[6], B. Picard[1], T. Jaud[1], M. Authier[1], C. Cotté[1,7], A. C. Dragon[1], E. Diamond[2,3], D. Antoine[2,3], P. Lovell[8], S. Blain[9,10], F. D'Ortenzio[2,3], and H. Claustre[2,3]

[1]Centre d'Etudes Biologiques de Chizé-CNRS, Villiers en Bois, France
[2]Laboratoire d'Océanographie de Villefranche, Villefranche-sur-Mer, France
[3]Université Pierre et Marie Curie (Paris-6), Unité Mixte de Recherche 7093, Laboratoire d'Océanographie de Villefranche, Villefranche-sur-Mer, France
[4]Ocean University of China, Qingdao, China
[5]Unité Biostatistique et Processus Spatiaux-INRA, Avignon, France
[6]Muséum national d'histoire naturelle, DMPA USM 402/LOCEAN, Paris, France
[7]Université Pierre et Marie Curie (Paris-6), DMPA USM 402/LOCEAN, Paris, France
[8]Sea Mammal Research Unit, University of St. Andrews, St. Andrews, Scotland
[9]Laboratoire d'Océanographie Microbienne, Université Paris VI, Banyuls sur mer, France
[10]Université Pierre et Marie Curie (Paris-6), Unité Mixte de Recherche 7621, Laboratoire d'Océanographie Microbienne, Banyuls-sur-Mer, France
Correspondence to: C. Guinet (guinet@cebc.cnrs.fr)

Abstract. In situ observation of the marine environment has traditionally relied on ship-based platforms. The obvious consequence is that physical and biogeochemical properties have been dramatically undersampled, especially in the remote Southern Ocean (SO). The difficulty in obtaining in situ data represents the major lim-itations to our understanding, and interpretation of the coupling between physical forcing and the biogeochem-ical response. Southern elephant seals (*Mirounga leonina*) equipped with a new generation of oceanographic sensors can measure ocean structure in regions and seasons rarely observed with traditional oceanographic platforms. Over the last few years, seals have allowed for a considerable increase in temperature and salinity profiles from the SO, but we were still lacking information on the spatiotemporal variation of phytoplank-ton concentration. This information is critical to assess how the biological productivity of the SO, with direct consequences on the amount of CO_2 "fixed" by the biological pump, will respond to global warming. In this research programme, we use an innovative sampling fluorescence approach to quantify phytoplankton concen-tration at sea. For the first time, a low energy consumption fluorometer was added to Argos CTD-SRDL tags, and these novel instruments were deployed on 27 southern elephant seals between 25 December 2007 and the 4 February 2011. As many as 3388 fluorescence profiles associated with temperature and salinity measurements were thereby collected from a vast sector of the Southern Indian Ocean. This paper addresses the calibration issue of the fluorometer before being deployed on elephant seals and presents the first results obtained for the Indian sector of the Southern Ocean. This in situ system is implemented in synergy with satellite ocean colour radiometry. Satellite-derived data is limited to the surface layer and is restricted over the SO by extensive cloud cover. However, with the addition of these new tags, we are able to assess the 3-dimension distribution of phy-toplankton concentration by foraging southern elephant seals. This approach reveals that for the Indian sector of the SO, the surface chlorophyll *a* (chl *a*) concentrations provided by MODIS were underestimated by a factor 2 compared to chl *a* concentrations estimated from HPLC corrected in situ fluorescence measurements. The scientific outcomes of this programme include an improved understanding of both the present state and variability in ocean biology, and the accompanying biogeochemistry, as well as the delivery of real-time and open-access data to scientists.

1 Introduction

Polar marine ecosystems, and in particular the Southern Ocean (SO hereafter), are among the most vulnerable ecosystems to climate change. However, there is conflicting evidence on how the biological productivity of these Polar Oceans will respond to global warming. The SO plays an important role in the carbon cycle and it is one of the largest sink for anthropogenic CO_2 through the formation of deep water around Antarctica and intermediate water in the vicinity of the subantarctic zone (Caldeira et al., 2000; Lo Monaco et al., 2005). Furthermore, by contributing to roughly half of the biosphere's primary production, photosynthesis by oceanic phytoplankton is a vital link between living and inorganic stocks of carbon (Field et al., 1998; Berhenfeld et al., 2006), but our current understanding of the variability of SO's primary productivity is hampered by the lack of in situ observations available for this logistically difficult region, and much of the existing observations are heavily biased towards the austral summer.

Furthermore, the degree of confidence for primary production derived from satellite-based estimates of phytoplankton biomass is still debated. This is especially true in SO, where satellite measurements tend to under-estimate chlorophyll a (chl a hereafter) concentrations (Dierssen and Smith, 2000; Holm-Hansen et al., 2004; Garcia et al., 2005; Dierssen, 2010; Kahru and Mitchell, 2010). Satellites scan the sea surface and are unable to provide subsurface chlorophyll profiles. Deep fluorescence maxima have been found within the frontal zone of the Antarctic Circum Current (ACC hereafter, Quéguiner and Brzezinski, 2002; Holm-Hansen et al., 2004) or in the vicinity of the ice edge (Waite and Nodder, 2001). Persistent cloud cover and fragmented sea-ice also constitute towards a major limitation of satellite ocean colour measurements in the SO (Arrigo et al., 1998; Buesseler et al., 2003).

Evaluation of the distribution of chl a throughout the water column is one of the most important biological parameters in the ocean because it is an indicator of the spatial and temporal variability of primary productivity (Behrenfeld and Falkowski, 1997). The limitations of satellite assessments of primary production combined with a lack of primary productivity measurements in the field requires to complement remotely sensed ocean colour data with year-round surveys of the in situ optics as well as the physical oceanographic measurements for a description of spatial (horizontal and vertical) and temporal (seasonal, inter-annual) distribution of phytoplankton, but also give insights on its advection and fate. In turn, this data will contribute to our understanding of how primary production within SO may respond to climatic changes.

Subsurface chl a measurements are traditionally performed from research vessels, using profiling fluorometers and water samples collected by Niskin bottles. Alternatively, chl a profiles can be obtained from fluorometers deployed on fixed moorings or autonomous platforms like Argo floats (Roemmich et al., 2004), or autonomous underwater vehicles (Yu et al., 2002). Rapid technological advances in ocean observation have nevertheless been achieved during the last decade, particularly with respect to physical climate variables. Developing such in situ observation systems is an essential step towards a better understanding of biogeochemical cycles and ecosystem dynamics, especially at spatial and temporal scales that have been unexplored until now. However, with regard to the carbon cycle, the establishment of in situ observing systems in the under-sampled SO remains challenging due to its remoteness, harsh weather conditions and the presence of sea-ice.

Here we present the development of an original synergy between biologist's efforts to understand the marine life of top predators, physical and biogeochemical oceanographic studies through development of new bio-logging devices deployed on southern elephant seals (*Mirounga leonina*), SES hereafter. This device incorporates high accuracy temperature and salinity sensors, as well as a fluorometer and provides a range of new behavioural and physiological data on free ranging marine animals for biologists, while simultaneously gathering vertical profiles of temperature, salinity and fluorescence for oceanographers. Profiles sampled in the remote SO are of great interest as they can fill a niche within the ocean observing system, where such measurements are lacking (e.g., Charrassin et al., 2008; Nicholls et al., 2008; Roquet et al., 2009; Wunch et al., 2009). One important aspect of this methodology is the near real-time delivery of CTD-Fluo profiles using the Argos satellite system (Argos, 1996). SESs provide an ideal "platform" for such investigation as they dive nearly continuously and at great depths (Hindell et al., 1991). Moreover, they undertake long foraging trips each year, exploring large areas of the SO (Biuw et al., 2007).

However, to make most of the use of these fluorescence data, it is essential to develop effective means for calibration, quality control and postprocessing to provide a consistent dataset to oceanographers and for climatologies. Therefore, the first objective of this paper is to report the calibration and the profile qualification procedure on a unique 3-yr fluorescence dataset collected by SES within the Indian sector of the Southern Ocean. From this data, we will assess how in situ measurements compare with surface chl a concentration measured by ocean colour satellites. This new approach allows sustained acquisition of chl a fluorescence profiles (proxy for chl a concentration) in areas where data scarcity is the rule and how they complement satellite ocean colour data.

2 Materials and methods

2.1 Instrumentation

A thorough technical description of CTD-SRDLs can be found in Boehme et al. (2009), which we briefly summarise here (see also Fedak et al., 2002). CTD-Fluoro-SRDLs have been designed as miniaturised platforms to record behavioural data and log in situ CTD profiles. They can be deployed on a range of marine mammals (e.g., Lydersen et al., 2002; Boehme et al., 2008; Nicholls et al., 2008; Roquet et al., 2009). The devices contain (1) a Platform Terminal Transmitter (PTT) to transmit compressed data through the Argos satellite system, (2) a micro-controller coordinates the different functions e.g., sensor data acquisition (data processing and transmission based on the internal setup and energy budget, Boehme et al. (2009) and (3) a miniaturised CTD (Valeport LTD, Totnes, UK).

The specifications of the miniaturised CTD (Valeport Ltd, Totnes, UK) result from a trade-off between the need for miniaturisation, energy consumption, stability and sensor performance. The pressure measurements are made by a Keller series-PA7 piezoresistive pressure transducer1 (Keller AG, CH) with a given accuracy of better than 1 % of the full-scale reading (± 20 dbar at 2000 dbar). However, laboratory experiments have shown a performance of better than 0.25 % of the actual reading (Boehme et al., 2009). The temperature probe is a fast response Platinum Resistance Thermometer (PRT) made by Valeport (range: $-5\,°C$ to $+35\,°C$, accuracy: $\pm 0.005\,°C$, time constant: 0.7 s) and an inductive conductivity sensor by Valeport (range: 0 to 80 mS cm^{-1}, accuracy: better than ± 0.01 mS cm^{-1}).

Implementation of a fluorometer to estimate chl a concentration

In vivo fluorescence F is a widely used technique to estimate chl a concentration in aquatic environments and can be expressed as:

$$F = Ea^*[\text{chl } a]\varphi_f \tag{1}$$

Where E (mole quanta m^{-2} s^{-1}) is the intensity of the exciting source, a^* is the chl-specific absorption coefficient (m^2 mg [chl a]$^{-1}$) where [chl a] is the chl a concentration ([chl-a] hereafter) (mg chl a m^{-3}) and φ_f is the quantum yield for fluorescence [mole of emitted quanta (mole absorbed quanta^{-1})].

The fluorescence-chl a relationship for a given fluorometer varies according to environmental conditions such as the phytoplankton taxonomic composition and physiological adaptative mechanisms (e.g., Falkowski and Kolber, 1995; Babin et al., 1996, 2008).

The Cyclops 7 is a compact cylinder (110×25 mm after removal of the end cap), low energy consumption single channel fuorescence detector that can be used for many different applications. It delivers a voltage output that is proportional to the concentration of the chl a particle, or compound of interest. For chl a detection a 460 nm exciting wavelength and a 620–715 nm fluorescence detection photodiode are used. According to Turner Design specifications the minimum detection limit is 0.025 μg L^{-1} of chl a. The Cyclops 7 can be

set on a different level of sensitivity for chl a detection allowing detection of maximum chl a concentration ranging generally from low (i.e., detection range 0–500 μg L^{-1}) to medium (0–50 μg L^{-1}) and high (0–5 μg L^{-1}) sensitivities. For our application according to chl a climatologies available, the initial detection range was set between 0–2.5 μg L^{-1}, a range well matching the chl a concentration generally encountered within the oceanic waters of the SO (Reynolds et al., 2001; Marrari et al., 2006; Uitz et al., 2009).

The Cyclops 7 was integrated in a new CTD-Fluo Satellite Relay Data Loggers (Tags hereafter). They were built by the Sea Mammal Research Unit (SMRU) (University of St. Andrews, Scotland). Fluorescence was sampled continuously between the surface and 180 m. As Argos messages are restricted in length, we had to reduce the resolution of fluorescence data. Therefore, values were averaged for eighteen 10 m vertical sections. For each section the mean fluorescence value was allocated to the mid-depth point of the corresponding section.

Fluorometer calibrations, relying essentially on chl a solutions or on phytoplankton cultures, are generally provided by manufacturers. Most of the time these calibrations are established for large range of chl a concentrations not always representative of in situ ones. Therefore, it is highly desirable to confirm or adjust through in situ calibration on natural samples (see Xing et al., 2012). As part of this programme, a thorough calibration and testing procedure was undertaken for the CTD-Fluo SRDL. Pre-deployment calibrations of the tags and at-sea validating tests were conducted prior to SES deployment. This procedure was followed for most deployments in this study. Before being taken into the field, devices were calibrated at Valeport, Service Hydrographique de la Marine (Brest, France), and had temperature (T) and conductivity (C) resolutions of 0.001 °C and 0.002 mS cm^{-1}, respectively (see Roquet et al., 2011 for details).

2.2   Calibration procedure

The fluoremeters were inter-calibrated by implementing a Bayesian procedure using all information available regarding predeployment tests as well as post-deployment information collected.

2.2.1   Fluorometer inter-calibration and conversion in chl a concentration

Pre-deployment tests

Five consecutive sessions of CTD-Fluo SRDL deployments on SES (ft01, ft02, ft03, ft04 and ft06) were conducted as part of this study. The first two tags (ft01) were deployed on a seal without any pre-deployment test. For the second deployment (ft02), 8 tags were simultaneously tested at sea at Kerguelen along a 100 m-cast.

For the following deployment (ft03, ft04, ft06) and previous to their operational deployments on SES at Kerguelen

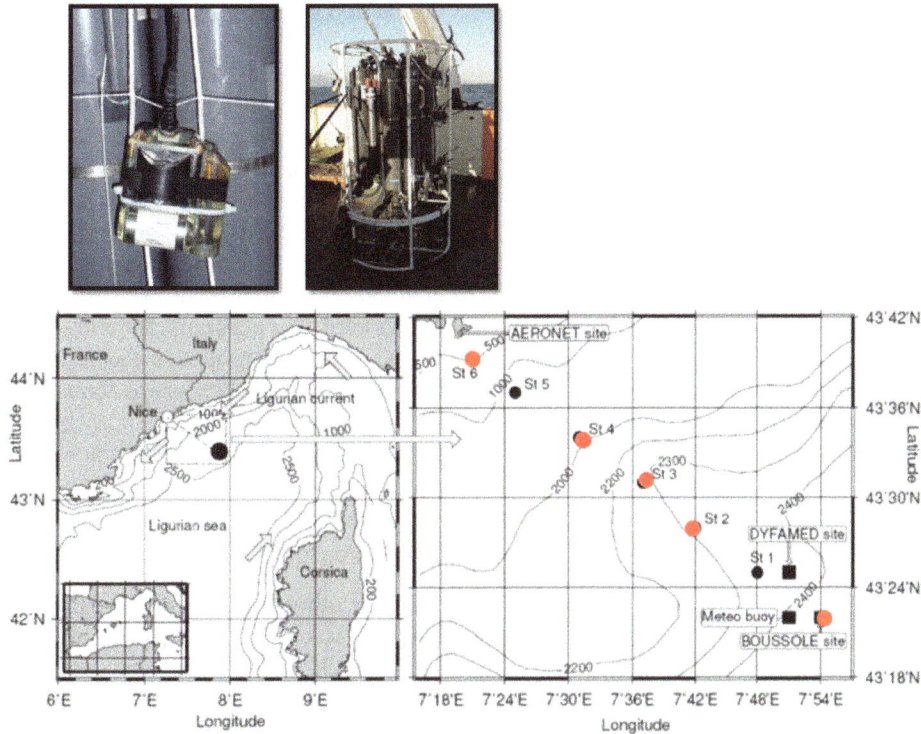

Figure 1. Top: CTD-FLUO-SRDL were fixed on the external part of the CTD-cage. Bottom: the Boussole at-sea test set up. Fluorescence profiles were conducted on stations located on the transect between Villefranche sur mer and the BOUSSOLE site (right). A fluorescence profile combined with water sampling for HLPC assessment of Chlorophyll a concentration was performed at the BOUSSOLE site located 70 miles off Villefranche sur mer (left).

Island the tags were tested in the Mediterranean sea during shipboard experiments. At sea tests were performed during the BOUSSOLE oceanographic cruises on the SSV "Tethys II" (Resp. D. Antoine, LOV). Each cruise consisted of a transect between the Nice harbour and the BOUSSOLE mooring site located in the north western Mediterranean sea (43°20′ N, 7°54′ E) with up to 6 oceanographic casts performed in between (Fig. 1). As part of the BOUSSOLE programme and associated cruises (Antoine et al., 2008) in the Ligurian Sea (Western Mediterranean), each series of tags were indeed attached to a CTD rosette generally immersed at Boussole site.

Water samples were collected for 10 different depths, filtered onboard and immediately frozen in liquid nitrogen before being stored at − 80°C back in the laboratory. High Performance Liquid Chromatography (HPLC) analysis of filtered samples was performed according to Ras et al. (2008) for the accurate determination of total chl a and accessory pigments (other chlorophylls and carotenoids).

The in situ calibration procedures for each tag subsequently include the deep offset fluorescence correction. Offset is detected in the profile through the fluorescence value (Fluo) in deep waters (like $z > 200$ m). Chl a fluorescence is considered as null at these depths because

HPLC [chl a] is below the detection limit (DL) of the method (DL $= 0.05$ mg m^{-3}). For each tag and each cast, the fluorescence-offset was calculated as difference between 0 and the fluorescence value provided by the fluorometer for every depth greater than 200 m. The mean and standard deviation (SD) of the fluorescence offset was then calculated for each tag. The mean offset value calculated for a given cast and a given fluorometer was retrieved to the fluorescence values provided by this fluorometer. The at-sea test offset calculation for ft02 was restrained between 80 and 100 m, for which fluorescence for all fluorometers reached constant and minimum values which were consistent with the offset values generally found during the other at-sea test.

Only the ascent values were used as (i) the water samples were only collected during the ascent phase and (ii) the CTD-Fluoro SRDL tags were programmed to sample the fluorescence only during the ascent phase when deployed on an elephant seal.

For this calibration procedure, tags were set on a sampling protocol and programmed to record pressure, temperature, conductivity and fluorescence every 2 s continuously while in the water. They were deployed to the side of the water-sampling rosette which was equipped with 11, 12 L Niskin bottles (General Oceanics) and a SBE 21 CTD profiler (Seabird Electronics) and with a CHELSEA Aqua-Track chl *a* fluorometer (hereafter named reference fluorometer). Chl *a* concentration was determined by HPLC on a sample of 2.27 L of water collected by the Niskin bottles for one and sometime two casts during the cruise. Water samples were usually collected from 12 depths (5, 10, 20, 30, 40, 50, 60, 70, 80, 150, 200 m).

The multiple casts conducted, without water sampling, allow us to (1) assess the deep fluorescence offset for each tag and to assess its inter-cast variability and (2) to assess if discrepancies between different fluorometers were consistent among and between casts.

chl *a* concentrations of the water samples collected on BOUSSOLE site were determined using standard fluorometric analysis of acetone extracts of the filtered samples. Water samples collected using Niskin bottles were filtered (2250 cm^3) onto glass fiber filters (Whatman GF/F, nominally 0.7 μm) using positive pressure. The filters were placed in a test tube, wrapped in aluminium foil and frozen in the dark. Back in the Villefranche laboratory chl *a* was extracted from the filter with 7 mL of HPLC grade acetone for 24 h in the dark. The pigment concentration was then analysed by the fluorometric method (Yentsch and Menzel, 1963) with a blanked and calibrated fluorometer (Turner Designs 10-AU).

Despite the off-set correction and a good agreement in the general shape of the fluorescence profiles provided by each fluorometer for a given cast, differences in absolute fluorescence values are clearly noticeable between fluorometers (see Fig. 2) which means that the calibration parameters provided by the manufacturer were not precise enough and/or that the integration of the fluorometer into the CTD SRDL tag degraded the fluorometer calibration. Therefore, the fluorometers of the CTD-Fluo tags needed to be re-calibrated in situ again.

To do so, for a given cast, the regression coefficient was calculated (without constant) for depth ranging between 0 and 200 m between the offset corrected fluorescence values provided by each CTD-Fluo SRDL and the reference fluorometer. This was performed for ft03/ft04 and ft06 deployments. Several casts were conducted for a given at-sea test allowing to estimate intra and inter-fluorometer variability.

Post-deployment procedure

The second step was to proceed to the inter-calibration of the fluorometers between all the at-sea tests and this was essential for the ft01 and ft02 for which no proper complete at-sea calibration procedure was performed previous to seal deployment. For ft02, the simultaneous testing of all the tags provided information about the proportionality between the fluorescence measurements provided by the different fluorometers and that information was used.

To intercalibrate the tags between each deployment we used all the information provided by at-sea tests as well as the proportionality found between surface values provided by

the tag fluorometer within a deployment and the corresponding chl *a* surface values provided by MODIS. IMODIS values were not used as an absolute measurement of [chl *a*], but as a relative measurement to better assess the proportionality between each fluorometer. An 8-day composite 9 km scale resolution MODIS data was the highest usable resolution to investigate the relationship between in situ surface fluorescence chl *a* and those provided by MODIS. Indeed too few MODIS values were available to investigate this relationship at a higher temporal (daily) and spatial (1 km) resolution at the tag level. The tag's surface fluorescence values used were offset and quenching corrected and saturated values retrieved (see below). The relationships found between the MODIS surface fluorescence values for each deployment was used to proceed to the production of a homogeneous fluorescence dataset.

Conversion to chl *a* concentration

The inter-calibrated fluorescence values were then converted into a chl *a* concentration value by using the relationship between the chl *a* concentration provided by the reference fluorometer and the [chl *a*] provided from HPLC measurements. This relationship was estimated over 70 test profiles ranging between 0 to 200 m and performed between 2002 and 2009. As all these profiles were performed during daylight hours, therefore, only fluorescence values deeper than 30 m were used to avoid any quenching effect and fluorescence values provided by the reference fluorometer were offset corrected.

2.2.2 Deployment on elephant seals

Instruments were deployed on SES either at the end of their moult in late summer to cover their pre-breeding, winter foraging trips or in October on post-breeding females. Animals were anesthetised with intravenous injection of tiletamine and zolazepam 1 : 1, and then instruments were attached to the fur on their head by using a two component industrial epoxy. Seals dove repeatedly with CTD-Fluo data being collected every 2 s during the ascent phase of dive and processed onboard before being transmitted via the Argos satellite system when animals were at the surface. On average, 1.8 ± 0.5 vertical temperature (T), conductivity (C) and fluorescence profiles were transmitted daily. Because of the narrow bandwidth of Argos transmitters, each profile was transmitted in a compressed form consisting of 18 fluorescence and 24 T and C data points. The 18 first T and C corresponded to the fluorescence measurements for the 0–180 m depth range. Fluorescence, T and C measurements were averaged over 10 m bin sampled for the upper 180 m of the dive. The 6 additional T and C corresponded to the most important inflection points determined onboard over between 180 m and the deepest part of the dive by using a "broken stick algorithm" (Roquet et al., 2011).

BOUSSOLE SITE

STATION 2

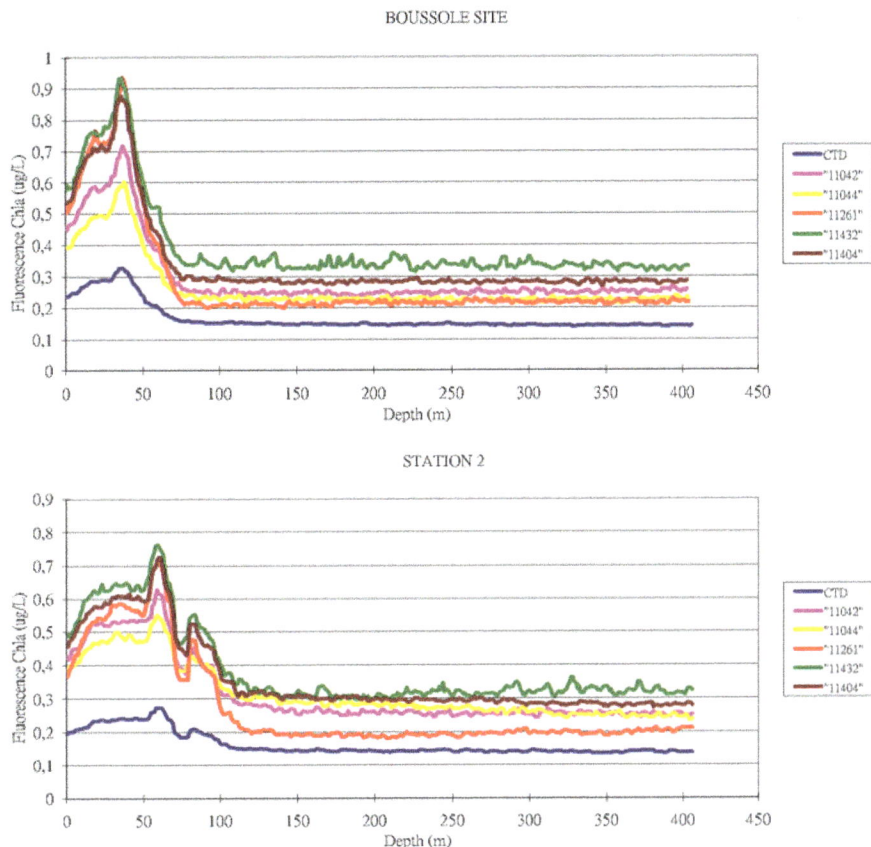

Figure 2. Example of fluorescence profiles provided by different CTD-FLUO-SRDLs and the reference fluorometer (in blue) obtained at two different stations. These profiles exhibit the existence of a fluorescence offset differing between CTD-FLUO-SRDLs. Fluorometers were providing consistent data between each other, but differed in the absolute amount of fluorescence produced.

2.2.3   Post deployment issues and correction processes

Chl a saturated values

According to chl a measurements available for the study area ft02, ft03, ft04 CTD-Fluoro SRDL Tags the Cyclops 7 fluorometer gain was set to monitor chl a concentration ranging between 0 and $2.5\,\mu g\,L^{-1}$. In situation of high in situ chl a concentration, some raw profiles exhibited saturated values. Therefore, these profiles were flagged accordingly and retained as saturated one in the data base. For the ft06 Tags the gain of the Cyclops 7 fluorometer was set for a dynamical range of 0 to $4\,\mu g\,L^{-1}$ and saturated chl a profiles were exceptionally encountered and flagged accordingly. Unsaturated profiles were flagged as "1" while saturated profiles were flagged as "2".

Offset correction

Initially each in situ profiles was corrected according to the mean offset calculated for each tag from the at-sea test conducted prior to deployments which exhibited very little variability for a given fluorometer between cast (see result part).

However, post deployment values revealed that this offset could vary over time and the reasons for such variations are not yet fully understood, but are thought to be related to the water masses encountered and in particular the amount of non-phytoplanktonic particles. The use of a constant offset correction was leading to positively or negatively bias profiles values depending on situations. Therefore, a profile-by-profile-offset-correction method was implemented and three cases were distinguished and flagged accordingly. In the first situation the 4 deepest chl a values corresponded (i) to the minimum values of the profiles, with a standard deviation (SD) on chl a value lower than 10 %. In this situation the lowest chl a value among the 4 deepest values was used as the offset value and the whole profile was corrected and flagged as "1"; (ii) when the deepest chl a values did not correspond to the minimum values, or when the SD was higher than 10 % the profile was flagged as "2" and was corrected according to the mean offset calculated for that Tag over the whole deployment period; (iii) a third situation was observed in a few cases, and for which chl a concentration was increasing at depth. No offset correction was applied to these profiles which were flagged as "3" and these flag 3 profiles are not integrated in the current data base.

Quenching correction

In both, laboratory and field studies, a daily rhythm of in vivo fluorescence that is not correlated with diel changes in the concentration of chl a have been reported in a number of studies. During periods of high irradiance, fluorescence tends to be lower than the value at night (Kiefer, 1973; Loftus and Seliger, 1975; Falkowski and Kolber, 1995; Dandonneau and Neveux, 1997; Behrenfeld and Kolber, 1999; Kinkade et al., 1999). The photo-inhibition of phytoplankton by an excess of light, result in a decrease of the fluorescence quantum yield (i.e., the ratio of photons emitted as fluorescence to those absorbed by photosynthetic pigments). This is often ascribed to a set of processes generally termed non-photochemical fluorescence quenching (NPQ; Falkowski and Kolber, 1995; Krause and Jahns, 2004). Quenching is commonly observed during daytime with a maximum intensity at midday (Kiefer, 1973; Dandonneau and Neveux, 1997) and it is referred as daytime fluorescence quenching. Quenching poses a problem and, therefore, need to be accounted for. We applied the new method of quenching correction developed, tested and successfully applied to fluorescence data collected by elephant seals (Xing et al., 2012). In short, for mixed type waters, the maximum fluorescence values within the mixed layer is extended to the surface (i.e., all upper points are replaced by this maximum value). However, for stratified waters, which usually have a thin mixed layer, obviously, the quenching effect will pass through the mixed layer into the stratified layer. In the stratified layer, we were unable to perform such correction and only night time fluorescence profiles were used. However, the stratified layer situations were only encountered for 16 % of the fluorescence profiles sampled, and nearly half of them were obtained at night. In this process, we assume the maximum value in the mixed layer as the non-quenching value and all above points are corrected to the same value (see Xing et al., 2012). After implementing such corrections, no differences could be detected between day time fluorescence corrected profiles and the proximate night profiles (just before or after the day profiles).

2.2.4 Estimation of the tag specific chl a correction coefficient: a Bayesian approach

Using a Bayesian adjustment framework and by combining each at-sea test, the information available for each profile and the HPLC values available, a chl a calibration coefficient was calculated for each CTD-Fluoro SRDL tag. A Bayesian framework is especially suited as it guarantees that between-casts variability is taken into account in estimating these inter-fluorometer coefficients. An attractive feature of this framework is the relative ease with which different datasets are combined in a single analysis, allowing information transfer between disparate sources. Another advantage stems from the easy computation of confidence intervals for parameters of interest and prediction errors. Models were fitted with *WinBUGS* (Spiegelhalter et al., 2003) called from *R* (R Development Core Team, 2009) with the package *R2WinBUGS* (Sturtz et al., 2005). Weakly informative and robust priors were favoured. We used uniform priors (Gelman, 2006) for standard deviation parameters. Because all slopes were expected to be positive, we used Student-t priors with mean 0, scale 10 and 7 df on a log scale (Gelman et al., 2008). Batches of tag-specific coefficient were assumed to follow a bivariate Gaussian distribution with covariance matrix $\sum$. We used the prior described in Tokuda et al. (2011) for the covariance matrix $\sum$.

We built a model to predict HPLC [chl a] from either in situ fluorescence (Fluo) measured by instrumented SES or from MODIS chl. Specifically we first used long-term data at the Boussole site to estimate the relationship between HPLC [chl a] and measurement from the reference fluorometer:

$$\text{HPLC [chl } a] = \delta \cdot \text{Fluo[reference]} + \varepsilon 1 \qquad (2)$$

where $\varepsilon 1$ is a Gaussian residual error term.

Secondly, we used data from the intercalibration experiment to estimate the relationship between a specific tag j and the reference fluorometer:

$$\text{Fluo[reference]} = \alpha j \cdot (\text{Fluo } j - \text{Offset} j) + \varepsilon 2 \qquad (3)$$

where $\varepsilon 2$ is a Gaussian residual error term. When measurements from the reference fluorometer were unavailable, but measurements from the different tags at a specific location were available, we rewrote Eq. (3) as:

$$\text{Fluo[reference]} - (\alpha j \cdot (\text{Fluo } j - \text{Offset} j) + \varepsilon 2) = 0 \qquad (4)$$

and put a weakly informative prior (Half-Student-t with mean 0, scale 5 and 4 df, Gelman et al., 2008) on the value of Fluo[reference]. We then used the *WinBUGS* "zero-trick" (Spiegelhalter et al., 2003) to incorporate these data into the model. Note that the error term $\varepsilon 2$ is the same for Eqs. (3) and (4).

Finally, we evaluated the relationship between in situ fluorescence as measured from a specific tag j and MODIS:

$$(\text{Fluo } j - \text{Offset} j) = \beta j \cdot \text{MODIS} + \varepsilon 3 \qquad (5)$$

where $\varepsilon 3$ is a Student residual error term. Equation (5) is a regression with heteroskedastic noise to account for possible outliers, or extreme observations, when evaluating the relationship between MODIS and in situ fluorescence. To make robust inferences, the parameter ν, that is the degrees of freedom of the Student distribution, was estimated from the data (with a Gamma (shape = 2, scale = 4) prior for ν). All data points available and corresponding to a match between chl a provided by MODIS and the fluorometer for a given tag were used in estimating βj.

Table 1. Details of the 23 CTD-fluo tags deployments. Deployment date, Sampling period, number of fluoro-profiles obtained, and correction coefficient applied to each tag.

Deployment number	CTD-Fluo SRDL number	Deployment date	End transmission	number of fluorescence profiles collected	number of fluorescence profiles usable	Daylight hours profile	night hours profile	nb of saturated profile	nb of quenching corrected profile	nb prof flag 1 offset profile	nb of prof flag 2 Offset profile	Mean offset (predeployment sea test)
FT01	10863	12/22/2007	6/5/2008	241	241	166	75	32	76	212	29	–
FT02	10946	1/20/2009	12/20/2009	331	331	187	144	1	145	248	83	1.28
FT02	11034	1/20/2009	3/12/2009	73	73	45	28	0	15	40	33	0.36
FT02	11035	1/28/2009	9/21/2009	404	404	202	202	0	148	292	112	0.44
FT02	11039	1/25/2009	7/13/2009	289	289	169	120	51	122	215	74	0.72
FT02	11040	1/20/2009	2/16/2009	41	41	26	15	1	13	30	11	0.43
FT02	11042	12/24/2008	6/1/2009	236	229	118	111	33	113	146	83	0.71
FT02	11044	1/10/2009	6/19/2009	267	267	139	128	92	66	238	29	0.51
FT03	11038	10/17/2009	12/31/2009	141	130	82	48	63	50	71	59	0.20
FT03	11259	10/19/2009	1/4/2010	134	125	72	53	35	50	85	40	0.17
FT03	11260	10/20/2009	1/9/2010	156	155	101	54	42	82	122	33	0.23
FT03	11262	10/23/2009	1/18/2010	169	169	114	55	45	100	144	25	0.28
FT03	11263	10/21/2009	1/9/2010	157	156	114	42	12	91	116	40	0.16
FT04	11042	3/8/2010	5/25/2010	97	51	0	51	0	0	38	13	0.24
FT04	11044	2/20/2010	8/19/2010	314	128	0	128	2	0	95	33	0.23
FT04	11261	2/15/2010	2/27/2010	13	6	0	6	1	0	4	2	0.21
FT04	11404	2/20/2010	9/25/2010	418	125	1	124	1	0	82	43	0.28
FT04	11432	3/12/2010	9/15/2010	309	112	0	112	0	0	87	25	0.33
FT06	10946	11/4/2010	1/19/2011	150	101	61	40	7	44	101	0	
FT06	11035	11/4/2010	1/9/2011	120	102	84	18	35	53	102	0	0.30
FT06	11038	11/2/2010	1/25/2011	162	88	66	22	70	48	88	0	
FT06	11262	10/22/2010	12/30/2010	136	6	3	3	0	2	6	0	0.24
FT06	11263	9/9/2010	11/24/2010	140	59	26	33	1	26	59	0	0.16
Total				4498	3388	1776	1612	524	1244	2621	767	

Combining Eqs. (2) and (3) yields:

$$\text{HPLC [chl } a] = \delta \cdot \text{Fluo[reference]} = \delta \cdot \alpha j \cdot (\text{Fluo } j - \text{Offset } j) \tag{6}$$

where $\delta \cdot \alpha j$ is a tag specific calibration coefficient allowing to predict HPLC [chl a] from in situ fluorescence. Combining Eqs. (5) and (6) further yields:

$$\text{HPLC [chl } a] = \delta \cdot \alpha j \cdot \beta j \text{ MODIS} \tag{7}$$

which allows us to predict for each tag the likely value of HPLC [chl a] from MODIS measurement.

3 Results

3.1 At-sea trials prior to deployment

The at-sea testing previous to deployment revealed two types of issues which were tag's fluorometer dependant: (1) a fluorescence offset was, therefore, an instrument-specific clean water background, equivalent to the lowest values observed in the deepest part of the fluorescence profiles was subtracted from the raw fluorescence values; (2) for offset-corrected fluorescence profiles, differences between tags in the absolute amount of fluorescence were detected.

The variability of the fluorescence offset for a given tag was one order of magnitude smaller than the offset differences between tags (Table 1). The mean fluorescent offset for the tested tags during the BOUSSOLE cruise was $0.24 \pm 0.05\,\mu g\,L^{-1}$ (range 0.16–0.33, $n = 14$). For a given tag, the standard deviation of inter-cast fluorescent offset ranged from (0.0007 to 0.0620, mean $= 0.0148\,\mu g\,L^{-1}$).

3.1.1 Tag's fluorometer intercalibration

The multiple cast performed during the at-sea test prior to ft03, ft04 and ft06 deployments allowed assessing the intra tag-fluorometer variability. This variability ranged from to a minimum of 0.03 % to a maximum of 5.61 % with a mean of 2.08 ± 1.84 %. However, the difference between fluorometers was one order of magnitude larger than the within fluorometer variability and one fluoromenter provided fluorescence values which were on average 2.61 time higher than the minimal values. The mean inter-fluorometer variation in fluorescence values was 69 % (range 0.005 to 261 %).

On average αj indicated that the tested fluorometers provided [chl a] 4 times greater (i.e., 0.24^{-1}) than the reference fluorometer.

3.1.2 Reference fluorometer-HPLC relationship

Over the 2002–2009 period, HPLC values were found to be linearly related to the chl a estimates provided by the reference fluorometer. δ was estimated to be 2.53 (i.e., chl a concentration provided by HPLC were found to be 2.53 higher than those estimated from the reference fluorometer (Fig. 3, Table 1). The tag fluorometers provided chl a concentration which were on average 1.64 time greater than chl *a* concentration provided by concomitant HPLC measurements.

3.2 Post deployment data

From December 2007 to July 2010, a total of 27 SES, were fitted tags, of which 23 provided usable fluorescence data (Table 1). The 23 SES tracks provided a broad geographic and seasonal coverage ranging from Antarctic to subtropical waters, but with most individuals concentrating east of Kerguelen Island within the Kerguelen plume (Fig. 4). A total of 4662 fluorescence profiles were transmitted, but 1274 of them either incomplete, with constant values or presenting obvious anomalies were discarded (i.e., 27 %). Among the remaining 3388 profiles 1776 were collected during the day and 1612 at night and year round (Fig. 5). The summary of the different profiles and flagging situation are provided in Table 1. This dataset (doi:10.7491/MEMO.1) is freely available at http://www.cebc.cnrs.fr/ecomm/Fr_ecomm/ecomm_memoOCfd.html.

524 profiles exhibited saturated fluorescence values (i.e., 18 %) and were excluded for the comparison with the corresponding weekly chl *a* MODIS data. However, due to heavy cloud cover only 884 surface values of CTD-fluo profiles (i.e., 30.9 %) could be matched with the weekly MODIS data, and 126 with the MODIS data collected on the same day (i.e., 4.4 %).

Among these 23 tags deployed on SES, 8 were recovered when SES came back onshore after at sea periods ranging between 3 to 8 months. At recovery the optical face of Cyclops 7 was clean with no bio-fouling most likely because elephant seals are typically deep divers spending very short periods within the euphotic zone. Furthermore, they spent most time at low temperatures.

On the 3388 fluorescence profiles, 40 % exhibited day time fluorescence quenching (i.e., 70 % of daytime profiles – Table 1). Therefore these profiles were corrected according to the procedure proposed by Xing et al. (2012). The recovery of fluorescence from quenching is obvious when we compare day profiles with night profiles obtained for the same location at the same date (Fig. 6). Individual seal collected chl *a* transects along their track monitoring latitudinal/longitudinal changes for a given time period as well as seasonal change within a given area (Fig. 7).

As quenching could only be corrected in well mixed water and not in stratified ones, we were (1) unable to proceed to quenching correction of fluorescence profiles in 283 profiles (i.e., about 16 % of daytime profiles) and therefore unable to assess properly deep maximum fluorescence for these daytime fluorescence profiles. To address this question we only used the 1612 night time profiles. Among those 352 had a maximum value at surface, while 742 exhibited a maximum deeper than 30 m. However, the vast majority the maxima was not exceeding the surface values by more than 30 % (i.e., 1423 profiles). Only 148 night profiles (i.e., 9 % of the to-

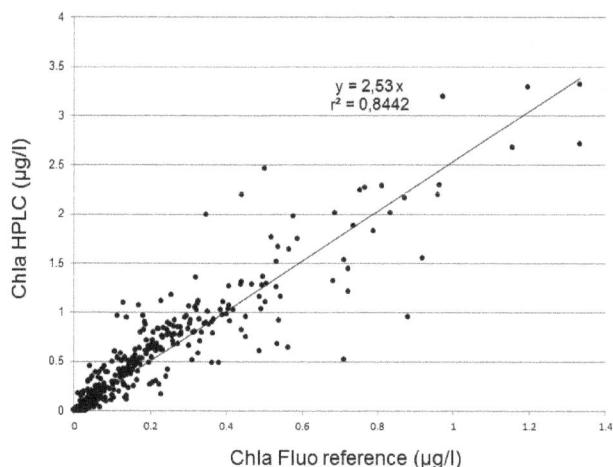

Figure 3. Relationship between the chl *a* concentration provided by the reference fluorometer and the chl *a* concentration estimated from HPLC measurements. This relationship was estimated over 70 test profiles ranging between 30 to 200 m and performed between 2002 and 2009.

tal number of night profiles) had a deep maximum exceeding 30 % of the surface value (maximum: 180 %). The depth class distribution of the maxima values exceeding 30 % of the surface value are shown in Fig. 8 and exhibit a clear bimodal distribution with a 30–40 m and a 70–80 m modes.

No obvious spatial structuring in the distribution of these profiles could be identified through the range of the SES.

Deep offset and quenching corrected surface fluorescence values provided by intercalibrated tags according to the reference fluorometer (ft03, ft04 and ft06) were related to the corresponding MODIS chl *a* value along SES tracks. This relationship was implemented within the Bayesian framework to correct tag fluorescence values for which no at-sea test was performed previous to the deployment. For the ft02 deployment the proportionality found between tags, all tested simultaneously at sea was used. Following the Bayesian procedure previously described the correction coefficients ($\delta \cdot \alpha j$, mean value) applied to each tag were calculated and are given in Table 2.

The surface [chl *a*] values derivated from by offset and quenching corrected profiles were found to be related to the 8-day-9 km MODIS chl *a* values. On average MODIS values were 3.04 times (βj, range: 1.90–8.74) lower than the corresponding tag fluorometer, and we found that on average, MODIS tended to underestimate HPLC related [chl *a*] by a mean 1.99 ($\delta \cdot \alpha j \cdot \beta j$, range: 1.04–3.21) factor compared to the in situ estimates provided by the inter-calibrated fluorescence tag (Table 2, Fig. 9). The variability of βj (i.e., the Fluoremeter/MODIS relationship) was on average larger than the inter-fluorometer one, suggesting that a large part of the error is likely due to the poor relationship found between [chl *a*] provided by individual fluorometers and MODIS. However, this work emphasises that despite the fact that all

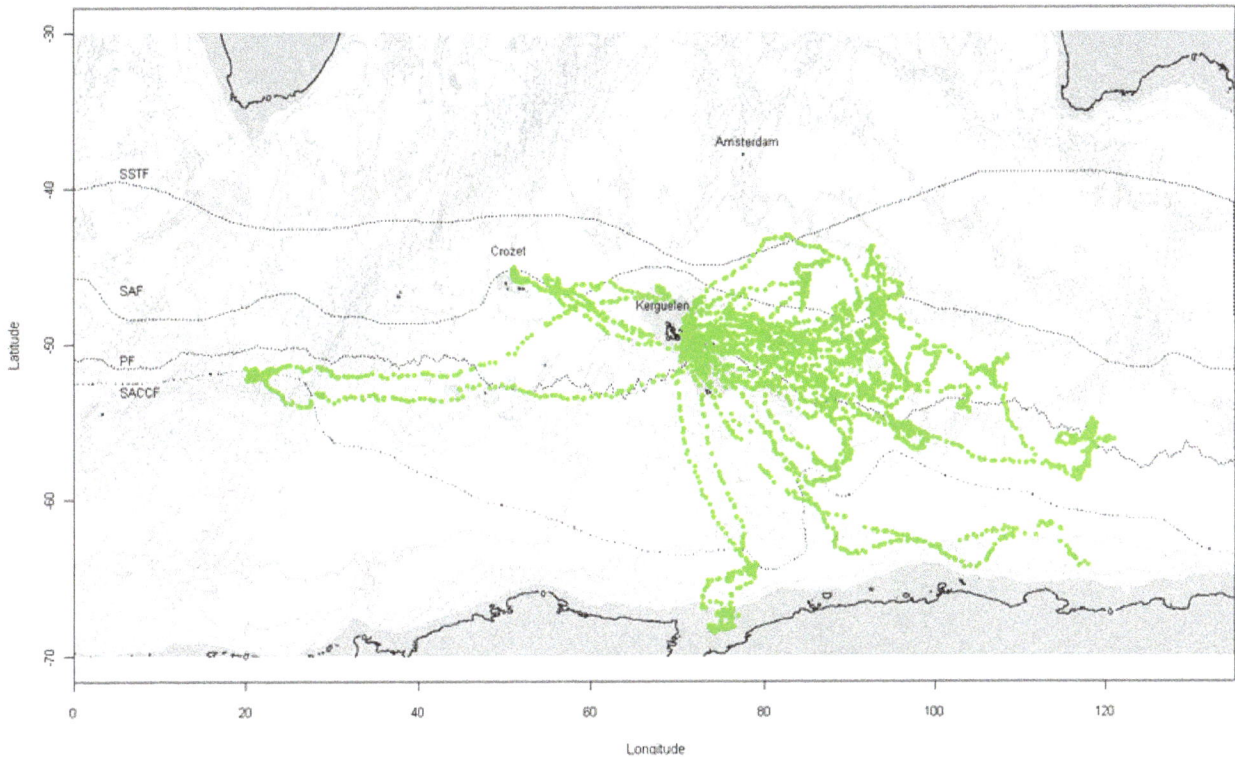

Figure 4. Location of the fluorescence profiles along the tracks of 24 SES successfully equipped with a CTD-FLUO-SRDL and collected within the scope of this study (December 2007–January 2011).

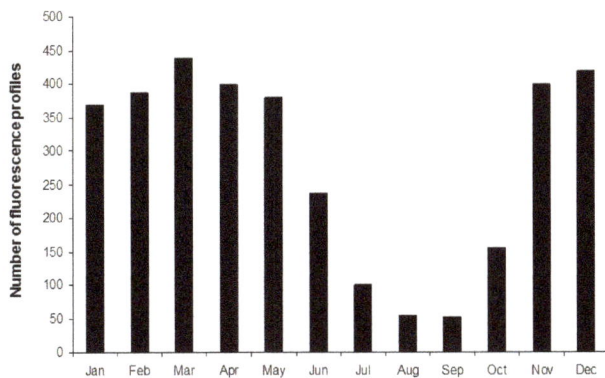

Figure 5. Monthly distribution of the fluorescence profiles.

the fluorometers were identical, nevertheless some large differences could be observed between fluorometers with αj ranging between 0.11 and 0.36 (mean: 0.24) and those differences in themselves require those fluorometers to be intercorrected between each other. For the SO total absence of chl a was never detected by MODIS with the lowest value observed of $0.06\,\mu g\,L^{-1}$ in our case, while in situ fluorescence measurements suggest that total absence ($\pm 0.025\,\mu g\,L^{-1}$, i.e., the detection limit of the Cyclops 7 fluorometer) of chl a can be observed.

4   Discussion and conclusion

The fluorescence profiles collected by the SES within the SO, provides 3-D information on otherwise poorly sampled area such as the Antarctic sea-ice zone, an area were the ocean satellites are blinded by sea-ice and/or cloud cover. SES provided an unequalled dataset of fluorescence profiles associated with temperature and salinity measurements (i.e., density) for a broad sector of the Indian part of the SO and this dataset represents a significant contribution in to understanding of the seasonal variation of phytoplankton biomass.

The significant contribution of this study was to propose a detailed and step-by-step procedure to intercalibrate the fluorometer to provide consistent chl a estimates. To summarise the first step requires the correction of the deep fluorescence offset, the second step requires to have at least a common reference fluorometer between the performed tests. The third step, i.e., to proceed to the HPLC calibration, requires eliminating quenching affected fluorescence surface values obtained during daytime or better to perform night fluorescence profiles. Furthermore, we suggest that this HPLC intercalibration should not to be performed on a profile-to-profile basis, but instead according to the fluorescence/chl a relationship established from multiple at-sea tests. We found, in this study, that while the fluorescence-HPLC chl a relationship can vary from one at-sea test to the other, however,

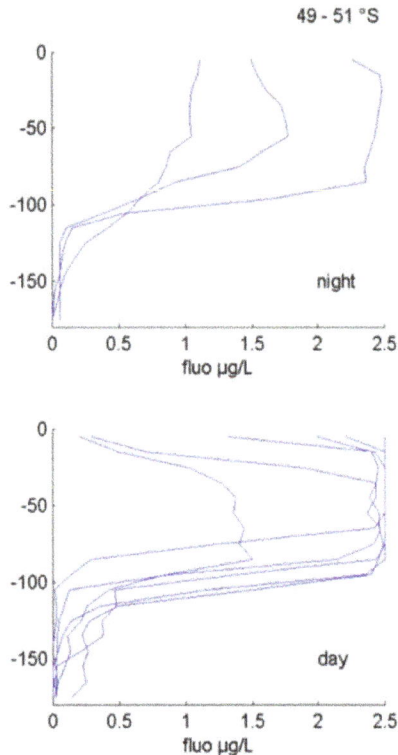

Figure 6. Example of unquenched night fluorescence profiles (top) and quenched day ones (bottom) collected on the same location and same period by a CTD-FLUO-SRDL deployed on a male SES foraging over the Kerguelen plateau.

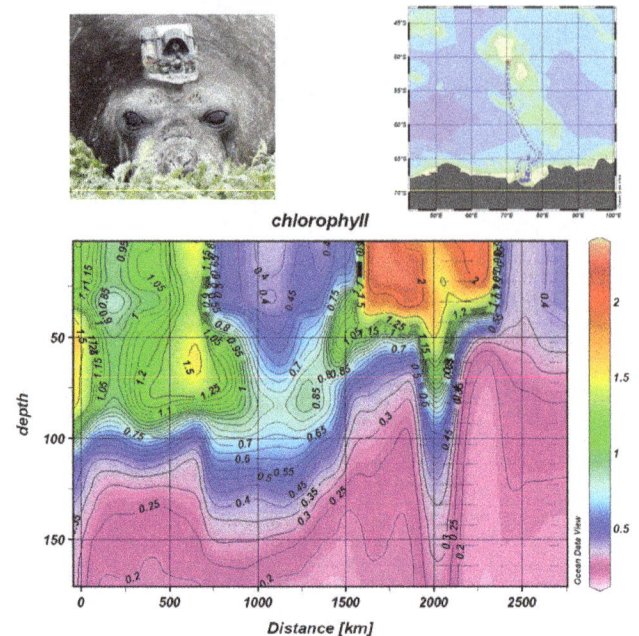

Figure 7. Top: female SES equipped with a CTD-FLUO-SRDL and section to the track followed by a juvenile SES female between January–April 2009. This female left Kerguelen on 12 January and reached the Antarctic shelf on 6 February. This female left the Antarctic shelf on 14 March, the female then remained associated with the marginal ice zone and the Antarctic divergence. Bottom: interpolated quenching corrected fluorescence profiles provided by this SES along its track. This tag exhibited a fluorescence offset of $0.2\,\mu g\,L^{-1}$ of chl a and Antarctic values higher than $2\,\mu g\,L^{-1}$ were saturated. The abrupt change in chl a concentration (km 2400) coincides with sea-ice formation which took place in mid march. These data show both the latitudinal change in the phytoplankton concentration along a north south transect performed during the inward trip (i.e., 1600 km south of Kerguelen 500 m isobath) and the transition in phytoplankton concentration in Antarctic waters from summer to fall (from km 1600 to 2400).

the long-term relationship established over several years and encompassing numerous at-sea test exhibit a very good linear relationship (Fig. 3). Therefore, we suggest using this global relationship to transform the inter-calibrated fluorescence values provided by the fluorometers to an actual estimation of [chl a] value. In a last step when there are sufficient surface fluorescence measurements coinciding with MODIS one, MODIS data can be used as a common but weak relative (not absolute) reference between fluorometers as many issues are affecting the quality of the relationship: low number of corresponding values, a poor temporal and spatial correspondence when using 9 km weekly data.

Due to logistical constraints, all the inter-calibrations were performed form at-sea test conducted in the north-west Mediterranean Sea. This could result in some differences in the absolute amount of chl a estimated from fluorescence data in the SO and this point should be investigated in greater details in future studies. This procedure, nevertheless, presents the major advantage of producing a dataset in which all fluorometers are inter-calibrated with each other. Furthermore the long-term relationship established between the reference fluorometer and HPLC in the north-west Mediterranean Sea is likely to be robust as it was established over

a broad range of years and seasons encompassing different phytoplankton assemblages.

One important result of this study was to show that MODIS may underestimate surface [chl a] compared to in situ measurements provided by the inter-calibrated fluorometers (Fig. 9). In situ chl a concentration provided by the fluorometer was correlated with the MODIS, but data points were highly dispersed. This is not surprising due to the low spatial and temporal resolution of the MODIS data used to investigate this relationship while surface values collected by the tags were associated with a unique location within that 9×9 km sector and, therefore, small scale variation which can be measured by the fluorometer are likely to be overlooked by the 9×9 km MODIS and weekly data.

This finding is consistent with several studies showing that standard satellite ocean colour algorithms tend to underestimate chl a concentrations in the SO compared to in situ

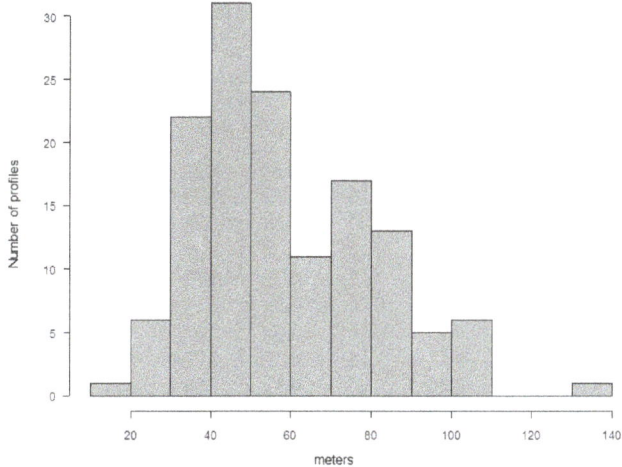

Figure 8. Depth distribution of chl *a* maxima exceeding 30 % of the surface values.

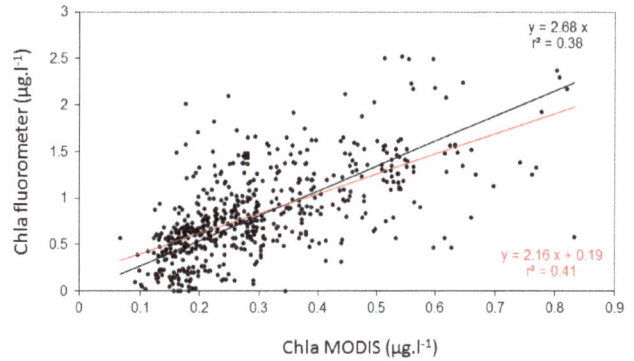

Figure 9. Relationship between the offset, quenching corrected HPLC inter-calibrated fluorometers with the corresponding 9 km weekly MODIS data. Both regressions with (in red) and without (in black) intercept are presented.

HPLC measurement or from calibrated fluorometer (Mitchell and Holm-Hansen, 1991; Dierssen and Smith, 2000; Holm-Hansen et al., 2004; Garcia et al., 2005) with standard algorithms typically underestimating chl *a* by 2–3 times (Kahru and Mitchell, 2010) which is consistent with our study, and up to 5 times (Dierssen, 2010). These errors are typically transferred to the estimates of primary production and carbon fluxes.

Therefore, we are confident that the current MODIS data underestimate by a large extent the in situ [chl *a*]. In our study the MODIS underestimation was found over the whole study area and, therefore, encompassed a broad range of phytoplankton assemblage. But future studies should investigate in greater details if the relationships between chl *a* surface concentration estimated from in situ fluorescence measurements and MODIS do vary according to the biogeographic regions of the SO visited by the SES. Furthermore, the in situ chl *a*/MODIS relationship is characterised by an intercept which is different from 0. A recent study reveals that within the SO the current ocean colour algorithms significantly underestimate and overestimate chl *a* at high and low concentrations, respectively (Johnson et al., 2013).

Global estimates of ocean primary production are now based on satellite Ocean Colour data (Longuhurst et al., 1995; Antoine et al., 1996; Behrenfeld and Falkowski, 1997; Behrenfeld et al., 2005). Time series have been built, from which climate-relevant trends can be extracted (Antoine et al., 2005; Polovina et al., 2008; Martinez et al., 2009). In situ and satellite data are highly complementary. Whereas in situ data extend the satellite information into the ocean interior (unseen by the remote sensor) and provide indispensable sea truth data, satellite provides the synoptic coverage.

We also found that when taking into account the quenching effect, deep maxima chl *a* concentration exceeding 30 % of the surface value were found only in 9 % of the night pro-

file. According to this result the decoupling of surface and deep phytoplankton biomass observed for only 9 % of the profiles is unlikely to be a major issue when estimating primary production from surface data. However, the real issue is the quenching effect observed during daylight hours and requiring to be properly dealt with (Xing et al., 2012).

Models can only provide useful answers if there are sufficient data to constrain the underlying processes and validate the model output. New approaches to assimilate biological and chemical data into these models are advancing rapidly (Brasseur et al., 2009). Notably, the progressive integration of biogeochemical variables in the next generation of operational oceanography systems is one of the long-term objectives of the GODAE OceanView international programme. Nevertheless, and in view of refining these models for improving their representativeness and predictive capabilities, the datasets currently available remain too scarce. There is an obvious and imperative need to reinforce biological and biogeochemical data acquisition and to organise databases and SES equipped with CTD-Fluo SRDL tags are contributing efficiently to this need.

In situ acquisition of fluorescence data by SES represents a significant contribution to the observation of biogeochemical and ecosystem variables within this undersampled Ocean. These new data implemented in tight synergy with two other essential bricks of an integrated ocean observation system: modelling and satellite observation should represent a significant contribution towards the resolution of important scientific questions relative to the overall phytoplankton biomass and primary production and ultimately changes in carbon fluxes within the SO in relation to climate variability and longer term changes.

Table 2. Correction coefficient calculated for each fluorometer tag (see methods for details). The bold values refer to the correction coefficient applied to the corresponding fluorometer tag, to estimate [chl a].

Deployment number	CTD-Fluo SRDL number	αj			βj			$\delta \cdot \alpha j$			$\delta \cdot \alpha j \cdot \beta j$		
		Mean	Lower	Upper	Mean	Lower	Upper	Mean	Lower	Upper	Mean	Lower	Upper
FT01	10863	0.26	0.13	0.46	2.96	2.20	3.77	**0.65**	0.32	1.16	1.91	0.92	3.52
FT02	10946	0.24	0.12	0.45	4.74	4.44	5.04	**0.61**	0.30	1.14	2.90	1.41	5.38
FT02	11034	0.33	0.18	0.50	2.08	1.65	2.51	**0.83**	0.47	1.26	1.73	0.93	2.72
FT02	11035	0.29	0.17	0.42	1.90	1.74	2.06	**0.74**	0.44	1.08	1.39	0.83	2.08
FT02	11039	0.14	0.09	0.20	8.74	7.77	9.75	**0.36**	0.22	0.51	3.12	1.88	4.49
FT02	11040	0.24	0.14	0.35	3.49	2.90	4.07	**0.61**	0.36	0.89	2.11	1.22	3.20
FT02	11042	0.19	0.12	0.27	3.50	3.32	3.67	**0.48**	0.30	0.68	1.67	1.05	2.36
FT02	11044	0.22	0.14	0.32	1.81	1.44	2.21	**0.57**	0.35	0.81	1.02	0.59	1.50
FT03	11038	0.11	0.10	0.12	3.58	2.74	4.46	**0.28**	0.25	0.31	1.00	0.75	1.28
FT03	11259	0.23	0.21	0.26	2.42	1.77	3.10	**0.59**	0.52	0.66	1.42	1.02	1.86
FT03	11260	0.22	0.19	0.24	2.07	1.69	2.44	**0.54**	0.48	0.61	1.13	0.89	1.36
FT03	11262	0.22	0.19	0.25	3.79	3.38	4.18	**0.56**	0.49	0.63	2.10	1.77	2.45
FT03	11263	0.27	0.23	0.30	2.29	1.90	2.71	**0.67**	0.59	0.76	1.54	1.23	1.89
FT04	11042	0.30	0.22	0.38	3.42	2.99	3.87	**0.76**	0.57	0.96	2.61	1.89	3.41
FT04	11044	0.35	0.26	0.45	2.61	2.24	3.01	**0.89**	0.66	1.15	2.33	1.65	3.09
FT04	11261	0.24	0.18	0.30	2.39	1.56	3.28	**0.61**	0.46	0.75	1.45	0.88	2.13
FT04	11404	0.25	0.19	0.31	5.07	4.51	5.64	**0.63**	0.48	0.80	3.21	2.37	4.14
FT04	11432	0.24	0.18	0.31	2.83	2.51	3.18	**0.62**	0.46	0.78	1.75	1.28	2.25
FT06	10946	0.24	0.12	0.45	4.06	3.68	4.45	**0.61**	0.31	1.13	2.49	1.24	4.59
FT06	11035	0.36	0.32	0.39	3.11	2.48	3.74	**0.90**	0.81	0.99	2.80	2.19	3.42
FT06	11038	0.25	0.12	0.46	4.24	2.99	5.60	**0.62**	0.30	1.16	2.63	1.20	5.23
FT06	11262	0.36	0.32	0.39	2.72	1.58	3.91	**0.90**	0.82	0.99	2.45	1.44	3.55
FT06	11263	0.21	0.19	0.23	1.95	1.44	2.45	**0.53**	0.48	0.58	1.04	0.76	1.32
Mean		0.24	0.20	0.29	3.04	2.53	3.57	0.61	0.51	0.73	1.99	1.44	2.33

Acknowledgements. This work was supported by CNES-TOSCA (Southern Elephant Seal as oceanographer Fluorescence measurements), The MEMO observatory as part of the SOERE CTD-02, the ANR VMC IPSOS-SEAL (2008–2011) and the Total Foundation. We are indebted to IPEV (Institut Polaire Français), for financial and logistical support of Antarctic research program 109 (Seabirds and Marine Mammal Ecology lead by H. Weimerskirch). Special to the BOUSSOLE cruise and team for their help in proceeding at-sea tests and to F. Roquet who proceeded to the correction/validation of the temperature/salinity data. Finally, we would like to thank all the colleagues and volunteers involved in the field work on southern elephant seals at Kerguelen Island, with the special acknowledgement of the invaluable field contribution of N. ElSkaby, G. Bessigneul, A. Chaigne and Q. Delorme.

Edited by: F. Schmitt

References

Antoine, D., André, J. M., and Morel, A.: Oceanic primary production 2. Estimation at global scale from satellite (coastal zone color scanner) chlorophyll, Global Biogeochem. Cy., 10, 57–69, 1996.

Antoine, D., Morel, A., Gordon, H. R., Banzon, V. F., and Evans R. H.: Bridging ocean color observations of the 1980s and 2000s in search of long-term trends, J. Geophys. Res., 110, C06009, doi:10.1029/2004JC002620, 2005.

Antoine, D., D'Ortenzio, F., Hooker, S. B., Bécu, G., Gentili, B., Tailliez, D., and Scott, A. J.: Assessment of uncertainty in the ocean reflectance determined by three satellite ocean color sensors (MERIS, SeaWiFS, and MODIS-A) at an offshore site in the Mediterranean Sea, J. Geophys. Res., 113, C07013, doi:101029/102007JC004472, 2008.

Argos: User's manual, CLS/Service Argos, Toulouse, 1996.

Arrigo, K. R., Worthen, D. L., Schnell, A., and Lizotte, M. P.: Primary production in Southern Ocean waters, J. Geophys. Res., 103, 587–600, 1998.

Babin, M.: Phytoplankton fluorescence: theory, current litterature and in situ measurements, in: Real-time coastal observing systems for marine ecosystem dynamics and harmful algal blooms, edited by: Babin, M., Roesler, C., and Cullen, J. J., 237–280, Unesco, Paris, 2008.

Babin, M., Morel, A., and Gentili, B.: Remote sensing of sea surface sun-induced chlorophyll fluorescence: Consequences of natural variations in the optical characteristics of phytoplankton and the quantum yield of chlorophyll a fluorescence, Int. J. Remote Sens., 17, 2417–2448, 1996.

Behrenfeld, M. J. and Falkowski, P. G.: Photosynthetic rates derived from satellite-based chlorophyll concentration, Limnol. Oceanogr., 42, 1–20, 1997.

Behrenfeld, M. J. and Kolber, Z. S.: Widespread iron limitation of phytoplankton in the south Pacific Ocean, Science, 283, 840–843, 1999.

Behrenfeld, M. J., Boss, E., Siegel, D. A., and Shea, D. M.: Carbon-based ocean productivity and phytoplankton physiology from space, Global Biogeochem. Cy., 19, GB1006, doi:10.1029/2004GB002299, 2005.

Behrenfeld, M. J., O'Malley, R. T., Siegel, D. A., McClain, C. R., Sarmiento, J. L., Feldman, G. C., Milligan, A. J., Falkowski, P. G., Letelier, R. M., and Boss, E. S.: Climate-driven trends in contemporary ocean productivity, Nature, 444, 752–755, 2006.

Biuw, M., Boehme, L., Guinet, C., Hindell, M., Costa, D., Charrassin, J. B., Roquet, F., Bailleul, F., Meredith, M., Thorpe, S., Tremblay, Y., McDonald, B., Park, Y.-H., Rintoul, S., Bindoff, N., Goebel, M., Crocker, D., Lovell, P., Nicholson, J., Monks, F., and Fedak, M.: Variations in behaviour and condition of a Southern Ocean top predator in relation to in-situ oceanographic conditions, P. Natl. Acad. Sci. USA, 104, 13705–13710, 2007.

Boehme, L., Meredith, M. P., Thorpe, S. E., Biuw, M., and Fedak, M.: The Antarctic Circumpolar Current frontal system in the South Atlantic: Monitoring using merged Argo and animal-borne sensor data, J. Geophys. Res., 113, C09012, doi:10.1029/2007JC004647, 2008.

Boehme, L., Lovell, P., Biuw, M., Roquet, F., Nicholson, J., Thorpe, S. E., Meredith, M. P., and Fedak, M.: Technical Note: Animal-borne CTD-Satellite Relay Data Loggers for real-time oceanographic data collection, Ocean Sci., 5, 685–695, doi:10.5194/os-5-685-2009, 2009.

Brasseur, P., Gruber, N., Barciela, R., Brander, K., Doron, M., El Moussaoui, A., Hobday, A. J., Huret, M., Kremeur, A.-S., Lehodey, P., Matear, R., Moulin, C., Murtugudde, R., Senina, I., and Svendsen, E.: Integrating Biogeochemistry and Ecology Into Ocean Data Assimilation Systems, Oceanography, 22, 206–215, 2009.

Buesseler, K. O., Barber, R. T., Dickson, M. L., Hiscock, M. R., Moore, J. K., and Sambrotto, R. N.: The effect of marginal ice-edge dynamics on production and export in the Southern Ocean along 170-W, Deep-Sea Res. Pt. II, 50, 579–603, 2003.

Caldeira, K., Hoffert, M. I., and Jain, A.: Simple ocean carbon cycle models, in: The Carbon Cycle, edited by: Wigley, T. M. L. and Schimel, D. S., Cambridge University Press, Cambridge, United Kingdom, 199–211, 2000.

Charrassin, J. B., Hindell, M., Rintoul, S. R., Roquet, F., Sokolov, S., Biuw, M., Costa, D., Boehme, L., Lovell, P., Coleman, R., Timmermann, R., Meijers, A., Meredith, M., Park Y.-H., Bailleul, F., Goebel, M., Tremblay, Y., Bost, C.-A., McMahon, C. R., Field, I. C., Fedak, M. A., and Guinet, C.: Southern ocean frontal structure and sea-ice formation rates revealed by elephant seals, P. Natl. Acad. Sci. USA, 105, 11634–11639, 2008.

Dandonneau, Y. and Neveux, J.: Diel variations of in-vivo fluorescence in the eastern equatorial Pacific an unvarying pattern, Deep-Sea Res., 44, 1869–1880, 1997.

Dierssen, H. M.: Perspective on empirical approaches for ocean color remote sensing of chlorophyll in a changing climate, P. Natl. Acad. Sci. USA, 107, 17073–17078, 2010.

Dierssen, H. M. and Smith, R. C.: Bio-optical properties and remote sensing ocean color algorithms for Antarctic Peninsula waters, J. Geophys. Res., 105, 26301–26312, 2000.

Falkowski, P. G. and Kolber, Z.: Variations in chlorophyll fluorescence yields in phytoplankton in the world oceans, Aust. J. Plant Physiol., 22, 341–355, 1995.

Fedak, M., Lovell, P., McConnell, B., and Hunter, C.: Overcoming the Constraints of Long Range Radio Telemetry from Animals: Getting More Useful Data from Smaller Packages, Integr. Comp. Biol., 42, 3–10, doi:10.1093/icb/42.1.3, 2002.

Field, C. B., Behrenfeld, M. J., Randerson, J. T., and Falkowski, P. G.: Primary production of the biosphere: Integrating terrestrial and oceanic components, Science, 281, 237–240, 1998.

Garcia, C. A. E., Garcia, V. M. T., and McClain, C. R.: Evaluation of SeaWiFS chlorophyll algorithms in the Southwestern Atlantic and Southern Oceans, Remote Sens. Environ., 95, 125–137, 2005.

Gelman, A.: Prior Distributions for Variance Parameters in Hierarchical Models (Comment on Article by Browne and Draper), Bayesian Analysis, 1, 515–534, 2006.

Gelman, A., Jakulin, A., Grazia Pittau, M., and Su, Y.-S.: A Weakly Informative Default Prior Distribution for Logistic and Other Regression Models, Ann. Appl. Stat., 2, 1360–1383, 2008.

Hindell, M., Slip, D., and Burton, H.: The diving behaviour of adult male and female southern elephant seals, *Mirounga leonina* (Pinnipedia: Phocidae), Aust. J. Zool., 39, 595–619, 1991.

Holm-Hansen, O., Kahru, M., Hewes, C. D., Kawaguchi, S., Kameda, T., Sushin, V. A., Krasovski, I., Priddle, J., Korb, R., Hewitt, R. P., and Mitchell, B. G.: Temporal and spatial distribution of chlorophyll-a in surface waters of the Scotia Sea as determined by both shipboard measurements and from satellite data, Deep-Sea Res. Pt. II, 51, 1323–1331, 2004.

Johnson, R., Strutton, P. G., Wright, S., McMinn, A., and Meiners, K. M.: Three Improved Satellite Chlorophyll Algorithms for the Southern Ocean, J. Geophys. Res., submitted, 2013.

Kahru, M. and Mitchell, G. B.: Blending of ocean colour algorithms applied to the Southern Ocean, Remote Sens. Lett., 1, 119–124, 2010.

Kiefer, D. A.: Fluoresence properties of natural phytoplankton populations, Mar. Biol., 22, 263–269, 1973.

Kinkade, C. S., Marra, J., Dickey, T. D., Langdon, C., Sigurdson, D. E., and Weller, R.: Diel bio-optical variability observed from moored sensors in the Arabian Sea, Deep-Sea Res. Pt. II, 46, 1813–1831, 1999.

Krause, G. H. and Jahns, P.: Non-photochemical energy dissipation determined by chlorophyll fluorescence quenching: characterization and function, in: Chlorophyll *a* Fluorescence: A Signature of Photosynthesis, edited by: Papageorgiou, G. C., and Govindjee, Springer, Dordorecht, The Netherlands, 463–495, 2004.

Lo Monaco, C., Goyet, C., Metzl, N., Poisson, A., and Touratier, F.: Distribution and inventory of anthropogenic CO_2 in the Southern Ocean: Comparison of three data-based methods, J. Geophys. Res., 110, C09S02, doi:10.1029/2004JC002571, 2005.

Loftus, M. E. and Seliger, H.: Some limitations of the in vivo fluorescence technique, Chesapeake Sci., 16, 79–92, 1975.

Longhurst, A., Sathyendranath, S., Platt, T., and Caverhill, C.: An estimate of global primary production in the ocean from satellite radiometer data, J. Plankton Res., 17, 1245–1271, 1995.

Lydersen, C., Nøst, O. A., Lovell, P., McConnell, B. J., Gammelsrød, T., Hunter, C., Fedak, M. A., and Kovacs, K. M.: Salinity and temperature structure of a freezing Arctic fjord monitored by white whales (Delphinapterus leucas), Geophys. Res. Lett., 29, 2119, doi:10.1029/2002GL015462, 2002.

Marrari, M., Hu, C., and Daly, K. L.: Validation of SeaWiFS chlorophyll *a* concentrations in the Southern Ocean: A revisit, Remote Sens. Environ., 105, 367–375, 2006.

Martinez, E., Antoine, D., D'Ortenzio, F., and Gentili, B.: Climate-driven basin-scale decadal oscillations of oceanic phytoplankton, Science, 36, 1253–1256, 2009.

Mitchell, B. G. and Holm-Hansen, O.: Bio-optical properties of Antarctic Peninsula waters: Differentiation from temperate ocean models. Deep-Sea Res., 38, 1009–1028, 1991.

Nicholls, K. W., Boehme, L., Biuw, M., and Fedak, M. A.: Wintertime ocean conditions over the southern Weddell Sea continental shelf, Antarctica, Geophys. Res. Lett., 35, L21605, doi:10.1029/2008GL035742, 2008.

Polovina, J. J., Howell, E. A., and Abecassis, M.: Ocean's least productive waters are expanding, Geophys. Res. Lett., 35, L03618, doi:10.1029/2007GL031745, 2008.

Quéguiner, B. and Brzezinski, M. A.: Biogenic silica production rates and particulate organic matter distribution in the Atlantic sector of the Southern Ocean during austral spring, Deep-Sea Res. Pt. II, 49, 1765–1786, 2002.

R Development Core Team: R: A Language and Environment for Statistical Computing, R Foundation for Statistical Computing, Vienna, Austria, ISBN 3-900051-07-0, 2002.

Ras, J., Claustre, H., and Uitz, J.: Spatial variability of phytoplankton pigment distributions in the Subtropical South Pacific Ocean: comparison between in situ and predicted data, Biogeosciences, 5, 353–369, doi:10.5194/bg-5-353-2008, 2008.

Reynolds, R. A., Darius, S, and Mitchell, B. G.: A chlorophyll-dependent semianalytical reflectance model derived from field measurements of absorption and backscattering coefficient within the Southern Ocean, J. Geophys. Res., 10, 7125–7138, 2001.

Roemmich, D., Riser, S., Davis, R., and Desaubies, Y.: Autonomous profiling floats: Workhorse for broad-scale ocean observations, Mar. Technol. Soc. J., 38, 31–39, 2004.

Roquet, F., Park, Y. H., Guinet, C., and Charrassin, J. B.: Observations of the Fawn Trough Current over the Kerguelen Plateau from instrumented elephant seals, J. Marine Syst., 78, 377–393, 2009.

Roquet, F., Charrassin, J. B., Marchand, S., Boehme, L., Fedak, M., Reverdin, G., and Guinet, C.: Validation of hydrographic data obtained from animal-borne satellite-relay data loggers, J. Atmos. Ocean Technol., 28, 787–801, 2011.

Spiegelhalter, D., Best, T., Best, N., and Lunn, D.: Winbugs user manual version 1.4, 2003.

Sturtz, S., Ligges, U., and Gelman, A.: R2winbugs: a Package for Running *WinBUGS* from *R*, J. Stat. Softw., 12, 1–16, 2005.

Tokuda, T., Goodrich, B., Van Mechelen, I., Gelman, A., and Tuerlinckx, F.: Visualizing Distributions of Covariance Matrices, Technical report, University of Leuwen, Belgium and Columbia University, USA, available at: http://www.stat.columbia.edu/~gelman/research/unpublished/Visualization.pdf, 2011.

Uitz, J., Claustre, H., Griffiths, B., Ras, J., Garcia, N., and Sandroni, V.: A phytoplankton class-specific primary production model applied to the Kerguelen Isalnds region (Southern Ocean), Deep-Sea Res. Pt. I, 56, 541–560, 2009.

Waite, A. M. and Nodder, S. D.: The effect of in situ iron addition on the sinking rates and export flux of Southern Ocean diatoms, Deep-Sea Res. Pt. II, 48, 2635–2654, 2001.

Wunsch, C., Heimbach, P., Ponte, R., Fukumori, I., and the ECCO-Consortium members: The global general circulation of the oceans estimated by the ECCO-Consortium, Oceanography, 22, 89–103, 2009.

Xing, X., Claustre, H., Blain, S., D'Ortenzio, F., Antoine, D., Ras, J., and Guinet, C.: Quenching correction for in vivo chlorophyll fluorescence measured by instrumented elephant seals in the Kerguelen region (Southern Ocean), Limnol. Oceanogr.-Meth., 10, 483–495, 2012.

Yentsch, C. S. and Menzel, D. W.: A method for the determination of phytoplankton chlorophyll and phaeophytin by fluorescence, Deep-Sea Res., 10, 221–231, 1963.

Yu, X., Dickey, T., Bellingham, J., Manov, D., and Streitlien, K.: The application of autonomous underwater vehicles for interdisciplinary measurements in Massachusettsand Cape Cod Bays, Cont. Shelf Res., 22, 2225–2245, 2002.

Permissions

The contributors of this book come from diverse backgrounds, making this book a truly international effort. This book will bring forth new frontiers with its revolutionizing research information and detailed analysis of the nascent developments around the world.

We would like to thank all the contributing authors for lending their expertise to make the book truly unique. They have played a crucial role in the development of this book. Without their invaluable contributions this book wouldn't have been possible. They have made vital efforts to compile up to date information on the varied aspects of this subject to make this book a valuable addition to the collection of many professionals and students.

This book was conceptualized with the vision of imparting up-to-date information and advanced data in this field. To ensure the same, a matchless editorial board was set up. Every individual on the board went through rigorous rounds of assessment to prove their worth. After which they invested a large part of their time researching and compiling the most relevant data for our readers.

The editorial board has been involved in producing this book since its inception. They have spent rigorous hours researching and exploring the diverse topics which have resulted in the successful publishing of this book. They have passed on their knowledge of decades through this book. To expedite this challenging task, the publisher supported the team at every step. A small team of assistant editors was also appointed to further simplify the editing procedure and attain best results for the readers.

Apart from the editorial board, the designing team has also invested a significant amount of their time in understanding the subject and creating the most relevant covers. They scrutinized every image to scout for the most suitable representation of the subject and create an appropriate cover for the book.

The publishing team has been an ardent support to the editorial, designing and production team. Their endless efforts to recruit the best for this project, has resulted in the accomplishment of this book. They are a veteran in the field of academics and their pool of knowledge is as vast as their experience in printing. Their expertise and guidance has proved useful at every step. Their uncompromising quality standards have made this book an exceptional effort. Their encouragement from time to time has been an inspiration for everyone.

The publisher and the editorial board hope that this book will prove to be a valuable piece of knowledge for researchers, students, practitioners and scholars across the globe.

List of Contributors

E. T. Buitenhuis
Tyndall Centre for Climate Change Research and School of Environmental Sciences, University of East Anglia, Norwich NR4 7TJ, UK

W. K. W. Li
Fisheries and Oceans Canada, Bedford Institute of Oceanography, Dartmouth, Nova Scotia, Canada

D. Vaulot and F. Partensky
CNRS and UPMC, Paris 06, UMR7144, Station Biologique, 29680 Roscoff, France

M. W. Lomas
Bermuda Institute of Ocean Sciences, St. George's GE01, Bermuda, USA

M. R. Landry
Scripps Institution of Oceanography, University of California San Diego, La Jolla, California, USA

D. M. Karl
Department of Oceanography, University of Hawaii, Honolulu, HI 96822, USA

O. Ulloa
Department of Oceanography, University of Concepci´on, Casilla 160-C, Concepción, Chile

L. Campbell
Department of Oceanography, Texas A&M University, College Station, TX 77843, USA

S. Jacquet
INRA, UMR CARRTEL, 75 Avenue de Corzent, 74200 Thonon-les-Bains, France

F. Lantoine
UPMC Univ Paris 06, CNRS, LECOB, Observatoire Océanologique, 66650, Banyuls/Mer, France

F. Chavez
MBARI, 7700 Sandholdt Rd, Moss Landing, CA 95039, USA

D. Macias
Department of Coastal Ecology and Management, Instituto de Ciencias Marinas de Andalucía (ICMAN-CSIC), Avd. Republica Saharaui s/n, CP11510, Puerto Real, Cádiz, Spain

M. Gosselin
Institut des sciences de la mer de Rimouski, Université du Québec á Rimouski, 310 Allée des Ursulines, Rimouski, Québec G5L 3A1, Canada

G. B. McManus
Department of Marine Sciences, University of Connecticut, Groton, CT 06340, USA

A. J. Sutton and S. Musielewicz
Joint Institute for the Study of the Atmosphere and Ocean, University of Washington, Seattle, Washington, USA
Pacific Marine Environmental Laboratory, National Oceanic and Atmospheric Administration, Seattle, Washington, USA

C. L. Sabine, S. Maenner-Jones, N. Lawrence-Slavas, C. Meinig, R. A. Feely, J. T. Mathis, R. Bott and P. D. McLain
Pacific Marine Environmental Laboratory, National Oceanic and Atmospheric Administration, Seattle, Washington, USA

H. J. Fought
Battelle Memorial Institute, Columbus, Ohio, USA

A. Kozyr
Carbon Dioxide Information Analysis Center, Oak Ridge National Laboratory, Department of Energy, Oak Ridge, Tennessee, USA

R. Moriarty
School of Earth, Atmospheric and Environmental Sciences, University of Manchester, Williamson Building, Oxford Road, Manchester M13 9PL, UK

T. D. O'Brien
National Marine Fisheries Service, 1315 East-West Highway, Silver Spring, Maryland, USA

R. Sander
Air Chemistry Department, Max-Planck Institute of Chemistry, 55020 Mainz, Germany

A. A. P. Pszenny and A. H. Young
University of New Hampshire, Durham, NH, USA

E. Crete
University of New Hampshire, Durham, NH, USA
The Earth Institute, Columbia University, NY, USA

W. C. Keene and J. R. Maben
Department of Environmental Sciences, University
of Virginia, Charlottesville, VA 22904, USA

M. S. Long
Department of Environmental Sciences, University
of Virginia, Charlottesville, VA 22904, USA
Harvard University, Cambridge, MA, USA

B. Deegan
Mount Washington Observatory, North Conway,
New Hampshire, USA
97 Raymond St., Fairhaven, MA, USA

U. Schuster, A. J. Watson and A. Louwerse
College of Life and Environmental Sciences,
University of Exeter, Exeter, EX4 4PS, UK
School of Environmental Sciences, University of
East Anglia, Norwich Research Park, Norwich, NR4
7TJ, UK

D. C. E. Bakker, G. A. Lee and O. Legge
Centre for Ocean and Atmospheric Science, School
of Environmental Sciences, University of East
Anglia, Norwich Research Park, Norwich, NR4
7TJ, UK

A. M. de Boer
Department of Geological Sciences and Bolin
Centre for Climate Research, Stockholm University,
Stockholm, Sweden
School of Environmental Sciences, University of
East Anglia, Norwich Research Park, Norwich, NR4
7TJ, UK

E. M. Jones
Alfred Wegener Institute for Polar and Marine
Research, Climate Sciences, Postfach 120161, 27515
Bremerhaven, Germany
School of Environmental Sciences, University of
East Anglia, Norwich Research Park, Norwich, NR4
7TJ, UK

J. Riley
International CLIVAR Project Office, National
Oceanography Centre, Southampton, Waterfront
Campus, European Way, Southampton, SO14 3ZH,
UK

S. Scally
School of Environmental Sciences, University of
East Anglia, Norwich Research Park, Norwich, NR4
7TJ, UK

M. Vogt, C. O'Brien and J. Peloquin
Institute for Biogeochemistry and Pollutant
Dynamics, Universit¨atsstrasse 16, 8092 Z¨urich,
Switzerland

V. Schoemann and L. Peperzak
Royal Netherlands Institute for Sea Research, 1790
AB Den Burg (Texel), The Netherlands

E. Breton
Universit´e Lille Nord de France, ULCO, CNRS,
LOG UMR8187, 32 Avenue Foch, 62930 Wimereux,
France

M. Estrada
Institut de Ci`encies del MAR (CSIC), Passeig
Maritim de la Barceloneta, 3749, 08003 Barcelona,
Catalunya, Spain

J. Gibson
Tasmanian Aquaculture and Fisheries Institute,
University of Tasmania, Private Bag 50, Hobart
Tasmania 7001, Australia

D. Karentz
University of San Francisco, College of Arts and
Sciences, 2130 Fulton Street, San Francisco, CA
94117, USA

M. A. Van Leeuwe and J. Stefels
University of Groningen, Centre for Ecological
and Evolutionary Studies, Department of Plant
Ecophysiology, 9750AA Haren, The Netherlands

C. Widdicombe
Plymouth Marine Laboratory, Prospect Place, The
Hoe, Plymouth PL1 3DH, UK

**R. Sauzède, H. Claustre, J. Uitz, C. Schmechtig
and F. D'Ortenzio**
Laboratoire d'Océanographie de Villefranche,
CNRS, UMR7093, Villefranche-Sur-Mer, France
Université Pierre et Marie Curie-Paris 6, UMR7093,
Laboratoire d'océanographie de Villefranche,
Villefranche-Sur-Mer, France

H. Lavigne
Istituto Nazionale di Oceanografia e di Geofisica
Sperimentale, Sgonico (OGS), Italy

C. Guinet
Centre d'Etudes Biologiques de Chizé, CNRS, Villiers en Bois, France

S. Pesant
MARUM, Center for Marine Environmental Sciences, Universität Bremen, Bremen, Germany PANGAEA, Data Publisher for Earth and Environmental Science, Bremen, Germany

S. Torres Valdés, S. C. Painter, A. P. Martin and R. Sanders
Ocean Biogeochemistry and Ecosystems Research Group. National Oceanography Centre. European Way, Southampton, SO14 3ZH, UK

J. Felden
Center for Marine Environmental Sciences. Universität Bremen. Leobener Strasse, POP 330 440, 28359 Bremen, Germany

B. Nechad and K. Ruddick
Operational Directorate Natural Environment, Royal Belgian Institute for Natural Sciences (RBINS/ODNE), 100 Gulledelle Brussels, 1200, Belgium

T. Schroeder and K. Oubelkheir
Commonwealth Scientific and Industrial Research Organisation (CSIRO), Land and Water, Environmental Earth Observation Program, Brisbane, QLD 2001, Australia

D. Blondeau-Patissier
Charles Darwin University, 0815 Darwin, Australia

N. Cherukuru, V. Brando, A. Dekker and L. Clementson
Commonwealth Scientific and Industrial Research Organisation (CSIRO), Canberra, ACT, Australia

A. C. Banks
Hellenic Centre for Marine Research (HCMR), Institute of Oceanography, Heraklion 71003, Crete, Greece
European Commission – Joint Research Centre (JRC), Institute for Environment and Sustainability, Via Enrico Fermi 2749, Ispra (Va) 21027, Italy

S. Maritorena
Earth Research Institute (ERI), University of California, Santa Barbara, CA 93106-3060, USA

P. J. Werdell
NASA Goddard Space Flight Center, Greenbelt, MD 20771, USA

C. Sá and V. Brotas
Marine and Environmental Sciences Centre (MARE), Faculdade de Ciências da Universidade de Lisboa, Campo Grande, 1749-016 Lisbon, Portugal

I. Caballero de Frutos
Institute of Marine Sciences of Andalucia (ICMAN-CSIC) Puerto Real-Cádiz, 11519, Spain

Y.-H. Ahn
Korea Ocean Research & Development Institute (KORDI), Ansan, 425–600, South Korea

S. Salama
Faculty of Geo-information Science and Earth Observation (ITC), Department of Water Resource, University of Twente, Hengelosestraat 99, 7500 AA Enschede, the Netherlands

G. Tilstone and V. Martinez-Vicente
Plymouth Marine Laboratory, Prospect Place, The Hoe, Plymouth PL1 3DH, UK

D. Foley
National Oceanic and Atmospheric Administration (NOAA), Southwest Fisheries Science Center, 110 Shaffer Road, Santa Cruz, CA 95060, USA

M. McKibben and J. Nahorniak
College of Earth, Ocean and Atmospheric Sciences (CEOAS), Oregon State University, Corvallis, OR, USA

T. Peterson
Center for Coastal Margin Observation and Prediction and Institute of Environmental Health, Oregon Health and Science University, 3181 SW Sam, Jackson Park Road, Portland, Oregon 97239, USA

A. Siliò-Calzada
Environmental Hydraulics Institute of the University of Cantabria, Cantabria, Spain

R. Röttgers
Institute of Coastal Research, Helmholtz-Zentrum Geesthacht, Centre for Materials and Coastal Research, Max-Plank-Str. 1, 21502 Geesthacht, Germany

Z. Lee
School for the Environment, University of Massachusetts Boston, Boston, MA 02125, USA

M. Peters and C. Brockmann
Brockmann Consult, Max-Planck-Str. 2, 21502 Geesthacht, Germany

M.-P. Gosselin
British Antarctic Survey, High Cross, Madingley Road, Cambridge CB3 0ET, UK
Environment Agency, North West – Water Resources, Ghyll Mount, Gillan Way, Penrith 40 Business Park, Penrith CA11 9BP, UK

R. Moriarty and C. Le Quéré
British Antarctic Survey, High Cross, Madingley Road, Cambridge CB3 0ET, UK
School of Environmental Sciences, University of East Anglia, Norwich Research Park, Norwich NR4 7TJ, UK
School of Earth, Atmospheric and Environmental Sciences, University of Manchester, Williamson Building, Oxford Road, Manchester M13 9PL, UK
Tyndall Centre for Climate Change Research, School of Environmental Sciences, University of East Anglia, Norwich Research Park, Norwich NR4 7TJ, UK

E. T. Buitenhuis
School of Environmental Sciences, University of East Anglia, Norwich Research Park, Norwich NR4 7TJ, UK
Tyndall Centre for Climate Change Research and School of Environmental Sciences, University of East Anglia, Norwich NR4 7TJ, UK

W. K. W. Li
Fisheries and Oceans Canada, Bedford Institute of Oceanography, Dartmouth, Nova Scotia, Canada

M. W. Lomas
Bermuda Institute of Ocean Sciences, St. George's GE01, Bermuda

D. M. Karl
Department of Oceanography, University of Hawaii, Honolulu, HI 96822, USA

M. R. Landry
Scripps Institution of Oceanography, University of California San Diego, La Jolla, California, USA

S. Jacquet
INRA, UMR CARRTEL, 75 Avenue de Corzent, 74200 Thonon-les-Bains, France

S. de Villiers, K. Siswana and K. Vena
Oceans and Coastal Research, Department of Environmental Affairs, Cape Town, South Africa

F. A. C. Le Moigne, S. A. Henson and R. J. Sanders
Ocean Biogeochemistry and Ecosystems, National Oceanography Centre, Southampton, UK

E. Madsen
School of Applied Sciences, Cranfield University, Cranfield, UK

N. Bednarŝek
NOAA Pacific Marine Environmental Laboratory, 7600 Sand Point Way NE, Seattle, WA 98115, USA

J. Možina
University of Nova Gorica, Laboratory for Environmental Research, Vipavska 13, Rožna Dolina, 5000 Nova Gorica, Slovenia

M. Vogt and C. O'Brien
Institute for Biogeochemistry and Pollutant Dynamics, ETH Zürich, Universitaetstrasse 16, 8092 Zürich, Switzerland

G. A. Tarling
British Antarctic Survey, Natural Environment Research Council, High Cross, Madingley Road, Cambridge CB3 0ET, UK

A. R. Fay
Space Science and Engineering Center, University of Wisconsin–Madison, 1225 W. Dayton St., Madison, Wisconsin 53706, USA

G. A. McKinley
Department of Atmospheric and Oceanic Sciences, University of Wisconsin–Madison, 1225 W. Dayton St., Madison, Wisconsin 53706, USA

C. Guinet, B. Picard, T. Jaud, M. Authier and A. C. Dragon
Centre d'Etudes Biologiques de Chizé-CNRS, Villiers en Bois, France

C. Cotté
Centre d'Etudes Biologiques de Chizé-CNRS, Villiers en Bois, France
Université Pierre et Marie Curie (Paris-6), DMPA USM 402/LOCEAN, Paris, France

E. Diamond, D. Antoine, F. D'Ortenzio and H. Claustre
Laboratoire d'Océanographie de Villefranche, Villefranche-sur-Mer, France
Université Pierre et Marie Curie (Paris-6), Unité Mixte de Recherche 7093, Laboratoire d'Océanographie de Villefranche, Villefranche-sur-Mer, France

X. Xing
Laboratoire d'Océanographie de Villefranche, Villefranche-sur-Mer, France
Université Pierre et Marie Curie (Paris-6), Unité Mixte de Recherche 7093, Laboratoire d'Océanographie de Villefranche, Villefranche-sur-Mer, France
Ocean University of China, Qingdao, China

E.Walker and P. Monestiez
Unité Biostatistique et Processus Spatiaux-INRA, Avignon, France

S. Marchand
Muséum national d'histoire naturelle, DMPA USM 402/LOCEAN, Paris, France

P. Lovell
Sea Mammal Research Unit, University of St. Andrews, St. Andrews, Scotland

S. Blain
Laboratoire d'Océanographie Microbienne, Université Paris VI, Banyuls sur mer, France
Universit´e Pierre et Marie Curie (Paris-6), Unité Mixte de Recherche 7621, Laboratoire d'Océanographie Microbienne, Banyuls-sur-Mer, France

Index

www.ingramcontent.com/pod-product-compliance
Lightning Source LLC
Chambersburg PA
CBHW061241190326
41458CB00011B/3547